*Elements of*

# Digital Geometry, Mathematical Morphology, and Discrete Optimization

*Elements of*

# Digital Geometry, Mathematical Morphology, and Discrete Optimization

**Christer Oscar Kiselman**
Uppsala University, Sweden

World Scientific

NEW JERSEY • LONDON • SINGAPORE • BEIJING • SHANGHAI • HONG KONG • TAIPEI • CHENNAI • TOKYO

*Published by*

World Scientific Publishing Co. Pte. Ltd.
5 Toh Tuck Link, Singapore 596224
*USA office:* 27 Warren Street, Suite 401-402, Hackensack, NJ 07601
*UK office:* 57 Shelton Street, Covent Garden, London WC2H 9HE

Library of Congress Control Number: 2021060593

**British Library Cataloguing-in-Publication Data**
A catalogue record for this book is available from the British Library.

**ELEMENTS OF DIGITAL GEOMETRY, MATHEMATICAL MORPHOLOGY, AND DISCRETE OPTIMIZATION**



ISBN 978-981-124-829-0 (hardcover)
ISBN 978-981-124-830-6 (ebook for institutions)
ISBN 978-981-124-831-3 (ebook for individuals)

For any available supplementary material, please visit
https://www.worldscientific.com/worldscibooks/10.1142/12584#t=suppl

# Preface

DEAR READER,

Please feel welcome to learn from this book on three distinct but related branches of science: digital geometry, mathematical morphology, and discrete optimization. They are united by a common mindset as well as by the many applications where they can serve together. In addition to being useful, they are also intellectually challenging.

The word *Elements* in the title is a humble reference to Euclid's *Elements*,[1] as well as to the *Éléments de mathématique* by Nicolas Bourbaki,[2] but is also a mild warning that the book does not present a complete building but rather provides a reliable foundation on which to build.

### *What is different with this book?*

Among the special traits offered here I would like to mention the following three.

The concepts of *upper* and *lower inverse* of a mapping between preordered sets are completely elementary—yet they unify and generalize many procedures used in several fields of knowledge, starting with the Galois connections in algebra two centuries ago and with significant later instances, maybe the most prominent being the Fenchel transformation in convexity theory. By a systematic study of the properties of the inverses, the book offers a common foundation for all these phenomena.

[1]Euclid's Greek name was Εὐκλείδες (*Eukleídes*). His work Στοιχεῖα (*Stoikheĩa*), in Latin *Elementa*, is the most successful mathematics book ever written. He was at the Μουσεῖον (*Mouseĩon*) in Alexandria, and was known as ὁ στοιχειωτής (*ho stoikheiōtḗs*), the Teacher of Elements.

[2]This is the pseudonym of a group of mathematicians, predominantly French alumni of the *École normale supérieure*, founded in 1935.

In later chapters the inverses are used to define dualities for convolution equations. This is essential since there is no general associative convolution multiplication.

*Difference operators* are the simplest convolution operators and yet serve as a most useful tool in the study of digital straightness and digital convexity, also using the duality just mentioned.

*Infimal convolution* is a tropical counterpart of ordinary convolution—and it is actually more elementary to define. In fact, it is just an instance of Minkowski addition, which in turn generalizes usual vector addition from vectors to sets of vectors. However, the use of infimal convolution provides us with a valuable tool in our studies of mathematical morphology. It is also quite natural in discrete optimization.

### *To read this book*

A modest knowledge of mathematics will suffice in order to understand the contents of this book. It is aimed at readers with developing interests like my own: from pure mathematics to a fascination of the digital world, both for its applications and for its intrinsic beauty.

The book is organized as follows.

We start by some (necessarily incomplete) motivations for the topics in Chapter 1. Then various fundamental concepts are recalled in Chapter 2.

Morphological operations in a set-theoretical setting are introduced in Chapter 3. Infimal convolution is presented as a useful tool.

In Chapter 4 lower and upper inverses of a mapping between preordered sets, mentioned above, are defined and their basic properties are presented. Inverses and quotients of mappings are then studied in detail and some of their many uses are explained in Chapter 5, and Chapter 6 takes a look at the structure of the studied operators.

Chapter 7 contains an elementary discussion of digitization, followed by a presentation of digital straightness and digital convexity in Chapter 8.

Convexity properties in vector spaces are recalled in Chapter 9, followed by a discussion of analogous definitions and results in a discrete setting in Chapter 10, with a careful look at two dimensions in Chapter 11. It is shown in Chapter 12 that there are three fundamental properties in classical convexity theory which pose problems in discrete optimization. Solutions to these problems are then presented.

A general framework for many phenomena in mathematics is provided by the concept of duality, to be considered in Chapter 13.

A survey of topology is given in Chapter 14, to be followed in Chapter 15 by a presentation of topologies which are of special interest in a digital setting; first of all the one named for Efim Khalimsky.

Chapter 16 presents distance transforms in both $\mathbf{R}^n$ and $\mathbf{Z}^n$. Skeletonizing is the topic in Chapter 17.

Every chapter has a section called *Notes*, where the history of the presented concepts are briefly mentioned, as well as some notes on etymology and alternative choices of terms.

Almost all chapters contain exercises, intended as a test ground for readers' understanding. In the *Appendix* there are solutions to some of the exercises—but please try not to look at them too soon.

### *The speed of time, or Hofstadter's Law*

While the speed of light is defined to be exactly 299,792,458 m/s, the speed of time is defined by Hofstadter's Law. This important law is self-referencing and contains an infinite (not necessarily convergent) recursion:

> It always takes longer than you think it will take, even if you take into account Hofstadter's Law. (Hofstadter 1985:47)

Since I started to lecture on these topics in 2002 and 2004, I have had some experience in this direction ... However, please do not believe that this text is from 2002 or 2004. I have followed the development and done research myself ever since then and up till now, as should be evident from the notes and the references.

2021 December 02

*Christer Oscar Kiselman*

# Acknowledgments

It is impossible to mention all persons who have helped me to arrive at the writing of this book, but let me mention some of them.

My navigating father Sam Svensson started my interest in astronomy when I was very young and taught me spherical trigonometry even before I learnt plane trigonometry and Euclidean geometry from my remarkable math teacher Bertil Broström. In high school, I had an equally remarkable teacher Karl Dagerholm, whose advisor had been Torsten Carleman.

Also my interest in languages can be explained: my most inspiring teacher was Karl Axnäs, who had a PhD in Slavic languages and was my teacher of German. Henrik Birnbaum, later to become one of the world's leading Slavists, replaced him for a short time. These two outstanding intellectuals made a long-lasting impression on me and led to early studies of Russian and much later to studies of Persian, Classical Greek, and Yiddish. Also Inga-Lisa Edstrand, my first teacher of English, was important in this respect, as was, on a general level, my philosophy teacher Gustav Boström.

Olof Hanner taught me calculus in one and several variables in my first year at Stockholm University College. Göran Björck and Benny Brodda were also among my first and definitely memorable teachers. From Stig Kanger I learnt logic. Aina Elvius taught me astrophysics; Gunnar Larsson-Leander and Bertil Lindblad celestial mechanics. Bo Döös showed me how in a remarkable way partial differential equations provide models in dynamic meteorology. Other meteorologists worth remembering were Bert Bolin, Christer Morales (weather), Leo Rannaleet (analysis methods in meteorology), Artturi Similä (oceanography), and Georg Witt (thermodynamics and noctilucent clouds).

Lars Hörmander, a world leader in the theory of partial differential equations, was my advisor in my work towards a PhD. Thanks to him,

I could spend the academic year 1965–1966 at the Institute for Advanced Study in Princeton, NJ. He had a profound influence on me; see my note (2015b).

An equally profound influence had André Martineau, who invited me to a year in Nice, which proved to be rich in consequences—scientifically and culturally as well as personally. This was for the academic year 1967–1968; see my essay (2020).

Much more recently, Chris Ciesielski helped me by clarifying essential distinctions in logic; Patrik Wahlberg gave me advice on books in logic.

Göran Björck, Jan Boman and Ragnar Sigurðsson, my mathematical brothers, have been a constant inspiration over several decades.

As to convexity, let me mention that Werner Fenchel wrote me a personal letter in 1977 carefully explaining his views on the history of the field where he himself had been a pioneer; see my paper (2019a). Recently, Jean-Baptiste Hiriart-Urruty commented helpfully on Chapter 9.

So far, the fields that I had worked in were partial differential equations and complex analysis in several variables, including complex geometry, even in infinite dimensions, and convexity in real vector spaces. Some of the results were described by Mikael Passare in his survey (2009). Pierre Lelong was very important for me as one of the two fathers of plurisubharmonic functions; see my essay (2014b).

A new field came to my attention with Gunilla Borgefors and Ewert Bengtsson (the latter already from 1979): digital geometry. I invited Gunilla to give a talk at my seminar in 1995 and with her and several others I organized in 2001 the First Franco-Nordic Summer School in Mathematics (*Première École d'Été Franco-Nordique de Mathématiques*), which was initiated by Jean-Marie Guastavino, Attaché for Science and Technology at the French Embassy in Sweden.

One of the three main lecturers at this summer school was Jean Serra, who taught us about mathematical morphology, the new science that he and Georges Matheron had created in the 1960s.

Thanks to these contacts I was welcomed in 2014 as a guest professor at the Department for Information Technology of Uppsala University—an invitation for which I am most grateful.

Jean has continued to be an inspiration ever since. But not only: he helped me with the understanding of mathematical morphology, with explanations of concepts, and with references.

In addition to these significant and happy events, I have learnt a lot in many conferences, in particular from three series of international conferences called *Discrete Geometry for Computer Imagery* (DGCI); *International Symposium for Mathematical Morphology* (ISMM);[3] and *International Workshop on Combinatorial Image Analysis* (IWCIA), where I have often been a referee or a member of the program committee.

I have participated also in some of the national symposia organized by the Swedish Society for Automated Image Analysis (SSBA).

In connection with these conferences and other meetings, I have learnt a lot from Jesús Angulo, Valérie Berthé, Srečko Brlek, Alfred Marcel Bruckstein, Rémy Malgouyres and Jean-Pierre Reveillès as well as from many other outstanding scientists.

During two visits to Singapore I have had the best possible experience from all points of view, intellectually as well as personally. I was invited in February and March 2011 to Nanyang Technical University; a second time in November and December of 2013 to the Institute for Mathematical Sciences at the National University of Singapore.

Maybe not all of these studies and activities have been a necessary preparation for the writing of an elementary book. This is, however, hard to prove or disprove.

Torbjörn Lodén explained elements of ancient Chinese philosophy; Bo Göran Johansson similarly Arabic thoughts on a discrete spacetime. Rikard Enberg introduced me to some of the intriguing problems of a discrete spacetime in modern physics. Fred A. Hamprecht gave me thoughtful comments to Chapter 1, Lawrence Gruman sent me comments and corrections to almost all chapters, and Boguslaw Obara made encouraging comments to parts of the manuscript. Mary Bass gave me advice concerning American English.

Last but by no means least, I have learnt a lot from the eighteen persons from seven countries whom I have advised as doctoral students, especially from Mikael Passare, who during his last years worked on tropical geometry, a field with strong links to digital geometry—see my paper (2014a) on mathematical questions inspired by him.

Of these eighteen former PhD students, five have worked in the general area of the present book, viz. Ola Weistrand (PhD in 2005), Erik Melin (PhD in 2008), Hania Uscka-Wehlou (PhD in 2009), Shiva Samieinia (PhD in 2011), and Adama Arouna Koné (PhD in 2016).[4]

---

[3]A joint symposium DGCI + ISMM, called *International Conference on Discrete Geometry and Mathematical Morphology* (DGMM), was held in Uppsala for four days, 2021 May 24–27.

[4]Shiva got her PhD at Stockholm University and Adama at Université des Sciences,

Several of these five later doctoral students participated in one of the courses I gave in 2002 and 2004. And all of them have contributed to the present book. In particular, Erik, Hania and Adama have provided several illustrations.

It is with profound gratitude that I acknowledge the multifaceted and most important influence that all these persons have had on my life and work.

### *Financial support*

It is a pleasure to acknowledge the support of the Swedish Research Council for Engineering Sciences (*Tekniska forskningsrådet*) and the Swedish Research Council (*Vetenskapsrådet*). The first-mentioned agency granted us the sum of 1,459,635 SEK for the years 1998–2000; the second, 1,859,100 SEK for 2004–2006. The grants enabled us to employ Céline Fouard as a postdoc and led to improved collaboration between departments within Uppsala University and between Uppsala University and the Swedish University for Agricultural Sciences (SLU).

### *World Scientific*

Kok Khoo Phua of World Scientific Publishing Co. in Singapore accepted to publish my book manuscript on 2021 March 11. With sincere gratitude I received this most welcome decision—a goal reached after eighteen years.

During the refereeing and production processes, Kim Tan and Tan Rok Ting have been most helpful. I am most grateful for their most valuable assistance, as well as for the truly professional help with LaTeX provided by D. Rajesh Babu. And I highly appreciate the cover design, artistically created by Pauline Tan Poon Ling.

2021 December 02

Christer Oscar Kiselman

---

des Techniques et des Technologies de Bamako (USTTB), Bamako I; the other sixteen at Uppsala University.

# Contents

# List of Figures

# List of Tables

# Chapter 1

# Introduction

DIGITAL GEOMETRY is, simply put, the geometry of the computer screen. Mathematical morphology is, in equally simple words, the theory and practice of transformations of sets and functions with an emphasis on their shape. In many cases these transformations have been known for a long time, but they have come into focus for the same reason as digital geometry: the operations can actually be performed on a computer.

Discrete optimization is a natural companion to digital geometry: we look for the best solution to a problem among several possibilities which form a discrete set.

In this introductory chapter we shall present the motives behind the development of these fields of knowledge.

## 1.1 Why digital geometry?

There is a universal answer (*Why not?*) to all such questions, but I shall nevertheless try to give some more specific motivation for this relatively new field and why I think it is worthwhile as a mathematical theory.[1]

Discretization is an important procedure in contemporary mathematics.

Discrete objects, like carpets and mosaics, have been around for thousands of years, but the advent of computers and digital cameras has made them ubiquitous.

A carpet can consist of thousands of knots, but these are only finite in number.

[1]An explanation—or is it an excuse?—is given in a study of the vertebrates: "The primates are notable by reason of their activity and restlessness, and especially because of their curious concern with what goes on around them." (Colbert 1966:272).

Figure 1.1. Persian carpet.

A mosaic is made up of finitely many little stones, tessellas.[2]

Carpets, embroidery work, and mosaics are examples of artifacts that are discrete. All three can present pictures quite well—the human eye and the human brain work together successfully to create nice images. As Eric Kandel writes:

> [. . . ] we have learned that the brain is not a camera but a Homeric storyteller [. . . ] (Kandel 2012:351)

So all this makes digital geometry into a kind of geometry of growing interest. Discretization of sets and functions is studied from many points of view.

### *What is discretization?*

We replace a slope by a staircase. That is discretization! And an old one at that. There are many ways to do it; see Chapter 7.

[2]The Latin word *tessella* means just a small stone with four sides, used to build up mosaics; the word being related to the Classical Greek τέσσαρες (*téssares*) 'four'.

Figure 1.2. Ravenna mosaic.

A transition to discrete calculations experienced by many people, not just engineers, is that from the slide rule to the electronic calculator.

### *What is a discrete set?*

A discrete set need not be the result of a discretization: it can be there from the beginning. Intuitively, a discrete set is a set where every point is a bit away from all the other points.

To make this precise, we need some kind of distance. So denote by $d(x, y)$ the distance between two points $x$ and $y$ in some kind of space, like the plane or three-space. Then a set $A$ is said to be ***discrete*** if for every element $a$ of $A$, there is a positive number $r$ such that $d(a, b) \geqslant r$ for all elements $b$ of $A$ with $b \neq a$.

Let us first mention the real line as a negative example, with $d(x, y) = |x - y|$, $x, y \in \mathbf{R}$. Then for every number $x$ there are other numbers $y$ that are as close as we like to $x$. So $\mathbf{R}$ with this metric is not discrete.

The subset $\mathbf{Z}$ of integers is discrete, because now $d(x, y) \geqslant 1$ for all points $y \neq x$.

But beware! Every set is discrete if we define the distance as

$$d_{\text{discr}}(x, y) = 1 \text{ when } x \neq y \text{ and } d_{\text{discr}}(x, x) = 0. \tag{1.1.1}$$

So also $\mathbf{R}$ is discrete with this distance. This means that the distance

Figure 1.3. Embroidery with many stitches.

between $\pi$ and

$$3.14159265358979323846264338327950288 4197$$

is 1 with this metric, whereas it is rather small in the unusal metric.

Mathematical models based on real or complex numbers have been very successful—think of celestial mechanics and the theory of electric circuits.

However, we should not think of real numbers as being closer to reality than discrete models, or than complex numbers. The attribute *real* is misleading. Imaginary numbers and integers can serve as well as reals in many models.

The idea that matter consists of atoms goes back some time: Leucippus (Λεύκιππος) and Democritus (Δημόκριτος) are regarded as the initiators of atomic theory. Epicurus (Επίκουρος) was a follower of the two. But today nobody believes that atoms are indivisible.[3] And even the protons and neutrons in their nuclei are considered to be divisible—into quarks. How far can we go? Are even neutrinos infinitely divisible? I have seen in scientific publications claims to the effect that "reality" is infinitely divisible, just like $\mathbf{R}^3$ (references withheld). For me, $\mathbf{R}^3$ and $\mathbf{Z}^3$ are models of three-space, often quite successful—each with its limitations.

We may ask: is space granular, with a smallest possible volume?

But it is not only about space: time and space come together.

[3]The adjective ἄτομος (*átomos*) in Classical Greek means the same as the Latin *individuus*, viz. 'indivisible'.

We can go back to Zeno of Elea (Ζήνων ὁ Ἐλεάτης). He formulated four paradoxes about movement—paradoxes behind which there is apparently a conflict between a smoothly flowing time and a discrete time. For these paradoxes, see Nils Bertil Thelin's book (2014:204–207), and more generally all of his Sections 4.6–4.8 (2014:194–236) for a survey of the thoughts on motion and time in the history of philosophy which these paradoxes have aroused during more than two millennia. It is not only a problem for physics, but a problem for semiotics, grammar, and theories of knowledge as well.[4]

We may wonder whether Euclid's geometry was affected by Zeno's paradoxes. The answer is in the negative as far as I can judge. The constructions are not made by mathematicians: they are already finished. Mathematicians just look at the eternal pictures. Euclid's geometry contains no movements.[5] See my paper (2015a, Section 5) for more on what constructions meant to the Greek.

A straight line is an entity of its own: it does not consist of points. Instead points are special entities, like the stars in a constellation. This is clear from Michel Federspiel's studies. A point in Euclid's geometry is what in French is rendered by a single term, *repère*, see (Federspiel 1992:388, 389; 1998:67) and also (Kiselman 2015a: Subsection 4.4). This term corresponds in English to more than one expression: *a reference mark*, *a benchmark*, *a distinctive sign.*

That time is discrete was a theory developed in Spain and northern Africa in the Middle Ages:

> As regards the theoretical and philosophical analysis of time, the most important and original contribution of medieval Islamic thinkers was their theory of discontinuous, or atomistic, time. (Whitrow 1990:79)

Moshe ben Maimun, who was born in 1135 or 1138 in Córdoba and died in 1204 near Cairo, was a Jewish philosopher working in Muslim cultural circles, and is better known under his Greek name Maimonides (Μαϊμωνίδης). He wrote about the Muslim ideas on discrete space and discrete time—the two must come together.

---

[4]And it lasts: up to this day, people argue with force that Zeno was wrong, and claim with emphasis that movement is possible—since we can talk and write. "We cannot deny the existence of motion. [. . . ] Hence, this discussion proves that motion has reality and it exists." (Chattopadhyay 2019:21).

[5]There is an exception to this in his *Data*, Δεδομένα (*Dedoména*), as explained by Taisbak (2003:18), where "we are the audience in the class-room of the mathematician at work." This is indeed a rare deviation from Euclid's usual style.

In his book *Guide for the Perplexed*, Maimonides states twelve propositions, common to all Mutakallemim,[6] the followers of Kalām, of which the first three are the following.

> Proposition I. All things are composed of atoms.
> Proposition II. There is a vacuum.
> Proposition III. Time is composed of time-atoms.
> (Maimonides 1904, Chapter LXXIII, p. 120)

The first of them is explained as follows.

> "The Universe, that is, everything contained in it, is composed of very small parts [atoms] which are indivisible on account of their smallness; [...]" (Maimonides 1904, Chapter LXXIII, p. 120)

Similarly for time in the third proposition:

> "Time is composed of time-atoms," i.e., of many parts, which on account of their short duration cannot be divided. This proposition also is a logical consequence of the first.
> [...]
> An hour is, e.g., divided into sixty minutes, the minute into sixty seconds, the second into sixty parts, and so on; at last after ten or more successive divisions by sixty, time-elements are obtained, which are not subjected to division, and in fact are indivisible, just as is the case with space.
> (Maimonides 1904, Chapter LXXIII, p. 121)

These are the time atoms, later to be called chronons, a term introduced by Robert Lévi (1927), in analogy with *electron*, *photon*.

However, after having surveyed these ideas, Maimonides sharply criticizes them, and writes for instance:

> They held that locomotion consisted in the translation of each atom of a body from one point to the next one; accordingly the velocity of one body in motion cannot be greater than that of another body. When, nevertheless, two bodies are observed to move during the same time through different spaces, the cause of this difference is not attributed by them to the fact that the body which has moved through a larger distance had a greater velocity, but to the circumstance that motion which in ordinary language is called slow, has been interrupted by more moments of rest, while the motion which ordinarily is called quick has been interrupted by fewer moments of rest.[7]
> [...]
> But we ask them: "Have you observed a complete revolution of a millstone? Each point in the extreme circumference of the stone describes a large circle

[6] More often called Mutakallemūn, plural of Mutakallim. This term is used in the later translation by Schlomo Pines (Maimonides 1963).

[7] This sounds exactly as an equation for a digital line with a slope smaller than one—see Chapter 8.

> in the very same time in which a point nearer the centre describes a small circle: the velocity of the outer circle is therefore greater than that of the inner circle. You cannot say that the motion of the latter was interrupted by more moments of rest; for the whole moving body, i.e., the millstone, is one coherent body." (Maimonides 1904, Chapter LXXIII, pp. 121–122)

This remark is of utmost interest also today. Indeed, rotation, so natural in $\mathbf{R}^2$ with a Euclidean norm, is problematic in $\mathbf{Z}^2$, and many papers are now devoted to studying rotation in digital spaces, for instance by Lin et al. (2008), Yohan Thibault et al. (2008), Andres & Largeteau-Skapin (2016), Ngo et al. (2017), Pluta et al. (2017), Andres et al. (2019), Ngo et al. (2019), Passat et al. (2019), Avkan et al. (2020).

Centuries later, Vasily Yanchilin (2013) cries out at us that discrete movement is the key to understanding quantum measurements. In his book (2003:123) he writes about discrete motion and chaotic motion: "So, the electron disappears from one point of its cloud and, in an infinitesimal period of time, appears at another point of the cloud." What is real? What is just a mathematical model of reality?

For a detailed analysis of ancient concepts of spatial magnitude, time and motion, see Michael J. White's book (1992). He presents three concepts: an Aristotelian, a quantum, and a Stoic model, of which the second, being discrete and in the Epicurian tradition, is the one of interest here, the two others being of what he calls the continuous kind.

As to other cultures, let us quote John L. Bell, who writes: "From a very early period atomism played a role in Indian philosophical thought." (Bell 2006:50). By contrast, "Chinese natural philosophy seems to have inclined more to synechism than to atomism" (Bell 2006:48).[8]

Albert Einstein wrote

> Es scheint mir nun in der Tat, daß die Beobachtungen über die „schwarze Strahlung“, Photolumineszenz, die Erzeugung von Kathodenstrahlen durch ultraviolettes Licht und andere die Erzeugung bez. Verwandlung des Lichtes betreffende Erscheinungsgruppen besser verstandlich erscheinen unter der Annahme, daß die Energie des Lichtes diskontinuierlich im Raume verteilt sei.
>
> [...]
>
> es besteht dieselbe aus einer endlichen Zahl von in Raumpunkten lokalisierten Energiequanten, welche sich bewegen, ohne sich zu teilen und nur als Ganze absorbiert und erzeugt werden können. (Einstein 1905:133)

"These simple and clear lines are the real birth certificate of quantum the-

---

[8]Here the term *synechism*, from the Greek adjective συνεχής (*synekhēs*) 'continuous', refers to divisibility; cf. the verb συνέχειν (*synékhein*) 'to hold together'.

ory," writes Rovelli (2016:94). But acceptance did not come easily:

> At first, Einstein's idea that light could be made up of photons is regarded by his colleagues as no more than youthful waywardness. Everyone commends him for his theory of relativity, but everybody judges the notion of photons to be outlandish. (Rovelli 2016:95)

This kind of judgment still prevails in some quarters. From time to time I have experienced a resistance towards digital geometry and mathematical morphology, a resistance—or hostility—which is hard to document since it has always been covert.

There are small groups researching digital geometry all over the planet, but even in our present society—which is indeed going digital—mathematicians often reject discrete approaches.

However, physicists are studying discrete models—and did not start just recently.

Erwin A. K. Brüning & Shigeaki Nagamachi (2004) have established a quantum field theory with a positive fundamental length $l$: in distances smaller than $l$, space-time melts down. They use a kind of hyperfunctions, tempered ultra-hyperfunctions, to construct this theory.

For a popular presentation of discrete models of spacetime, see (Rovelli 2016); for a somewhat less popular one, see (Hossenfelder 2013) and (Rovelli & Vidotto 2015). Volume in a discrete spacetime can be measured by counting, and one expects the density to be near the Planck densitiy of $7 \cdot 10^{142}\,\mathrm{m}^{-3}\mathrm{s}^{-1}$ (Bombelli et al. 1987:522).

What is real? In the world of real numbers, the diagonal of a square is $\sqrt{2}$ times longer than a side, but in the real world, maybe there do not exist three points $a$, $b$, $c$ with the distance between $a$ and $c$ equal to $\sqrt{2}$ times that between $a$ and $b$.

Often we do not go from models based on real numbers to discrete models, but from one very fine discrete model to another model, also discrete, but less fine. An example is when we go from water molecules to droplets in the air, or from molecules of a mineral to sand. More abstract is the description of the animal kingdom as consisting of individual beings, then as being subject to the equivalence relation $x$ and $y$ belong to the same species; then $x$ and $y$ belong to the same genus, the same family, the same order, and so on. Or going from bricks, to houses, blocks, and towns.

Points, straight lines and planes have been studied for well over two millenia, and certain curves, like ellipses and hyperbolas, have been an object of our curiosity for almost as long. Other, less well-known curves, like lemniscates and cardioids, have been studied for several centuries. In the study of these curves we rely very much on the fact that we can draw them on paper. But with the advent of computers we have acquired a new method of drawing pictures. On the computer screen we see images, and the images consist of little picture elements, pixels that our eyes put together to form geometric objects. A straight line is therefore not what Euclid understood by a straight line, but rather a finite collection of dots on the screen, which the brain nevertheless perceives as a connected line segment. Also in the time dimension: when in a cinema we see 24 pictures a second but in spite of this, persons and even cars seem to move smoothly—until stroboscopic phenomena set in.

Is there a geometry for these images on the screen? The answer is in the affirmative. We shall not be content with considering the images as being more or less accurate approximations of ideal straight lines or curves: we can treat these finite sets of points with the same accuracy as Euclid had in his geometry.

This is digital geometry.

Although the ideas touched upon are indeed ancient, digital geometry is young in comparison with Euclid's: the concept of a digital straight line was clarified in 1974 by Azriel Rosenfeld. We can also talk about curves in the digital plane. In fact, we can take any concept in Euclidean geometry, try to translate it to digital geometry, and see if a certain result in Euclidean geometry becomes true in a digital interpretation. The Jordan curve theorem is an instructive example; see Chapter 15.

For an elementary discussion of digital geometry and in particular of the Khalimsky Jordan curve theorem, see (Kiselman 2003, 2008b).

The plane may be divided into squares, hexagons, or triangles.[9] These are the most common tessellations of the plane (Figures 1.4–1.6). The centers of the pixels form in the first two cases a square or a triangular pattern, respectively. However, in all three cases we may use a pair of integer coordinates to indicate the location of a pixel. This is obvious when we tesselate by squares, but it is useful also in the other cases, although we need to be careful about the metric then. (In the case of triangles, one can use $(x, \sqrt{3}y)$ with $(x, y) \in \mathbf{Z}^2$ as addresses.)

For a study of more exotic tessellations, including a description of connections between discrete geometry and other areas by means of combinatorics on words, see Blondin-Massé et al. (2009).

[9]For a study of simple closed curves in the triangular grid, a grid not so well studied as the square grid, see, e.g., the article by Lidija Čomić (2019).

One often speaks of $\mathbf{Z}^2$ as the "set of pixels," although, speaking more precisely, a pair of integers $a = (a_1, a_2) \in \mathbf{Z}^2$ is just the address of a pixel, and need not be the barycenter of the picture element. When we come to Voronoi cells in Chapter 7, we can say that the pixels in the first two cases arise as the Voronoi cells of $\mathbf{Z}^2$, indicated by dots in Figures 1.4 and 1.5, while this is not so in Figure 1.6. But for identification, this is of no consequence: one can put the address at a corner, so that the pixel with address $(a_1, a_2)$ is $[a_1, a_1 + 1] \times [a_2, a_2 + 1]$.

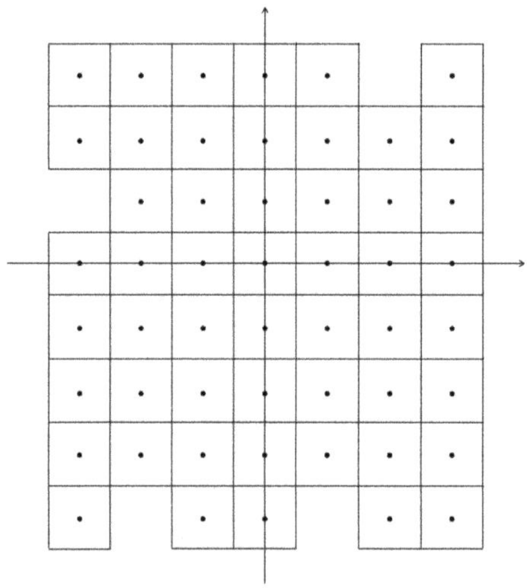

Figure 1.4. The plane tessellated by squares. (Courtesy Hania Uscka-Wehlou)

Maurits Cornelis Escher was inspired by the tilings, tessellations as well as by the geometrically presented groups of movements on the walls of Alhambra. But while the elements there are exclusively geometric, not depicting any living being, Escher filled his two-dimensional planes and three-dimensional spaces also with animals of ever-changing forms (Escher 1967; Locher, Ed. 1991; Ernst 2018). The last-mentioned book offers a background and an account of the genesis of Escher's work. In turn, his art inspired Douglas Hofstadter in his *Gödel, Escher, Bach: an Eternal Golden Braid* (1979, 1999).

Non-periodic tessellations have attracted attention with the discoveries by Roger Penrose and Dan Shechtman. In spite of the fact that quasicrystals are of great interest in chemistry as well as in mathematics, we leave them aside here.

## 1.2 Why mathematical morphology?

The branch of science that has become known as mathematical morphology was created in the 1960s by two persons: Georges Matheron and Jean Serra. For its fascinating history, which started with the analysis of minerals, see their story (2002). See also Serra's words in (2005b) about Matheron and the survey by Ronse & Serra (2008, 2010).

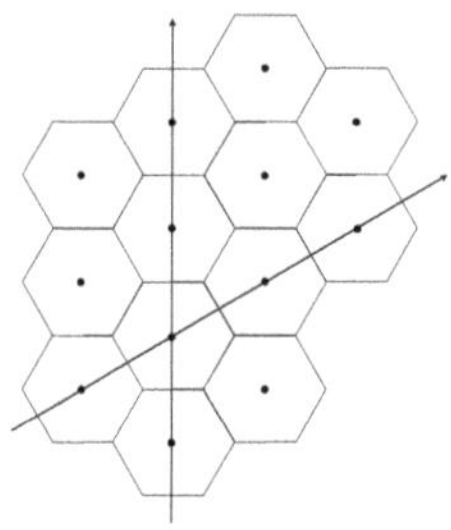

Figure 1.5. The plane tessellated by hexagons. (Courtesy Hania Uscka-Wehlou)

In a sense, mathematical morphology is quite old—it is about operations on sets and functions that have been around for a long time, but which are now being systematized and studied under a new angle, precisely because it is possible to actually perform operations on the computer and see on the screen what happens.

My personal view is that morphology has its origin in our trying to understand a complicated world. The world is so complex that the human mind—and the human eye—cannot perceive all its minute details, but needs a simplified image, a simplified structure. The need to simplify a complicated object is, in this view of things, a basic impulse behind mathematical morphology, and this is what mathematical morphology does. Related to this is the fact that an image may contain a lot of disturbances, or rather, it almost always does. Therefore, most images need to be tidied up. Hence another need to process images; it is related to the first, for the border line between dirt and other kind of disturbances is not too clear.

Somewhat simplified we can say that mathematical morphology is a tool for vision, while Fourier analysis and synthesis is a tool for hearing. Let me explain.

If an orchestra is playing, you can hear the tones produced by a musician even if she or he sits behind another musician. This is because the wavelength of the sound is much larger than the wavelength of light, typically more than a million times larger.

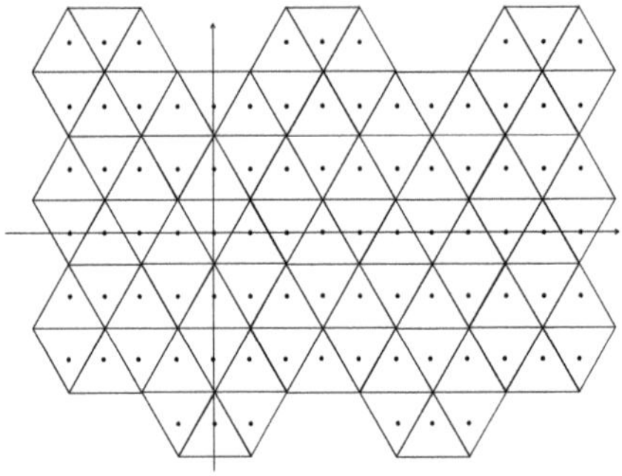

Figure 1.6. The plane tessellated by triangles. (Courtesy Hania Uscka-Wehlou)

As a consequence of this comparison, we can say that vision is modelled by the equation $\max(a, x) = b$, where one cannot reconstruct the unknown quantity $x$ from knowledge of $a$ and $b$ if $x \leqslant a$, since $x$ is then hidden behind $a$, i.e., $\max(a, x) = a$ for all $x \leqslant a$. By constrast, from the sum $a + x = b$ we can find $x$ if $a$ and $b$ are known, since it equals $b - a$. Similarly, if two tones are played, we can hear both of them.

The example mentioned indicates that the operations $(x, y) \mapsto x \wedge y = \min(x, y)$ and $(x, y) \mapsto x \vee y = \max(x, y)$ are more important for vision than the operations $+$ and $-$, and this motivates the study of ordered structures and lattices.

Jean Serra discusses the nature of mathematical morphology in his article (1978:151). He writes that this science is not faithful to any single branch of mathematics: it uses different mathematical tools for different purposes. Unfaithful as mathematical morphology might be, it is in itself a valuable and most useful way to look at the world around us.

Serra continues this line of thought in a later book chapter, where he compares with processes in Gestalt psychology, mentions books by Wolfgang Köhler and Kurt Koffka and observes "that vision tends to operate more in gestalt mode."(1988f:245).

For a readable introduction to the ideas behind this branch of science, see the article by Henk J. A. M. Heijmans (1995).

Mathematical morphology uses complete lattices as opposed to lattices which are not complete. This means that we are allowed to form freely suprema and infima of arbitrary families of elements, not just of finite families, and so amounts to the existence of actual infinities, not just potential infinities. This touches discussions that have lasted thousands of years.

Segmentation is a significant activity in image processing. Random closed sets and random functions, as pioneered by Georges Matheron, is also of great significance. This book mentions these branches only cursively.

There are ideas common to the fields of digital geometry, mathematical morphology and tropical geometry. The latter is a rapidly growing branch of science; see the book by Diane Maclagan and Bernd Sturmfels (2015). The articles by Petros Maragos et al. (2021) and Arthur Renaudineau (2021) may serve as helpful introductions to mathematical morphology.

Concerning the relation between tropical geometry and mathematicial morphology, see for instance (Maragos 2019) and (Maragos & Theodosis 2019). Intuitively, we can say that in digital geometry we replace an integral by a sum, while in mathematical morphology and tropical geometry we replace an integral by a supremum. The three operations $\int$, $\sum$ and sup have properties in common—to be investigated.

## 1.3 How is image processing at all possible?

Let us think first of Euclidean geometry and consider cardinalities. The set $\mathbf{N}$ of nonnegative integers is infinite, and its cardinality is denoted by $\mathbf{card}(\mathbf{N}) = \aleph_0$, the smallest infinite cardinal called *aleph nought.*

The family of all subsets of any given set $X$ always has a strictly larger cardinality than has $X$, in symbols

$$\mathbf{card}(\mathscr{P}(X)) = 2^{\mathbf{card}(X)} > \mathbf{card}(X),$$

where $\mathscr{P}(X)$ denotes the family of all subsets of $X$.

The set of real numbers $\mathbf{R}$ has the same cardinality as the set of all subsets of $\mathbf{N}$, thus

$$\mathbf{card}(\mathbf{R}) = \mathbf{card}(\mathscr{P}(\mathbf{N})) = 2^{\aleph_0}.$$

The points in the Euclidean plane have the same cardinality: $\mathbf{card}(\mathbf{R}^2) = \mathbf{card}(\mathbf{R})$. But the set of all subsets of the line or the plane has the larger cardinality, viz. $2^{2^{\aleph_0}}$, which is too much for our brains to keep track of.

This is because there are a lot of highly irregular sets. They are simply too many—and most of them are not interesting: we need to restrict attention to some not too large subclass of this huge class, a subclass consisting of more regular sets. For instance, the set of all disks has a much smaller cardinality, because three numbers suffice to determine a disk in the plane: its radius and the two coordinates of its center. Similarly, four numbers suffice to specify a rectangle $[a_1, b_1] \times [a_2, b_2]$ with sides parallel to the axes; a fifth is needed if we want to rotate it. This leads to the idea of simplifying a general, all too wild set, to some reasonable, more well-behaved set.

It is true that one can think of a Euclidean line as containing only denumerably many points. We can define a line as the set of solutions in $\mathbf{Q}^2$ of an equation $a_1x_1 + a_2x_2 + a_3 = 0$ with integer coefficients. Then two lines which are not parallel intersect in a point with rational coordinates. The cardinality of the set of all subsets of $\mathbf{Q}^2$ is $2^{\aleph_0}$, so there are fewer sets to keep track of than in the real case. However, the conclusion is more or less the same: there are too many subsets to grasp even of the plane $\mathbf{Q}^2$ of rational points.

If we consider digital geometry we can make a similar discussion about cardinality. On a computer screen with, say, 1,024 pixels in a horizontal row and 768 pixels in a vertical column there are $1{,}024 \cdot 768 = 786{,}432$ pixels. On such a screen a rectangle with sides parallel to the axes is the Cartesian product

$$R(a, b) = [a_1, b_1]_{\mathbf{Z}} \times [a_2, b_2]_{\mathbf{Z}}, \qquad a, b \in \mathbf{Z}^2,$$

of two intervals; we need to consider all pairs of integers $(a_j, b_j)$ such that $0 \leqslant a_1 \leqslant b_1 < 1{,}024$ and $0 \leqslant a_2 \leqslant b_2 < 768$. We can form

$$\frac{1{,}024 \cdot 1{,}025}{2} \cdot \frac{768 \cdot 769}{2} \approx 1.55 \cdot 10^{11}$$

such rectangles.

Let us for instance ask if a given subset $A$ is close to a rectangle $R(a, b)$. We can measure the distance between two sets $A$ and $B$ in some way, for instance simply by computing

$$d(A, B) = \mathbf{card}(A \smallsetminus B) + \mathbf{card}(B \smallsetminus A),$$

the number of pixels that belong to one of the sets but not to the other. Then the problem is to compute $d(A, R(a, b))$ for all permissible values of $a$ and $b$, i.e., in $1.55 \cdot 10^{11}$ cases. At least one of the rectangles is closest to $A$ in this metric. Thus to find the best-fitting rectangle to an arbitrary set is a finite optimization problem—but some finite problems are not small.

But the number of rectangles is very small compared with the number of arbitrary sets. To describe a subset of the screen we need to specify for each pixel whether it belongs to the set or not. This means that there are $2^{786,432} \approx 10^{236,740}$ different subsets of the screen, or binary images. A binary image is a black-and-white image, i.e., we only specify whether a point belongs to the set or to the background. (If we want to consider gray-level images, or color images, the cardinality goes up of course.) The really pure mathematician then says: there are only finitely many binary images. But the number of binary images must be compared with other finite numbers. Some astronomers estimate the mass of the universe at $10^{53}$ kg, which is $6 \cdot 10^{79}$ proton masses or $10^{83}$ electron masses. One sometimes talks about "astronomical numbers," which Webster (1983) defines as "enormously or inconceivably large" numbers. This metaphor has not only faded; it is actually misleading[10] in the world of image analysis, since $10^{83} \ll 10^{236,740}$.

Thus, although the number of binary images on a computer screen is finite, it is so huge that the conclusion must be the same as in the case of the infinite cardinal $2^{2^{\aleph_0}}$: there are too many; we cannot search through the whole set; we must simplify. This leads, again, to image processing and mathematical morphology, this time of subsets of $\mathbf{Z}^2$, or, a little more generally, of $\mathbf{Z}^n$, the set of all $n$-tuples of integers.

When we discuss mathematical morphology we want to keep both cases in mind, i.e., both the vector space $\mathbf{R}^n$ of all $n$-tuples of real numbers (the addresses of points in space) and the digital space $\mathbf{Z}^n$ (the addresses of pixels or voxels).[11] The latter covers the case of rectangular pixels or voxels, but, as already pointed out, also triangular and hexagonal pixels in the plane. What is common to $\mathbf{R}^n$ and $\mathbf{Z}^n$ is that they form an abelian group. Therefore we shall always start the discussion assuming that the space, called ***image carrier***, is just an abelian group. When necessary we shall then specialize to $\mathbf{R}^n$ and $\mathbf{Z}^n$. In many ways, $\mathbf{R}^n$ will guide us in the perhaps less familiar study of $\mathbf{Z}^n$.

---

[10]In case you think that this conclusion depends on the universe having a low density, please calculate the mass of a fictitious universe with a radius of $14 \cdot 10^9$ light years $\approx 1.3 \cdot 10^{26}$ m and the density of a neutron star, say $10^{17}$ kg/m$^3$. The conclusion is the same.

[11]In addition to the terms *pixel* and *voxel*, also *spel* and *surfel* are used for 'spatial element' and 'surface element', respectively. See, e.g., (Herman 1998:58).

## 1.4 Why discrete optimization?

Figure 1.7. Judgment of King Solomon.

Optimization is an important branch of mathematics. The word means 'to search for the best', from Latin's *optimum* 'optimum, the best'. The very word *optimism* is said to come from Leibniz' idea that the existing world is the best possible—God optimizes.

To solve an optimization problem one must first decide what is best: is the best price the lowest or the highest price?

Next one must decide about the acceptable states or values. One usually cannot buy half an apple, half a house, a fifth of an automobile: one has to buy either zero, one, two, ...; only integer values are admissible. On the contrary, an optimal length can be 3.14 meters or $\pi = 3.14159265\ldots$ meters: in such a situation, all real numbers are possible. I will present some significant differences between integer and real optimization, and show how to find suitable classes of functions with integer values for which desirable results hold.

We learn early in life to calculate with the natural numbers and much later with real numbers, like $\sqrt{5}$, $\pi$, e. So one might think that natural numbers are easier than the more general real numbers, sometimes difficult even to define. However, from a logical point of view, the opposite is true: the elementary theory of real numbers and hence for the elementary theory of lines and planes in Euclidean geometry is decidable, while the theory of Peano arithmetic is not decidable and even incomplete, hence much more difficult as a theory. These results are associated with the names of Alfred Tarski, Giuseppe Peano, Alonzo Church, and Kurt Gödel: they show that, while natural numbers are the ones we learn first, their theory is quite difficult. This will be mentioned with some explanation in Chapter 7.

We learn in the Bible what King Solomon said when two women both claimed to be the mother of a child: "Divide the living child in two, and give half to the one, and half to the other" (Bible, 1 Kings 3:16–28). The value $\frac{1}{2}$ was accepted by one of the women but not by the other.

King Solomon understood the difference between real and discrete optimization more than 2,900 years ago.

## 1.5 Notes

### 1.5.1 *History of discreteness*

I have quoted above passages on discreteness from (Maimonides 1904). His book *Guide for the Perplexed* contains an exemplarily clear account of his own thinking as well as that of the Islamic thinkers of his time.

Gerald James Whitrow (1990) has written a fascinating account on how time has been understood over the ages, and John L. Bell (2006) likewise for space and time in mathematics and philosophy in many different cultures. Carlo Rovelli, seemingly moved by Hamlet's "The time is out of joint" (but himself not necessarily the one who was "born to set it right") goes further in his book (2019) with ideas in today's physics.

### 1.5.2 *Digital geometry and mathematical morphology*

A general reference for digital geometry is the book by Klette & Rosenfeld (2004).

Mathematical morphology was established by Jean Serra in his book (1982) and then even more firmly by him as an editor in (Serra, Ed. 1988), which contains contributions by several leading researchers in the field. For some early work by Matheron, see (Matheron 1965, 1978a).

Among later books on mathermatical morphology, let me mention those by Marchand-Maillet & Sharaiha (2000) and Soille (2004), as well as the ones edited by Najman & Talbot (2008, 2010).

### 1.5.3 *Origin the terms in the title of this book*

• *Element.* The noun *element* is of Latin origin (*elementum*₍ₙ₎) and has several meanings. In the title it refers to the most basic principles of a science, like in Euclid's *Elements*, Στοιχεῖα (*Stoikheĩa*). Another meaning is that of a smallest constituent, like hydrogen H and oxygen O being the elements in water, $H_2O$. And we say: "$x$ is an element of $A$" and write it $x \in A$.

• *Digit.* The noun *digit* and the adjective *digital* come from Latin's *digitus* (m), meaning 'finger, toe' and is related to the Greek δάκτυλος (m) (*dáktylos*) 'finger'. Since most of us have ten fingers and ten toes, the reason behind the modern meaning 'symbol for a number' is easy to guess, but in digital geometry the emphasis is on discreteness rather than on counting.

The Greek word is also the origin of the name of a fruit, *date.* If you look at dates, not in a box, but growing high up in a palm *Phoenix dactylifera*, you will appreciate the similarity—*dactylifera* means 'carrying fingers'. So digital geometry is about counting on fingers and toes (perhaps implying a system of base twenty) but if that seems to you not to be sweet enough, think of counting dates.

What I call *digital geometry* in this book could, theoretically, have been called *discrete geometry*, but by tradition the latter expression has come to be used mainly for research of a different character, an observation supported by the contents of journals like *Discrete & Computational Geometry*; *Discrete Mathematics*; *The Art of Discrete and Applied Mathematics.* In the combination *discrete optimization* the same cultural phenomenon is not present.

• *Geometry.* The science geometry, in Classical Greek γεωμετρία (f) (*geōmetría*), started, as we are told, from the measuring of land after the inundations of the river Nile. Anyway, the origin is clearly the Classical Greek γῆ (f) (*gế*) 'earth' and μετρεῖν (*metreĩn*) 'to measure'; μέτρον (n) (*métron*) 'measure'.

• *Mathematics.* The word *mathematics* has its origin in the Classical Greek noun μαθηματική (f) (*mathēmatikḗ*) 'mathematics', from the verb μανθάνειν (*manthánein*) 'to learn', giving rise to the noun μαθεματικός (m) (*mathe-*

*matikós*) 'mathematician', which is also an adjective 'mathematical'. The Indo-European root for these words is **mundh-* 'to pay attention to, to be alert, to be lively', and has also given rise to the Swedish word *munter* 'cheerful, blithe'. An encouraging start for any student of mathematics.

• *Morphology.* The term *morphology* is of Greek origin: μορφή$_{(f)}$ (*morfḗ*), meaning 'form, shape, appearance', and λόγος$_{(m)}$ (*lógos*), which has several meanings, among them 'word', 'thought', and is now highly productive in the naming of sciences. So mathematical morphology is the study and description of shapes. The epithet *mathematical* is needed in order to distinguish this theory from the morphology in biology, archeology, linguistics as well as in other fields of knowledge.

• *Discrete.* The adjectives *discrete* and *discreet* have their origin in the Latin *cretus*, past participle of the verb *discernere* 'to sift, to discern, to distinguish' and equipped with the prefix *dis-* 'apart'. The verb is related to the Classical Greek verb κρίνειν (*krínein*) 'to separate, to discern, to judge'.

• *Optimization.* To optimize means to search for the best, from the Latin comparison *bonus, melior, optimus* 'good, better, best'.[12]

—

The sources for the etymology here and in later chapters are the dictionaries by Millén (1853), Linder & Walberg (1862), The Cabinet Dictionary (1871), Cavallin (1885), Ahlberg et al. (1966), Schwartzman (1994), Webster (1983, 1994), Solvang (2002), Kiselman & Mouwitz (2008), where Ebbe Vilborg provided the etymology, Lo Bello (2013), and the web-based Oxford English Dictionary.

## 1.6 Exercises

1.1. Find some discretizations in ordinary life which can be modelled as an operation from $(b\mathbf{Z})^n$ to $(c\mathbf{Z})^n$, where $b$ and $c$ are constants with $0 < b < c$. Can you find a discretization which can be modelled as an operation from $\mathbf{R}^n$ to $\mathbf{Z}^n$?

1.2. As an ambitious project, try to find other philosophical ideas on discrete spacetime than those mentioned in this chapter.

1.3. Find optimizations in actual life where there is a significant difference

[12]The English language possesses a strengthened superlative, *bestest.* However, few optimizers use the letter closing *Bestest regards*—it seems to be strictly limited to researchers in several complex variables.

between discrete optimization and optimization over $\mathbf{R}^n$.

1.4. Radioactive decay is often modelled on the function $t \mapsto 2^{-t/h}$, where $t$ is time and $h$ the half-life of the process. If you imagine that a piece of radiactive substance consists of a finite number $N_0$ of atoms at time $t = 0$, then the number of remaining atoms will be represented by the equation $N(t) = 2^{-t/h}N_0$, a quantity which is always positive, indicating a process going on forever. But $N_0$ atoms can disappear no more than $N_0$ times.

Can you reconcile the finitenes of $N_0$ with this model using an exponential function?

1.5. What would Zeno of Elea have said if you had told him—in Classical Greek—that, if there is a midnight and if there is one hour past midnight, then the time can never be $\pi$ hours after midnight? Could it be "I told you so!" or perhaps even "Time does not exist!"?

1.6. (a) Prove that to store the information contained in one chrono-synclastic infundibulum in another chrono-synclastic infundibulum, the latter has to be twice as large as the first. *Hint:* See *The Syrens of Titan* by Kurt Vonnegut, Jr.

(b) Give a lower bound for the cardinality of the set of all chrono-synclastic infundibula in the visible universe. Is there an upper bound?

# Chapter 2

# Sets, mappings, and order relations

## 2.1 Introduction

SETS AND MAPPINGS belong to all branches of mathematics. In this chapter we start by mentioning these fundamental concepts and presenting standard notation for them.

Next we come to order relations, which are important for all what follows. A typical order is the inclusion relation $\subset$ between sets.

## 2.2 Sets

The ***empty set***, the unique set with no elements, will be denoted by $\varnothing$. If $A$ and $B$ are sets such that every element of $A$ belongs also to $B$, then $A$ is said to be a ***subset*** of $B$ and $B$ a ***superset*** of $A$. This is written $A \subset B$ and $B \supset A$.

The family of all subsets of a set $W$ is called the ***power set*** of $W$, and will denoted by $\mathscr{P}(W)$. Thus $A \in \mathscr{P}(W)$ if and only if $A \subset W$. We denote by $\mathscr{P}_{\text{finite}}(W)$ the family of all finite subsets of $W$.

We shall use the usual symbols for the ***intersection*** and ***union*** of a family $(A_j)_{j \in J}$ of sets:

$$\bigcap_{j \in J} A_j \text{ and } \bigcup_{j \in J} A_j, \qquad A_j \in \mathscr{P}(W). \tag{2.2.1}$$

When the index set $J$ has only two elements, we write

$$A_1 \cap A_2 \text{ and } A_1 \cup A_2. \tag{2.2.2}$$

The set of all elements in $A$ which are not elements of $B$ is written $A \smallsetminus B$. When $A$ is equal to the whole set $W$ we can write $\complement B$ for $W \smallsetminus B$, the ***complement*** of a subset $B$ of $W$.

The ***Cartesian product*** of two sets $X$ and $Y$ is the set of all pairs $(x, y)$ with $x \in X$ and $y \in Y$. The Cartesian product of $n$ sets $X_j$, $j = 1, \dots, n$, denoted by

$$X_1 \times \cdots \times X_n = \prod_{j=1}^{n} X_j,$$

is the set of all $n$-tuples $(x_1, \dots, x_n)$ with $x_j \in X_j$.

## 2.3 Notation for sets of numbers

We write $\mathbf{N} = \{0, 1, 2, \dots\}$ for the set of natural numbers, following (Bourbaki 1963:67).[1] We also write $\mathbf{Z}$, $\mathbf{Q}$, $\mathbf{R}$ and $\mathbf{C}$ for, respectively, the ring of integers, the field of rational numbers, the field of real numbers, and the field of complex numbers.

We use a dot to indicate that zero is removed, thus $\dot{\mathbf{N}} = \{1, 2, \dots\}$.

We shall also use a notation for the ***extended real line***

$$\mathbf{R}_! = \mathbf{R} \cup \{-\infty, +\infty\} = [-\infty, +\infty],$$

adding two infinities. We also add these infinities to the integers,

$$\mathbf{Z}_! = \mathbf{Z} \cup \{-\infty, +\infty\} = [-\infty, +\infty]_{\mathbf{Z}}.$$

Similarly we write $Y_! = Y \cup \{-\infty, +\infty\}$ for any subset $Y$ of $\mathbf{R}$.

## 2.4 Counting with infinities

How shall we define a sum like $(+\infty) + (-\infty)$? Can addition $\mathbf{R} \times \mathbf{R} \ni (x, y) \mapsto x + y \in \mathbf{R}$ be extended to an operation $\mathbf{R}_! \times \mathbf{R}_! \ni (x, y) \mapsto x + y \in \mathbf{R}_!$ in a reasonable way?

$+\dot{+}$A convenient solution is to define two extensions, ***upper addition*** and ***lower addition***. The first is an upper semicontinuous mapping from $\mathbf{R}_! \times \mathbf{R}_!$ into $\mathbf{R}_!$; the second a lower semicontinuous mapping. They are denoted by $\dot{+}$ and $\underset{\cdot}{+}$ and are defined by the requirements of being commutative and to satisfy

$$\begin{aligned} &x \dot{+} (+\infty) = +\infty \text{ for all } x \in \mathbf{R}_!; \\ &x \dot{+} (-\infty) = -\infty \text{ for all } x \in [-\infty, +\infty[; \text{ and} \\ &\quad x \underset{\cdot}{+} y = -\big((-x) \dot{+} (-y)\big) \text{ for all } x, y \in \mathbf{R}_!. \end{aligned} \tag{2.4.1}$$

[1] Also Alfred Tarski (1956:121) calls 0 a natural number.

When there are several terms we may use the summation symbol with a dot:

$$\sum_{j=1}^{m}{}^{\bullet}\, t_j = t_1 \dotplus \cdots \dotplus t_m, \qquad t_j \in \mathbf{R}_!. \tag{2.4.2}$$

A convenient rule is the following.

**Lemma 2.4.1.** *For any element $c \in \mathbf{R}_!$ and any function $f\colon X \to \mathbf{R}_!$ defined on an arbitrary set $X$ we have*

$$\inf_{x\in X} \left(c \dotplus f(x)\right) = c \dotplus \inf_{x\in X} f(x). \tag{2.4.3}$$

Note that there are no exceptions to this formula.

**Proof.** We just need to check all possibilities where our intuition is less reliable than usual, i.e., when $c = \pm\infty$ or $X$ is empty. □

We also note the equivalence

$$a \dotplus b \geqslant c \Leftrightarrow a \geqslant c \mathbin{\underset{\cdot}{+}} (-b), \qquad a, b, c \in \mathbf{R}_!. \tag{2.4.4}$$

## 2.5 The floor and ceiling functions

The ***floor function*** and the ***ceiling function*** $\mathbf{R} \ni t \mapsto \lceil t \rceil \in \mathbf{Z}$ map the real numbers to the integers. They are defined by the inequalities

$$t - 1 < \lfloor t \rfloor \leqslant t \leqslant \lceil t \rceil < t + 1, \qquad t \in \mathbf{R}. \tag{2.5.1}$$

Note that

$$\lfloor s + t \rfloor \geqslant \lfloor s \rfloor + \lfloor t \rfloor, \qquad s, t \in \mathbf{R},$$

and that

$$\lfloor s + k \rfloor = \lfloor s \rfloor + k, \qquad s \in \mathbf{R},\ k \in \mathbf{Z}.$$

We extend the floor and ceiling functions to $\mathbf{R}_!$ in a natural way.

We note that for any index set $J$ and any family $(t_j)_{j\in J}$ of elements of $\mathbf{R}_!$ we have

$$\lfloor \inf t_j \rfloor = \inf \lfloor t_j \rfloor, \tag{2.5.2}$$

whereas

$$\lfloor \sup t_j \rfloor \geqslant \sup \lfloor t_j \rfloor, \tag{2.5.3}$$

where strict inequality can occur. By symmetry,

$$\lceil \inf t_j \rceil \leqslant \inf \lceil t_j \rceil, \tag{2.5.4}$$

where strict inequality can occur, while

$$\lceil \sup t_j \rceil = \sup \lceil t_j \rceil. \tag{2.5.5}$$

## 2.6 Relations and mappings

A ***relation between two sets*** $X$ ***and*** $Y$ is a subset of the Cartesian product $X \times Y$.

A ***mapping*** $f$ ***from a set*** $X$ ***into a set*** $Y$ is a triple $(X, Y, \rho)$ where $\rho$ is a relation between $X$ and $Y$ such that, for every $x \in X$, there is one and only one $y \in Y$ with $(x, y) \in \rho$. We write $f \colon X \to Y$ to indicate this. Following (Bourbaki 1954: Chapter II, § 5, No. 1, p. 101) we shall denote the set of all mappings from $X$ into $Y$ by $\mathscr{F}(X, Y)$. If $Y$ is a set of numbers or $\mathbf{R}_!$, a mapping is often called a ***function***.

The set $X$ is called the ***domain of definition*** of $f$; the set $Y$, the ***target set*** or ***codomain*** of $f$, and the set of all images $\{f(x);\ x \in X\}$, the ***range*** or ***image*** of $f$, denoted by $\mathbf{im}(f)$. The element $f(x) \in Y$ is said to be the ***image*** of the element $x \in X$.

The ***identity mapping*** in a set $X$ will be denoted by $\mathbf{id}_X$ and is defined by $\mathbf{id}_X(x) = x$ for all $x \in X$.

If we have three sets $X, Y, Z$ and two mappings $f \colon X \to Y$ and $g \colon Y \to Z$, we can form the ***composition*** $h = g \circ f \colon X \to Z$ of $f$ followed by $g$. It is the mapping defined by $h(x) = g(f(x))$.

If $Y$ is an abelian group like $\mathbf{Z}^n$ or $\mathbf{R}^n$, we shall call the set where $f$ is nonzero the ***support*** of $f$, to be denoted by $\operatorname{supp} f$. (It is actually the support in the usual sense if the domain is given the discrete topology.) We shall write $\mathscr{F}_{\text{finite}}(X, Y)$ for the set of all mappings with finite support.

**Definition 2.6.1.** To a mapping $f \colon X \to Y$ we associate two mappings on a higher level:

$$f^* \colon \mathscr{P}(Y) \to \mathscr{P}(X), \text{ defined by } f^*(B) = \{x \in X;\ f(x) \in B\}, \tag{2.6.1}$$

and

$$f_* \colon \mathscr{P}(X) \to \mathscr{P}(Y), \text{ defined by } f_*(A) = \{f(x) \in Y;\ x \in A\}. \tag{2.6.2}$$

We say that $f^*(B)$ is the ***preimage*** or ***inverse image*** of $B$ and that $f_*(A)$ is the ***(direct) image*** of the set $A$. Thus $f_*(\{x\}) = \{f(x)\}$. □

We note that $f_* \circ f^* \leqslant \mathbf{id}_{\mathscr{P}(Y)}$ and that $f^* \circ f_* \geqslant \mathbf{id}_{\mathscr{P}(X)}$.

Given any mapping $f \colon X \to Y$ and a subset $A$ of $X$, we define the ***restriction*** of $f$ to $A$ as the function $g \colon A \to Y$ with domain of definition equal to $A$ and taking the same values as $f$ there. We shall write $f|_A$ for the restriction, and also say that $f$ is an ***extension of*** $g$.

A mapping $f \colon X \to Y$ is said to be ***injective*** if $f(x) = f(x')$, where $x, x'$ belong to $X$, only if $x = x'$. It is called ***surjective*** if every $y \in Y$ is equal to $f(x)$ for some $x \in X$. Finally, it is called ***bijective*** if it is both injective and surjective.

If $Y = X$ we can speak about the fixed points of $f$: we say that $p \in X$ is a ***fixed point*** of $f\colon X \to X$ if $f(p) = p$. The set of all fixed points is also called the ***invariance set*** of $f$; we denote it by $\mathbf{invar}(f)$, thus

$$\text{(2.6.3)} \qquad \mathbf{invar}(f) = \{x \in X; f(x) = x\}, \qquad f\colon \mathscr{P}(X) \to \mathscr{P}(X).$$

The invariance set of a mapping carries essential information about the mapping. Within certain classes of mappings, the invariance set determines the mapping uniquely as we shall see in Section 4.3.

To every mapping $f\colon X \to Y$ we associate its ***graph***, which is the set

$$\text{(2.6.4)} \qquad \mathbf{graph}(f) = \{(x, y) \in X \times Y;\ y = f(x)\}.$$

### 2.6.1 *Linear and affine mappings*

When we have vector spaces it is of interest to study linear and affine mappings. A mapping $f\colon E \to G$ from one vector space into another is said to be ***linear*** if it satisfies

$$\text{(2.6.5)} \qquad f(x + ty) = f(x) + tf(y) \text{ for all } x, y \in E \text{ and all } t \in \mathbf{R}.$$

It follows from this inequality that, taking $x = \sum_{j=1}^{m} t_j x^{(j)}$,

$$\text{(2.6.6)} \qquad f(x) = \sum_{j=1}^{m} t_j f(x^{(j)}) \text{ for all } x^{(j)} \in E \text{ and all } t_j \in \mathbf{R}.$$

Here $m$ can be any natural number. If $G = \mathbf{R}$ we say that $f$ is a ***linear form.***

We say that $g$ is ***affine*** if it is of the form $f + c$, where $f$ is linear and $c$ is a constant in $G$. It follows that affine mappings satisfy (2.6.6) when the sum of all the $t_j$ is equal to 1.

## 2.7 Preorders

For the morphological operations on the computer screen we need to consider sets of sets. The set of all subsets of a given set is ordered by the inclusion relation, which is an example of an abstract order relation. It is therefore convenient to introduce concepts that will be useful in the general theory of order relations. In this section we shall do so.

Let us define first the concept of preorder, which is even more fundamental than that of order.

**Definition 2.7.1.** A ***preorder*** in a given set $X$ is a relation (a subset of $X \times X = X^2$) which is reflexive and transitive. A ***preordered set*** is a set together with a preorder. □

The definition means, if we denote the relation by $\leqslant$, that

(2.7.1) $$x \leqslant x, \qquad x \in X; \text{ and that}$$

(2.7.2) $$x \leqslant y \text{ and } y \leqslant z \text{ implies } x \leqslant z, \qquad x, y, z \in X.$$

We shall write $x < y$ if $x \leqslant y$ and $x \neq y$. We shall also write $x \geqslant y$ and $x > y$ for $y \leqslant x$ and $y < x$, respectively.

**Definition 2.7.2.** A preorder in a set $X$ is said to be ***directed***, if for any two elements $x, y \in X$ there is an element $z \in X$ such that $x \leqslant z$ and $y \leqslant z$. □

**Definition 2.7.3.** An ***order*** is a preorder which is antisymmetric, meaning that the relation shall satisfy

(2.7.3) $$x \leqslant y \ \&\ y \leqslant x \ \Rightarrow\ x = y, \qquad x, y \in X.$$

An ***ordered set*** is a set together with an order.

Two elements $x$ and $y$ are said to be ***comparable*** if either $x \leqslant y$ or $y \leqslant x$.

An order is said to be ***total*** if any two elements are comparable. □

A total order is always directed.

As already mentioned, a basic example of an ordered set is the power set of a set $W$ with the order relation given by inclusion, thus $A \leqslant B$ being defined as $A \subset B$ for $A, B \in \mathscr{P}(W)$.

If $A$ is a subset of a preordered set, we shall say that $b$ is a ***minorant*** of $A$ and that $c$ is a ***majorant*** of $A$ if $b \leqslant a \leqslant c$ for all $a \in A$.

An ordered set is said to be a ***causal set*** if, for each $a$ and $b$, there are only finitely many elements $x$ satisfying $a \leqslant x \leqslant b$. This concept is of interest in discrete models of spacetime (Bombelli 1987:522).

A subset $A$ of a preordered set will be called a ***down-set*** if from $a \in A$ and $b \leqslant a$ it follows that $b \in A$, and an ***up-set*** if $c \in A$ for all $c$ such that there is an element $a$ of $A$ with $c \geqslant a$.

Suppose that we have two preorders defined in a set $X$; denote them by $\leqslant$ and $\preccurlyeq$. The preorder $\leqslant$ is said to be ***finer*** than the preorder $\preccurlyeq$, and $\preccurlyeq$ is said to be ***coarser*** than $\leqslant$, if $x \leqslant y$ implies $x \preccurlyeq y$ for all $x, y$.

There is a finest preorder in a set, viz. when we define $x \leqslant y$ to mean that $x = y$. This preorder is of course an order; let us call it the ***discrete***

***order***. There is also a coursest preorder in any set $X$, when we declare that $x \leqslant y$ for all $x, y \in X$. Let us call this the ***chaotic preorder***. The set of all preorders on any set is thus an ordered set with a largest and a smallest element.

Given any preorder we define $[a, \rightarrow[$ as the set of all $x$ such that $a \leqslant x$ and $]\leftarrow, a]$ as the set of all $y$ such that $y \leqslant a$ (cf. Bourbaki 1963:26). The set $[a, \rightarrow[$ determines $a$ for all $a \in X$ if and only if $\leqslant$ is an order; similarly for $]\leftarrow, a]$.

In a preordered set $L$, given $a, b \in L$, the ***interval*** $[a, b]$ is the set $\{x \in L;\ a \leqslant x \leqslant b\}$, in particular to be used when $L = \mathbf{R}_!$. We shall write $[a, b]_{\mathbf{Z}}$ for an interval of integers even though $a$ and $b$ are non-integers, thus

$$[a, b]_{\mathbf{Z}} = [a, b] \cap \mathbf{Z} = \{x \in \mathbf{Z};\ a \leqslant x \leqslant b\}, \qquad a, b \in \mathbf{R}. \tag{2.7.4}$$

**Definition 2.7.4.** An ***equivalence relation*** is a preorder which is symmetric, i.e., such that $x \preccurlyeq y$ if and only if $y \preccurlyeq x$. □

This means that $x \leqslant y$ implies $y \leqslant x$ for all $x, y \in X$.

Given a preorder $\leqslant$ in $X$, we can introduce an equivalence relation $\simeq$ in $X$ by defining $x \simeq y$ to mean that $x \leqslant y$ and $y \leqslant x$. If $\leqslant$ is an order, then $\simeq$ is just equality. We can form the quotient space $X/\simeq$ of all equivalence classes of $X$ modulo $\simeq$. The equivalent classes are just

$$[a, \rightarrow[\ \cap\ ]\leftarrow, a] = \{x \in X;\ a \leqslant x \leqslant a\}, \qquad a \in X.$$

In this quotient space, $\leqslant$ induces an order (Exercise 2.14).

## 2.8 Mappings between preordered sets

In preordered spaces, the increasing mappings form a subfamily of importance:

**Definition 2.8.1.** If $f\colon X \to Y$ is a mapping from a preordered set $X$ to a preordered set $Y$, then we say that $f$ is ***increasing*** if

$$\text{for all } x, x' \in X, \text{ the relation } x \leqslant_X x' \text{ implies } f(x) \leqslant_Y f(x').$$

We shall write $\mathbf{incr}(X, Y)$ for the set of all increasing mappings $X \to Y$.

We shall say that $f$ is ***decreasing*** if

$$\text{for all } x, x' \in X, \text{ the relation } x \leqslant_X x' \text{ implies } f(x) \geqslant_Y f(x').$$

We shall call a mapping $f\colon X \to Y$ ***coincreasing*** if for all $x, x' \in X$ the relation $f(x) \leqslant_Y f(x')$ implies $x \leqslant_X x'$. □

The increasing mappings play the same role in the context of ordered sets as the linear mappings in the theory of vector spaces and as the continuous mappings in the theory of topological spaces.

A preorder $\leqslant$ is finer than another preorder $\preccurlyeq$ if and only if the identity mapping $(X, \leqslant) \to (X, \preccurlyeq)$ is increasing.

Increasing does not imply coincreasing; coincreasing does not imply increasing; cf. Exercise 2.14. If $X$ is ordered, then a mapping is coincreasing if and only if it is injective and the mapping $f(x) \mapsto x$ is increasing.

To emphasize the symmetry between the two concepts increasing and coincreasing, let us define, given any mapping $f\colon X \to Y$ between preordered sets, a preorder $\leqslant_f$ in $X$ by the requirement that $x \leqslant_f x'$ if and only if $f(x) \leqslant_Y f(x')$. Then $f$ is increasing if and only if $\leqslant_X$ is finer than $\leqslant_f$, and $f$ is coincreasing if and only if $\leqslant_f$ is finer than $\leqslant_X$.

If $f, g\colon X \to Y$ are increasing, then so are $f \wedge g$ and $f \vee g$. If also $h\colon Y \to Z$ is increasing, then so is $h \circ f\colon X \to Z$. In particular, when $X = Y$, we have the three operations $(f, g) \mapsto f \wedge g, f \vee g, g \circ f$, which all preserve the property of being increasing.

If $X$ has the chaotic preorder and $Y$ is ordered, then all increasing mappings are constant. If on the other hand $X$ has the discrete order, where $x \leqslant y$ means $x = y$, then $\mathbf{incr}(X, Y)$ consists of all mappings $X \to Y$, i.e., $\mathbf{incr}(X, Y) = \mathscr{F}(X, Y)$. The conclusion is the same if $Y$ is equipped with the chaotic preorder.

If $Y$ has the discrete order, then $\mathbf{incr}(X, Y)$ consists of those mappings $X \to Y$ that are constant on any chain of comparable elements, thus $f(x_j)$ is constant if $x_1 \leqslant x_2 \geqslant x_3 \leqslant \cdots \leqslant x_{2n}$.

A comparison with topology is in order here. If $f\colon X \to Y$ is a mapping of a topological space $X$ into a topological space $Y$ with topologies (families of open sets) $\tau_X$ and $\tau_Y$, we can define a new topology $\tau_f$ in $X$ as the family of all sets $\{x \in X; f(x) \in V\}$, $V \in \tau_Y$. Then $f$ is continuous if and only if $\tau_X$ is finer than $\tau_f$.

**Definition 2.8.2.** Given a preordered set $L = (L, \leqslant)$ with the relation denoted by $\leqslant$, we also consider the preordered set $(N, \leqslant) = (L, \geqslant)$, thus with the relation $\leqslant_N$ equal to $\geqslant_L$.

Given a mapping $f\colon L \to M$ we define a mapping $f^{\mathrm{op}}\colon L \to (M, \geqslant)$ taking the same values as $f$ but with the target set provided with the opposite preorder. Similarly, $f_{\mathrm{op}}\colon (L, \geqslant) \to M$, and

$$(f^{\mathrm{op}})_{\mathrm{op}} = (f_{\mathrm{op}})^{\mathrm{op}}\colon (L, \geqslant) \to (M, \geqslant)$$

with both $L$ and $M$ provided with the opposite preorders. □

We note that $(f^{\mathrm{op}})^{\mathrm{op}} = (f_{\mathrm{op}})_{\mathrm{op}} = f$. It is permissible to write just $f^{\mathrm{op}}_{\mathrm{op}}$ for $(f^{\mathrm{op}})_{\mathrm{op}} = (f_{\mathrm{op}})^{\mathrm{op}}$.

For mappings $f\colon X \to X$ with target set equal to the domain we can form the iterations $f \circ f$, $f \circ f \circ f$ and so on, and among these mappings those which satisfy $f \circ f = f$ are of interest:

**Definition 2.8.3.** We shall say that a mapping $f\colon X \to X$ is ***idempotent*** if $f \circ f = f$, i.e., if $f(f(x)) = f(x)$ for all $x \in X$. A mapping which is both increasing and idempotent will be called an ***ethmomorphism***. □

We have $\mathbf{invar}(f) \subset \mathbf{im}(f)$ with equality if and only if $f$ is idempotent.

These concepts were the starting point for Georges Matheron's development in his book (1975) and in many other publications.

**Lemma 2.8.4.** *Let $f$ and $g$ be two mappings $X \to X$ in a set $X$. Assume that $f$ is idempotent. Then* $\mathbf{invar}(f) \subset \mathbf{invar}(g)$ *if and only if* $g \circ f = f$.

**Proof.** Assume first that $\mathbf{invar}(f)$ is contained in $\mathbf{invar}(g)$. For any $x \in X$, $f(x)$ belongs to $\mathbf{invar}(f)$, hence also to $\mathbf{invar}(g)$, i.e., $g(f(x)) = f(x)$.

Conversely, if $g \circ f = f$, and $x$ belongs to $\mathbf{invar}(f)$, then $g(f(x)) = f(x) = x$, proving that $g(x) = g(f(x)) = f(x) = x \in \mathbf{invar}(g)$. □

**Definition 2.8.5.** We shall say that a mapping $f\colon X \to X$ is ***extensive*** if it is larger than the identity, i.e., $f(x) \geqslant x$ for all elements $x \in X$. We shall say that it is ***antiextensive*** if it is smaller than the identity, i.e., $f(x) \leqslant x$ for all $x \in X$. □

For extensive mappings $f$ the invariance set is decreasing in $f$, while it is increasing in $f$ for antiextensive mappings.

## 2.9 Epigraphs and hypographs

**Definition 2.9.1.** To a mapping $f\colon X \to Y$ where $Y$ possesses a preorder $\leqslant$, we associate its ***epigraph***

$$\mathbf{epi}(f) = \{(x, y) \in X \times Y;\ y \geqslant f(x)\}; \tag{2.9.1}$$

and its ***strict epigraph***

$$\mathbf{epi}_{\text{strict}}(f) = \{(x, y) \in X \times Y;\ y > f(x)\}. \tag{2.9.2}$$

If we take $Y = \mathbf{R}_! = \mathbf{R} \cup \{-\infty, +\infty\}$, we define its ***finite epigraph*** as

$$\mathbf{epi}^{\text{finite}}(f) = \{(x, y) \in X \times \mathbf{R};\ y \geqslant f(x)\}; \tag{2.9.3}$$

and its ***strict finite epigraph*** as

(2.9.4) $$\mathbf{epi}^{\text{finite}}_{\text{strict}}(f) = \{(x, y) \in X \times \mathbf{R};\ y > f(x)\}.$$

Analogously we define the ***hypograph*** of a function:

(2.9.5) $$\mathbf{hypo}(f) = \{(x, y) \in X \times Y;\ y \leqslant f(x)\},$$

and of course the ***strict hypograph***, $\mathbf{hypo}_{\text{strict}}(f)$, the ***finite hypograph***, $\mathbf{hypo}^{\text{finite}}(f)$, and the ***strict finite hypograph***, $\mathbf{hypo}^{\text{finite}}_{\text{strict}}(f)$. □

It is easy to pass from the finite epigraph to the strict finite epigraph as well as in the other direction:

(2.9.6)
$$\mathbf{epi}^{\text{finite}}_{\text{strict}}(f) = \bigcup_{c>0} \mathbf{epi}^{\text{finite}}(f + c); \qquad \mathbf{epi}^{\text{finite}}(f) = \bigcap_{c>0} \mathbf{epi}^{\text{finite}}_{\text{strict}}(f - c).$$

Similarly for $\mathbf{epi}_{\text{strict}}$ and $\mathbf{epi}$.

The finite epigraph of a function $f\colon \mathbf{R}^n \to \mathbf{R}_!$ is contained in the closure (taken for the usual topology in $\mathbf{R}^n \times \mathbf{R}$) of $\mathbf{epi}^{\text{finite}}_{\text{strict}}(f)$, maybe strictly.

The intersection of $\mathbf{epi}(f)$ and $\mathbf{hypo}(f)$ contains the graph of $f$. If $Y$ is not only preordered but ordered, the intersection is equal to the graph.

If $X$ and $Y$ are given, any mapping $X \to Y$ is determined by its graph, and, if $Y$ is an ordered set, also by its epigraph as well as by its hypograph. It is often convenient to express properties of mappings in terms of their epigraphs or hypographs; for examples, see Proposition 5.7.2 and formulas (5.8.3).

## 2.10 Calculating with sets and functions

Calculating with sets is equivalent to calculating with functions. To every set $A \subset X$ we associate its ***characteristic function*** $\chi_A$, defined by

(2.10.1) $$\chi_A(x) = 1 \text{ for } x \in A, \quad \chi_A(x) = 0 \text{ for } x \in X \smallsetminus A.$$

We also define its ***indicator function*** $\mathbf{indf}_A$, defined by

(2.10.2) $$\mathbf{indf}_A(x) = 0 \text{ for } x \in A, \quad \mathbf{indf}_A(x) = +\infty \text{ for } x \in X \smallsetminus A.$$

We have $\chi_{A\cap B} = \chi_A \wedge \chi_B$ and $\chi_{A\cup B} = \chi_A \vee \chi_B$ and similar formulas for the indicator function.

In the other direction, given a function $f\colon X \to \mathbf{R}$ we associate to it its epigraph, a subset of $X \times \mathbf{R}$, defined in (2.9.1), and any calculation on functions can be done on their epigraphs.

## 2.11 Cleistomorphisms and anoiktomorphisms

**Definition 2.11.1.** A ***cleistomorphism*** in an ordered set $X$ is an ethmomorphism (see Definition 2.8.3) $\kappa\colon X \to X$ which is extensive (see Definition 2.8.5); in other words, which satisfies the following three conditions.

$$(2.11.1)\qquad \begin{aligned} &x \leqslant y \text{ implies } \kappa(x) \leqslant \kappa(y);\\ &\kappa(\kappa(x)) = \kappa(x);\\ &x \leqslant \kappa(x), \end{aligned}$$

for all $x, y \in X$. □

The element $\kappa(x)$ is said to be the ***closure*** of $x$. Elements $x$ such that $\kappa(x) = x$ are called ***invariant*** or ***closed*** (for this operator). An element is closed if and only if it is the closure of some element (and then it is the closure of itself).

Sometimes we shall consider operators that are defined only on a subset of $X$.

Let $f\colon X_1 \to X$ be a mapping, where $X_1 \subset X$. Then it is not obvious what idempotency shall mean, for the composition $f(f(x))$ need not be defined for $x \in X_1$. We solve this problem by a modified definition; cf. (Kiselman 1969:336).[2]

**Definition 2.11.2.** We shall say that $\kappa\colon X_1 \to X$, where $X_1$ is a subset of an ordered set $X$, is a ***partial cleistomorphism*** if it the two conditions

$$(2.11.2)\qquad \begin{aligned} &x \leqslant \kappa(x) && x \in X_1;\\ &x \leqslant \kappa(y) \text{ implies } \kappa(x) \leqslant \kappa(y), && x, y \in X_1, \end{aligned}$$

are satisfied. □

It follows from this that $\kappa$ is increasing and that $\kappa(\kappa(x)) = x$ for every $x \in X_1$ such that $\kappa(x)$ happens to belong to $X_1$. It is also clear that if $X_1 = X$, then the new definition agrees with Definition 2.11.1.

Any partial cleistomorphism $\kappa$ in the sense of Definition 2.11.2 can be extended to an idempotent operator $\kappa_1\colon Z \to X$ by taking $Z = X_1 \cup \mathbf{im}(\kappa)$ and defining $\kappa_1(x) = x$ when $x \in Z \smallsetminus X_1$.

In many applications the set $X$ is the the power set $\mathscr{P}(W)$ of some set $W$. Then the cleistomorphism is given as an intersection:

$$\overline{A} = \bigcap_Y (Y; Y \text{ is closed and } Y \supset A).$$

[2]Matheron (1975:186) required instead that $f(x)$ belong to $X_1$ whenever $x \in X_1$.

In Chapter 4 we shall return to these constructions.

When $W$ is a topological space, a basic example is the topological closure operator which associates to a set in a topological space its topological closure, i.e., the smallest closed set containing the given set, denoted by $A \mapsto \overline{A}$. In fact a cleistomorphism in $\mathscr{P}(W)$ defines a topology in $W$ if and only if it satisfies, in addition to (2.11.1) above, two extra conditions, viz.

$$\overline{\varnothing} = \varnothing \text{ and } \overline{A \cup B} = \overline{A} \cup \overline{B} \text{ for all } A, B \subset W, \tag{2.11.3}$$

where $\varnothing$ denotes the empty set. In Chapter 14 we shall return to these concepts.

Another cleistomorphism of great importance is the operator which associates to a set $A$ in $\mathbf{R}^n$ its convex hull, the smallest convex set containing the given set, denoted by $\mathbf{cvxh}(A)$. The composition $A \mapsto \overline{\mathbf{cvxh}(A)}$ is a cleistomorphism, whereas the composition in the other order, $A \mapsto \mathbf{cvxh}\left(\overline{A}\right)$ is not idempotent if $n \geqslant 2$ (Exercise 2.14). We see that the composition of two cleistomorphisms is sometimes, but not always, a cleistomorphism.

**Definition 2.11.3.** With Willian Andrew Coppel (1998:7) we shall say that a cleistomorphism $\kappa\colon \mathscr{P}(W) \to \mathscr{P}(W)$ is ***algebraic***, if for any $A \subset X$ and any $x \in \kappa(A)$, there is a finite subset $B$ of $A$ such that $x$ belongs to $\kappa(B)$. □

The operation of taking the convex hull is algebraic whereas that of taking the closure in $\mathbf{R}^n$, $n \geqslant 1$, is not.

Dual to the concept of cleistomorphism is the concept of anoiktomorphism.

**Definition 2.11.4.** An ethmomorphism $\alpha\colon X \to X$ is said to be an ***anoiktomorphism*** if it is antiextensive; in other words, if the following three conditions

$$x \leqslant y \text{ implies } \alpha(x) \leqslant \alpha(y), \qquad x, y \in X;$$
$$\alpha \circ \alpha = \alpha;$$
$$\alpha(x) \leqslant x, \qquad x \in X.$$

are satisfied. □

Just as for cleistomorphisms, we can extend this definition to a more general setting:

**Definition 2.11.5.** Let $\alpha\colon X_1 \to X$ be a mapping from a subset $X_1$ of $X$ into $X$. We shall say that $\alpha$ is a ***partial anoiktomorphism*** if it satisfies the following two conditions

$$(2.11.4) \qquad \begin{aligned} &\alpha(x) \leqslant x; \\ &\alpha(x) \leqslant y \text{ implies } \alpha(x) \leqslant \alpha(x), \end{aligned}$$

for all $x, y \in X_1$. □

The composition of a cleistomorphism and an anoiktomorphism is always idempotent:

**Proposition 2.11.6.** *Let $\alpha, \kappa \colon L \to L$ be an anoiktomorphism and a cleistomorphism. Then $\eta = \alpha \circ \kappa$ and $\theta = \kappa \circ \alpha$ are ethmomorphisms.*

**Proof.** That $\eta$ and $\theta$ are increasing is obvious. Since $\kappa$ is extensive, we get

$$\eta \circ \eta = \alpha \circ \kappa \circ \alpha \circ \kappa \geqslant \alpha \circ \alpha \circ \kappa = \alpha \circ \kappa = \eta.$$

Since $\alpha$ is antiextensive, we get

$$\eta \circ \eta = \alpha \circ \kappa \circ \alpha \circ \kappa \leqslant \alpha \circ \kappa \circ \kappa = \alpha \circ \kappa = \eta,$$

so $\eta$ is idempotent. The proof for $\theta$ is similar. □

**Example 2.11.7.** A typical example is when we take $\alpha$ and $\kappa$ as the operations of taking the interior and the topological closure in a topological space, respectively; $\alpha(A) = A^\circ$, $\kappa(A) = \overline{A}$. Then a fixed point of the composition $\alpha \circ \kappa$ is called a *regular open set* and a fixed point of $\kappa \circ \alpha$ is called a *regular closed set.* These operations are in general neither extensive nor antiextensive. □

**Proposition 2.11.8.** *The infimum of a family of cleistomorphisms in a complete lattice $L$ is a cleistomorphism. The supremum of any family of anoiktomorphisms in $L$ is an anoiktomorphism.*

**Proof.** Let $\kappa_j$, $j \in J$, be cleistomorphisms and define $\kappa = \bigwedge_{j \in J} \kappa_j$, meaning that the value of $\kappa$ at $x \in L$ equals $\bigwedge_{j \in J} \kappa_j(x)$. Clearly $\kappa$ is increasing and larger than the identity. It follows that $\kappa \circ \kappa \geqslant \kappa$. To prove the opposite inequality, we note that $\kappa \circ \kappa \leqslant \kappa_j \circ \kappa_j = \kappa_j$. Taking the infimum over all $j$ we get what we want.

The result for anoiktomorphisms follows by duality. □

**Corollary 2.11.9.** *To any mapping $f \colon L \to L$ of a complete lattice $L$ into itself there exists a smallest cleistomorphism $f^\kappa$ majorizing $f$ and a largest anoiktomorphism $f_\alpha$ minorizing $f$.*

Thus if $\alpha$ is an anoiktomorphism and $\kappa$ is a cleistomorphism with $\alpha \leqslant f \leqslant \kappa$, then $\alpha \leqslant f_\alpha \leqslant f \leqslant f^\kappa \leqslant \kappa$.

**Proof.** We just take the infimum of all cleistomorphisms larger than or equal to $f$. □

## 2.12 How to measure distribution of sizes?

Matheron (1975:24) and (1975: Chapter 7, pp. 186–216) introduced granulometries with the help of anoiktomorphisms as follows. A ***granulometry*** is a family $(\alpha_s)_s$ of anoiktomorphisms indexed by nonnegative numbers or natural numbers and satisfying

(1) $\alpha_0 = \mathbf{id}$; [(3).]
(2) *The family is decreasing:* $s \leqslant t$ *implies* $\alpha_s \geqslant \alpha_t$;
(3) *They form a semigroup with multiplication table* $\alpha_s \circ \alpha_t = \alpha_{\max(s,t)}$.

If we apply the $\alpha_s$ to a component $C$ of a space $X$, the result shrinks when $s$ increases and may become empty for $s$ large enough. The infimum of all $s$ such that $\alpha_s(C)$ vanishes measures the size of $C$. Therefore, a measure of the size of the components of $\alpha_s(X)$, like the Lebegue measure in $\mathbf{R}^n$ or the cardinality in $\mathbf{Z}^n$, yields information on the distribution of sizes among the components.

Jean Serra (2007, 2010) studied these semigroups in the context of subsets of $\mathbf{Z}^n$ which are the Minkowski sum of finitely many sets, each of which is a segment of the form $\{ja;\ j \in [-m, m]_\mathbf{Z}\}$ where the coordinates $a_j$ of the vector $a = (a_1, \ldots, a_n)$ are relatively prime.

A typical granulometry is obtained by taking $\alpha_s$ given by

$$\alpha_{B_s}(A) = \{x \in X;\ x + B_s \subset A\},$$

where the $B_s$ increase with increasing $s$. See (3.2.2).

These constructions are at the very origin of mathematical morphology since it was a task to estimate the distribution of sizes in minerals like Caucasian dolomites with the formula $\mathrm{CaMg(CO_3)_2}$; see (Serra 1982: Chapter X), especially (1982:338).

## 2.13 Notes

Another notation for the complement $\complement B$ of a set $B$ is $B^c$.

Upper and lower addition, denoted by $\dot{+}$ and $\underset{\cdot}{+}$, were used early by Jean-Jacques Moreau in (1970).

For the definitions of preorder and order we have essentially followed (Bourbaki 1963:7). The first of these fundamental relations is often called a *partial order.* Garret Birkhoff (1940:5) uses *partially ordered system*; in (1948:49) *partly ordered.* Birkhoff & Mac Lane (1957:349) mention *partial ordering* and *partially ordered system.* A set provided with an order is sometimes called a "poset," an abbreviation for *partially ordered set.*

The idea of a total order goes back to Leibniz according to Garret Birkhoff (1940:5, footnote *).

For the concept of preorder, Bourbaki (1963:9) uses just this term. Birkhoff (1940:7) uses *quasi-ordered system*; in (1995:20) *quasi-ordering* and *quasi-ordered set.* Other terms for the concept of preorder are *preordering, quasiorder,* and *pseudoordering.* Nowadays *preorder* is more common (e.g., Gierz et al. 2003:1).

The terms *increasing* and *decreasing* are widely used; see for example (Bourbaki 1963:15). There a mapping which is either increasing or decreasing is said to be *monotone.* Garret Birkhoff (1995:2), Blyth & Janowitz (1972:6), Blyth (2005:5), and Grätzer (1998:20) call an increasing mapping *order-preserving* or *isotone*; the latter was used also by Garret Birkhoff (1948:49). Blyth & Janowitz (1972:6) and Blyth (2005:5) call a decreasing mapping *order-inverting* or *antitone.* Gierz et al. use *order preserving* and *monotone* (2003:5) as well as *antitone* (2003:35).

Garret Birkhoff used the term *extensive* for a mapping which is larger than the identity (1948:49). Pfaltz & Šlapal (2013) used *expansive* for this concept and *contractive* for the opposite.

The terms *ethmomorphism, cleistomorphism,* and *anoiktomorphism,* which were introduced by the author in (2007a; 2010b), are formed in analogy with the many terms *homomorphism, isomorphism, automorphism, homeomorphism* etc. Their origins are in the Classical Greek words ἠθμός₍ₘ₎ (*ethmós*) 'strainer, colander'; κλείς₍f₎ (*kleis*) 'key', κλειστός (*kleistós*) 'closed', 'that can be shut or closed'; ἄνοιξις₍f₎ (*ánoixis*) 'opening', ανοικτός (*anoiktós*) 'opened'. Ebbe Vilborg advised me concerning the choice of terms. In English we have other terms of the same origin such as *ethmoid bone, cleistogamy.*

Other terms for ethmomorphism are *morphological filter*[3] (Serra 1988d: 101, 104), *projection operator* and *projection* (Gierz et al. 2003:26).

For cleistomorphisms, Ore (1944:494) and Garret Birkhoff (1948:49) used the term *closure operation*; the latter attributed the concept to Moore,

[3]Not to be abbreviated to "filter"; cf. Definition 4.2.2.

who wrote about *closure* and *closure properties* (Moore 1910:53). Everett (1944) and Gierz et al. (2003:26) used *closure operator.* Similar terms are *closure mapping* (Blyth & Janowitz 1972:9), *closing* (Matheron 1975:187; Serra 1982:56), *hull operator* (Singer 1997:8, Coppel 1998:7), *closure map* (Grätzer 1998:69).

In French we have *fermeture*$_{(f)}$ *de Moore* (Dubreil & Dubreil-Jacotin 1964:177) and *application enveloppante* (Kiselman 1969:336; a term proposed by André Hirschowitz).

Georges Matheron used the term *closing* in his book (1975:18). In his earlier book (1967:18) he used the French term *fermeture*$_{(f)}$. Jean Serra informs me that the first appearance of the French terms *ouverture*$_{(f)}$ and *fermeture* was in a note by Georges Matheron (1965); the operations were called "associées à une transformation de Serra" there.

For anoiktomorphism there are several other terms: Tucker (1936:94) used the term *aperture* for the dual of a closure. Other choices have been *dual closure mapping* (Blyth & Janowitz 1972:9), *opening* (Matheron 1975:18, 187; Serra 1982:56), *kernel operator* (Gierz et al. 2003:26). In French, Matheron (1967:18) used the term *ouverture.*

Matheron (1988b:129) denotes the cleistomorphism $f^\kappa$ defined in Corollary 2.11.9 by $\hat{f}$ and the anoiktomorphism $f_\alpha$ by $\check{f}$.

## 2.14 Exercises

2.1. As a nasty little exercise, try to find the exact conditions under which the equality

$$\inf_x \left(c \dotplus f(x)\right) = c \dotplus \inf_x f(x)$$

holds (cf. Lemma 2.4.1).

2.2. Verify that any preorder in a set $X$ induces an order in the quotient $X/\simeq$, where $x \simeq y$ denotes the equivalence relation $x \leqslant y \leqslant x$ (cf. Definitions 2.7.1 and 2.7.3).

2.3. Consider the following four properties of a mapping $f\colon X \to Y$ between two preordered sets (see Definition 2.8.1).

(A) $f$ is increasing: $x \leqslant y \Rightarrow f(x) \leqslant f(y)$;

(B) $f$ is strictly increasing: $x < y \Rightarrow f(x) < f(y)$, where $x < y$ means that $x \leqslant y$ and $x \neq y$;

(C) $f$ is injective: $f(x) = f(y) \Rightarrow x = y$;

(D) $f$ is coincreasing: $f(x) \leqslant f(y) \Rightarrow x \leqslant y$.

Prove that (B) implies (A); that (A) & (C) implies (B); and that, if $X$ is ordered, (D) implies (C).

Prove by examples that none of the implications

$$(\mathrm{B}) \Rightarrow (\mathrm{C}); \quad (\mathrm{B})\,\&\,(\mathrm{C}) \Rightarrow (\mathrm{D}); \quad (\mathrm{D}) \Rightarrow (\mathrm{A})$$

holds, not even for mappings $f\colon X \to X$.

If we assume that $X$ is ordered and that (A) holds, we thus have (D) $\Rightarrow$ (C) and (C) $\Rightarrow$ $(B)$, but the converse implications do not hold.

2.4. Let $f$ denote the mapping $\mathscr{P}(\mathbf{R}^2) \ni A \mapsto \overline{A} \in \mathscr{P}(\mathbf{R}^2)$ of taking the topological closure of a set $A$ and $g$ the mapping $A \mapsto \mathbf{cvxh}(A)$ of taking the convex hull of $A$. Are $f \circ g$ and $g \circ f$ cleistomorphisms? Prove or disprove.

2.5. Give an example which shows that the inclusion relation for anoiktomorphisms $\alpha_j$,

$$\bigcup_{j \in J} \mathbf{invar}(\alpha_j) \subset \mathbf{invar}\left(\bigvee_{j\in J} \alpha_j\right),$$

can be strict.

# Chapter 3

# Morphological operations: Set-theoretical duality

## 3.1 Groups and semigroups

OUR PURPOSE here is to describe morphological operations in $\mathbf{Z}^2$ as well as in $\mathbf{R}^2$ and more generally in $\mathbf{Z}^n$ and $\mathbf{R}^n$. It will therefore be an advantage if we can choose a common framework for these two cases. Since $\mathbf{Z}^2$ is not a vector space, we cannot assume that we live in vector space. What $\mathbf{Z}^n$ and $\mathbf{R}^n$ have in common is that they are abelian groups. It turns out that this concept is rich enough for a theory of morphological operations to be set up.

In an abelian group $G$ we have an associative and commutative operation, written as addition, and a ***neutral element***, denoted by 0 (i.e., such that $x+0=x$ for all $x \in G$); finally every element $x$ has an inverse, written $-x$ and satisfying $x+(-x)=0$. This generality causes no problem in the definitions and proofs. However, the reader can think of the special abelian groups $\mathbf{Z}^2$ and $\mathbf{R}^2$ all the time.

For some definitions and results it is not necessary to assume the group to be commutative; sometimes it is not even necessary to assume that we have a group: a semigroup will suffice. A ***semigroup*** is a set together with an associative operation, written as juxtaposition. If it is commutative, we call it an ***abelian semigroup***, but in general we do not assume commutativity, nor the existence of a neutral element. If there is one, it will be denoted by 1 in the multiplicative notation ($x1 = 1x = x$), by 0 in the additive one ($x + 0 = 0 + x = x$). A semigroup with a neutral element is also called a ***monoid***. If every element has an inverse, the monoid is called a ***group***.

The reason for assuming so little is not generality in the first place but a kind of Occam's razor: by not assuming too much we make the constructions more transparent and clarify the dependence of a result on the hypotheses.

**Definition 3.1.1.** Let $A$ and $B$ be subsets of a semigroup $G$. Then we define their ***product*** as the set $AB = \{ab;\ a \in A,\ b \in B\}$. If the semigroup is commutative, we shall write $+$ for the operation and define the ***Minkowski sum*** of $A$ and $B$ as $A + B = \{a + b;\ a \in A,\ b \in B\}$. □

The operation makes the power set $\mathscr{P}(G)$ of all subsets of $G$ into a semigroup. In that semigroup the empty set $\emptyset$ is a ***zero***: $A\emptyset = \emptyset A = \emptyset$; and $\{1\}$ is a neutral element if $G$ happens to have a neutral element 1: $A\{1\} = \{1\}A = A$.

In the commutative case the operation $(A, B) \mapsto A + B$ is called ***Minkowski addition***. We have $A + \emptyset = \emptyset$ and $A + \{0\} = A$ if $G$ has a neutral element 0.

If $B$ is finite, as is often the case in $\mathbf{Z}^n$, only finitely many checks are needed to decide whether a point $x$ belongs to $A + B$: we check whether $x - b$ belongs to $A$ for the finitely many $b \in B$. Therefore dilation by a finite set satisfies Serra's third principle (see Section 7.2).

If $A$ consists of only one point $a$ we shall write $AB = \{a\}B = aB = S_a(B)$, where the last equality defines an operation $S_a\colon \mathscr{P}(G) \to \mathscr{P}(G)$, called ***left translation by*** $a$; all points are moved by a fixed amount. Similarly we define ***right translation by*** $b$ as the operation $T_b$, $T_b(A) = A\{b\} = Ab$. In the commutative case the two translations are equal, and we write $S_b(A) = b + A = A + b = T_b(A)$. If $G$ is a group, $S_a$ and $T_a$ have inverses, viz. $(S_a)^{-1} = S_{a^{-1}}$ and $(T_b)^{-1} = T_{b^{-1}}$; in the commutative case $(S_a)^{-1} = (T_a)^{-1} = S_{-a} = T_{-a}$.

In the sequel we shall often mention only commutative semigroups—for most applications this is sufficient.

**Definition 3.1.2.** A ***zonotope*** is the Minkowski sum of finitely many segments in $\mathbf{R}^n$. If $n = 3$, it is also called a ***zonohedron***. □

A zonotope $A$ is symmetric with respect to the point $c$ equal to the sum of the midpoints of the segments defining it, i.e., $-(A - c) = A - c$.

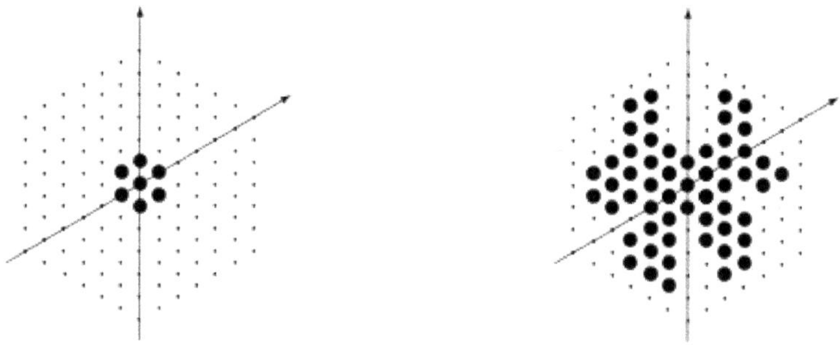

Figure 3.1. To the left a set $S$ consisting of seven pixels with Cartesian coordinates $(0,0)$, $(1,0)$, $(0,1)$, $(-1,1)$, $(-1,0)$, $(0,-1)$, $(1,-1)$ and to be used as a structuring element. To the right an image $A$ with 55 elements, to be subject in the next figures to the dilation $\delta = \delta_S$, the corresponding erosion $\varepsilon$, as well as the cleistomorphism $\kappa = \varepsilon \circ \delta$ and the anoiktomorphism $\alpha = \delta \circ \varepsilon$.

## 3.2 Dilations and erosions: Set-theoretical duality

Using translations, we see that

$$A + B = \bigcup_{y \in B} (A + y) = \bigcup_{x \in A} (x + B).$$

Thus the sum is a union of translates of $A$, and equally a union of translates of $B$. We can view it in either way: we move $A$ around in a manner determined by $B$ or vice versa. Often $A$ is a complicated set and $B$ is a simple, small set, called the ***structuring element***. Then we say that $A + B$ is $A$ ***dilated*** by $B$. Think of $G = \mathbf{R}^2$ and with $B$ equal to a small disk. We fix $B$ and apply the operation $A \mapsto A + B$ to a lot of sets $A$:

**Definition 3.2.1.** Let $G$ be a semigroup and let $B$ be a subset of $G$. We define two mappings $\delta_B, \varepsilon_B \colon \mathscr{P}(G) \to \mathscr{P}(G)$ as

$$\delta_B(A) = AB, \quad \varepsilon_B(A) = \{x;\ xB \subset A\}, \qquad A \in \mathscr{P}(G);$$

in the abelian case

$$\delta_B(A) = A + B, \quad \varepsilon_B(A) = \{x;\ x + B \subset A\}, \qquad A \in \mathscr{P}(G).$$

The first is called ***dilation*** by $B$. The second, ***erosion*** by $B$. □

To emphasize the similarity between the two definitions we may rewrite them as

$$\delta_B(A) = \bigcup_x (\{x\} + B;\ \{x\} \subset A), \qquad \varepsilon_B(A) = \bigcup_x (\{x\};\ \{x\} + B \subset A).$$

We note that dilation commutes with translation:

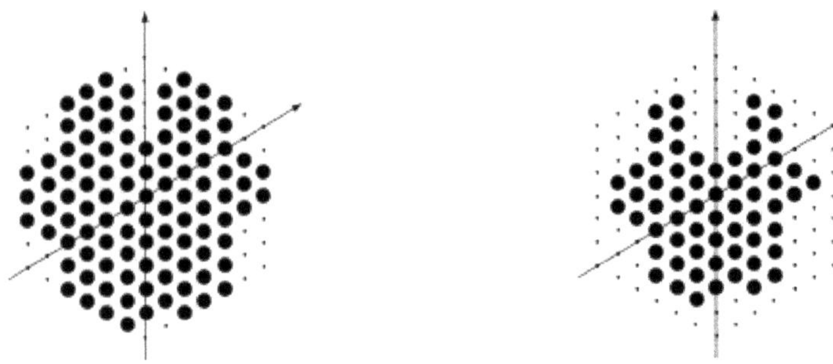

Figure 3.2. To the left the dilated set $\delta(A) = \delta_S(A)$, $S$ being as in Figure 3.1 to the left and $A$ as in Figure 3.1 to the right. To the right the closed set $\kappa(A) = \varepsilon(\delta(A))$. We have $\mathbf{card}(\kappa(A) \smallsetminus A) = 61 - 55 = 6$. We note that the lower, narrow gulf has been filled in, while the upper, wider gulf has been preserved.

$$S_x(\delta_B(A)) = x + (A + B) = (x + A) + B = \delta_B(S_x(A)), \qquad A \in \mathscr{P}(G),$$

which may be written as $S_x \circ \delta_B = \delta_B \circ S_x$. Similarly, if $G$ is a group, erosion commutes with translation:

$$\begin{aligned} S_z(\varepsilon_B(A)) &= \{z + x;\ x + B \subset A\} = \{y;\ (y - z) + B \subset A\} \\ &= \{y;\ y + B \subset z + A\} = \varepsilon_B(S_z(A)), \end{aligned} \tag{3.2.1}$$

thus $S_z \circ \varepsilon_B = \varepsilon_B \circ S_z$.

Dilation commutes also with the formation of unions:

$$\bigcup (A_j B) = \left(\bigcup A_j\right) B; \qquad \bigcup (A_j + B) = \left(\bigcup A_j\right) + B,$$

where $(A_j)_{j \in J}$ is a finite or infinite family of subsets of an arbitrary semigroup $G$. There is a converse to this statement:

**Proposition 3.2.2.** *Let $G$ be a semigroup with neutral element* 1. *Any mapping $f\colon \mathscr{P}(G) \to \mathscr{P}(G)$ which commutes with left translations and the formation of infinite unions is a dilation, in fact the dilation by $f(\{1\})$.*

**Proof.** We write $A = \bigcup_{x\in A}\{x\} = \bigcup\{x\}$ so that

$$\begin{aligned} f(A) &= f\left(\bigcup\{x\}\right) = \bigcup f(\{x\}) = \bigcup f(S_x\{1\}) \\ &= \bigcup S_x(f(\{1\})) = Af(\{1\}) = \delta_{f(\{1\})}(A). \end{aligned}$$

□

Analogously, erosion commutes with the formation of arbitrary intersections:

$$\varepsilon_B\left(\bigcap A_j\right) = \{x;\ xB \subset \bigcap A_j\} = \bigcap\{x;\ xB \subset A_j\} = \bigcap \varepsilon_B(A_j).$$

See Proposition 3.2.7. We shall take this property of commuting with the formation of unions and intersections as the starting point for a more general theory; see Chapter 4.

We now note a significant relation between the two operations.

**Proposition 3.2.3.** *Let $A$, $B$ and $C$ be three subsets of a semigroup. Then the following three statements are equivalent.*

(a) *$AB \subset C$ (in the commutative case $A + B \subset C$);*
(b) *$\delta_B(A) \subset C$;*
(c) *$A \subset \varepsilon_B(C)$.*

**Proof.** That (a) and (b) are equivalent is obvious from the definition of dilation. The definition of erosion can be written

$$x \in \varepsilon_B(C) \Leftrightarrow xB \subset C.$$

If (a) holds, the statement to the right holds for all $x \in A$. Hence $A \subset \varepsilon_B(C)$, i.e., (c) is true. Conversely, if (c) holds, then the statement to the left holds for all $x \in A$, hence $AB \subset C$. □

**Example 3.2.4.** The processing of binary images is the main source of inspiration—examples abound. A binary image consists of pixels, e.g., spots on a computer screen, and each pixel can be given an address which is a pair of integers. Thus we let $\mathbf{Z}^2$ or $\mathbf{Z}^n$ be the set of addresses. (This does not mean that each pixel must be a square; in fact a hexagonal or triangular tessellation can be given addresses in $\mathbf{Z}^2$ just like the pixels in a rectangular tessellation; see Figures 1.4–1.6.) Let us consider the dilation $\delta = \delta_S$ defined by the Minkowski sum,

$$\delta(A) = A + S = \{x + y;\ x \in A, y \in S\}, \qquad A \in \mathscr{P}(G),$$

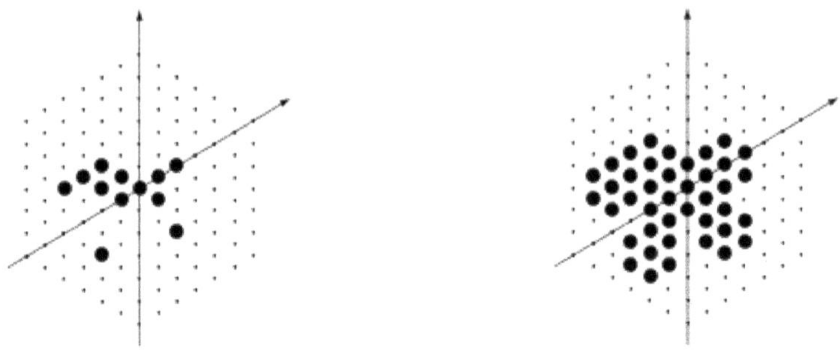

Figure 3.3. To the left the eroded set $\varepsilon(A) = (\delta_S)^{[-1]}(A)$, $S$ being as in Figure 3.1 to the left and $A$ as in Figure 3.1 to the right. To the right the open set $\alpha(A) = \delta(\varepsilon(A))$. We have $\mathbf{card}(A \smallsetminus \alpha(A)) = 55 - 41 = 14$. We note that the narrow arms (the two upper arms and the arm to the right) have disappeared, while the thicker arm to the left has survived.

where $S$ is a fixed subset of $G$, the structuring element. Here $G$ is any abelian group, for instance $\mathbf{Z}^n$ or $\mathbf{R}^n$, in particular $\mathbf{Z}^2$. We also consider the erosion $\varepsilon = \varepsilon_S$ related to this dilation:

$$\varepsilon(B) = \bigcup_A (A;\ \delta(A) \subset B) = \{x;\ x + S \subset B\}, \qquad B \in \mathscr{P}(G).$$

We can now form the compositions $\kappa = \varepsilon \circ \delta$, a cleistomorphism, and $\alpha = \delta \circ \varepsilon$, an anoiktomorphism. The effect of these four operations on a finite set $A$ is illustrated in Figures 3.1–3.3. □

If we assume that $G$ is not only a semigroup but a group, we can use inverses to define sets. To any given subset $A$ of a group $G$ we define its ***opposite set*** as $\check{A} = \{x^{-1};\ x \in A\}$; in the commutative case $\check{A} = \{-x;\ x \in A\}$. Then we can also define ***Minkowski difference*** by the formula

$$A - B = A + \check{B} = \{a - b;\ a \in A,\ b \in B\}.$$

Note that $\check{B} = \{0\} - B = 0 - B$; we may also write it as $-B$. We should not mix this up with the ***set-theoretical difference***:

$$A \smallsetminus B = A \cap \complement B = \{a \in A;\ a \notin B\}.$$

In a group, erosion is dual to dilation in a natural sense. Equivalently, an erosion of a set $A$ is an intersection of translates of $A$.

To any mapping $f\colon \mathscr{P}(X) \to \mathscr{P}(Y)$ we may define its ***set-theoretically dual mapping*** $f^{\text{dual}}$ by

$$f^{\text{dual}}(A) = \complement f(\complement A), \qquad A \in \mathscr{P}(X).$$

We shall now see that, in a group, erosion by $B$ is dual to dilation by $\check{B}$.

**Theorem 3.2.5** (Set-theoretical duality). *Let $G$ be an abelian group and $B$ a subset of $G$. Then, for all $A$,*

$$\varepsilon_B(A) = \bigcap_{y\in B} Ay^{-1} = \complement \bigcup_{y\in B} \complement\left(Ay^{-1}\right) = \complement \bigcup_{y\in B} \left((\complement A)\, y^{-1}\right) = \complement\delta_{\check{B}}\left(\complement A\right).$$

*With additive notation this becomes*

$$\varepsilon_B(A) = \bigcap_{y\in B} (A-y) = \complement \bigcup_{y\in B} \complement\left(A-y\right) = \complement \bigcup_{y\in B} \left((\complement A) - y\right) = \complement\delta_{\check{B}}\left(\complement A\right).$$

In the commutative case we can write

(3.2.2) $$\varepsilon_B(A) = \{x;\ x+B \subset A\} = \complement\left((\complement A) + \check{B}\right).$$

**Proof.** To prove the first equality, note that $x$ belongs to $\varepsilon_B(A)$ if and only if $xy \in A$ for all $y \in B$, which is equivalent to $x \in Ay^{-1}$ for all $y \in B$, i.e., to $x$ belonging to $\bigcap_{y\in B} Ay^{-1}$. The second equality is just one of De Morgan's laws. The third equality depends on the fact that the formation of complement commutes with right translation; see the next lemma. The last equality follows from the definition of $\delta_{\check{B}}$. We are done. □

In the proof of the theorem we needed the following result.

**Lemma 3.2.6.** *Let $G$ be a group. Then the formation of complement commutes with left and right translation:*

$$\complement(zA) = z(\complement A) \ \text{ and } \ \complement(Az) = (\complement A)z, \qquad A \in \mathscr{P}(G), \quad z \in G,$$

*briefly $\complement \circ S_z = S_z \circ \complement$ and $\complement \circ T_z = T_z \circ \complement$.*

**Proof.** If we analyze the relations $x \in \complement(zA)$ and $x \in z(\complement A)$ we find that they are both equivalent to $z^{-1}x \notin A$. Similarly, $y \in \complement(Az)$ and $y \in (\complement A)z$ are equivalent to $yz^{-1} \notin A$. □

Thus Theorem 3.2.5 says that

$$\varepsilon_B = (\delta_{\check{B}})^{\text{dual}} = \complement \circ \delta_{\check{B}} \circ \complement,$$

from which it follows that $\delta_B = (\varepsilon_{\check{B}})^{\text{dual}}$.

From Proposition 3.2.2 and Theorem 3.2.5 we now deduce a result on erosions using this duality.

**Proposition 3.2.7.** *Let $G$ be a group. Any mapping $g\colon \mathscr{P}(G) \to \mathscr{P}(G)$ which commutes with left translations and the formation of infinite intersections is an erosion, viz. the erosion by $\complement\left[g\left(\complement\{1\}\right)\right]^{\smile}$.* □

## 3.3 Convolution

For comparison with infimal convolution, to be defined in the next section, we recall here the definition of classical convolution.

**Definition 3.3.1.** Given two functions $F, G\colon \mathbf{R}^n \to \mathbf{R}$, we define their ***convolution product*** $H = F * G$ by the formula

$$H(x) = (F * G)(x) = \int_{\mathbf{R}^n} F(x-y)G(y)\mathrm{d}y, \qquad x \in \mathbf{R}^n, \tag{3.3.1}$$

whenever the integral can be given a sense. □

We then define a discrete counterpart:

**Definition 3.3.2.** Given a semigroup $G$ and a vector space $E$ as well as two functions $f, g\colon G \to E$, we define their ***discrete convolution product*** $h = f * g$ by

$$h(z) = (f * g)(z) = \sum_{xy=z} f(x)g(y), \qquad z \in G, \tag{3.3.2}$$

whenever the sum can be given a sense. In case $G$ is commutative, we write

$$h(z) = (f * g)(z) = \sum_{x+y=z} f(x)g(y), \tag{3.3.3}$$

where the sum goes over all elements $x, y$ with sum equal to the argment of $h$. □

If $G$ is an abelian group, there is only one element $y$ for given $z$ and $x$, viz. $y = x^{-1}z$ or $y = z - x$.

The Kronecker delta $\delta_0$ is a ***neutral element*** for the discrete convolution: $f * \delta_0 = f$ for all functions $f$. Here the ***Kronecker delta*** $\delta_a$ placed at $a$ is defined by

$$\delta_a(a) = 1, \quad \delta_a(x) = 0 \text{ for } x \neq a. \tag{3.3.4}$$

When $G = \mathbf{Z}$ or $G = \mathbf{R}$, an important function is the ***Heaviside function*** $Y_a$, defined by

$$Y_a(x) = 0 \text{ for } x < a; \quad Y_a(x) = 1 \text{ for } x \geqslant a. \tag{3.3.5}$$

We note the formula $Y_a * \delta_b = Y_{a+b}$ for $a, b \in G$. For $G = \mathbf{Z}$ we have $(\delta_0 - \delta_1) * Y_a = \delta_a$, $a \in \mathbf{Z}$.

Discrete convolution will be studied in detail in Chapter 13.

## 3.4 Infimal convolution

The product $AB$ of two subsets can be put into the wider framework of infimal convolution.

**Definition 3.4.1.** Given two functions $f, g\colon G \to \mathbf{R}_!$ defined on a semigroup $G$ and with values in the extended real line, we define a new function $h = f \sqcap g$, called the ***infimal convolution*** of $f$ and $g$, as

$$(3.4.1) \qquad (f \sqcap g)(z) = h(z) = \inf_{x,y \in G} \big(f(x) \dot{+} g(y);\ xy = z\big), \quad z \in G.$$

And of course

$$(3.4.2) \qquad (f \sqcap g)(z) = h(z) = \inf_{x,y \in G} \big(f(x) \dot{+} g(y);\ x + y = z\big), \quad z \in G,$$

when $G$ is commutative. □

The infimum is taken over all elements $x, y \in G$ such that their product is $z$, the argument of $h$. If there are no such elements, the infimum is plus infinity; that element is by definition the infimum over the empty set. For the definition of upper addition $\dot{+}$ used here, see formula (2.4.1).

The points where $f$ or $g$ takes the value $+\infty$ play no role in the formation of the infimum: the definition of upper addition guarantees this. Removing these points therefore yields an equivalent definition:

$$(3.4.3) \qquad (f \sqcap g)(z) = \inf_{x,y \in G} \big(f(x) + g(y);\ xy = z,\ f(x), g(y) < +\infty\big),$$

defined for all $z \in G$. □

The ***effective domain***, written $\mathbf{dom}(f)$, of a function $f\colon X \to \mathbf{R}_!$ defined on an arbitrary set $X$ is the set where it is strictly less than plus infinity:

$$(3.4.4) \qquad \mathbf{dom}(f) = \{x \in X;\ f(x) < +\infty\}.$$

With this notation we can write (3.4.3) as

$$(f \sqcap g)(z) = \inf_{\substack{x \in \mathbf{dom}(f) \\ y \in \mathbf{dom}(g) \\ xy = z}} \big(f(x) + g(y)\big), \qquad z \in G.$$

Here there is no doubt about the meaning of the sum; the disadvantage with this approach is that we have to remember, each time we take an infimum, over which set the variables range.

Intuitively, plus infinity corresponds to vacuum and $-\infty$ to an infinitely dense neutron star. This may appear to be upside down. However, we should think of the density as $\mathrm{e}^{-f(x)}$, and then of course $\mathrm{e}^{-(+\infty)} = 0$. Infimal convolution of $f$ and $g$ is related to supremal convolution, defined in (5.8.1) below, of the functions $\mathrm{e}^{-f}, \mathrm{e}^{-g}$, viz.

$$\sup_y \left(\mathrm{e}^{-f(y)}\mathrm{e}^{-g(x-y)}\right) = \mathrm{e}^{-(f\sqcap g)(x)}.$$

The supremum is often comparable to integration in $\mathbf{R}^n$, which means that we sometimes have a remarkably good approximation

$$\begin{aligned} \mathrm{e}^{(f\sqcap g)(x)} &= \sup_{y\in\mathbf{R}^n} \left[\mathrm{e}^{-f(y)}\mathrm{e}^{-g(x-y)}\right] \approx \int_{\mathbf{R}^n} \mathrm{e}^{-f(y)}\mathrm{e}^{-g(x-y)}dy \\ &= \left[\mathrm{e}^{-f} * \mathrm{e}^{-g}\right](x), \qquad x \in \mathbf{R}^n, \end{aligned} \tag{3.4.5}$$

where the asterisk denotes usual convolution (Definition 3.3.1).

**Proposition 3.4.2.** *Infimal convolution is associative:*

$$(f \sqcap g) \sqcap h = f \sqcap (g \sqcap h).$$

**Proof.** We start calculating:

$$\left[(f \sqcap g) \sqcap h\right](z) = \inf_{\substack{y,\, x_3 \\ yx_3 = z}} \Big[\Big( \inf_{\substack{x_1,\, x_2 \\ x_1x_2 = y}} \big(f(x_1) \dotplus g(x_2)\big)\Big) \dotplus h(x_3)\Big].$$

According to Lemma 2.4.1 this expression is equal to

$$\begin{aligned} &\inf_{\substack{y,\, x_3 \\ yx_3 = z}} \inf_{\substack{x_1,\, x_2 \\ x_1x_2 = y}} \left[\big(f(x_1) \dotplus g(x_2)\big) \dotplus h(x_3)\right] \\ &= \inf_{\substack{x_1,\, x_2,\, x_3 \\ x_1x_2x_3 = z}} \left[\big(f(x_1) \dotplus g(x_2)\big) \dotplus h(x_3)\right]. \end{aligned}$$

A similar calculation shows that also $(f \sqcap (g \sqcap h))(z)$ can be transformed to the last expression. This proves associativity. □

If $G$ is a group, we know that $xy = z$ if and only if $y = x^{-1}z$, which in turn is equivalent to $x = zy^{-1}$, so the definition of $f \sqcap g$ can be written

$$(f \sqcap g)(z) = \inf_{x\in G} \left[f(x) \dotplus g(x^{-1}z)\right] = \inf_{y\in G} \left[f(zy^{-1}) \dotplus g(y)\right].$$

In the case of an abelian group this formula reads

$$(f \sqcap g)(z) = \inf_{x\in G} \left[f(x) \dotplus g(z-x)\right] = \inf_{y\in G} \left[f(z-y) \dotplus g(y)\right].$$

Now why is infimal convolution more general that the formation of products of sets or Minkowski addition? This is because of the formula

$$\mathbf{dom}(f \sqcap g) = \mathbf{dom}(f) + \mathbf{dom}(g), \tag{3.4.6}$$

which is easily proved.

A special case of this formula is obtained when we consider indicator functions, defined in (2.10.2). It is clear that $\mathbf{dom}(\mathbf{indf}_A) = A$. We then have $\mathbf{indf}_A \sqcap \mathbf{indf}_B = \mathbf{indf}_{AB}$ for all subsets $A, B$ of a semigroup $G$, or, in the commutative case, $\mathbf{indf}_A \sqcap \mathbf{indf}_B = \mathbf{indf}_{A+B}$. Hence the Minkowski sum may be defined in terms of infimal convolution as $A+B = \mathbf{dom}(\mathbf{indf}_A \sqcap \mathbf{indf}_B)$.

If $G$ has a neutral element 1, then $\mathbf{indf}_{\{1\}}$ is a neutral element for infimal convolution; when using addition, $\mathbf{indf}_{\{0\}}$.

More generally, if $f(x) = 0$ when $x \in A$ and $f(x) \geqslant 1$ otherwise and similarly with $g$ and $B$, then $(f \sqcap g)(z) = 0$ when $z \in AB$ and $(f \sqcap g)(z) \geqslant 1$ when $z \notin AB$.

Thus multiplication of sets and Minkowski addition can be expressed in terms of infimal convolution. But we can go also in the other direction and express any infimal convolution as a product of sets, albeit at the expense of adding one more dimension. This is done using strict epigraphs.

If $X = G$ is a semigroup, then we make $G \times \mathbf{R}$ into a semigroup by defining $(x,s)(y,t) = (xy, s+t)$; in the commutative case $(x,s)+(y,t) = (x+y, s+t)$. It is not difficult to show that

$$\mathbf{epi}^{\text{finite}}_{\text{strict}}(f \sqcap g) = (\mathbf{epi}^{\text{finite}}_{\text{strict}} f)(\mathbf{epi}^{\text{finite}}_{\text{strict}} g).$$

This means that the function $f \sqcap g$ can be defined as the function whose strict finite epigraph is the product $(\mathbf{epi}^{\text{finite}}_{\text{strict}} f)(\mathbf{epi}^{\text{finite}}_{\text{strict}} g)$. With the additive notation we have

$$\mathbf{epi}^{\text{finite}}_{\text{strict}}(f \sqcap g) = \mathbf{epi}^{\text{finite}}_{\text{strict}} f + \mathbf{epi}^{\text{finite}}_{\text{strict}} g. \tag{3.4.7}$$

For the finite epigraph we always have

$$\mathbf{epi}^{\text{finite}}(f \sqcap g) \supset \mathbf{epi}^{\text{finite}} f + \mathbf{epi}^{\text{finite}} g,$$

but here the inclusion relation may be strict (see Exercise 3.13).

Let us also note that the marginal function, to be defined in Definition 9.21.1, is actually an infimal convolution: If we define $G(x,y) = 0$ when $x = 0$ and as $+\infty$ otherwise, then the marginal function of $F$ is

$$\begin{aligned} F_{\text{marg}}(x) &= \inf_{y' \in \mathbf{R}^m} F(x,y') = \inf_{(x',y') \in \mathbf{R}^n \times \mathbf{R}^m} \big(F(x',y') \dotplus G(x-x', y-y')\big) \\ &= (F \sqcap G)(x,y) \end{aligned}$$

for all $(x,y) \in \mathbf{R}^n \times \mathbf{R}^m$. (Here $F \sqcap G$ is a function of $(x,y)$ but actually independent of $y$.)

## 3.5 Approximation using infimal convolution

Infimal convolution is useful when we want to approximate a given function by functions with better regularity, for instance with a finite Lipschitz constant—for the definition of the Lipschitz constant, see (7.3.5).

**Proposition 3.5.1.** *Define $V_k(x) = k\|x\|$ for $x \in \mathbf{R}^n$ and $k > 0$, and $f_k = f \sqcap V_k$. If $f$ is lower semicontuous, if $f \geqslant -V_1$, and if $f(p) < +\infty$ for at least one point $p$, then $f \sqcap V_k$, which has Lipschitz constant at most $k$, tends to $f$ as $k$ tends to plus infinity. For functions on $\mathbf{Z}^n$ the same result holds without any assumption of semicontinuity.*

**Proof.** Since $f(p) < +\infty$ we see that $(f \sqcap V_k)(x) \leqslant f(p) \dotplus V_k(x-p) < +\infty$ for all $x$. Clearly $f \sqcap V_k$, as the infimum of a family of functions with Lipschitz constant $\leqslant k$, also has Lipschitz constant $\leqslant k$. It is the largest minorant of $f$ with this property.

If $f(x) > c$, then $f(y) > c$ for all points $y$ close to $x$, say for all $y$ with $\|y - x\| \leqslant s$. Then $(f \sqcap V_k)(x)$ is the minimum of

$$\inf_{\|y-x\| \leqslant s} \big(f(y) + V_k(x-y)\big) \text{ and } \inf_{\|y-x\| > s} \big(f(y) + V_k(x-y)\big).$$

The first expression is $\geqslant c$; the second majorizes

$$-V_1(y) + V_k(x-y) \geqslant (k-1)\|x-y\| - \|x\| > (k-1)s - \|x\| > c$$

when $k$ is large (depending on $x$).

The same proof holds for functions $\mathbf{Z}^n \to \mathbf{R}_!$. It is easy to prove similar results for other metric spaces, taking $V_k(x) = kd(x, 0)$. □

We can easily get better regularity by substituting another function for $V_k$ in this approximation: see Proposition 9.10.3 on page 177.

## 3.6 Iterated infimal convolutions

Let us see what happens if we iterate infimal convolutions several times. The result will serve in the construction of distances.

**Proposition 3.6.1.** *Let $f_j \colon G \to \mathbf{R}_!$, $j \in J$, be arbitrary functions on an abelian group $G$ indexed by an index family $J$. Define, for any finite subset $K$ of $J$, $h_K$ as the infimal convolution of all the $f_j$ with $j \in K$. If $f_j \geqslant s_j > 0$, then $h_K \geqslant c_K$, where $c_K = \sum_{j \in K} s_j$.*

*If the $f_j$ are nonnegative, then the infimum*

$$h = \inf_{K \in \mathscr{P}_{\text{finite}}(J)} h_K$$

*is nonnegative. If moreover $f_j(0) = 0$ and $f_j(x) \geqslant s_j > 0$ for $x \neq 0$, then the $h_K$ satisfy $h_K \geqslant \inf_{j\in K} s_j = t_K > 0$ outside the origin.*

*The infimum $h$ is subadditive.*

*If all the $f_j$ are equal to a single subadditive function $f$, then $h_K = h = f$ for all $K \in \mathscr{P}_{\text{finite}}(G)$.*

**Proof.** The proof is easy, except perhaps for the subadditivity of $h$.

The function $h_K$ is well defined since infimal convolution is commutative and associative. We have

$$h_K(x) = \inf_{\sum x_j = x} \sum{}^{\bullet} f_j(x_j), \qquad x \in G,$$

where the summation symbol with a dot is defined in (2.4.2). It is easy to see that

$$h_K(x) \dotplus h_L(y) \geqslant h_{K\cup L}(x+y),$$

if $K$ and $L$ are two finite and disjoint subsets of $J$. This is because

$$\sum_{z\in K} z = x \;\&\; \sum_{z\in L} z = y \quad \Rightarrow \quad \sum_{z\in K\cup L} z = x+y,$$

so that the infimum of all upper sums of $f(z)$ with $z \in K \cup L$ satisfying the left-hand side of this implication is larger than the infimum of all $f(z)$ satisfying the right-hand side. We now take the infima over all $K$ and all $L$ in the inequality

$$h_K(x) \dotplus h_L(y) \geqslant h(x+y)$$

to see that $h$ is subadditive. □

We see that if $\sum s_j = +\infty$, then $c_K \nearrow +\infty$ as $K$ grows.

We also note that any function $d_0$ such that $d(x,y) = d_0(x-y)$ is a metric is subadditive and therefore $d_0 \sqcap \cdots \sqcap d_0 = d_0$. (Cf. Lemmas 16.2.1 and 16.5.1.)

## 3.7 A more general infimal convolution

However, it is possible here to use another operation than $\dotplus$, for instance $f \vee g$, following Jesús Angulo (2015, 2017). More generally, we can define

(3.7.1) $$h(x) = (f \sqcap^{\diamond} g)(x) = \inf_{y+z=x} \big(f(y) \diamond g(z)\big), \qquad x \in G,$$

and then investigate the conditions to be imposed on $\diamond$ to get some desirable properties of this more general operation $\sqcap^{\diamond}$. It seems that $\diamond = \vee$ and $\diamond = \dot{+}$ are the most often studied ones, but a general study may be interesting.

A related topic is taken up by Angulo (2019), who studies a discrete Laplacian on an ultrametric space. See also (Zhang et al. 2019).

## 3.8 Restricted Minkowski addition

We define a restricted Minskowski addition by restricting the points in Definition 3.1.1 by means of a relation:

**Definition 3.8.1.** Let $\rho$ be a relation between elements in an abelian semigroup $G$, written $a\,\rho\,b$ for $a, b \in G$. The ***restricted Minkowski sum with respect to*** $\rho$ of $A$ and $B$ is

$$A +^{\rho} B = \{a + b;\ a \in A,\ b \in B,\ a\,\rho\,b\}, \tag{3.8.1}$$

thus imposing a well-defined restriction on the admissible elements in the formation of the sum. □

If all pairs of elements of $G$ have the relation $\rho$, then there is no restriction, and we get $A +^{\rho} B = A + B$, usual Minkowski addition. The other extreme case is when $\rho$ is equality. Then $A +^{\rho} B = 2(A \cap B)$, twice the intersection of the two sets.

Given a subset $J$ of $\{1, \dots, n\}$ we define a relation $\rho$ in $\mathbf{R}^n$ by saying that $x\,\rho\,y$ if and only if $x_j = y_j$ for all $j \in J$. We shall then denote $A +^{\rho} B$ by $A +_J B$.

When the sets are strict finite epigraphs in $\mathbf{R}^n \times \mathbf{R}$ of functions $\mathbf{R}^n \to \mathbf{R}_!$, three choices of subsets of $\{1, 2, \dots, n, n+1\}$ are of interest: $J = \{1, \dots, n\}$, $J = \varnothing$, and $J = \{n+1\}$. With the first choice we have

$$\mathbf{epi}^{\text{finite}}_{\text{strict}}(f) +_{\{1,\dots,n\}} \mathbf{epi}^{\text{finite}}_{\text{strict}}(g) = \mathbf{epi}^{\text{finite}}_{\text{strict}}(f \dot{+} g) \subset \mathbf{R}^n \times \mathbf{R},$$

thus usual upper addition of the function values.

With the second choice we get, as already noted above,

$$\begin{aligned}\mathbf{epi}^{\text{finite}}_{\text{strict}}(f) + \mathbf{epi}^{\text{finite}}_{\text{strict}}(g) &= \mathbf{epi}^{\text{finite}}_{\text{strict}}(f) +_{\varnothing} \mathbf{epi}^{\text{finite}}_{\text{strict}}(g) \\ &= \mathbf{epi}^{\text{finite}}_{\text{strict}}(f \sqcap^{\dot{+}} g) = \mathbf{epi}^{\text{finite}}_{\text{strict}}(f \sqcap g),\end{aligned}$$

usual infimal convolution.

The operation $(A, B) \mapsto A +_{\{n+1\}} B$ is sometimes called ***parallel addition***; it appears in the third choice mentioned above. We see that parallel addition yields

$$\mathbf{epi}^{\text{finite}}_{\text{strict}}(f) +_{\{n+1\}} \mathbf{epi}^{\text{finite}}_{\text{strict}}(g) = \mathbf{epi}^{\text{finite}}_{\text{strict}}(f \sqcap^{\vee} g),$$

a variant of infimal convolution mentioned above in Section 3.7.

Thus restricted Minkowski additions $\mathbf{epi}^{\text{finite}}_{\text{strict}}(f) +_J \mathbf{epi}^{\text{finite}}_{\text{strict}}(g)$ for some choices of the set $J$ define important operations on functions.

## 3.9 Combining dilations and erosions

Having defined dilations and erosions, we shall now combine them.

It is clear that $(A+B)+C = A+(B+C)$, so the composition of two dilations is a dilation: $\delta_C \circ \delta_B = \delta_{B+C}$. Similarly,

$$\varepsilon_C(\varepsilon_B(A) = \varepsilon_{B+C}(A),$$

so the composition of two erosions is an erosion: $\varepsilon_C \circ \varepsilon_B = \varepsilon_{B+C}$. (In the non-commutative case the conclusions are the same.) But what about the composition of a dilation and an erosion? Let us consider $\varphi = \varepsilon_C \circ \delta_B$, defined by $\varphi(A) = \varepsilon_C(A+B)$, where $B$ and $C$ are fixed subsets of an abelian group. This operation is always increasing, and it is extensive if and only if $C \subset B$, which in turn is equivalent to $\varphi(\{0\}) \supset \{0\}$. In this section we shall study this operator when $C = B$ (this is the only really interesting case) and see that it is idempotent then (cf. Exercise 3.13). We shall also see that, in typical cases, it is neither a dilation nor an erosion.

**Definition 3.9.1.** Given two subsets $A$ and $B$ of a semigroup $G$ we define a cleistomorphism $\kappa_B = \varepsilon_B \circ \delta_B$:

$$\kappa_B(A) = \varepsilon_B(\delta_B(A)) = \{x \in G;\ xB \subset AB\}, \qquad A \in \mathscr{P}(G); \tag{3.9.1}$$

in the commutative case

$$\kappa_B(A) = \{x \in G;\ x + B \subset A + B\}. \tag{3.9.2}$$

Analogously we define an anoiktomorphism $\alpha_B = \delta_B \circ \varepsilon_B$:

$$\alpha_B(A) = \delta_B(\varepsilon_B(A)) = \bigcup_{x \in G}(xB;\ xB \subset A), \qquad A \in \mathscr{P}(G); \tag{3.9.3}$$

and

$$\alpha_B(A) = \bigcup(x + B;\ x + B \subset A). \tag{3.9.4}$$

in the commutative case. □

For $\kappa_B$, this means that we perform on $A$ first a dilation by $B$, then an erosion by the same set. We shall see that this is actually a cleistomorphism. We call $\kappa_B(A)$ the ***B-closure*** of $A$, and a set $A$ is called ***B-closed*** if $\kappa_B(A) = A$. For $\alpha_B$ we perform first an erosion by $B$ and then a dilation by $B$. We call $\alpha_B$ the $B$-***anoiktomorphism***, and a set $A$ is called $B$-***open*** if $\alpha_B(A) = A$.

Extreme special cases are when $B = \varnothing$ or $B = G$:

$$\delta_\varnothing(A) = \alpha_\varnothing(A) = \varnothing; \quad \varepsilon_\varnothing(A) = \kappa_\varnothing(A) = G \text{ for all } A;$$

$$\delta_G(A) = \kappa_G(A) = G \text{ when } A \neq \varnothing; \quad \delta_G(\varnothing) = \kappa_G(\varnothing) = \varnothing;$$

$$\varepsilon_G(A) = \alpha_G(A) = \varnothing \text{ when } A \neq G; \quad \varepsilon_G(G) = \alpha_G(G) = G.$$

**Theorem 3.9.2.** *For any subset $B$ of an arbitrary semigroup $G$ the mapping $\kappa_B = \varepsilon_B \circ \delta_B$ is a cleistomorphism and the mapping $\alpha_B = \delta_B \circ \varepsilon_B$ is an anoiktomorphism.*

The theorem will follow on combining Proposition 3.9.3 and Corollary 3.9.5 below.

**Proposition 3.9.3.** *For any subset $B$ of a semigroup $G$ the mappings $\kappa_B = \varepsilon_B \circ \delta_B$ and $\alpha_B = \delta_B \circ \varepsilon_B$ are increasing; the first is extensive and the second is antiextensive.*

**Proof.** It is obvious that the mappings are increasing. Formula (2.11.1) shows that $\kappa_B(A)$ contains $A$; similarly we see from (2.11.4) that $\alpha_B(A)$ is contained in $A$. □

**Corollary 3.9.4.** *For any subset $B$ of a semigroup $G$ we have*

$$\delta_B \circ \varepsilon_B \circ \delta_B = \delta_B \text{ and } \varepsilon_B \circ \delta_B \circ \varepsilon_B = \varepsilon_B.$$

**Proof.** Let us omit the subscript $B$ for brevity. Then $\delta \circ \varepsilon \circ \delta \leqslant \delta$ since $\delta \circ \varepsilon \leqslant \mathbf{id}$. But we also have $\delta \circ \varepsilon \circ \delta \geqslant \delta$ since $\varepsilon \circ \delta \geqslant \mathbf{id}$ and $\delta$ is increasing.

Similarly, $\varepsilon \circ \delta \circ \varepsilon \leqslant \varepsilon$ since $\delta \circ \varepsilon \leqslant \mathbf{id}$ and $\varepsilon$ is increasing. But we also get $\varepsilon \circ \delta \circ \varepsilon \geqslant \varepsilon$ since $\varepsilon \circ \delta \geqslant \mathbf{id}$. □

**Corollary 3.9.5.** *For any subset $B$ of a semigroup $G$ the mappings $\kappa_B = \varepsilon_B \circ \delta_B$ and $\alpha_B = \delta_B \circ \varepsilon_B$ are idempotent.*

**Proof.** We have $\kappa \circ \kappa = (\varepsilon \circ \delta)(\varepsilon \circ \delta) = (\varepsilon \circ \delta \circ \varepsilon) \circ \delta = \varepsilon \circ \delta = \kappa$. Similarly $\alpha \circ \alpha = (\delta \circ \varepsilon) \circ (\delta \circ \varepsilon) = (\delta \circ \varepsilon \circ \delta) \circ \varepsilon = \delta \circ \varepsilon = \alpha$. □

**Corollary 3.9.6.** *For any subsets $A, B$ of a semigroup $G$, $\delta_B(A) = AB$ is $B$-open and $\varepsilon_B(A)$ is $B$-closed.*

**Proof.** We have $\alpha \circ \delta = \delta \circ \varepsilon \circ \delta = \delta$, so $\alpha_B(\delta_B(A)) = \delta_B(A)$, showing that $\delta_B(A)$ is $B$-open. Similarly $\kappa \circ \varepsilon = \varepsilon \circ \delta \circ \varepsilon = \varepsilon$ and $\kappa_B(\varepsilon_B(A)) = \varepsilon_B(A)$; $\varepsilon_B(A)$ is $B$-closed. □

For any fixed $B$, the family of all sets $\varepsilon_B(A)$, $A \subset G$, is equal to the family of all $B$-closed sets. This is the same as the family of all sets of the form $\kappa_B(C)$, $C \subset G$.

If $G$ is a group, $\varepsilon_B$ is the mapping set-theoretically dual $\delta_{\check{B}}$, i.e., $\varepsilon_B = (\delta_{\check{B}})^{\text{dual}} = \complement \circ \delta_{\check{B}} \circ \complement$, so we have

$$\alpha_B = \delta_B \circ \varepsilon_B = \delta_B \circ \complement \circ \delta_{\check{B}} \circ \complement = \complement \circ \varepsilon_{\check{B}} \circ \delta_{\check{B}} \circ \complement = \complement \kappa_{\check{B}} \complement = (\kappa_{\check{B}})^{\text{dual}},$$

so that $\alpha_B$ and $\kappa_{\check{B}}$ are dual to each other. This implies that $\complement(\alpha_B(A)) = \kappa_{\check{B}}(\complement A)$ and shows that a set is $B$-open if and only if its complement is $\check{B}$-closed.

It may be convenient to put dilations, erosions, cleistomorphisms, and anoiktomorphisms under a common roof; see Section 6.1.

## 3.10 Commuting with translations

In an abelian group the mappings that commute with translations are of special significance. We can characterize cleistomorphisms which commute with translations in terms of Minkowski addition as follows.

**Theorem 3.10.1.** *Let $G$ be an abelian group and $\mathscr{H}$ a subfamily of $\mathscr{P}(G)$ such that $\mathscr{H}$ contains all singleton sets $\{x\}$, $x \in G$, and such that $A+B \in \mathscr{H}$ for all $A, B \in \mathscr{H}$. Let $f\colon \mathscr{H} \to \mathscr{P}(G)$ be a partial cleistomorphism (see Definition 2.11.2). Then the following conditions are equivalent.*

(A) *$f$ commutes with all translations: $f \circ T_x = T_x \circ f$ for all $x \in G$;*
(B) *$f(A+x) \supset f(A)+x$ for all $A \in \mathscr{H}$ and all $x \in G$;*
(C) *$f(A+B) \supset f(A)+B$ for all $A, B \in \mathscr{H}$ (thus $f \circ \delta_B \supset \delta_B \circ f$ for all $B \in \mathscr{H}$);*
(D) *$f(A+B) \supset f(A)+f(B)$ for all $A, B \in \mathscr{H}$.*

**Proof.** That (A) and (B) are equivalent is easily proved.

Assume that (B) holds. Then $f(A+B) \supset f(A+b) \supset f(A)+b$ for all $b \in B$. Now (C) follows on taking the union over all $b$.

If (C) holds we know, after interchanging $A$ and $B$, that $f(A+B) \supset A+f(B)$, which implies that $f(A+B) \supset f(A+f(B))$ by (2.11.2). Applying (C) a second time we see that the latter set contains $f(A)+f(B)$; hence (D) holds.

Finally, if (D) holds, then $f(A+x) \supset f(A)+f(\{x\}) \supset f(A)+x$. Thus (B) holds. □

**Remark 3.10.2.** In the case of a vector space and a mapping which commutes with homotheties, thus $f(\lambda A) = \lambda f(A)$ for nonnegative $\lambda$, it follows from (D) that

(3.10.1) $f(\lambda A+\mu B) \supset \lambda f(A)+\mu f(B)$ for all $\lambda, \mu \geqslant 0$ and all $A, B \in \mathscr{H}$,

which is a property analogous to concavity. □

## 3.11 Matheron's structural theorems

We are now acquainted with dilations, erosions, cleistomorphisms, and anoiktomorphisms. They are examples of morphological mappings, but we have so far no idea how general they are. Are they just some special mappings that we have come across? There are obviously many more mappings $\mathscr{P}(G) \to \mathscr{P}(G)$, even if we restrict ourselves to mappings which commute with translations, which is reasonable to do. More precisely, we ask the following two questions.

How special are the dilations $\delta_B$ and erosions $\varepsilon_B$ that we have studied so far in the family of all increasing mappings which commute with translations?

How special are the cleistomorphisms $\kappa_B$ and the anoiktomorphisms $\alpha_B$ we have constructed in the family of all cleistomorphisms and anoiktomorphisms which commute with translations?

Georges Matheron proved two results which gave very neat answers to these two questions.

Matheron's first structural theorem describes the structure of increasing mappings in $\mathbf{R}^n$ which commute with translations in terms of dilations and erosions. Actually the result holds in every abelian group. It is important because it underlines the fact that the mappings we already know are the building blocks of a much more general class of mappings.

For any mapping $f\colon \mathscr{P}(G_1) \to \mathscr{P}(G_2)$ of the family of all subsets of an abelian group $G_1$ into those of another, $G_2$, we define its ***kernel*** as

$$\mathbf{ker}(f) = \{A \in \mathscr{P}(G_1);\ 0 \in f(A)\}.$$

The concept is due to Matheron (1975:217). A mapping which commutes with all translations is completely determined by its kernel.

We mention some examples of kernels.

(1) The kernel of the identity $\mathbf{id}\colon \mathscr{P}(G) \to \mathscr{P}(G)$ is
$$\mathbf{ker}(\mathbf{id}) = \{A \in \mathscr{P}(G);\ 0 \in A\}.$$
(2) The kernel of the mapping $f^{\text{dual}}$ dual to $f$ is
$$\mathbf{ker}(f^{\text{dual}}) = \{A;\ 0 \notin f(\complement A)\}.$$
(3) The kernel of a translation $T_b(A) = A + b$ is
$$\mathbf{ker}(T_b) = \{A;\ -b \in A\}.$$
(4) A dilation $\delta_B\colon A \mapsto A + B$ has kernel
$$\mathbf{ker}(\delta_B) = \{A;\ A \cap \check{B} \neq \varnothing\}.$$
(5) The kernel of an erosion $\varepsilon_B$ is
$$\mathbf{ker}(\varepsilon_B) = \{A;\ A \supset B\}.$$
(6) The kernel of a cleistomorphism $\kappa_B = \varepsilon_B \circ \delta_B$ is
$$\mathbf{ker}(\kappa_B) = \{A;\ B \subset A + B\} = \bigcap_{y \in B} \mathbf{ker}(\delta_{B-y}).$$
(7) An anoiktomorphism $\alpha_B = \delta_B \circ \varepsilon_B$ has the kernel
$$\mathbf{ker}(\alpha_B) = \{A;\ \exists y \in B \text{ such that } A \supset B - y\} = \bigcup_{y \in B} \mathbf{ker}(\varepsilon_{B-y}).$$

**Theorem 3.11.1** (Matheron's first structural theorem, 1975: Proposition 8-1-3). *Let $G$ be an abelian group and $f\colon \mathscr{P}(G) \to \mathscr{P}(G)$ an increasing mapping which commutes with translations. Then $f$ is a union of erosions as well as an intersection of dilations:*
$$f(A) = \bigcup_{B \in \mathbf{ker}(f)} \varepsilon_B(A) = \bigcap_{B \in \mathbf{ker}(f^{\text{dual}})} \delta_{\check{B}}(A) = \bigcap_{B \in \mathbf{ker}(f^{\text{dual}})} (A - B),$$
*for all $A \in \mathscr{P}(G)$, where $f^{\text{dual}}(A) = \complement f(\complement A)$ is the mapping set-theoretically dual to $f$.*

**Proof.** If $B \in \mathbf{ker}(f)$, then $0 \in f(B)$ and, since $f$ commutes with all translations, $x \in f(x+B)$ for all $x \in G$. Therefore, given any subset $A$ of $G$, $x \in \varepsilon_B(A)$ implies that $x+B \subset A$ and $x \in f(x+B) \subset f(A)$ ($f$ is increasing). Since $x$ is arbitrary in $\varepsilon_B(A)$, we have proved that $\varepsilon_B(A) \subset f(A)$ for all $A$ and all $B \in \mathbf{ker}(f)$. Letting $B$ vary in $\mathbf{ker}(f)$ we see that the union of all the $\varepsilon_B(A)$ is contained in $f(A)$.

To prove the inclusion in the other direction, take an arbitrary element $x$ of $f(A)$. Then $0 = x - x \in f(A) - x = f(A - x)$, i.e., $A - x \in \mathbf{ker}(f)$. If we now define $B$ as $A - x$, then $x \in \varepsilon_B(A)$. This means that there exists a $B \in \mathbf{ker}(f)$ such that $x \in \varepsilon_B(A)$, and we have proved that $f(A)$ is contained in the union of all the $\varepsilon_B(A)$.

The second representation follows on applying the first to $f^{\text{dual}}$. $\square$

Matheron's second structural theorem requires a preliminary study of extensions of increasing mappings which we shall now undertake. We first include a definition from (Matheron 1975:187):

**Definition 3.11.2.** Let $X$ be an arbitrary set and $\mathscr{H}$ a subset of $\mathscr{P}(X)$. For any mapping $f\colon \mathscr{H} \to \mathscr{P}(X)$ we define

$$f_\vee(A) = \bigcup_{\substack{B \in \mathscr{H} \\ B \subset A}} f(B) \text{ and } f^\wedge(A) = \bigcap_{\substack{B \in \mathscr{H} \\ B \supset A}} f(B),$$

for $A \in \mathscr{P}(X)$. □

**Lemma 3.11.3.** *Let $X$ be an arbitrary set, $\mathscr{H}$ any subset of $\mathscr{P}(X)$, and $f\colon \mathscr{H} \to \mathscr{P}(X)$. The mappings $f_\vee$ and $f^\wedge$ are increasing and dual to each other in the sense that $(f_\vee)^{\text{dual}} = (f^{\text{dual}})^\wedge$ and $(f^\wedge)^{\text{dual}} = (f^{\text{dual}})_\vee$. The mapping $f_\vee$ is the smallest increasing mapping $h$ such that $h|_{\mathscr{H}} \geqslant f$ and $f^\wedge$ is the largest increasing mapping $k$ such that $k|_{\mathscr{H}} \leqslant f$.*

*If $f\colon \mathscr{H} \to \mathscr{P}(X)$ is increasing, then $f_\vee$ is the smallest increasing extension of $f$ to all of $\mathscr{P}(X)$, and $f^\wedge$ is the largest increasing extension of $f$ to all of $\mathscr{P}(X)$.*

The proof is easy.

**Proposition 3.11.4.** *Let $X$ be any set, $\mathscr{H}$ a subset of $\mathscr{P}(X)$, and let $\alpha\colon \mathscr{H} \to \mathscr{P}(X)$ be a partial anoiktomorphism. Assume that $\alpha(A) \in \mathscr{H}$ for every $A \in \mathscr{H}$. Then $\alpha_\vee\colon \mathscr{P}(X) \to \mathscr{P}(X)$ is an anoiktomorphism.*

**Proof.** That $\alpha_\vee$ is increasing and antiextensive is obvious. It remains to be proved that $\alpha_\vee$ is idempotent. We know that $\alpha_\vee$ is an extension of $\alpha$, which implies that $\alpha_\vee(\alpha(B)) = \alpha(\alpha(B)) = \alpha(B)$ for all $B \in \mathscr{H}$, since by hypothesis $\alpha(B) \in \mathscr{H}$. Hence $\alpha_\vee(\alpha_\vee(A)) \supset \alpha_\vee(\alpha(B)) = \alpha(B)$ if $B \in \mathscr{H}$ and $B \subset A$. On taking the union over all $B \in \mathscr{H}$ contained in $A$ we get

$$\alpha_\vee(\alpha_\vee(A)) \supset \bigcup_{\substack{B \in \mathscr{H} \\ B \subset A}} \alpha(B) = \alpha_\vee(A) \supset \alpha_\vee(\alpha_\vee(A)),$$

showing that $\alpha_\vee$ is idempotent. □

**Remark 3.11.5.** The hypothesis that $\alpha$ maps $\mathscr{H}$ into itself cannot be dispensed with. Indeed, take four sets $A_j$, $j = 0, 1, 2, 3$ with $A_0 \subset A_1 \subset A_2 \subset A_3$; $X = A_3$ with $A_1 \neq A_2 \neq A_3$, and define $\alpha(A_j) = A_{j-1} = [0, j-1]$, $j = 1, 3$, and $\mathscr{H} = \{A_1, A_3\}$. Then $\alpha\colon \mathscr{H} \to \mathscr{P}(X)$ is a partial anoiktomorphism; in particular the second part of (2.11.4) holds: $\alpha(A_j) \subset A_k$

implies $\alpha(A_j) \subset \alpha(A_k)$ for $j, k = 1, 3$. But $\alpha_\vee$ is not idempotent: $A_0 = \alpha_\vee(\alpha_\vee(A_3)) \neq \alpha_\vee(A_3) = A_2$. □

**Proposition 3.11.6.** *If $\alpha\colon \mathscr{H} \to \mathscr{P}(X)$ is a partial anoiktomorphism, where $\mathscr{H}$ is closed under the formation of unions, then also* $\mathbf{invar}(\alpha)$ *is closed under the formation of unions (in particular $\emptyset \in \mathbf{invar}(\alpha)$), and $\alpha = (\mathbf{id}_{\mathbf{invar}(\alpha)})_\vee \big|_{\mathscr{H}}$, i.e., $\alpha$ is the smallest increasing extension to $\mathscr{P}(X)$ of the identity on* $\mathbf{invar}(\alpha)$.

*Conversely, let a class $\mathscr{C} \subset \mathscr{P}(X)$ be given. Then $(\mathbf{id}_{\mathscr{C}})_\vee$, the smallest increasing extension of the identity on $\mathscr{C}$, is an anoiktomorphism $\nu$, and* $\mathbf{invar}(\nu)$ *is the class closed under the formation of unions generated by $\mathscr{C}$.*

**Proof.** If $A_j \in \mathscr{H}$, $j \in J$, are $\alpha$-invariant, then also $\bigcup A_j \in \mathscr{H}$ and $\alpha(\bigcup A_j) \supset \alpha(A_k) = A_k$ for all $k$, so $\alpha(\bigcup A_j) \supset \bigcup A_k$. Since on the other hand we always have $\alpha(\bigcup A_j) \subset \bigcup A_j$, it follows that $\bigcup A_j \in \mathbf{invar}(\alpha)$.

That $\alpha$ is the smallest increasing extension of the identity on $\mathbf{invar}(\alpha)$ follows from

$$(\mathbf{id}_{\mathbf{invar}(\alpha)})_\vee (A) = \bigcup_{\substack{B \in \mathscr{H} \\ B \subset A}} B = A \text{ if } A \in \mathscr{H}.$$

Since the inclusion mapping $\mathscr{C} \to \mathscr{P}(X)$ is a partial anoiktomorphism satisfying the condition in Proposition 3.11.4, it follows that $(\mathbf{id}_{\mathscr{C}})_\vee$ is an anoiktomorphism.

To prove the last assertion in the statement of the proposition, we note that by the definition of $\nu = (\mathbf{id}_{\mathscr{C}})_\vee$,

$$A \in \mathbf{invar}(\nu) \text{ implies that } A = \nu(A) = \bigcup(B;\ B \in \mathscr{C}, B \subset A),$$

which means that $A$ belongs to the set generated by $\mathscr{C}$ and closed under union formation. □

In case $X = G$ is an abelian group, let us say that a subset $\mathscr{B}$ of $\mathbf{invar}(\alpha)$ is a ***basis*** for $\mathbf{invar}(\alpha)$, if $\mathbf{invar}(\alpha)$ is the class which is closed under the formation of unions and translations and generated by $\mathscr{B}$. This means that $\mathbf{invar}(\alpha)$ is the family of all sets of the form $\bigcup_{(B,x)\in M}(B + x)$ for some subset $M$ of $\mathscr{B} \times G$.

We are now ready to state and prove Matheron's second structural theorem, which characterizes general cleistomorphisms and anoiktomorphisms in terms of the elementary cleistomorphisms and anoiktomorphisms $\kappa_B$ and $\alpha_B$ by a given subset $B$. His result, which shows that the latter are not so special as one could imagine, is as follows.

**Theorem 3.11.7** (Matheron's second structural theorem, 1975: Proposition 7-1-3). *Let $G$ be an abelian group and let $g\colon \mathscr{P}(G) \to \mathscr{P}(G)$ be a mapping which commutes with translations. We define $\alpha_B$ and $\kappa_B$ as in Definition 3.9.1. Then $g$ is an anoiktomorphism if and only if it admits a representation*

$$g(A) = \bigcup_{B \in \mathscr{B}} \alpha_B(A)$$

*for some class $\mathscr{B} \subset \mathscr{P}(G)$. If this is so, the class of sets invariant under $g$ is the family of all elements of $\mathscr{B}$ and all translates and unions of these. If $g$ is an anoiktomorphism, the dual mapping $g^{\text{dual}}$ is a cleistomorphism and has the representation*

$$g^{\text{dual}}(A) = \bigcap_{B \in \mathscr{B}} \kappa_B(A).$$

**Proof.** First assume that $g$ is an anoiktomorphism. The smallest extension of the identity on $\mathbf{invar}(g)$ is given by

(3.11.1)
$$g(A) = \left(\mathbf{id}_{\mathbf{invar}(g)}\right)_{\vee}(A) = \bigcup_{\substack{C \in \mathbf{invar}(g) \\ C \subset A}} C = \bigcup_{B \in \mathscr{B}} \bigcup_{x \in G} (B + x;\ B + x \subset A),$$

if $\mathscr{B}$ is a basis for $\mathbf{invar}(g)$ as defined before the statement of the theorem. In fact, given a set $C \in \mathbf{invar}(g)$ contained in $A$, there is a subset $M$ of $\mathscr{B} \times G$ such that

$$C = \bigcup_{(B,x) \in M} (B + x;\ B + x \subset A) \subset \bigcup_{(B,x) \in \mathscr{B} \times G} (B + x;\ B + x \subset A) = C',$$

but the union will not be changed if we add $C'$, and (3.11.1) follows. Let us now note that

$$\alpha_B(A) = \bigcup_{x \in G} (B + x;\ B + x \subset A).$$

Therefore the last expression in (3.11.1) is equal to $\bigcup_{B \in \mathscr{B}} \alpha_B(A)$.

Conversely, if $g_j$, $j \in J$, are anoiktomorphisms, so is $g$ defined by $g(A) = \bigcup g_j(A)$. Indeed, $g$ is certainly increasing and antiextensive, hence $g \circ g \leqslant g$. Idempotency then follows from

$$g(g(A)) = \bigcup_{j \in J} g_j\Big[\bigcup_{k \in J} g_k(A)\Big] \supset \bigcup_{j \in J} g_j(g_j(A)) = \bigcup_{j \in J} g_j(A) = g(A).$$

□

## 3.12 Notes

### 3.12.1 *The most fundamental concepts*

Hugo Hadwiger defined dilation and erosion in an article (1950:212) and in a book (1957:146–147), although he did not use these words. This seems to be the first appearance of the two concepts. In his 1957 book he coined the terms *Außenmenge*$_{(f)}$ *von A relativ zu B der Spanne* $\rho$ for $\delta_{\rho B}(A)$ and *Innenmenge*$_{(f)}$ *von A relativ zu B der Spanne* $\rho$ for $\varepsilon_{\rho B}(A)$. Here $A$ and $B$ are nonempty sets in a Euclidean space and $\rho$ an arbitrary positive number. However, as remarks Jean Serra (private communication 2019-12-05), Hadwiger does not do anything with the erosion, nor does he define the concept of anoiktomorphism.

Jean Serra (private communication 2019-12-05) obtained in the year 1965 patents for the analysis of textures in several countries: France, Canada, Great Britain, the United States, Germany, Austria, Sweden, and Japan. These include the operation of dilation. The concepts of these patents reappear in publications by Serra (1966) and Matheron (1967), as well as by Haas et al. (1967).

Jean Serra emphasizes that a most decisive step in the development of mathematical morphology was the passage from functionals defined on families of sets and with numbers as values, to functionals with values in families of sets (private letter of 2019-12-05). This took place around 1965 and permitted the formation of compositions, like the composition of a cleistomorphism as a dilation followed by an erosion, $\kappa = \varepsilon \circ \delta$, and many others. Indeed, most of Hadwiger's book is about functionals defined for sets and yielding numbers, like diameter, surface area and volume. Only a small part is devoted to operations yielding sets.

Definition 3.11.2, Lemma 3.11.3, and Proposition 3.11.4 are taken from (Matheron 1975:187).

In a book chapter in the 1988 book edited by Jean Serra, he studies the properties of compositions $\alpha_k \circ \kappa_k \circ \alpha_{k-1} \circ \cdots \alpha_2 \circ \kappa_2 \circ \alpha_1$, thus with alternating anoiktomorphisms and cleistomorphisms (Serra 1988e). They can be used to clean up images.

For properties of dilations on topological spaces, see (Matheron 1988a).

For more recent developments, see the book by Chan & Shen (2005)—on modeling, model analysis, computation and simulation in image processing and analysis—and the article by Welk et al. (2019).

### 3.12.2 *Addition and subtraction*

As seen above, we use addition on three different levels: the sum of two elements $a + b$; the sum of an element and a set

$$a + B = \{a + b;\ b \in B\};$$

and the sum of two sets

$$A + B = \{a + b;\ a \in A,\ b \in B\}.$$

We can express this by saying that we simplify the addition of sets $\{a\} + B$ to $a + B$ and $\{a\} + \{b\}$ to $a + b$. But sometimes a need is felt to distinguish them: $A + B$ can then be written $A \oplus B$; see (Matheron 1975:16), (Serra 1982; 2001) and (Gonzalez & Woods 1993:519); note however that the sign $\oplus$ is otherwise used for a direct sum.

Restricted Minkowski addition (see Section 3.8) was used in (Kiselman & Samieinia 2010; 2017).

We can also form $a - b$, $a - B$ and

$$A - B = A + \check{B} = A + (0 - B) = \{a - b;\ a \in A,\ b \in B\},$$

the latter being Minkowski subtraction.

There is another minus sign used in an abelian group, written $\ominus$:

$$A \ominus B = \varepsilon_B(A) = \{x;\ x + B \subset A\}. \tag{3.12.1}$$

Note that $\varepsilon_\varnothing(A) = A \ominus \varnothing = G$, that $\varepsilon_G(A) = A \ominus G = \varnothing$ provided $A \neq G$, and that $\varepsilon_{\{0\}}(A) = A \ominus \{0\} = A$ if $G$ has a neutral element 0.

To illustrate the difference in meaning between these three minus signs $-$, $\smallsetminus$, and $\ominus$ as defined in (3.12.1) above, note that $A \smallsetminus A = \varnothing$ for all sets $A$; that $A - A \neq \varnothing$ if $A$ is a nonempty subset of an abelian group; and that, finally, $A \ominus A$ is a subsemigroup if $A$ is a subset of an abelian group $G$. If $A$ is a bounded subset of $\mathbf{R}^n$ or $\mathbf{Z}^n$ for example, then $A \ominus A = \varepsilon_A(A) = \{0\}$.

Gonzalez & Woods (1993:521) use the sign $\ominus$ as in (3.12.1) above. Serra (1982:43) defines instead $A \ominus B$ as $\bigcap_{y \in B}(A + y) = \varepsilon_{\check{B}}(A)$, thus with structuring element equal to $\check{B} = 0 - B$, the set opposite to $B$.

### 3.12.3 *Terms used, and their origin*

The word *dilation* is of Latin origin, the prefix *di-*, which is a variation of *dis-* 'apart, separate' and the adjective *latus* 'wide, broad', thus *to dilate* means 'to pull apart'.

Both forms *dilation* and *dilatation* occur, but it seems that the former is the most common term today; in his English book, Matheron (1975:17) calls it *dilatation*, a term which is often reserved for mappings $x \mapsto \lambda x$.[1]

Also the word *erosion* is of Latin origin: the prefix *e-* is a variation of *ex-* 'from, away' and the remainder from the verb *rodere* 'to gnaw', thus *to erode* 'to gnaw away'.

A structuring element (see Serra 1988c:73) is sometimes called a *structural element*, for example in (Klette & Rosenfeld 2004:486).

The property mentioned in formula (3.10.1) was introduced and used to study cleistomorphisms appearing in the theory of partial differential equations by Kiselman (1969); however, I did not know at the time that it is implied by property (A) in Theorem 3.10.1.

### 3.12.4 *Notation*

Matheron (1975) and Serra (1982, 2001) use the notation $A^B$ for $\kappa_B(A)$; Gonzalez & Woods (1993:524) write $A \bullet B$.

Matheron (1975) and Serra (1982, 2001) use the notation $A_B$ for $\alpha_B(A)$, Gonzalez & Woods (1993:524) use $A \circ B$. Note, however, that $\circ$ is most often used to denote composition of functions or relations.

## 3.13 Exercises

3.1. Dilate and erode $A$ by $B$ when

(a) $A$ is a disk in $\mathbf{R}^2$, $B$ a pair of points. Here the word *disk* shall be understood with regard to one of the norms $\|\cdot\|_p$, $p = 1, 2, \infty$.

(b) $A$ is an open disk in $\mathbf{R}^2$ of radius $R$, $B$ a closed disk of radius $r$.

(c) $A$ is a bounded set in $\mathbf{R}^2$ and $B$ is a translate of $A$.

(d) $A$ is a bounded set in $\mathbf{R}^2$ and $B$ is a translate of $\check{A}$.

3.2. We know that $\delta_B(A) \subset C$ if and only if $A \subset \varepsilon_B(C)$.

(a) Give examples to show that we do not have an equivalence $\delta_B(A) = C$ if and only if $A = \varepsilon_B(C)$. Is one of the implications true?

(b) Show by examples that it is not true that $\varepsilon_B(A) \subset C$ if and only if $A \subset \delta_B(C)$. Is one of the implications true?

---

[1] In French, *dilatation* (*f*) is the usual term; see, e.g., (Najman & Talbot 2008:23).

However, the equivalence holds for certain families of sets ...

3.3. Prove that, in an abelian group, erosion by $B$ satisfies $\varepsilon_B(A) \subset A$ for all sets $A$ if and only if $0 \in B$.

3.4. Calculate the zonotope generated by two segments

$$[a, b] = \{(1-t)a + tb;\ 0 \leqslant t \leqslant 1\}, \qquad a, b \in \mathbf{R}^2,$$

and $[c, d]$ in $\mathbf{R}^2$, where $a, b, c, d$ are four arbitrary points in $\mathbf{R}^2$.

3.5. Calculate the sum of three arbitrary segments in $\mathbf{R}^2$.

3.6. Calculate the zonohedron generated by three orthogonal segments in $\mathbf{R}^3$.

3.7. Calculate in $\mathbf{R}^3$ the Minkowski sum of four segments, three of which are contained in a plane while the fourth is not.

3.8. Prove that a triangle cannot be the Minkowski sum of a finite number of segments.

3.9. A heptagon in $\mathbf{R}^2$ can never be the sum of a finite number of segments.

3.10. A tetrahedron in $\mathbf{R}^3$ cannot be the sum of a finite number of segments.

3.11. Give an example of a quadrilateral in $\mathbf{R}^2$ that is not the sum of two segments.

3.12. (a) Calculate in $\mathbf{R}^3$ the sum of the four segments $[(0,0,0),(1,0,0)]$, $[(0,0,0),(0,1,0)]$, $[(0,0,0),(0,0,1)]$, and $[(0,0,0),(1,1,1)]$. The result is a polyhedron. How many vertices, edges, and faces does it have?

(b) Calculate in $\mathbf{R}^3$ the sum of the four unit normals to a regular tetrahedron.

3.13. The Minkowski sum of two filled squares in the plane with side lengths $s_1$ and $s_2$ is either a square with side length $s_1 + s_2$ or an octagon with four sides of length $s_1$ and four of length $s_2$. To be proved.

3.14. Let $T_\gamma$ denote the filled triangle which has vertices $(\cos\theta, \sin\theta) \in \mathbf{R}^2$ with $\theta = \gamma, \gamma + 2\pi/3, \gamma + 4\pi/3$. Determine the Minkowski sum of two triangles $r_0 T_0$ and $r_\gamma T_\gamma$, $0 \leqslant \gamma \leqslant \pi$, where $r_0$ and $r_\gamma$ are two positive numbers.

3.15. We keep the notation for triangles in Exercise 3.13. Denote by $P$ the tetrahedron with one vertex in $(0,0,0)$ and with the others those in $T_{\pi/3} \times \{1\}$, a triangle placed at level $x_3 = 1$.

(a) How many vertices, edges, and faces are there in the convex hull of $(T_0 \times \{0\}) \cup P$?

(b) The same question for the Minkowski sum $(T_0 \times \{0\}) + P$.

3.16. The vector sum of two triangles in $\mathbf{R}^2$ is a convex polygon. Find examples that show that it can have three, four, five and six vertices.

3.17. Find examples that show that the sum of a triangle and a square can have five, six and seven vertices.

3.18. Find examples (for instance in $\mathbf{Z}^2$) that show that we do not always have

$$\kappa_B(A_1 \cup A_2) = \kappa_B(A_1) \cup \kappa_B(A_2) \text{ or}$$
$$\kappa_B(A_1 \cap A_2) = \kappa_B(A_1) \cap \kappa_B(A_2);$$

in other words, $\kappa_B = \varepsilon_B \circ \delta_B$ is in general neither a dilation nor an erosion.

3.19. Define in an abelian group $G$ a mapping $\varphi = \delta_C \circ \varepsilon_B$, $A \in \mathscr{P}(G)$, where $B$ and $C$ are fixed subsets of $G$.

(a) Prove that $\varphi$ is antiextensive if and only if $C \subset B$, which in turn is equivalent to $\varphi(G \smallsetminus \{0\}) \subset G \smallsetminus \{0\}$.

(b) Prove that if $B$ and $C$ are nonempty bounded subsets of $\mathbf{Z}^n$ or $\mathbf{R}^n$ with $C \subset B$, then $\varphi$ is idempotent only if $B = C$, thus only if $\varphi(A) = \alpha_B(A)$. *Hint:* Calculate $\varphi(B)$ and $\varphi(\varphi(B))$.

3.20. Find an example which shows that the inclusion

$$\mathbf{epi}(f \sqcap g) \supset \mathbf{epi}(f) + \mathbf{epi}(g), \qquad f, g \in \mathscr{F}(\mathbf{R}^n, \mathbf{R}_!),$$

can be strict.

3.21. Find a semigroup $G$ such that left translation does not commute with erosions, i.e., such that $S_z(\varepsilon_B(A)) \neq \varepsilon_B(S_z(A))$ for some choices of $z \in G$ and $A, B \subset G$ (cf. (3.2.1) on page 42).

# Chapter 4

# Complete lattices

## 4.1 Introduction

LATTICE THEORY is a mature mathematical theory thanks to the pioneering work by Garrett Birkhoff, Øystein Ore and others in the first half of the twentieth century. A standard reference is still Birkhoff's book (1995), first published in 1940.

Lattice theory can (now) be viewed as a tropicalization of other mathematical theories. (As mentioned at the end of Section 1.2, tropicalization means that we replace sums and integrals by supremum operators.) Developments originate in several branches of mathematics, for instance algebra (Blyth & Janowitz 1972, Blyth 2005), logic (Stoltenberg-Hansen et al. 1994, Gierz et al. 2003), general topology and functional analysis (Gierz et al. 2003:xxx–xxxii), convexity theory (Singer 1997), and, most importantly as a background for this book, mathematical morphology with applications in image processing: books by Matheron (1975), Serra (1982), Serra, Ed. (1988), and Heijmans (1994); articles by Heijmans & Ronse (1990), Ronse (1990), Ronse & Heijmans (1991, 1998), Serra (2006), and Ronse & Serra (2008, 2010). Other areas where concepts from lattice theory are used include semantics (abstract interpretation) of programming, the theory of fuzzy sets, fuzzy logic, and formal concept analysis (Ganter & Wille 1999). For general lattice theory a standard reference is (Grätzer 1998).

This variety of sources for fundamental concepts has led to varying terminology and hence to difficulties in tracing history.

The two concepts of lattice and complete lattice must be carefully distinguished. This becomes obvious when we see that a complete lattice $L$ can contain another complete lattice $M$ with $M$ as a sublattice of $L$ ... but $M$ is not a sub-complete-lattice of $L$ (see Example 4.2.6).

Complete lattices, although introduced more than eighty years ago by Garrett Birkhoff, became useful in image processing some forty years ago, as they combined the properties of the three most fundamental exemples, to whit: power sets $\mathscr{P}(W)$; functions with values in $\mathbf{R}_!$; partitions of a given set. These three examples remain the main ones up till now.

Another, also quite important example is the family of functions with a common Lipschitz constant and with values in a bounded set. For the definition of the Lipschitz constant, see (7.3.5).

Often we require of the ordered sets studied that they shall be complete lattices. However, of the various phenomena brought together here, the Galois connection is the oldest, and it makes sense between preordered sets which are not necessarily complete lattices, or even lattices.

A goal will therefore be to study this general situation and relate it to the more special theories of residuation and adjunction. And this general situation comes forward in the definition and study of inverses and quotients to be undertaken in the next chapter.

Both inverses and quotients come in two versions, lower and upper. It turns out that anoiktomorphisms can be characterized as lower quotients of the form $f/_{\star}f$, and cleistomorphisms as upper quotients $f/^{\star}f$.

To define an inverse of a general mapping seems to be a hopeless task. However, if the mapping is between preordered sets, there is some hope of constructing mappings that can serve in certain contexts just like inverses do.

There is an analogy between lattice theory and the theory of vector spaces. The theory of topological vector spaces was developed to a large extent because of the theory of distributions, which in turn was motivated by applications in partial differential equations. Developments in image processing motivated a renewed interest in lattice theory, in particular in complete lattices. Lattice theory was applied to switching circuits, and it was then enough, because of general finiteness conditions, to form models using lattices, but in image processing it is more convenient to assume completeness; for a motivation, see (Ronse 1990).

While vector spaces are useful in modelling linear problems, lattices seem to be more adapted to nonlinear problems. Auditory phenomena are often additive: as remarked on page 12 all the instruments of an orchestra can be heard; while with visual phenomena this is not so: one object can block another from our view. This indicates that linear models may suffice for the first kind of phenomena (Fourier analysis and synthesis are successful for sound waves), while the visual ones are more in agreement with nonlinear operators, like maximum and minimum.

As we shall see, there are also analogies between topological spaces and preordered sets, in particular lattices. The continuous linear mappings in the first case correspond to increasing mappings in the second.

## 4.2 Definitions and examples

Let $L$ be an ordered set and $A$ a subset of $L$. An element $b \in L$ is said to be the ***infimum*** of all elements $a \in A$ if $b$ is the largest minorant of all $a \in A$. This means that $b \leqslant a$ for all elements $a \in A$, and that if $b' \leqslant a$ for all $a \in A$, then $b' \leqslant b$. The infimum, if it exists, is necessarily unique. The infimum of the empty set exists if and only if $L$ possesses a largest element, and if so, the infimum is this largest element.

We shall write

$$b = \inf_{a \in A} a = \inf(a;\ a \in A) = \bigwedge_{a \in A} a$$

for the infimum of all elements in $A$; if $A$ has only $n$ elements we write $b = a_1 \wedge \cdots \wedge a_n$, in particular $b = a_1 \wedge a_2$ if $n = 2$. If the infimum belongs to $A$, we call it a ***minimum***. As an example, the set of all positive real numbers has 0 as its infimum, but 0 is not a positive number and therefore not a minimum.

Similarly we define the ***supremum***

$$c = \sup_{a \in A} a = \sup(a;\ a \in A) = \bigvee_{a \in A} a$$

as the smallest majorant of all elements in $A$. If the supremum belongs to $A$, we call it a ***maximum***. The supremum of all elements in the empty set, $\sup_{x \in \varnothing} x$, exists if and only if $L$ has a smallest element.

**Definition 4.2.1.** Let $L$ be a nonempty set. If any subset consisting of two elements in $L$ has an infimum, we shall call $L$ an ***inf-semilattice***; similarly, if any two-set of $L$ has a supremum, we shall call $L$ a ***sup-semilattice***. If $L$ is both an inf-semilattice and a sup-semilattice we shall call $L$ a ***lattice***. □

**Definition 4.2.2.** A nonempty subset $I$ of a lattice $L$ is said to be an ***ideal*** if $I \neq L$ and $x, y \in I$, $z \in L$ implies that $x \vee y, x \wedge z \in I$.

A nonempty subset $F$ of a lattice $L$ is said to be a ***filter*** if $F \neq L$ and $x, y \in F$, $z \in L$ implies that $x \wedge y, x \vee z \in F$. □

An ideal is necessarily a down-set; a filter an up-set as defined in Section 2.7.

**Definition 4.2.3.** If any nonempty subset (finite or infinite) of a nonempty set $L$ has an infimum, $L$ will be said to be a ***complete inf-semilattice***; analogously we define ***complete sup-semilattice*** and ***complete lattice***. (see also the definition of a Moore family, Definition 4.3.7). □

A complete inf-semilattice has a smallest element, which is the infimum of all elements, to be denoted by $\mathbf{0}$, and a complete sup-semilattice has a largest element, the supremum of all elements, to be denoted by $\mathbf{1}$. These extremal elements could also be written as $-\infty$ and $+\infty$, just to emphasize the fact that we work with actual infinities, not just potential infinities. In some cases, like $\mathbf{R}_! = [-\infty, +\infty]$, we do so, but in general we shall keep $\mathbf{0}$ and $\mathbf{1}$. In the complete lattice $\mathscr{P}(W)$, the extremal elements are $\mathbf{0} = \varnothing$ and $\mathbf{1} = W$.

A complete inf-semilattice with a largest element $\mathbf{1}$ is also a complete lattice. Indeed, the supremum of any set of elements is equal to the infimum of all majorants of the set—this set is not empty since $\mathbf{1}$ is a majorant.

In a complete lattice $L$ we have

$$\sup_{x \in L} x = \inf_{x \in \varnothing} x = \mathbf{1}; \quad \inf_{x \in L} x = \sup_{x \in \varnothing} x = \mathbf{0};$$

the infimum of the empty set exists and is $\mathbf{1}$, and the supremum of the empty set is $\mathbf{0}$.

It is possible to define a lattice as a set with two binary operations $\wedge$ and $\vee$ satisfying certain axioms.

A ***sublattice*** is defined just like a subgroup with respect to the operations $\wedge$ and $\vee$: that $M$ is a sublattice of $L$ means that for all $x, y \in M$, $x \wedge y$ and $x \vee y$, when calculated in $L$, are elements of $M$. A sublattice is therefore something more than a subset with the induced order; see the following examples.

**Example 4.2.4.** The space of real-valued continuous functions on a topological space is a lattice with the usual order: $f \leqslant g$ if and only if $f(x) \leqslant g(x)$ for all $x$. The space $C^1(\mathbf{R}^n, \mathbf{R})$ of real-valued continuously differentiable functions on $\mathbf{R}^n$ is not a sublattice of $C(\mathbf{R}^n, \mathbf{R})$ if $n \geqslant 1$. It is not even a lattice on its own. (The functions $\mathbf{R} \ni x \mapsto \sqrt{t^2 + x^2}$, $t > 0$, converge to $x \mapsto |x|$ as $t \to 0$, but there is no infimum in $C^1(\mathbf{R}, \mathbf{R})$.) □

The example just given—just as all other counterexamples—is essential for our understanding: it signals the boundaries of a given concept.

**Example 4.2.5.** The family $\mathscr{P}(W)$ of all subsets of a set $W$ is a complete lattice, with $\bigwedge A_j = \bigcap A_j$ and $\bigvee A_j = \bigcup A_j$. The compact sets in $\mathbf{R}^n$ form a sublattice $\mathscr{K}(\mathbf{R}^n)$ of $\mathscr{P}(\mathbf{R}^n)$. This lattice is a complete inf-semilattice but not a complete sup-semilattice. The family $\mathscr{K}_{cvx}(\mathbf{R}^n)$, $n \geqslant 1$, of all convex compact sets is a lattice but not a sublattice of $\mathscr{K}(\mathbf{R}^n)$: the supremum of two convex compact sets is not always the same in the two lattices. □

**Example 4.2.6.** The family of all closed sets in $\mathbf{R}^n$, denoted by $\mathscr{C}(\mathbf{R}^n)$, is a sublattice of $\mathscr{P}(\mathbf{R}^n)$: the union and intersection of two closed sets are closed. But, although $\mathscr{C}(\mathbf{R}^n)$ is a complete lattice, it is not a sub-complete-lattice of the complete lattice $\mathscr{P}(\mathbf{R}^n)$ when $n \geqslant 1$. The union of a family of closed sets is not always closed, but there is a supremum, viz. the closure of the union. Thus, finite suprema agree with those in $\mathscr{P}(\mathbf{R}^n)$ while infinite suprema do not.[1] □

**Example 4.2.7.** The set $\mathscr{F}(\mathbf{R}^n, \mathbf{R}_!)$ of all functions defined on $\mathbf{R}^n$ and with values in the extended real line $\mathbf{R}_!$ is a lattice under the usual order for real numbers, extended in an obvious way to the two infinities. The subset of all convex functions is ordered in the same way, and is also a lattice under this order. However, the convex functions $CVX(\mathbf{R}^n, \mathbf{R}_!)$ do not form a sublattice of $\mathscr{F}(\mathbf{R}^n, \mathbf{R}_!)$ if $n \geqslant 1$. The supremum of two convex functions is equal to the pointwise supremum of them:

$$(f \vee g)(x) = f(x) \vee g(x) = \max(f(x), g(x)),$$

but the infima are different in the two lattices: the infimum in the lattice of convex functions is

$$f \wedge_{cvx} g = \sup\big(h \in CVX(\mathbf{R}^n, \mathbf{R}_!);\ h \leqslant f, g\big) \leqslant f \wedge g = \min(f, g),$$

where the supremum of all $h \leqslant f, g$ is calculated in $\mathscr{F}(\mathbf{R}^n, \mathbf{R}_!)$ and has a sense because that lattice is complete. That the two infima may be different is shown by easy examples like $f(x) = \mathrm{e}^x$, $g(x) = \mathrm{e}^{-x}$, $x \in \mathbf{R}$. Here $\min(f, g)(x) = \mathrm{e}^{-|x|}$, while $f \wedge_{cvx} g = 0$. □

**Example 4.2.8.** A ***partition*** of a set $X$ is a family $(P_i)_{i\in I}$ of subsets of $X$ such that the subsets are disjoint and their union is all of $X$. A partition $(Q_j)_{j\in J}$ is said to be finer than $(P_i)_{i\in I}$ if every $Q_j$ is contained in some of the $P_i$. With this order, the family of all partitions is a complete lattice, the finest being the family of all singleton sets; the least fine the one consisting of just $\{X\}$. □

---

[1]This and other examples indicate that it would be desirable to have a different word for *complete lattice*, to allow for a better term than *sub-complete-lattice*.

**Definition 4.2.9.** Let $L$ be a preordered set, $Z$ an arbitrary subset of $L$, and $M$ a complete lattice. We define for any mapping $f\colon Z \to M$ two increasing mappings $f_\vee$ and $f^\wedge$ as follows.

$$f_\vee(x) = \bigvee_{\substack{y \in Z \\ y \leqslant x}} f(y) \text{ and } f^\wedge(x) = \bigwedge_{\substack{y \in Z \\ y \geqslant x}} f(y) \tag{4.2.1}$$

for $x \in L$. □

This is an easy generalization of Definition 3.11.2. So $f_\vee$ is the smallest increasing mapping $h$ such that $h|_Z$ majorizes $f$, and $f^\wedge$ is the largest increasing mapping $k$ such that $k|_Z \leqslant f$. (It may happen that $f_\vee > f^\wedge$ at some points.)

If $f$ is increasing, $f_\vee$ and $f^\wedge$ are actually extensions of $f$; we shall call them the ***lower extension*** and the ***upper extension***.

If $Z = L$, then $f_\vee$ and $f^\wedge$ are just the largest increasing minorant and the smallest increasing majorant of $f$.

### 4.2.1 *Distributive, modular and complemented lattices*

A lattice $L$ is said to be ***distributive*** if we have for all $x, y, z \in L$,

$$x \vee (y \wedge z) = (x \vee y) \wedge (x \vee z) \tag{4.2.2}$$

and

$$x \wedge (y \vee z) = (x \wedge y) \vee (x \wedge z). \tag{4.2.3}$$

Actually (4.2.2) implies (4.2.3) and conversely, so that it suffices to require only one of the two (Dubreil & Dubreil-Jacotin 1964:188). In particular (4.2.2) says that

$$x \leqslant z \text{ implies that } x \vee (y \wedge z) = (x \vee y) \wedge z. \tag{4.2.4}$$

If this is so, the lattice is said to be ***modular***.

If we have a lattice $L$ with smallest element $\mathbf{0}$ and largest element $\mathbf{1}$, we say that $L$ is ***complemented*** if to every element $x$ there exists an element $y$ such that

$$x \wedge y = \mathbf{0} \text{ and } x \vee y = \mathbf{1}.$$

If $L$ is distributive, the element $y$ is uniquely determined by the two equations. A lattice which is distributive and complemented is said to be a ***Boolean lattice***. Obviously the power set $\mathscr{P}(W)$ of any set $W$ is Boolean. For a study of Boolean lattices in the context of mathematical morphology and skeletonizing, see (Serra 1988b, 1989).

## 4.3 Moore families and dual Moore families

The infimum of a family of complete inf-semilattices is a complete inf-semilattice. This implies that to any subset $K$ of a complete lattice $L$, there is a smallest complete inf-semilattice containing $K$. Similarly to any subset $K$ there is a smallest sup-semilattice containing $K$. We make this precise in the following definition.

**Definition 4.3.1.** To any subset $K$ of a complete lattice $L$ we define $K^{\uparrow}$ as the smallest complete sup-semilattice containing $K$ and $K^{\downarrow}$ as the smallest complete inf-semilattice containing $K$. We define, given any subset $K$ of $L$, two mappings, $\beta_{[K]}$ and $\lambda^{[K]}$ by

$$\beta_{[K]}(x) = \bigvee_{y \in K^{\uparrow}} (y;\ y \leqslant x), \qquad \lambda^{[K]}(x) = \bigwedge_{y \in K^{\downarrow}} (y;\ y \geqslant x)$$

for $x \in L$. □

The definition of $\lambda^{[K]}$ is a good one by hypothesis—indeed, $\lambda^{[K]}(x)$ is the infimum of all $y$ in the subset $K_x = \{y;\ y \geqslant x\}$ of $L$.[2]

**Theorem 4.3.2.** *The mapping $\beta_{[K]}$, just defined, is an anoiktomorphism and $\lambda^{[K]}$ is a cleistomorphism for any subset $K$ of the complete lattice $L$. The invariance set* **invar**$(\beta_{[K]})$ *of $\beta_{[K]}$ is $K^{\uparrow}$; the invariance set* **invar**$(\lambda^{[K]})$ *of $\lambda^{[K]}$ is $K^{\downarrow}$.*

**Proof.** Clearly $\lambda^{[K]}$ is increasing and larger than the identity. We note that $y \geqslant x$ if and only if $y \geqslant \lambda^{[K]}(x)$, which implies that

$$\lambda^{[K]}(\lambda^{[K]}(x)) = \bigwedge_{y \in K^{\downarrow}} (y;\ y \geqslant \lambda^{[K]}(x)) = \bigwedge_{y \in K^{\downarrow}} (y;\ y \geqslant x) = \lambda^{[K]}(x)$$

for all $x \in L$, proving that $\lambda^{[K]}$ is idempotent.

If $x$ belongs to $K^{\downarrow}$, then clearly $\lambda^{[K]}(x) = x$ (take $y = x$ in the definition). Conversely, if $\lambda^{[K]}(x) = x$, we invoke the definition of a complete inf-semilattice to conclude that $x \in K^{\downarrow}$. □

In particular we shall study $\beta_{[K]}$ and $\lambda^{[K]}$ when $K$ is the invariance set of a mapping $f$, and shall then denote them by $f_{\diamond}$ and $f^{\diamond}$, respectively, thus

$$f_{\diamond} = \beta_{[\mathbf{invar}(f)]} \text{ and } f^{\diamond} = \lambda^{[\mathbf{invar}(f)]}, \qquad f\colon L \to L. \tag{4.3.1}$$

[2] We note that $\lambda^{[K]}$ is an upper inverse in the sense of Definition 5.2.1 below, viz. $\lambda^{[K]} = g^{[-1]}$, where $g$ is the inclusion mapping $K \to L$. It is also equal to $g^{\wedge}$ as defined in Definition 4.2.9, taking $Z = K^{\downarrow}$.

We then have the following result.

**Corollary 4.3.3.** *If $\eta$ is an ethmomorphism in a complete lattice $L$, then $\eta_\diamond \leqslant \eta \leqslant \eta^\diamond$.*

**Proof.** In the definition of $\lambda^{[K]}$, where $K = K^{\downarrow} = \mathbf{invar}(\eta)$, the inequality $y \geqslant x$ for $y \in K^{\downarrow}$ and $x \in L$ implies that $y = \eta(y) \geqslant \eta(x)$, proving that $\eta^\diamond \geqslant \eta$. □

**Corollary 4.3.4.** *If $\eta$ in Corollary 4.3.3 is an anoiktomorphism $\alpha$, then $\eta_\diamond = \alpha_\diamond = \alpha$. If $\eta$ is a cleistomorphism $\kappa$, then $\eta^\diamond = \kappa^\diamond = \kappa$.*

**Proof.** In the definition of $\kappa^\diamond$ we can take $y = \kappa(x)$, proving that $\kappa^\diamond \leqslant \kappa$. Conversely, the inequality $y \geqslant x$ for any $y \in \mathbf{invar}(\kappa)$ implies that $y = \kappa(y) \geqslant \kappa(x)$, proving that $\kappa^\diamond \geqslant \kappa$. Hence the equality $\kappa^\diamond = \kappa$. □

**Remark 4.3.5.** We note that $K^{\downarrow}$, given as the intersection of all complete inf-semilattices containing $K$ can also be given as the set of all infima of elements in $K$: $K^{\downarrow}$ is the set of all infima of elements $y \in Y$, where $Y$ is an arbitrary subset of $K$, thus

$$K^{\downarrow} = \big\{ \textstyle\bigwedge_{y \in Y} y;\ Y \subset K \big\}.$$

Because of this representation we can actually define $\lambda^{[K]}$ without recourse to $K^{\downarrow}$:

$$\lambda^{[K]}(x) = \bigwedge_{Y \in \mathscr{P}(L)} \bigwedge_{y \in Y} (y;\ y \geqslant x), \qquad x \in L.$$

Similarly for $\beta_{[K]}$. □

Can we characterize the possible invariance sets of cleistomorphisms? The answer is easy to find:

**Theorem 4.3.6.** *If $\kappa$ is a cleistomorphism in a complete lattice $L$, then its invariance set is a complete inf-semilattice (see Definition 4.2.3).*

*Conversely, for any complete inf-semilattice $K$ in $L$, there exists a cleistomorphism with invariance set equal to $K$, viz. $\lambda^{[K]}$ (Definition 4.3.1).*

*For anoiktomorphisms, the corresponding result holds for complete sup-semilattices.*

**Proof.** Assume that $\kappa$ is a cleistomorphism and let $x_j$, $j \in J$, be any elements satisfying $\kappa(x_j) = x_j$. Define $x = \bigwedge_{j \in J} x_j$. We have $\kappa(x) \leqslant \kappa(x_j) = x_j$ for every $j$. Taking the infimum over all $j$ we see that $\kappa(x) \leqslant x$

and that $x$ belongs to $\mathbf{invar}(\kappa)$, proving that $\mathbf{invar}(\kappa)$ is a complete inf-semilattice for the operations inherited from $L$.

Conversely, let $K$ be a complete inf-semilattice contained in $L$. Then $\lambda^{[K]}$, as we have seen above, is a cleistomorphism with invariance set equal to $K$.

The result for complete sup-semilattices follows by duality. □

**Definition 4.3.7.** Let $L$ be a complete lattice. A ***Moore family*** in $L$ is a subset $M$ of $L$ with the property that for every subset $N$ of $M$, the infimum of all elements in $N$ belongs to $M$. By duality we define a ***dual Moore family*** using the supremum of elements in a subset $N$. □

By taking $N$ as the empty set we see that a Moore family is nonempty—the same holds for a dual Moore family.

By tradition, we call the set $M$ a "family" although it is just a subset of a complete lattice. The reason is that it was first defined when $L$ is the power set of some set, and then $M$ is a family of subsets of this set.

The definition means that $M$ is a complete inf-semilattice (see Definition 4.2.3) for the operations given in its supset $L$. 99999

To every Moore family $M$ we define a cleistomorphism $\lambda^{[M]}$ as in Definition 4.3.1 above.

**Proposition 4.3.8.** *Given a Moore family $M$, there exists one and only one cleistomorphism with invariance set equal to $M$.*

**Proof.** We have seen in Theorem 4.3.6 above that $\lambda^{[M]}$ is a cleistomorphism with invariance set equal to $M$. There can be no other, since $\lambda^{[\mathbf{invar}(\kappa)]} = \kappa^{\diamond} = \kappa$ for every cleistomorphism $\kappa$ as shown in Corollary 4.3.4. □

**Corollary 4.3.9.** *Every cleistomorphism is uniquely determined, within the family of all cleistomorphisms, by its invariance set. Analogously for anoiktomorphisms.* □

We have defined, given any ethmomorphism $\eta\colon L \to L$, two anoiktomorphisms: $\eta_\alpha$ and $\eta_\diamond = \beta_{[\mathbf{invar}(\eta)]}$, and likewise two cleistomorphisms: $\eta^\kappa$ and $\eta^\diamond = \lambda^{[\mathbf{invar}(\eta)]}$ (see Corollary 2.11.9 and Definition 4.3.1). By definition, $\eta_\alpha$ is the anoiktomorphism which is maximal among the anoiktomorphisms minimizing $\eta$, while $\eta_\diamond$ is the anoiktomorphism minimizing the invariance set among all anoiktomorphisms with invariance set containing that of $\eta$, thus the solutions to two different optimization problems. What is the relation between them? The answer is the following.

**Theorem 4.3.10.** *For any ethmomorphism $\eta\colon L \to L$, where $L$ is a complete lattice, we have*

$$\eta_\diamond \leqslant \eta_\alpha \leqslant \eta_{(1)} \leqslant \eta \leqslant \eta^{(1)} \leqslant \eta^\kappa \leqslant \eta^\diamond, \tag{4.3.2}$$

*where $\eta_{(1)} = \eta \wedge \mathbf{id}$ and $\eta^{(1)} = \eta \vee \mathbf{id}$, and where we have defined $\eta_\diamond = \beta_{[\mathbf{invar}(\eta)]}$ and $\eta^\diamond = \lambda^{[\mathbf{invar}(\eta)]}$; see* (4.3.1).

**Proof.** Both $\eta_\diamond$ and $\eta_\alpha$ are anoiktomorphisms minorizing $\eta$. But we know that $\eta_\alpha$ is the largest such minorant. Hence $\eta_\diamond \leqslant \eta_\alpha \leqslant \eta$. □

Looking at the invariance sets, the inequalities (4.3.2) yield

$$\mathbf{invar}(\eta) \subset \mathbf{invar}(\eta)^\uparrow = \mathbf{invar}(\eta_\diamond) \subset \mathbf{invar}(\eta_\alpha) \subset \mathbf{invar}(\eta_{(1)}),$$

$$\mathbf{invar}(\eta) \subset \mathbf{invar}(\eta)^\downarrow = \mathbf{invar}(\eta^\diamond) \subset \mathbf{invar}(\eta^\kappa) \subset \mathbf{invar}(\eta^{(1)}).$$

A natural question is whether the inequalities $\eta_\diamond \leqslant \eta_\alpha$ and $\eta^\kappa \leqslant \eta^\diamond$ can be strict; equivalently whether the inclusion relations $\mathbf{invar}(\eta_\diamond) \subset \mathbf{invar}(\eta_\alpha)$ and $\mathbf{invar}(\eta^\diamond) \subset \mathbf{invar}(\eta^\kappa)$ can be strict. The answer is in the affirmative:

**Example 4.3.11.** Let $L$ be the complete lattice with four elements $\mathbf{0}, a, b, \mathbf{1}$ and with the order relations $\mathbf{0} < a < \mathbf{1}$ and $\mathbf{0} < b < \mathbf{1}$, implying that $a \wedge b = \mathbf{0}$ and $a \vee b = \mathbf{1}$. Define $\gamma\colon L \to L$ by $\gamma(x) = x$ for $x \neq b$; $\gamma(b) = \mathbf{1}$. Then $\gamma$ is an ethmomorphism with invariance set $\mathbf{invar}(\gamma) = \{\mathbf{0}, a, \mathbf{1}\}$, and $\gamma_\diamond(b) = \mathbf{0} < b = \gamma_\alpha(b)$. To see this, we note that we must have $\gamma_\diamond(b) \leqslant b$ and also $\mathbf{invar}(\gamma_\diamond) = \mathbf{invar}(\gamma_\alpha) = \{\mathbf{0}, a, \mathbf{1}\}$, so that $\gamma_\diamond(b)$ must be different from $b$, and then the only possibility is $\gamma_\diamond(b) = \mathbf{0}$. Explicitly, we have

| $x$ | $\gamma_\diamond(x)$ | $\gamma_\alpha(x)$ | $\gamma_{(1)}(x)$ | $\gamma(x)$ |
|---|---|---|---|---|
| $\mathbf{1}$ | $\mathbf{1}$ | $\mathbf{1}$ | $\mathbf{1}$ | $\mathbf{1}$ |
| $b$ | $\mathbf{0}$ | $b$ | $b$ | $\mathbf{1}$ |
| $a$ | $a$ | $a$ | $a$ | $a$ |
| $\mathbf{0}$ | $\mathbf{0}$ | $\mathbf{0}$ | $\mathbf{0}$ | $\mathbf{0}$ |

Sure enough the invariance set of $\gamma_\alpha$ is strictly larger than that of $\gamma_\diamond$. □

## 4.4 Dilations and erosions in complete lattices

We have studied dilations in an abelian group, i.e., mappings of the form $A \mapsto A + B$. We also remarked (Proposition 3.2.2) that a mapping which commutes with translations and the formation of infinite unions is necessarily of this form. In lattice theory it is therefore natural to take the latter property as a definition:

**Definition 4.4.1.** We say that a mapping $\delta\colon L \to M$, where $L$ and $M$ are complete lattices, is a ***dilation*** if it commutes with the formation of suprema, i.e.,

$$\delta\big(\textstyle\bigvee_{x\in A} x\big) = \bigvee_{x\in A} \delta(x)$$

for all subsets $A$ of $L$. □

In particular we get $\delta(\mathbf{0}_L) = \mathbf{0}_M$ (take $A$ empty), while

$$\delta(\mathbf{1}_L) = \bigvee_{x\in L} \delta(x) \leqslant \mathbf{1}_M. \tag{4.4.1}$$

**Definition 4.4.2.** Similarly we shall say that $\varepsilon$ is an ***erosion*** if it commutes with the formation of infinite infima,

$$\varepsilon\big(\textstyle\bigwedge_{x\in A} x\big) = \bigwedge_{x\in A} \varepsilon(x)$$

for all subsets $A$ of $L$. □

We note that $\varepsilon(\mathbf{1}_L) = \mathbf{1}_M$ (take $A$ empty), while $\varepsilon(\mathbf{0}_L) = \bigwedge_{x\in L} \varepsilon(x) \geqslant \mathbf{0}_M$.

Dilations and erosions are always increasing. Indeed, we have $\delta(x \vee y) = \delta(x) \vee \delta(y)$. If $x \leqslant_L y$, this equation simplifies to $\delta(y) = \delta(x) \vee \delta(y) \geqslant_M \delta(x)$, which shows that $\delta$ is increasing, $\delta \in \mathbf{incr}(L, M)$. A similar argument shows that erosions are increasing.

We now recall Proposition 3.2.3, where it was stated that, in a semigroup, $\delta_B(A) \subset C$ if and only if $A \subset \varepsilon_B(C)$. In a lattice this may be written as $\delta(x) \leqslant y$ iff $x \leqslant \varepsilon(y)$, equivalently as $\mathbf{epi}\,\delta = (\mathbf{hypo}\,\varepsilon)^{\check{}}$, where the symbol $\check{}$ means that we swap the components: for a subset $A$ of a Cartesian product $X \times Y$ we define

$$A^{\check{}} = \{(y, x);\ (x, y) \in A\} \subset Y \times X. \tag{4.4.2}$$

May we use this as a model to define erosions from dilations and conversely in the more general lattice situation? Indeed this is the case, and we shall do so in the next chapter.

**Example 4.4.3.** The mapping $f^*\colon \mathscr{P}(Y) \to \mathscr{P}(X)$, defined by (2.6.1), is both a dilation and an erosion, while $f_*\colon \mathscr{P}(X) \to \mathscr{P}(Y)$, defined in (2.6.2), is a dilation but in general not an erosion. □

The ceiling function is a dilation between the two complete lattices $\mathbf{R}_!$ and $\mathbf{Z}_!$:

$$\sup_j \lceil t_j \rceil = \lceil \sup_j t_j \rceil,$$

but it is not an erosion, since it may happen that

$$\inf_j \lceil t_j \rceil > \lceil \inf_j t_j \rceil.$$

Similarly, the floor function $t \mapsto \lfloor t \rfloor$ is an erosion but not a dilation. When considered as mappings from $\mathbf{R}_!$ to $\mathbf{R}_!$, the ceiling and floor functions are ethmomorphisms; the first a cleistomorphism, the second an anoiktomorphism.

We also extend the ceiling and floor functions to functions $F\colon X \to \mathbf{R}_!$ defined on any set $X$ and with values in the set of extended reals $\mathbf{R}_!$. Then $\lfloor F \rfloor, \lceil F \rceil \in \mathscr{F}(X, \mathbf{Z}_!)$.

**Definition 4.4.4.** A mapping is said to be an ***antierosion*** if $f^{\mathrm{op}}$ is an erosion, i.e., if $f\left(\bigwedge_{j\in J} x_j\right) = \bigvee_{j\in J} f(x_j)$ for all families $(x_j)_{j\in J}$.

A mapping is said to be an ***antidilation*** if $f^{\mathrm{op}}$ is a dilation, i.e., if $f\left(\bigvee_{j\in J} x_j\right) = \bigwedge_{j\in J} f(x_j)$ for all families $(x_j)_{j\in J}$. □

## 4.5 Notes

### 4.5.1 *The sources*

The crucial distinction between lattices and complete lattices was first made by Garret Birkhoff (1933:442), although he called a lattice "finite lattice" and a complete lattice just "lattice." In his seminal book published seven years later, he used *lattice* and *complete lattice* as is the custom today (Birkhoff 1940:16, 17).

For lattice theory, see (Dubreil & Dubreil-Jacotin 1964), (Davey & Priestely 2003) and (Grätzer 1998).

Matheron studies lattices of several kinds (compact, distributive etc.) in his report (1996).

Petros Maragos studies in (2017) complete lattices with two operations in addition to $\wedge$ and $\vee$, called multiplication and dual multiplication.

A book with many chapters on applications of mathematical morphology to image processing is (Heijmans & Roerdink 1998, Eds.). For the algebraic foundations of mathematical morphology, see the chapter by Ronse & Serra (2008, 2010). See also Serra's Thèse d'État (1986a), his publication (1986b), and his report (1987).

For several studies of lattices by Jean Serra, see his publications (1998, 2002, 2005a, 2005c), the last one on segmentation of color images.

Jesús Angulo introduced in his study (2007) an algorithmic framework to apply mathematical morphology operators to color images. Similarly, Aptoula and Lefèvre (2007) extend mathematical morphology to multivalued images.

Najman et al. (2015) construct a variety of continuously varying operators in between erosions and dilations, and, perhaps more interestingly, between anoiktomorphisms and cleistomorphisms.

Roerdink (1996) studied a homogeneous space on which a transitive but not necessarily commutative transformation group is defined, and generalized dilations, erosions, anoiktomorphisms and cleistomorphisms in this non-commutive setting.

Theorem 4.3.2 is essentially Proposition 5.4 in (Serra 1988d:106).

For hierarchies of partitions, see (Kiran & Serra 2013).

### 4.5.2 *Terms used*

Matheron in his pioneering treatise (1975:17) used the terms *dilatation* and *erosion* for operations $\mathscr{P}(\mathbf{R}^n) \to \mathscr{P}(\mathbf{R}^n)$. Serra (1986, 1988a) defined dilations and erosions as here in the case of complete lattices with $L = M$; antierosions and antidilations were introduced by Serra (1987).

Singer (1997:172) uses the term *duality* for an antierosion. The study of dualities in the sense of Singer is therefore equivalent to that of dilations or erosions.

The Moore families are named for Eliakim Hastings Moore; cf. his publication (1910). Heijmans & Keshet (2002) present a framework for image processing based on complete inf-semilattices, called *cisl*, which, as we have seen, are just Moore families.

## 4.6 Exercises

4.1. Give an example of a dilation $\delta\colon L \to M$ where $\delta(\mathbf{1}_L) \neq \mathbf{1}_M$; cf. (4.4.1).

4.2. Prove all assertions in Example 4.3.11.

## Chapter 5

# Inverses and quotients of mappings

## 5.1 Introduction

In this chapter we shall study inverses and quotients of mappings between ordered sets which are analogous to inverses $1/y$ and quotients $x/y$ of positive numbers. The theory of lower and upper inverses defined in Section 5.2 generalizes the theory of Galois connections as well as residuation theory and the theory of adjunctions. We investigate in Sections 5.4, 5.5 and 5.6 to what extent a generalized inverse can serve as a left inverse, as a right inverse, and how an inverse of an inverse relates to the identity mapping. These inverses and quotients are then used in Section 6.1 to create a convenient formalism for a unified treatment of dilations $\delta\colon L \to M$ and erosions $\varepsilon\colon M \to L$ as well as of cleistomorphisms $\kappa = \varepsilon \circ \delta\colon L \to L$ and anoiktomorphisms $\alpha = \delta \circ \varepsilon\colon M \to M$.

In general a mapping $g\colon X \to Y$ between sets does not have an inverse. If $g$ is injective, we may define a left inverse $u\colon Y \to X$, thus with $u \circ g = \mathbf{id}_X$, where $\mathbf{id}_X$ denotes the identity mapping in $X$, defining $u(y)$ in an arbitrary way when $y$ is not in the image of $g$. If $g$ is surjective, we may define a right inverse $v\colon Y \to X$, thus with $g \circ v = \mathbf{id}_Y$. We then need to define $v(y)$ as an element of the preimage $\{x;\ g(x) = y\}$. In the general situation this has to be done using the axiom of choice. In a complete lattice, however, it could be interesting to define $v(y)$ as the supremum or infimum of all points $x$ such that $g(x) = y$, even though this supremum or infimum need not belong to the set of points that are mapped to $y$. At any rate, the preimage of $y$ is contained in the interval defined by the infimum and the supremum. However, for various reasons it is convenient to take instead the infimum of all $x$ such that $g(x) \geqslant y$ or the supremum of all $x$ such that $g(x) \leqslant y$. This yields better monotonicity properties. (The case

$g(x) = y$ is allowed, since we can take the preorder in $Y$ be the discrete order.)

If, given a mapping $g\colon L \to M$ from an ordered set $L$ into an ordered set $M$, we can find mappings $u, v\colon M \to L$ such that $\mathbf{hypo}\, u = (\mathbf{epi}\, g)^{\check{}}$ or $\mathbf{epi}\, v = (\mathbf{hypo}\, g)^{\check{}}$, we would be content to have some kinds of inverses to $g$. However, usually the best we can do is to study mappings satisfying either $\mathbf{hypo}\, u \supset (\mathbf{epi}\, g)^{\check{}}$ or $\mathbf{epi}\, v \supset (\mathbf{hypo}\, g)^{\check{}}$. This will be the approach in what follows, where we shall define not one but two inverses, viz. the lower (to be denoted by $g_{[-1]}$) and the upper (written as $g^{[-1]}$).

Given any mapping $g\colon X \to Y$ between preordered sets, one could for instance study the four compositions

$$g^{[-1]} \circ g,\ g_{[-1]} \circ g\colon X \to X \text{ and } g \circ g_{[-1]},\ g \circ g^{[-1]}\colon Y \to Y,$$

as well as those formed by the other inverses mentioned below in (5.2.3) and (5.2.4). When $X = Y$, one should also ask whether $g$, $g_{[-1]}$ and $g^{[-1]}$ are ethmomorphisms, cleistomorphisms, or anoiktomorphisms. The same questions apply to all of the compositions just mentioned.

## 5.2 Defining inverses of mappings

**Definition 5.2.1.** Let $L$ be a complete lattice, $M$ a preordered set, and $g\colon L \to M$ any mapping. We then define the ***lower inverse*** $g_{[-1]}\colon M \to L$ and the ***upper inverse*** $g^{[-1]}\colon M \to L$ as the mappings

(5.2.1) $$g_{[-1]}(y) = \bigvee_{x \in L} (x;\ g(x) \leqslant_M y) = \bigvee_{x \in L} (x;\ (x, y) \in \mathbf{epi}\, g);$$

(5.2.2) $$g^{[-1]}(y) = \bigwedge_{x \in L} (x;\ g(x) \geqslant_M y) = \bigwedge_{x \in L} (x;\ (x, y) \in \mathbf{hypo}\, g),$$

where $y \in M$. □

As a first observation, let us note that these inverses are always increasing. If there exists a largest element $\mathbf{1}_M$, then $g_{[-1]}(\mathbf{1}_M) = \mathbf{1}_L$. Similarly, if $M$ possesses a smallest element $\mathbf{0}_M$, then $g^{[-1]}(\mathbf{0}_M) = \mathbf{0}_L$. If $M$ has the chaotic preorder, then both inverses are constant, $g_{[-1]} = \mathbf{1}_L$ and $g^{[-1]} = \mathbf{0}_L$ identically. Here the lower inverse is larger than the upper inverse.

As we shall see, there are other examples where the lower inverse is not below the upper inverse: For the function $h(x) = x \vee c$ in Example 5.8.5 we have $h_{[-1]}(c) = c > h^{[-1]}(c) = \mathbf{0}$ and $h_{[-1]}(\mathbf{1}) = \mathbf{1} > h^{[-1]}(\mathbf{1}) = a$. We should rather think of the forming of lower inverses as an approach from below, or bottom-up, and upper inverses as approaching from above, or top-down.

We note that the upper inverse is equal to the mapping $((g_{\mathrm{op}}^{\mathrm{op}})_{[-1]})_{\mathrm{op}}^{\mathrm{op}}$, where the mapping $g_{\mathrm{op}}^{\mathrm{op}}$ is defined in Definition 2.8.2.

We can also define mappings

$$((g^{\mathrm{op}})_{[-1]})_{\mathrm{op}}(y) = ((g_{\mathrm{op}})^{[-1]})^{\mathrm{op}} = \bigvee_{x \in L} (x;\ y \leqslant_M g(x)); \tag{5.2.3}$$

$$((g^{\mathrm{op}})^{[-1]})_{\mathrm{op}}(y) = ((g_{\mathrm{op}})_{[-1]})^{\mathrm{op}} = \bigwedge_{x \in L} (x;\ y \geqslant_M g(x)), \tag{5.2.4}$$

where $y \in M$. The four inverses were introduced by Banon & Barrera (1993:311), who used them to construct decompositions of general mappings between complete lattices. They denoted them as follows:

$$g_{[-1]} = \overline{g\cdot},\quad g^{[-1]} = \underline{\cdot g},\quad ((g^{\mathrm{op}})_{[-1]})_{\mathrm{op}} = \overline{\cdot g},\quad ((g^{\mathrm{op}})^{[-1]})_{\mathrm{op}} = \underline{g\cdot}\ . \tag{5.2.5}$$

**Example 5.2.2.** If $L$ and $M$ are complete lattices and $g\colon L \to M$ is decreasing, then

$$g(\mathbf{1}_L) = \inf g \leqslant \sup g = g(\mathbf{0}_L),$$

so that $g_{[-1]}(y) = \mathbf{1}_L$ if $y \geqslant \inf g$ and $g_{[-1]}(y) = \mathbf{0}_L$ otherwise, while $g^{[-1]}(y) = \mathbf{0}_L$ if $y \leqslant \sup g$ and $g^{[-1]}(y) = \mathbf{1}_L$ otherwise.

Thus, the lower inverse forgets all information about $g$ except its infimum, and the upper inverse remembers only the supremum of $g$. □

This example shows that the lower and upper inverses are not so interesting for decreasing mappings. Similarly, if $g$ is increasing, the inverses $\overline{\cdot g}$ and $\underline{g\cdot}$ are not interesting. However, for other mappings the inverses may contain much more information.

**Example 5.2.3.** For any mapping $f\colon X \to Y$ we defined in (2.6.1) and (2.6.2) the mappings $f^*\colon \mathscr{P}(Y) \to \mathscr{P}(X)$ and $f_*\colon \mathscr{P}(X) \to \mathscr{P}(Y)$. We note that, for all $A \in \mathscr{P}(X)$ and all $B \subset \mathscr{P}(Y)$ we have $f_*(A) \subset B$ if and only if $A \subset f^*(B)$. It follows that $(f^*)^{[-1]} = f_*$ and $(f_*)_{[-1]} = f^*$. From this we see that $((f^*)^{[-1]})_{[-1]} = f^*$ and that $((f_*)_{[-1]})^{[-1]} = f_*$. □

## 5.3 First properties of inverses

As already noted, $g_{[-1]}$ and $g^{[-1]}$ are increasing, while $((g^{\mathrm{op}})_{[-1]})_{\mathrm{op}} = \overline{\cdot g}$ and $((g^{\mathrm{op}})^{[-1]})_{\mathrm{op}} = \underline{g\cdot}$ are decreasing.

We note that we always have

$$\mathbf{epi}\, g \subset (\mathbf{hypo}\, g_{[-1]})^{\smile}, \tag{5.3.1}$$

in other words, if $y \geqslant g(x)$, then $x \leqslant g_{[-1]}(y)$; and

$$\mathbf{hypo}\, g \subset (\mathbf{epi}\, g^{[-1]})^{\smile}, \tag{5.3.2}$$

in other words, if $y \leqslant g(x)$, then $x \geqslant g^{[-1]}(y)$. Here $R^{\smile}$ for a subset $R$ of $X \times Y$ is defined by (4.4.2). In general these inclusions are strict as we shall see.

Note that we do not require in (5.2.1) that the set of all $x$ such that $y \geqslant_Y g(x)$ shall have a largest element. In other words, the supremum in (5.2.1) is not necessarily a maximum; see Examples 5.3.2, 5.8.1 and 5.8.2. Using terms from (Blyth 2005:5, 26), the set of all $x$ such that $g(x) \leqslant y$ is not necessarily an ideal in $X$, although it is a down-set if $g$ is increasing. Similarly, the infimum in (5.2.2) is not necessarily a minimum; the set of all $x$ such that $g(x) \geqslant y$ is not necessarily a filter in $X$, although it is an up-set if $g$ is increasing.

The special situation when the supremum in (5.2.1) is actually a maximum is equivalent to the inequality $g(g^{[-1]}(y)) \leqslant_Y y$; if this holds for all $y \in M$, we have $g \circ g^{[-1]} \leqslant_Y \mathbf{id}_Y$. This can happen whether $g$ is increasing or not. Except for the simplest lattices there exist increasing mappings which do not have this property:

**Proposition 5.3.1.** *Let $L$ be a complete lattice which is not totally ordered, and let $M$ be a complete lattice with at least two elements. Then there exists an increasing mapping $g\colon L \to M$ such that $g\big(g^{[-1]}(\mathbf{0}_M)\big) \nleqslant \mathbf{0}_M$.*

**Proof.** Let $a, b$ be any two non-comparable elements of $L$ and define $g(x) = \mathbf{1}_M$ if $x \geqslant a \vee b$; $g(x) = \mathbf{0}_M$ otherwise. (In particular $g(a) = g(b) = \mathbf{0}_M$.) Then $g^{[-1]}(y) \geqslant a \vee b$ for all $y \in M$, so that $g(g^{[-1]}(y)) = \mathbf{1}_M$ for all $y$, in particular $g(g^{[-1]})(\mathbf{0}_M) = \mathbf{1}_M \nleqslant \mathbf{0}_M$. □

**Example 5.3.2.** Take $L = M = [0,1]$ and $g\colon [0,1] \to [0,1]$. Define $x_0 = g^{[-1]}(0) = \sup(x;\ g(x) \leqslant 0)$. The inequality $g(g^{[-1]}(0)) \leqslant 0$ means that $g(x_0) = 0$, which need not be the case, even if the supremum is attained and $g$ is increasing, for instance if $g = \chi_{[1/2,1]}$, the characteristic function of the interval $[1/2,1]$. (We have $x_0 = g^{[-1]}(0) = 1/2$ and $g(1/2) = 1$.) □

We next show that the lower and upper inverses are solutions to certain extremal problems.

**Proposition 5.3.3.** *Let $L$ be a complete lattice, $M$ a preordered set, and let $g\colon L \to M$, $u, v\colon M \to L$ be mappings. If* $\mathbf{hypo}\, v \subset (\mathbf{epi}\, g)\check{} \subset \mathbf{hypo}\, u$, *then $v \leqslant g_{[-1]} \leqslant u$ and*

$$\mathbf{hypo}\, v \subset (\mathbf{epi}\, g)\check{} \subset \mathbf{hypo}\, g_{[-1]} \subset \mathbf{hypo}\, u.$$

*Hence $g_{[-1]}$ is the smallest mapping $u$ which satisfies* $\mathbf{hypo}\, u \supset (\mathbf{epi}\, g)\check{}$.

*If on the other hand* $\mathbf{epi}\, v \supset (\mathbf{hypo}\, g)\check{} \supset \mathbf{epi}\, u$, *then $v \leqslant g^{[-1]} \leqslant u$ and*

$$\mathbf{epi}\, v \supset (\mathbf{epi}\, g)^{[-1]} \supset (\mathbf{hypo}\, g)\check{} \supset \mathbf{epi}\, u.$$

*Hence $g^{[-1]}$ is the largest mapping $v$ such that* $\mathbf{epi}\, v \supset (\mathbf{hypo}\, g)\check{}$.

The proof is straightforward.

**Corollary 5.3.4.** *With $g$, $u$ and $v$ given as in the proposition, assume that* $\mathbf{hypo}\, u = (\mathbf{epi}\, g)\check{}$. *Then $u = g_{[-1]}$. Similarly, if* $\mathbf{epi}\, v = (\mathbf{hypo}\, g)\check{}$, *then $v = g^{[-1]}$.*

*If also $M$ is a complete lattice, then* $\mathbf{hypo}\, u = (\mathbf{epi}\, g)\check{}$ *implies $u^{[-1]} = g$ in addition to $u = g_{[-1]}$. Similarly,* $\mathbf{epi}\, v = (\mathbf{hypo}\, g)\check{}$ *implies that $v_{[-1]} = g$ in addition to $v = g^{[-1]}$.* □

The corollary singles out the special case of adjunctions (for the definition of adjunctions, see Subsection 5.7.3) between $L$ and $M$ among all pairs $(g, g_{[-1]})$ and adjunctions between $M$ and $L$ among all pairs $(g^{[-1]}, g)$.

**Example 5.3.5.** If $f\colon \mathscr{P}(G) \to \mathscr{P}(G)$ is the dilation $f(A) = \delta_U(A) = A + U$, $A \in \mathscr{P}(G)$, where $G$ is an abelian group and $U$ a fixed subset of $G$, the structuring element, then

$$(\delta_U)_{[-1]}(C) = \bigcup_{A \in \mathscr{P}(G)} (A;\ \delta_U(A) \subset C) = \varepsilon_U(C), \qquad C \in \mathscr{P}(G),$$

the erosion associated to $\delta_U$. A most fundamental example.

The compositions $\alpha_U = \delta_U \circ \varepsilon_U$ and $\kappa_U = \varepsilon_U \circ \delta_U$ are the anoiktomorphism and the cleistomorphism associated to $\delta_U$, respectively.

When $X = Y = \mathbf{R}^n$, the upper inverse of $\delta_U$ is not interesting, since $(\delta_U)^{[-1]}(C) = \emptyset$ for all $C$ if $U$ has interior points, and equal to $C$ if $U = \{0\}$. □

An ideal inverse $u$ would satisfy $u \circ g = \mathbf{id}_L$, $g \circ u = \mathbf{id}_M$, and the inverse of $u$ would be $g$. It is therefore natural to compare $g_{[-1]} \circ g$ and $g^{[-1]} \circ g$ with $\mathbf{id}_L$; $g \circ g_{[-1]}$ and $g \circ g^{[-1]}$ with $\mathbf{id}_M$; and inverses of inverses of $g$ with $g$. This is what we shall do in the next three sections.

## 5.4 Left inverses

We shall now compare $g_{[-1]} \circ g$ and $g^{[-1]} \circ g$ with $\mathbf{id}_L$.

**Proposition 5.4.1.** *If $L$ is a complete lattice and $M$ a preordered set, then for all mappings $g\colon L \to M$ one has*

$$g^{[-1]} \circ g \leqslant \mathbf{id}_L \leqslant g_{[-1]} \circ g.$$

*Equality holds if $g$ is coincreasing in the sense of Definition 2.8.1. Conversely, if $g^{[-1]} \circ g \geqslant \mathbf{id}_L$ or $g_{[-1]} \circ g \leqslant \mathbf{id}_L$, then $g$ is coincreasing.*

**Proof.** It is clear that $g^{[-1]}(g(a)) = \bigwedge(x;\ g(x) \geqslant g(a)) \leqslant a$. If $g$ is coincreasing, then $\{x;\ g(x) \geqslant g(a)\}$ is contained in $\{x;\ x \geqslant a\}$, so that $g^{[-1]}(g(a)) \geqslant \bigwedge(x;\ x \geqslant a) = a$.

Conversely, if $g^{[-1]}(g(a)) \geqslant a$, then for all $x$, $g(x) \geqslant g(a)$ implies $x \geqslant a$. If this is true for all $a$, then $g$ is coincreasing. The assertions for $g_{[-1]}$ follows by duality. □

**Corollary 5.4.2.** *Let $L$ and $M$ be as in the proposition. Then, for any mapping $g\colon L \to M$, $g^{[-1]}(y) \leqslant g_{[-1]}(y)$ for all $y \in \mathbf{im}(g)$, and also for all $y$ majorizing or minorizing $\mathbf{im}(g)$. In particular, $g^{[-1]} \leqslant g_{[-1]}$ if $g$ is surjective.*

**Proof.** The statement for $y \in \mathbf{im}(g)$ follows directly from the proposition. If $y$ majorizes all elements in $\mathbf{im}(g)$, then $g_{[-1]}(y) = \mathbf{1}_L$, and if $y$ minorizes all elements in $\mathbf{im}(g)$, then $g^{[-1]}(y) = \mathbf{0}_L$. □

**Proposition 5.4.3.** *If $u, v$ are increasing mappings such that $u \circ g \leqslant \mathbf{id}_L \leqslant v \circ g$, then $u \leqslant g^{[-1]}$ and $v \geqslant g_{[-1]}$. Hence, in view of Proposition 5.4.1, $g^{[-1]}$ is the largest increasing mapping $u$ such that $u \circ g \leqslant \mathbf{id}_L$, and $g_{[-1]}$ is the smallest increasing mapping $v$ such that $v \circ g \geqslant \mathbf{id}_L$.*

**Proof.** If $u$ and $v$ are increasing and $u \circ g \leqslant \mathbf{id}_L \leqslant v \circ g$, then $\mathbf{epi}\, u \supset (\mathbf{hypo}\, g)^{\smallsmile}$ and $\mathbf{hypo}\, v \supset (\mathbf{epi}\, g)^{\smallsmile}$. We can now apply Proposition 5.3.3. □

**Theorem 5.4.4.** *Let $L$ be a complete lattice and $M$ a preordered set. Then the following six conditions are equivalent.*

(a) *$g$ is coincreasing;*
(b) $g_{[-1]} \circ g \leqslant \mathbf{id}_L$;
(c) $g_{[-1]} \circ g = \mathbf{id}_L$;

(d) $g^{[-1]} \circ g \geqslant \mathbf{id}_L$;
(e) $g^{[-1]} \circ g = \mathbf{id}_L$;
(f) $g_{[-1]} \leqslant g^{[-1]}$.

**Proof.** (a) implies (b), (c), (d) and (e). If $g$ is coincreasing we know already from Proposition 5.4.1 that (b) and (d) hold with equality.

(b), (c), (d) or (e) implies (a). This is contained in Proposition 5.4.1.

(a) implies (f). Assume that $g$ is coincreasing and fix any element $y \in M$. Let $x, x' \in L$ be such that $g(x) \leqslant y \leqslant g(x')$. Then $x \leqslant x'$. Letting $x$ vary we see that $g_{[-1]}(y) \leqslant x'$. Letting now also $x'$ vary, we see that $g_{[-1]}(y) \leqslant g^{[-1]}(y)$. Thus (f) holds.

(f) implies (a). If $x$ and $x'$ are given with $g(x) \leqslant g(x')$ we define $y = g(x)$. Then $x \leqslant g_{[-1]}(y)$ and $g^{[-1]}(y) \leqslant x'$. If we know that (f) holds, it follows that $x \leqslant x'$, proving that $f$ is coincreasing. □

## 5.5 Right inverses

Next we compose inverses of $g$ with $g$ in the other order: we shall see to what extent the inverses we have constructed can serve as right inverses. This will lead to a characterization of dilations—and, by duality, of erosions, antierosions and antidilations.

**Theorem 5.5.1.** *If $L$ and $M$ are complete lattices and $g\colon L \to M$ any mapping, then the following five properties are equivalent.*

(A) *$g$ is a dilation;*
(B) $\mathbf{epi}\, g \supset (\mathbf{hypo}\, g_{[-1]})\check{}$;
(C) $\mathbf{epi}\, g = (\mathbf{hypo}\, g_{[-1]})\check{}$;
(D) *$g$ is increasing and* $\mathbf{epi}\, g \supset (\mathbf{graph}\, g_{[-1]})\check{}$;
(E) *$g$ is increasing and* $g \circ g_{[-1]} \leqslant \mathbf{id}_M$.

*We get analogous characterizations for erosions by duality.*

Singer (1997:178, Proposition 5.3) proves that (A) and (E) above are equivalent (expressed for dualities in his sense, i.e., for antierosions).

**Proof.** (A) implies (B). Suppose that (A) holds. Then if $(y, x)$ belongs to $\mathbf{hypo}\, g_{[-1]}$, in other words if $x \leqslant g_{[-1]}(y)$, we obtain, since $g$ is increasing by hypothesis,

$$g(x) \leqslant g(g_{[-1]}(y)) = g\big(\bigvee(x;\ g(x) \leqslant y)\big) = \bigvee\big(g(x);\ g(x) \leqslant y\big) \leqslant y,$$

which means that $(x, y) \in \mathbf{epi}\, g$. Thus (B) holds.

(B) implies (A). We note first that $g$ is increasing if (B) holds. Indeed, if $x \leqslant x'$ and we define $y = g(x')$, then $g_{[-1]}(y) = g_{[-1]}(g(x')) \geqslant x' \geqslant x$ (see Proposition 5.4.1), which by (B) implies that $g(x) \leqslant y = g(x')$.

Let now any family $(x_j)_j$ of elements of $L$ be given and define $z = \bigvee g(x_j)$, $w = g(\bigvee x_j)$. Since $g$ is increasing we always have $z \leqslant w$. Is it true that $w \leqslant z$? We note that, by Proposition 5.4.1, $x_j \leqslant g_{[-1]}(g(x_j)) \leqslant g_{[-1]}(z)$. Taking the supremum over all $j$ we obtain $\bigvee x_j \leqslant g_{[-1]}(z)$, which by (B) implies that $w = g(\bigvee x_j) \leqslant z$. We have proved (A).

(B) is equivalent to (C). This is clear since we always have $\mathbf{epi}\, g \subset (\mathbf{hypo}\, g_{[-1]})^{\check{}}$.

(B) implies (D). We have seen that (B) implies that $g$ is increasing. That (B) implies $(\mathbf{graph}\, g_{[-1]})^{\check{}} \subset \mathbf{epi}\, g$ follows from the inclusion $\mathbf{graph}\, g_{[-1]} \subset \mathbf{hypo}\, g_{[-1]}$.

(D) implies (B). If $x \leqslant g_{[-1]}(y)$ we define $x' = g_{[-1]}(y)$ and note that $x \leqslant x'$ and that $(y, x') \in \mathbf{graph}\, g_{[-1]}$. If (D) holds, we conclude that $g(x') \leqslant y$. Hence, if $g$ is increasing, $g(x) \leqslant g(x') \leqslant y$, proving (B).

(D) and (E) are equivalent: (E) is just a rephrasing of (D). □

**Remark 5.5.2.** We may use (B), (C), (D) or (E) to define dilations $L \to M$ when $M$ is only a preordered set. □

**Corollary 5.5.3.** *If $L$ and $M$ are complete lattices and $g\colon L \to M$ and $u\colon L \to M$ are two mappings such that $\mathbf{epi}\, g = (\mathbf{hypo}\, u)^{\check{}}$, then $g$ is a dilation and $u$ is an erosion, and $g_{[-1]} = u$, $u^{[-1]} = g$; $(g^{\mathrm{op}}, u_{\mathrm{op}})$ and $(u_{\mathrm{op}}, g^{\mathrm{op}})$ are Galois connections (see Definition 5.7.1).*

**Proof.** It follows from $\mathbf{epi}\, g = (\mathbf{hypo}\, u)^{\check{}}$ that $g$ is increasing and that $g_{[-1]} = u$, hence that (D) in the theorem holds. Since (D) is equivalent to (A), we see that $g$ is a dilation. The rest follows by duality. □

**Corollary 5.5.4.** *Let $L$ and $M$ be complete lattices and $F\colon L \to M$ a mapping. Then the following conditions are equivalent.*

(A) *There exists a mapping $G\colon M \to L$ such that $(F, G)$ is a Galois connection;*
(B) *The mapping $g = F^{\mathrm{op}}$ is increasing and $g \circ g^{[-1]} \leqslant \mathbf{id}_Y$;*
(C) *$F$ is an antidilation.*

**Proof.** The corollary follows from the theorem and Proposition 5.7.2. □

**Theorem 5.5.5.** *If $L$ and $M$ are complete lattices and $\delta\colon L \to M$ a dilation, then $\delta_{[-1]}\colon M \to L$ is an erosion. Similarly, if $\varepsilon\colon M \to L$ is an erosion, then $\varepsilon^{[-1]}$ is a dilation.*

**Proof.** We know that $\delta_{[-1]}$ is increasing, so, writing for brevity just $g$ for $\delta_{[-1]}$, we have $g(\bigwedge y_j) \leqslant g(y_k)$ for all $k$; hence $g(\bigwedge y_j) \leqslant \bigwedge g(y_k)$. We need to prove the opposite inequality, $\bigwedge g(y_k) \leqslant g(y)$, where $y = \bigwedge y_j$. From (E) in Theorem 5.5.1 we learn that $\delta(g(y_k)) \leqslant y_k$ for all $k$, which implies that $\delta(\bigwedge g(y_j)) \leqslant y_k$, hence that $\delta(\bigwedge g(y_j)) \leqslant y$. Since $g$ is increasing we get

$$\bigwedge g(y_j) \leqslant g\left(\delta\left(\bigwedge g(y_j)\right)\right) \leqslant g(y),$$

where the first inequality follows from Proposition 5.4.1. This completes the proof. □

**Corollary 5.5.6.** *For any dilation $\delta\colon L \to M$ we have*

$$\delta \circ \delta_{[-1]} \circ \delta = \delta \text{ and } \delta_{[-1]} \circ \delta \circ \delta_{[-1]} = \delta_{[-1]}.$$

*In particular,*

$$\kappa_{[\delta]} = \delta_{[-1]} \circ \delta \text{ and } \alpha_{[\delta]} = \delta \circ \delta_{[-1]}$$

*are idempotent: $\kappa_{[\delta]} \in$ **ethmo**$(L)$ and $\alpha_{[\delta]} \in$ **ethmo**$(M)$. The first is a cleistomorphism in $L$, the second an anoiktomorphism in $M$.*

*Dually, $\varepsilon \circ \varepsilon^{[-1]} \circ \varepsilon = \varepsilon$ and $\varepsilon^{[-1]} \circ \varepsilon \circ \varepsilon^{[-1]} = \varepsilon^{[-1]}$ for any erosion $\varepsilon\colon M \to L$.*

*Also*

$$\alpha^{[\varepsilon]} = \varepsilon^{[-1]} \circ \varepsilon \text{ and } \kappa^{[\varepsilon]} = \varepsilon \circ \varepsilon^{[-1]}$$

*are idempotent; the first an anoiktomorphism, the second a cleistomorphism.*

**Proof.** We always have $g_{[-1]} \circ g \geqslant \mathbf{id}_L$ (Proposition 5.4.1); it follows that $g_{[-1]} \circ g \circ g_{[-1]} \geqslant g_{[-1]}$. If $g$ is increasing, we also get $g \circ g_{[-1]} \circ g \geqslant g$. Hence $\delta \circ \delta_{[-1]} \circ \delta \geqslant \delta$.

For dilations we have $\delta \circ \delta_{[-1]} \leqslant \mathbf{id}_M$ (Theorem 5.5.1), from which we deduce that $\delta \circ \delta_{[-1]} \circ \delta \leqslant \delta$ and $\delta_{[-1]} \circ \delta \circ \delta_{[-1]} \leqslant \delta_{[-1]}$. This shows what we want for dilations; the rest follows by duality. □

In many situations we have $\varepsilon \leqslant \mathbf{id} \leqslant \delta$, for erosions $\varepsilon$ and dilations $\delta$, for instance when they are defined by a structuring element $B$ containing the origin. Then the chain of inequatilities $\varepsilon \leqslant \alpha \leqslant \mathbf{id} \leqslant \kappa \leqslant \delta$ leads to

$$\varepsilon = \kappa \circ \varepsilon \leqslant \kappa \circ \alpha \leqslant \kappa \leqslant \kappa \circ \delta;$$

$$\varepsilon \circ \kappa \leqslant \alpha \circ \kappa \leqslant \kappa \leqslant \delta \circ \kappa = \delta;$$

$$\alpha \circ \varepsilon \leqslant \alpha \leqslant \alpha \circ \kappa \leqslant \alpha \circ \delta = \delta;$$

$$\varepsilon = \varepsilon \circ \alpha \leqslant \alpha \leqslant \kappa \circ \alpha \leqslant \delta \circ \alpha.$$

The first two are obtained by composition from the left and right with $\kappa$; the third and fourth by composition from the left and right with $\alpha$.

The equality $\kappa(\varepsilon(A)) = \varepsilon(A)$ implies that $\varepsilon(A)$ as well as $\kappa(\alpha(A))$, $\kappa(A)$ and $\kappa(\delta(A))$ are all closed. Since $\varepsilon(A) \subset A \subset \kappa(A)$, we can consider $\varepsilon(A)$, $\kappa(\alpha(A))$ and $\kappa(A)$ as closed approximants of $A$.

## 5.6 Inverses of inverses

**Proposition 5.6.1.** *If $L$ and $M$ are complete lattices and $g\colon L \to M$ any mapping, then quite generally*

$$(5.6.1) \qquad \big(g_{[-1]}\big)^{[-1]} \leqslant g \leqslant \big(g^{[-1]}\big)_{[-1]}.$$

*Equality holds at the first place if and only if $g$ is a dilation; at the second place if and only if $g$ is an erosion.*

If $\delta$ is a dilation and we define $\varepsilon$ as the lower inverse of $\delta$ we thus have

$$(5.6.2) \qquad \varepsilon^{[-1]} = \delta \leqslant \big(\delta^{[-1]}\big)_{[-1]}.$$

It follows that

$$(5.6.3) \qquad \delta(x) \leqslant y \;\Leftrightarrow\; x \leqslant \varepsilon(y).$$

Conversely, this equivalence implies that $\delta_{[-1]} = \varepsilon$ and that $\varepsilon^{[-1]} = \delta$.

If $\varepsilon$ is an erosion and we define $\delta = \varepsilon^{[-1]}$, then (5.6.1) becomes

$$(5.6.4) \qquad (\varepsilon_{[-1]})^{[-1]} \leqslant \varepsilon = \delta_{[-1]}.$$

**Proof.** We always have $\mathbf{epi}\, g \subset (\mathbf{hypo}\, g_{[-1]})\check{}$, i.e., $y \geqslant g(a)$ implies $g_{[-1]}(y) \geqslant a$. This yields

$$(g_{[-1]})^{[-1]}(a) = \bigwedge \big(y;\; g_{[-1]}(y) \geqslant a\big) \leqslant \bigwedge (y;\; y \geqslant g(a)) = g(a).$$

If $g$ is a dilation, then, as Theorem 5.5.1 shows, $\mathbf{epi}\, g = (\mathbf{hypo}\, g_{[-1]})\check{}$ and equality follows.

Conversely, let us note that, in view of (5.3.1) and (5.3.2), we always have

$$\mathbf{epi}\, g \subset (\mathbf{hypo}\, g_{[-1]})\check{} \subset \mathbf{epi}(g_{[-1]})^{[-1]}.$$

Now if $(g_{[-1]})^{[-1]} = g$, then these inclusions are equalities, and we conclude that $\mathbf{epi}\, g = (\mathbf{hypo}\, g_{[-1]})\check{}$, which according to Theorem 5.5.1 means that $g$ is a dilation. □

**Theorem 5.6.2.** *If $L$ and $M$ are complete lattices and $\delta\colon L \to M$ is a dilation, then the lower inverse $\delta_{[-1]}\colon M \to L$ is an erosion, equal to the residual of $\delta$ (see Subsection 5.7.2), and also to the upper adjoint in the adjunction $(\delta^{[-1]}, \delta)$ (see Subsection 5.7.3). Similarly, if $\varepsilon\colon L \to M$ is an erosion, then the upper inverse $\varepsilon^{[-1]}$ is a dilation.*

**Proof.** We know that $u = \delta^{[-1]}$ is increasing, so we have, for any family $(y_j)_{j\in J}$, $u(\bigwedge y_j) \leqslant u(y_k)$ for all $k$; hence $u(\bigwedge y_j) \leqslant \bigwedge u(y_k) = x$. We need to prove the opposite inequality, $x \leqslant u(y)$, where $y = \bigwedge y_j$. From (D) in Theorem 5.5.1 we learn that $\delta(u(y_k)) \leqslant y_k$ for all $k$, which implies that $\delta(x) = \delta(\bigwedge u(y_j)) \leqslant y_k$. By (5.3.1), this implies that $x \leqslant u(y)$. The proof is complete. □

## 5.7 Special cases of inverses

The inverses we have defined in Section 5.2 generalize a situation which has been known for a long time, although under different names and with different fields of applications. Our results generalize the theory of Galois connections, equivalently residuation theory and the theory of adjunctions, to a more general situation, a situation which appears even in very simple cases like Proposition 5.3.1 and Example 5.3.2; we shall see other examples in the following five subsections. In complete lattices the upper and lower inverses always exist, whereas only very special mappings are upper or lower adjoints.

It seems that this generalization of residuation theory has not been considered in the contexts of the branches of mathematics mentioned in Subsections 5.7.1–5.7.5 below. As a rare exception, Singer (1997:176) defines the *dual* $M \to L$ of a mapping $g\colon L \to M$, which, after a change of order in $L$, is the lower inverse as defined here. He notes the inclusion corresponding to (5.3.1) and proves that it is an equality if and only if $g$ is a duality (antierosion).

Let us now briefly recall three classical definitions, viz. of the concepts of a Galois connection, a residuation and an adjunction.

### 5.7.1 *Galois connections*

If we require both that $g \circ g^{[-1]} \leqslant \mathbf{id}_Y$ shall hold and that $g$ shall be increasing, then we are in a situation which has been studied for a long time.

**Definition 5.7.1.** Given two preordered sets $X$ and $Y$, a ***Galois connection*** is a pair $(F, G)$ of decreasing mappings $F\colon X \to Y$, $G\colon Y \to X$ such that $G \circ F \geqslant \mathbf{id}_X$ and $F \circ G \geqslant \mathbf{id}_Y$. □

It is well known and easy to prove that if $(F, G)$ is a Galois connection, then $G \circ F$ is in fact a cleistomorphism in $X$ (Ore 1944:496). In fact, we see that $F \circ G \circ F = F$ and $G \circ F \circ G = G$. From this follows that $G \circ F$ and $F \circ G$ are ethmomorphisms and in fact cleistomorphisms.

A special case is when $X = Y$ and $G = F$. Then $F^{\circ 2}$ is a cleistomorphism and

(5.7.1) $$F^{\circ k} = F \text{ for odd } k \geqslant 1;\ F^{\circ k} = F \circ F, \text{ for even } k \geqslant 2.$$

**Proposition 5.7.2.** *Given preordered sets $X$, $Y$, a pair $(F, G)$ of mappings $F\colon X \to Y$ and $G\colon Y \to X$ is a Galois connection if and only if* $\mathbf{hypo}\, F = (\mathbf{hypo}\, G)^{\smile}$. *In other words,*

(5.7.2) *for all $x \in X$, $y \in Y$ we have $y \leqslant F(x)$ if and only if $x \leqslant G(y)$.*

*It follows that the inverse $\overline{\cdot F}$ (defined in* (5.2.4) *and* (5.2.5)*) is equal to $G$.*

**Proof.** If $(F, G)$ is a Galois connection and $x \leqslant G(y)$, then $F(x) \geqslant F(G(y))$ since $F$ is decreasing. Since $F(G(y)) \geqslant y$, we get $F(x) \geqslant F(G(y)) \geqslant y$. We have proved that $(\mathbf{hypo}\, G)^{\smile} \subset \mathbf{hypo}\, F$. The opposite inclusion follows by symmetry.

Conversely, if $\mathbf{hypo}\, F = (\mathbf{hypo}\, G)^{\smile}$ and $y \leqslant F(x)$, we deduce that $x \leqslant G(y)$, in particular, taking $y = F(x)$, that $x \leqslant G(F(x))$. Also, if $x' \leqslant x \leqslant G(y)$ we deduce that $y \leqslant F(x)$ and $y \leqslant F(x')$. Since we now know that this holds for $y = F(x)$, we conclude that $F(x) \leqslant F(x')$, hence that $F$ is decreasing. That $G$ is decreasing and that $y \leqslant F(G(y))$ follows by symmetry. □

The preordered sets $X$ and $Y$ in the definition need not be lattices:

**Example 5.7.3.** Let $X = \{0, 1\}$ and let $Y$ be any preordered set. Then $(F, G)$, where $F\colon X \to Y$, $G\colon Y \to X$, is a Galois connection if and only if $Y$ possesses a largest element $\mathbf{1}_Y$, $F(0) = \mathbf{1}_Y$, and $G = \chi_B$, the characteristic function of the down-set $B = \{y \in Y;\ y \leqslant F(1)\}$. The equivalence class of $F(1)$ determines everything. □

Conversely, any cleistomorphism $\kappa$ in $X$ can be obtained in this way for a suitable choice of $Y$ and $G$ (this follows from Gierz et al. 1980:23:3.13; 2003:29: O-3.13). We simply define $Y$ as the set $\mathbf{invar}(\kappa)$ with the opposite order, $F = \kappa$, and $G$ as the inclusion mapping $Y \to X$. However, this rather formal result is, in a sense, completely uninteresting; it is more like a tautology. By contrast, there are many situations in mathematics where $\kappa\colon X \to X$ can be factorized as $\kappa = G \circ F$ for some $F\colon X \to Y$ and $G\colon Y \to X$ which are very interesting and give new insights. Examples include Galois theory, convex analysis (Chapter 9), and image processing (Examples 5.8.7, 5.8.10 and 5.8.11).

If we introduce increasing mappings $g = F^{\mathrm{op}}\colon X \to Y^{\mathrm{op}}$ and $h = G_{\mathrm{op}}\colon Y^{\mathrm{op}} \to X$, the equality $\mathbf{hypo}\, F = (\mathbf{hypo}\, G)^{\smile}$ can be written $\mathbf{epi}\, g = (\mathbf{hypo}\, h)^{\smile}$, which is just (5.3.1) with equality, taking $h = g^{[-1]}$.

**Example 5.7.4.** In functional analysis, the interplay between bounded sets and neighborhoods of the origin in a vector space is a good example of a Galois connection. Let us say that a subset $V$ of a real vector space ***absorbs*** a set $B$ if there exists a positive number $\lambda_0$ such that $\lambda V$ contains $B$ for all scalars $\lambda$ with $|\lambda| \geqslant \lambda_0$. We define two decreasing mappings $F, G\colon \mathscr{P}(\mathscr{P}(E)) \to \mathscr{P}(\mathscr{P}(E))$ by

$$F(\mathscr{V}) = \{B \subset E;\ \text{every } V \subset \mathscr{V} \text{ absorbs } B\}, \qquad \mathscr{V} \subset \mathscr{P}(E),$$

$$G(\mathscr{B}) = \{V \subset E;\ \text{every } B \in \mathscr{B} \text{ is absorbed by } V\},\ \mathscr{B} \subset \mathscr{P}(E).$$

Clearly $G \circ F \geqslant \mathbf{id}$ and $F \circ G \geqslant \mathbf{id}$, so $(F, G)$ is a Galois connection. Topological vector spaces such that $G(F(\mathscr{V})) = \mathscr{V}$, where $\mathscr{V}$ denotes the family of all balanced and convex neighborhoods of the origin, are called ***bornological*** (Schaefer 1971:61). □

### 5.7.2 *Residuation*

A mapping $g\colon X \to Y$ of an ordered set $X$ into an ordered set $Y$ is said to be *residuated* if for every $b \in Y$ there is an element $a \in X$ such that the inverse image of $\{y \in Y;\ y \leqslant b\}$ is equal to $\{x \in X;\ x \leqslant a\}$ (Blyth & Janowitz 1972:11; Blyth 2005:7). It is equivalent to require that there exists a mapping $h\colon Y \to X$ such that $(g^{\mathrm{op}}, h_{\mathrm{op}})$ is a Galois connection. The mapping $h$ is proved to be unique and is called the *residual* of $g$. It is equal to $g^{[-1]}$.

Residuation theory goes back at least to a paper by Ward & Dilworth (1939), then in an algebraic setting. Algebraically, residuation is a kind of dual to multiplication. In a lattice $L$ with a multiplication one fixes an element $c$ and assumes that, given $y$, the set of all $x$ such that $cx \leqslant y$ has a largest element, which is denoted by $y : c$ (we consider for simplicity only the commutative case). We see that this is $g^{[-1]}$ if $g\colon L \to L$ denotes the mapping $g(x) = cx$. Thus $cx \leqslant y$ if and only if $x \leqslant y : c$.

### 5.7.3 *Adjunctions*

Gierz et al. (2003:22) define a *Galois connection* or *adjunction* as a pair of increasing mappings $(h, k)$, $h\colon X \to Y$, $k\colon Y \to X$ satisfying $\mathbf{hypo}\, h = (\mathbf{epi}\, k)^{\smile}$. This aspect probably originates in logic.

If $(h, k)$ is an adjunction, Gierz et al. (2003:22), Heijmans & Ronse (1990:264), and Davey & Priestely (2003:156) call $h$ the *upper adjoint* and $k$ the *lower adjoint* in the adjunction. As follows from Proposition 5.7.2, in an adjunction $(h, k)$ we have $k = h^{[-1]}$ and $h = k_{[-1]}$, so each mapping determines the other.

The definition of an adjunction means that $(F, G) = (h^{\mathrm{op}}, k_{\mathrm{op}})$ is a Galois connection in the sense of Definition 5.7.1.

Gierz et al. (2003:22) require that $h$ and $k$ be increasing, but this is superfluous as we have seen in Proposition 5.7.2.

### 5.7.4 *Asplund distances*

Edgar Asplund introduced in his article (1960) a distance between subsets of the Euclidean plane as follows. Given two sets $M, B \in \mathscr{P}(\mathbf{R}^2)$, we consider positive numbers $s$ and $t$ such that $sM \subset B \subset tM$, and the infimum of all quotients $t/s$ such that the inclusions hold. Then $\inf \log(t/s)$ is a metric for convex bodies which are symmetric with respect to the origin; more generally, for sets that are bounded and have the origin as an interior point. He then goes on by considering classes of affinely equivalent sets, like ellipses, parallelograms and hexagons.

The Asplund distance is an example of inverse mappings. Indeed, if we define $f(s) = sM$ for positive numbers $s$, then

$$f_{[-1]}(B) = \sup_{s>0}(s;\ f(s) \subset B) \text{ and } f^{[-1]}(B) = \inf_{t>0}(t;\ f(t) \supset B),$$

for $B$ a subset of $\mathbf{R}^2$, so that $\log\left(f^{[-1]}(B)/f_{[-1]}(B)\right)$ is the Asplund distance between $B$ and $M$.

Noyel & Jourlin (2017, 2020) extended the Asplund distance to classes of functions. On this topic, see also (Noyel 2019).

### 5.7.5 *Rådström's definition of smooth functions on arbitrary sets*

Let $X$ be any set and define for a fixed $m \in \mathbf{N} \cup \{\infty\}$ two mappings $F_m$ and $G_m$ as follows. Let $\Gamma$ be any set of mappings from the real line $\mathbf{R}$ into $X$, and $\Phi$ any set of real-valued functions on $X$. Then we define $F_m(\Gamma)$ as the set of all functions $\varphi$ from $X$ into $\mathbf{R}$ such that the composition $\varphi \circ \gamma \colon \mathbf{R} \to \mathbf{R}$ is of class $C^m$ for all $\gamma \in \Gamma$. We define $G_m(\Phi)$ as the set of all mappings $\gamma$ from $\mathbf{R}$ into $X$ such that $\varphi \circ \gamma \colon \mathbf{R} \to \mathbf{R}$ is of class $C^m$ for all functions $\varphi \in \Phi$.

We can express this briefly as

$$F_m(\Gamma) = \{\varphi;\ \varphi \circ \gamma \in C^m(\mathbf{R}, \mathbf{R}) \text{ for all } \gamma \in \Gamma\}, \quad \Gamma \in \mathscr{P}(\mathscr{F}(\mathbf{R}, X));$$
$$G_m(\Phi) = \{\gamma;\ \varphi \circ \gamma \in C^m(\mathbf{R}, \mathbf{R}) \text{ for all } \varphi \in \Phi\}, \quad \Phi \in \mathscr{P}(\mathscr{F}(X, \mathbf{R})).$$

Thus $F_m$ and $G_m$ are mappings

$$F_m \colon \mathscr{P}(\mathscr{F}(\mathbf{R}, X)) \to \mathscr{P}(\mathscr{F}(X, \mathbf{R}));$$
$$G_m \colon \mathscr{P}(\mathscr{F}(X, \mathbf{R})) \to \mathscr{P}(\mathscr{F}(\mathbf{R}, X)).$$

The pair $(F_m, G_m)$ is obviously a Galois connection.

Hans Rådström defined smooth functions on a set $X$ by fixing a set of curves $\Gamma$ and then defining a function $\varphi \colon X \to \mathbf{R}$ to be smooth if it belongs to $F_\infty(\Gamma)$. As far as I know, he did not develop a theory, and he published nothing.

A basic question is whether we get the usual $C^\infty$ functions if $X$ is a differential manifold and $\Gamma$ is the set of $C^\infty$ curves; Rådström conjectured that this is so, and Jan Boman (1967) proved it. More precisely, Boman proved that, for finite $m \geqslant 1$,

$$C^m(\mathbf{R}^n, \mathbf{R}) \subset F_m(C^\infty(\mathbf{R}, \mathbf{R}^n)) \subset C^{m-1,1}(\mathbf{R}^n, \mathbf{R}).$$

Here $C^m(\mathbf{R}^n, \mathbf{R})$ denotes the space of all functions on $\mathbf{R}^n$ with real values whose derivatives of order at most $m$ exist and are continuous, while $C^{m-1,1}(\mathbf{R}^n, \mathbf{R})$ is the subspace of $C^{m-1}(\mathbf{R}^n, \mathbf{R})$ consisting of functions whose derivates of order $m-1$ are all Lipschitz continuous. Boman proved that the first inclusion here is strict; obviously so is the second. But taking the intersection over all finite $m \geqslant 1$ we see that $F_\infty(C^\infty(\mathbf{R}, \mathbf{R}^n)) = C^\infty(\mathbf{R}^n, \mathbf{R})$, proving Rådström's claim.

Jan Boman has also obtained an explicit description of $F_m(C^m(\mathbf{R}, \mathbf{R}^n))$ for all finite $m \geqslant 1$ (personal communication 2008-09-18).

For $m = \infty$, Petermann (1979) developed a formalism in the framework of category theory. Finally Michor (1984), Kriegl & Nel (1990), and Kriegl & Michor (1997) developed a theory for global analysis using smooth curves.

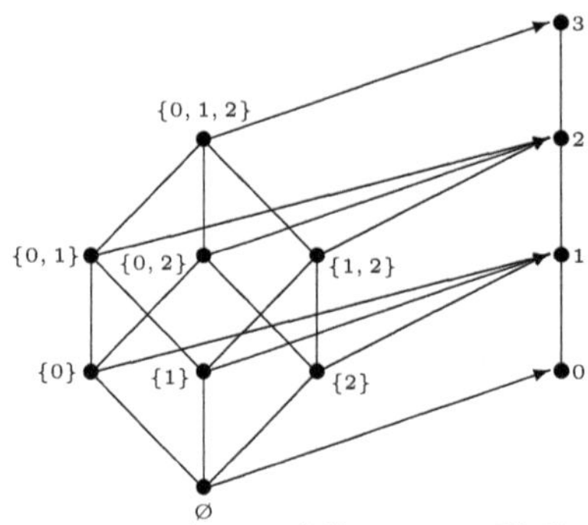


Figure 5.1. The mapping $A \mapsto g(A) = \mathbf{card}(A)$ in the case $n = 3$; see Example 5.8.1. One can see that $\mathbf{card}(\mathbf{epi}\, g) = 20$.

## 5.8 Examples of upper and lower inverses

The purpose of this section, which contains several examples, is twofold. The first of these are just intended to illustrate the concepts and show that some implications (that one could have expected to hold) do not hold. Examples 5.8.6–5.8.7 and 5.8.10–5.8.11 supplement Subsections 5.7.1 through 5.7.5 in showing that the definitions have a good meaning and yield some insight in a variety of situations.

We find a most fundamental example of inverses in convexity theory, to be discussed in detail later. We therefore postpone the discussion of this to Chapter 9.

**Example 5.8.1.** *Supremum not maximum.*
Take $X = \mathscr{P}(E)$, where $E$ is a finite set of cardinality $n \geqslant 2$, and $Y = [0, n]_{\mathbf{Z}} = \{y \in \mathbf{Z};\ 0 \leqslant y \leqslant n\}$, two complete lattices, and define $g(A) = \mathbf{card}(A)$, $A \in X$. Then $g_{[-1]}(0) = \varnothing = \mathbf{0}_X$ and $g_{[-1]}(y) = E = \mathbf{1}_X$ for all $y \in [1, n]_{\mathbf{Z}}$.

Thus $g_{[-1]}(1) = E = \mathbf{1}_X$, and the set $\{A \in X;\ g(A) \leqslant 1\}$ is not an ideal. It therefore comes as no surprise that, for $n = 2$, $\mathbf{card}(\mathbf{hypo}\, g_{[-1]}) = 25 > 20 = \mathbf{card}(\mathbf{epi}\, g)$; we have strict inequality in the inclusion in (5.3.1).

We also find that $g^{[-1]}(y) = \emptyset = \mathbf{0}_X$ when $y < n$ and $g_{[-1]}(y) = E = \mathbf{1}_X$ when $y = n$. See Figures 5.1 and 5.2, where $n = 3$. □

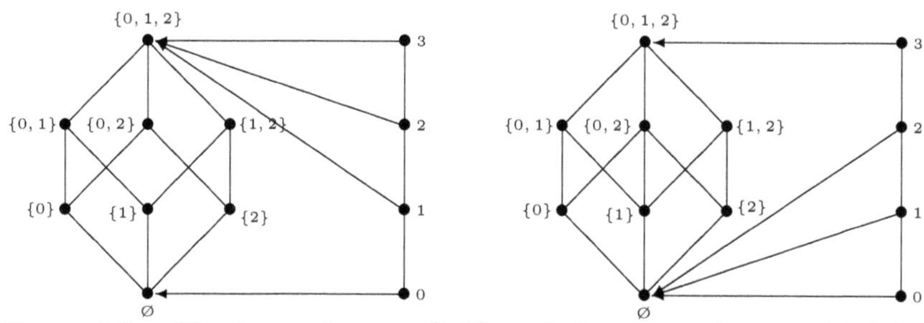


Figure 5.2. The lower inverse (left) and the upper inverse (right) of the mapping $g(A) = \mathbf{card}(A)$ in the previous figure. One can see that $\mathbf{card}(\mathbf{hypo}\, g_{[-1]}) = 25 > \mathbf{card}(\mathbf{epi}\, g) = 20$.

**Example 5.8.2.** *Non-residuated mappings.*
Take $L = M$, a complete lattice, in Definition 5.2.1, fix an element $c$ of $L$, and define a mapping $h\colon L \to L$ by $h(x) = x \vee c$, $x \in L$. In this case, the supremum in (5.2.1) is a maximum if $y \geqslant c$ but only then. Thus $h$ is not residuated unless $c = \mathbf{0}$; also, it is a dilation only if $c = \mathbf{0}$ (indeed, $h(\mathbf{0}) = c$ while $\bigvee_{x \in \emptyset} h(x) = \mathbf{0}$). But it is easy to determine its lower inverse: $h_{[-1]}(y) = y$ if $y \geqslant c$ and $h_{[-1]}(y) = \mathbf{0}$ otherwise. We have

$$\mathbf{epi}\, h = \{(x, y) \in L^2;\ y \geqslant x \vee c\},$$

while

$$\big(\mathbf{hypo}\, h_{[-1]}\big)^{\smile} = \mathbf{epi}\, h \cup \{(\mathbf{0}, y) \in L^2;\ y \ngeqslant c\},$$

so that

$$\big(\mathbf{hypo}\, h_{[-1]}\big)^{\smile} \smallsetminus \mathbf{epi}\, h = \{(\mathbf{0}, y) \in L^2;\ y \ngeqslant c\} \neq \emptyset \text{ if } c \neq \mathbf{0}.$$

We have $h(h^{[-1]}(y)) \leqslant y$ if and only if $y \geqslant c$.

For the upper inverse, we can only say that $h^{[-1]}(y) = \mathbf{0}$ if $y \leqslant c$ and that $h^{[-1]}(y) \leqslant y$ for $y \nleqslant c$. Both equality and strict inequality (see Example 5.8.4 below) can occur here. □

**Example 5.8.3.** We let $h$ be as in example 5.8.2 and assume in addition that $L$ is totally ordered. Then $h$ is an erosion. We have already determined $h_{[-1]}$ in Example 5.8.2, and we know that $h^{[-1]}(y) = \mathbf{0}$ for $y \leqslant c$. In the case of total order we have $h^{[-1]}(y) = y$ for all $y > c$. In the notation which Singer (1997:335) uses for $L = \mathbf{R}_!$, we can write $h^{[-1]}(y)$ as $h^{[-1]}(y) = y \top c$, $y \in L$. Thus $h_{[-1]}$ and $h^{[-1]}$ are equal except for $y = c$ if $c \neq \mathbf{0}$; there we get $h^{[-1]}(c) = \mathbf{0} \leqslant c = h_{[-1]}(c)$. Moreover, we have equality in (5.3.2):

$$\mathbf{hypo}\, h = \left(\mathbf{epi}\, h^{[-1]}\right)^{\smile} = \{(x, y) \in L^2;\ y \leqslant x \vee c\},$$

which, in view of the duality statement in Theorem 5.5.1 means that $h^{[-1]}$ is dually residuated with dual residual $h$; equivalently that the pair $(h^{\mathrm{op}}, (h^{[-1]})_{\mathrm{op}})$ is a Galois connection. □

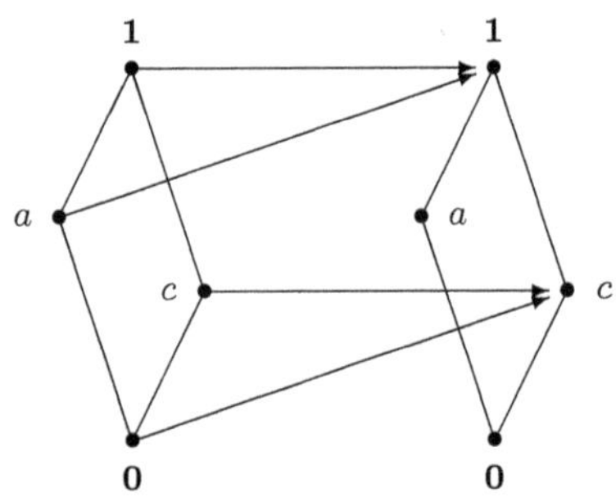


Figure 5.3. The mapping $h(x) = x \vee c$; see Example 5.8.5.

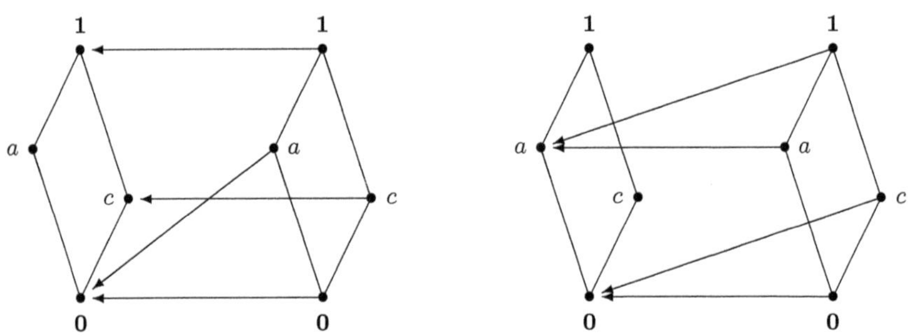


Figure 5.4. The lower inverse (left) and the upper inverse of the mapping $h(x) = x \vee c$ in the previous figure.

**Example 5.8.4.** Let now $L$ be $[0, 1]^2 \subset \mathbf{R}^2$, the Cartesian product of two intervals. The lower inverse is already known from Example 5.8.2. The upper inverse is

$$h^{[-1]}(y) = \begin{cases} \mathbf{0}, & y \leqslant c; \\ (0, y_2), & y_1 \leqslant c_1,\ y_2 > c_2; \\ (y_1, 0), & y_1 > c_1,\ y_2 \leqslant c_2; \\ y, & y_1 > c_1,\ y_2 > c_2. \end{cases}$$

Thus strict inequality in $h^{[-1]}(y) \leqslant y$ can occur. We have $\mathbf{hypo}\, h = (\mathbf{epi}\, h^{[-1]})^{\smile}$. □

**Example 5.8.5.** Let now $L$ be $\{0,1\}^2$ with the coordinatewise order, and let $h$ be as in Example 5.8.2, i.e., $h(x) = x \vee c$. We now choose $c = (1,0)$ and denote $(0,1)$ by $a$, so that $L$ consists of the four elements $\mathbf{0} = (0,0)$, $a = (0,1)$, $c = (1,0)$, and $\mathbf{1} = (1,1)$. From Example 5.8.2 we know that $h_{[-1]}(y) = y$ if $y \geqslant c$ and $h_{[-1]}(y) = \mathbf{0}$ otherwise. Thus

$$h_{[-1]}(y) = \begin{cases} \mathbf{0}, & y = \mathbf{0}; \\ \mathbf{0}, & y = a; \\ c, & y = c; \\ \mathbf{1}, & y = \mathbf{1}. \end{cases}$$

We find that

$$(\mathbf{hypo}\, h_{[-1]})^{\smile} \smallsetminus \mathbf{epi}\, h = \{(\mathbf{0}, \mathbf{0}), (\mathbf{0}, a)\} \neq \varnothing.$$

Thus $h$ is not residuated.

We also find that

$$h^{[-1]}(y) = \begin{cases} \mathbf{0}, & y = \mathbf{0}; \\ a, & y = a; \\ \mathbf{0}, & y = c; \\ a, & y = \mathbf{1}. \end{cases}$$

The infimum is in all cases a minimum, meaning that $h^{[-1]}$ is dually residuated, in other words, $(h^{\mathrm{op}}, (h^{[-1]})_{\mathrm{op}})$ is a Galois connection. We have equality in (5.3.2):

$$\mathbf{hypo}\, h = (\mathbf{epi}\, h^{[-1]})^{\smile} = L^2 \smallsetminus \{(\mathbf{0}, a), (\mathbf{0}, \mathbf{1}), (c, a), (c, \mathbf{1})\}.$$

We see in particular that $h_{[-1]}(c) = c > h^{[-1]}(c) = \mathbf{0}$ and that $h_{[-1]}(\mathbf{1}) = \mathbf{1} > h^{[-1]}(\mathbf{1}) = a$. □

**Example 5.8.6.** *Calculating with infinities.*

Let now $L$ be the complete lattice $\mathbf{R}_!$ of extended real numbers and define

$$S_c(x) = x \dotplus c, \quad s_c(x) = x \mathbin{\underset{\cdot}{+}} c, \qquad x \in \mathbf{R}_!,$$

where $c$ is a constant in $\mathbf{R}_!$. Here $\dot{+}$ and $\dot{+}$ denote lower and upper addition, defined in (2.4.1).

We note that $S_c = s_c$ for all real numbers $c$. The mapping $s_c$ is residuated for all $c \in \mathbf{R}_!$, and $(s_c)_{[-1]} = S_{-c}$. Also $(S_c)^{[-1]} = s_{-c}$ and the infimum is a minimum. But neither $S_{-\infty}$ nor $S_{+\infty}$ is residuated, although we have $(S_{-\infty})_{[-1]} = S_{+\infty}$ and $(S_{+\infty})_{[-1]} = S_{-\infty}$. Similarly, $(s_{-\infty})^{[-1]} = s_{+\infty}$ and $(s_{+\infty})^{[-1]} = s_{-\infty}$. □

**Example 5.8.7.** *Infimal and supremal convolution.*
Let $G$ be an abelian group—think of $\mathbf{Z}^n$ or $\mathbf{R}^n$. We have defined infimal convolution in Section 3.4; we now define similarly ***supremal convolution*** for arbitrary functions $f, g \in \mathscr{F}(G, \mathbf{R}_!)$,

$$(5.8.1) \qquad (f \sqcup g)(x) = \sup_{y \in G}(f(x-y) \underset{\cdot}{+} g(y)), \qquad x \in G,$$

dual to

$$(5.8.2) \qquad (f \sqcap g)(x) = \inf_{y \in G}(f(x-y) \dot{+} g(y)), \quad x \in G.$$

Here $\dot{+}$ and $\underset{\cdot}{+}$ are defined by (2.4.1).

Clearly $f \sqcup g = -((-f) \sqcap (-g))$, so from a theoretical point of view it is enough to study one of them; however, it is convenient to keep both operations.

We note that these convolutions are just Minkowski addition in another dimension: for $f, g \in \mathscr{F}(G, \mathbf{R}_!)$ we have

$$(5.8.3) \qquad \begin{aligned} \mathbf{epi}^{\text{finite}}_{\text{strict}}(f \sqcap g) &= \mathbf{epi}^{\text{finite}}_{\text{strict}}(f) + \mathbf{epi}^{\text{finite}}_{\text{strict}}(g), \\ \mathbf{hypo}^{\text{finite}}_{\text{strict}}(f \sqcup g) &= \mathbf{hypo}^{\text{finite}}_{\text{strict}}(f) + \mathbf{hypo}^{\text{finite}}_{\text{strict}}(g), \end{aligned}$$

where Minkowski addition is now in the abelian group $G \times \mathbf{R}$. □

We may study equations $f \sqcap s = g$ with $s$ and $g$ as given functions and $f$ as an unknown function, but in general such equations cannot be solved, and if so, they do not have a unique solution. However, if we pass to inequalities, something useful can be said:

**Proposition 5.8.8.** *Let $G$ be an abelian group, fix two functions $s, t \in \mathscr{F}(G, \mathbf{R}_!)$, and define two operations on functions $G \to \mathbf{R}_!$ by*

$$\delta(f) = f \sqcap s \quad \text{and} \quad \varepsilon(g) = g \sqcup t, \qquad f, g \in \mathscr{F}(G, \mathbf{R}_!).$$

*Then* $\mathbf{hypo}\,\delta = (\mathbf{epi}\,\varepsilon)^{\smile}$ *if and only if* $t(y) = -s(-y)$, $y \in G$.

**Proof.** We find that $\delta(f) \geqslant g$ if and only if $f(x-y) \dotplus s(y) \geqslant g(x)$ for all $x, y \in G$, and that $f \geqslant \varepsilon(g)$ if and only if $f(x) \geqslant g(x-y) \mathbin{\underset{\cdot}{+}} t(y)$ for all $x, y \in G$. We now invoke the equivalence (2.4.4), which shows that the condition $\delta(f) \geqslant g$ can be written $f(x-y) \dotplus (-g(x)) \geqslant -s(y)$ for all $x, y \in G$, equivalently

$$f(x) \dotplus (-g(x-y)) \geqslant -s(-y) \text{ for all } x, y \in G,$$

and that the condition $f \geqslant \varepsilon(g)$ can be written

$$f(x) \dotplus (-g(x-y)) \geqslant t(y) \text{ for all } x, y \in G.$$

Thus the equivalence holds for all $f, g \in \mathscr{F}(G, \mathbf{R}_!)$ if $t(y) = -s(-y)$ for all $y \in G$ and only then. □

We shall also need a similar result for nonnegative functions:

**Proposition 5.8.9.** *Let $G$ be an abelian group, fix two functions $s \in \mathscr{F}(G, [0, +\infty])$ and $t \in \mathscr{F}(G, [-\infty, 0])$, and define two operations on functions $G \to [0, +\infty]$ by*

$$\delta(f) = f \sqcap s \quad \textit{and} \quad \varepsilon^+(g) = (g \sqcup t)^+ = (g \sqcup t) \vee 0, \quad f, g \in \mathscr{F}(G, [0, +\infty]).$$

*Then* $\mathbf{hypo}\, \delta = (\mathbf{epi}\, \varepsilon^+)^{\smile}$ *if and only if* $t(y) = -s(-y)$ *for all* $y \in G$.

**Proof.** The proof is quite similar to the previous proof. Note that, for all $f \in \mathscr{F}(G, [0, +\infty])$, $f \geqslant \varepsilon(g)$ if and only if $f \geqslant \varepsilon^+(g)$. □

The mapping $\delta$ is an erosion for the order used. However, as we shall see in the next example, it is natural to use the order defined by the functions $\mathrm{e}^{-f}$ rather than by $f$, which explains that $\delta$ will be a dilation and $\varepsilon^+$ an erosion when operating on these functions.

**Example 5.8.10.** *Processing gray-scale images.*
The binary images on a set $X$ form a complete lattice $\mathscr{P}(X)$ isomorphic to $\mathscr{F}(X, \{0, 1\})$. A gray-scale image on $X$ is given by a function $\varphi \colon X \to P$ on $X$, where $P$ is a subset of $[0, 1]$ representing gray-levels in the image, for instance

$$P = \left\{0, \tfrac{1}{256}, \tfrac{2}{256}, \ldots, \tfrac{255}{256}, 1\right\}$$

with $2^8 + 1$ levels. If $P$ is closed, $\mathscr{F}(X, P)$ is a complete lattice, actually a sub-complete-lattice of $\mathscr{F}(X, [0, 1])$.

How to define dilations and erosions on gray-scale images on an abelian group $G$? A naive way would be to use convolution,

$$(\varphi * \sigma)(x) = \sum_{y \in G} \varphi(x-y)\sigma(y), \qquad x \in G,$$

possibly scaled to the interval $[0,1]$ by dividing by the supremum or taking just $(\varphi * \sigma) \wedge 1$ for any two functions $\varphi, \sigma\colon G \to [0,1]$, but these operations do not define dilations. So one could try instead with

$$\delta_\sigma(\varphi) = \sup_{y \in G} \varphi(x-y)\sigma(y), \qquad \varphi \in \mathscr{F}(G,[0,1]),$$

where $\sigma\colon G \to [0,1]$ is a given function. This does define a dilation. But if the set $P$ of values is finite, it yields a function with values in $P$ only if $\sigma$ takes values in $\{0,1\}$. This definition is one of several as we shall see; in the notation of Bloch & Maître (1995:1344) we have $\delta_\sigma = D3_\sigma$.

It will be convenient to formulate these arguments in terms of infimal and supremal convolution.

To any set $A$ we associated in (2.10.2) its indicator function $\mathbf{indf}_A = -\log \chi_A$; the characteristic function is $\chi_A = \mathrm{e}^{-\mathbf{indf}_A}$. Similarly, to any gray-scale image $\varphi$ we associate its ***indicator function*** $\mathbf{indf}_\varphi = -\log \varphi$.

We now note that the Minkowski sum $A+B$ of two sets $A$ and $B$ satisfies

$$\mathbf{indf}_{A+B} = (\mathbf{indf}_A) \sqcap (\mathbf{indf}_B).$$

This makes it natural to define a dilation $\delta$ by a fixed gray-scale image $\sigma$ and operating on arbitrary gray-scale images $\varphi$ by the formula

$$\text{(5.8.4)} \qquad \mathbf{indf}_{\delta(\varphi)} = (\mathbf{indf}_\varphi) \sqcap (\mathbf{indf}_\sigma), \quad \mathbf{indf}_\varphi \in \mathscr{F}(G,[0,+\infty]),$$

and, as we have seen in Proposition 5.8.9, its lower inverse $\delta^{[-1]} = \varepsilon^+$ is given by

$$\text{(5.8.5)} \quad \mathbf{indf}_{\varepsilon^+(\psi)} = (\mathbf{indf}_\psi) \sqcup ((-\mathbf{indf}_\tau)^+), \quad \mathbf{indf}_\psi \in \mathscr{F}(G,[0,+\infty]),$$

where $\tau(x) = \sigma(-x)$. And $\kappa = \varepsilon^+ \circ \delta$ is a cleistomorphism; $\alpha = \delta \circ \varepsilon^+$ an anoiktomorphism (Corollary 5.5.6). □

**Example 5.8.11.** *Fuzzy logic and fuzzy mathematical morphology.*
A subset $A$ of a set $E$, called a *crisp set* in this context, can be represented by its characteristic function $\chi_A\colon E \to \{0,1\}$. A fuzzy subset $A$ of a set $E$ can in the same way be represented by its *membership function* $\mu_A\colon E \to P$, where the value $\mu_A(x)$ represents the degree (or probability) that $x$ belongs to $A$. The intersection $A \cap B$ of two sets can be represented, respectively, by

$$\chi_{A\cap B} = \chi_A \wedge \chi_B \text{ and } \mu_{A\cap B} = \mu_A \wedge \mu_B;$$

similarly the union is represented by

$$\chi_{A\cup B} = \chi_A \vee \chi_B \text{ and } \mu_{A\cup B} = \mu_A \vee \mu_B.$$

Thus the family $\mathscr{P}(E)$ of all crisp sets is isomorphic to the complete lattice $\mathscr{F}(E, \{0,1\})$, and the family of all fuzzy sets is isomorphic to the lattice $\mathscr{F}(E, [0,1])$. But as shown by (5.8.3), the calculus of fuzzy sets can be reduced to calculus of crisp sets at the expense of adding one more dimension.

So far, the fuzzy sets are just like gray-scale images. However, fuzzy logic goes beyond the gray-scale images in generalizing also the minimum and maximum.

In addition to the operations of minimum and maximum there are many other operations on fuzzy sets. We say that a function $c\colon [0,1]^2 \to [0,1]$ is a ***conjunctor*** if its restriction to $\{0,1\}^2$ is Boolean conjunction; similarly we say that $d\colon [0,1]^2 \to [0,1]$ is a ***disjunctor*** if its restriction to $\{0,1\}^2$ is Boolean disjunction; moreover a ***negator*** is a function $n\colon [0,1] \to [0,1]$ such that $n(0) = 1$, $n(1) = 0$ (Sladoje 2005:18–19). Finally a function $i\colon [0,1]^2 \to [0,1]$ will be called an ***implicator*** if $i(0,0) = i(0,1) = i(1,1) = 1$ and $i(1,0) = 0$.

Deng & Heijmans (2000:5) use the terms *fuzzy conjunction* or just *conjunction* and *fuzzy implication* or *implication* for a conjunctor which is increasing in both arguments, and an implicator which is decreasing in the first and increasing in the second, respectively.

Examples of conjunctors and disjunctors on $[0,1]$ include

$$\begin{aligned} c_M(a,b) &= a \wedge b, & d_M(a,b) &= a \vee b; \\ c_A(a,b) &= ab, & d_A(a,b) &= a + b - ab; \\ c_L(a,b) &= (a+b-1) \vee 0, & d_L(a,b) &= (a+b) \wedge 1. \end{aligned}$$

Here $c_A$ and $d_A$ are called *algebraic* and $c_L$ and $d_L$ the *Łukasiewicz conjunctor* and *disjunctor.*

Given a conjunctor $c$ and a disjunctor $d$ we define the *c-intersection* $A \cap_c B$ and the *d-union* $A \cup_d B$ by their membership functions

$$\mu_{A\cap_c B}(x) = c(\mu_A(x), \mu_B(x)), \quad \mu_{A\cup_d B}(x) = d(\mu_A(x), \mu_B(x)), \quad x \in E.$$

It is customary to require more of the functions $c$ and $d$. We require $\cap_c$ and $\cup_d$ to be commutative and associative. This is guaranteed if

$$\text{(5.8.6)} \qquad c(a,b) = c(b,a), \quad d(a,b) = d(b,a), \qquad a, b \in [0,1], \text{ and}$$

$$\text{(5.8.7)} \qquad c(c(a,b),z) = c(a, c(b,z)), \quad d(d(a,b),z) = d(a, d(b,z))$$

for $a, b, z \in [0,1]$. We also require that 1 be a neutral element for $c$ and 0 a neutral element for $d$:

$$\text{(5.8.8)} \qquad c(a,1) = c(1,a) = a, \quad d(a,0) = d(0,a) = a, \qquad a \in [0,1].$$

Conditions (5.8.6)–(5.8.8) imply that the fuzzy sets form a commutative semigroup with neutral element under $\cap_c$ as well as under $\cup_d$. Finally, it is natural to require that

(5.8.9) *$c$ and $d$ are increasing in each variable.*

It follows from (5.8.6)–(5.8.9) that both semigroups have a zero element (the empty set $\varnothing$ for $\cap_c$ and the full set $E$ for $\cup_d$), and that $c$ and $d$ are uniquely determined on the boundary of $[0,1]^2$; in particular at the four corners $\{0,1\}^2$, which implies that $A \cap_c B$ and $A \cup_d B$ agree with $A \cap B$ and $A \cup B$ when $A$ and $B$ are crisp sets.

One says that a function $c$ or $d$ satisfying (5.8.6)–(5.8.9) is a ***T*-norm** or a ***T*-conorm**, respectively. Clearly the minimum is the largest of all $T$-norms, and the Łukasiewicz $T$-norm $c_L$ is the largest convex $T$-norm.

Translated to the language of fuzzy sets, (5.8.4) and (5.8.5) mean that the dilation $\delta_S$ by a fuzzy set $S$ with $s = \mathbf{indf}_S$ has lower inverse $(\varepsilon_T)^+$ if and only if $t = -\mathbf{indf}_T$ where $T = \{-s;\ s \in S\}$.

More general dilations can now be defined as

$$\delta_\sigma(\varphi) = \sup_{y \in G} c(\varphi(x-y), \sigma(y)), \qquad \varphi \in \mathscr{F}(G, [0,1]),$$

for any conjunctor $c$. In their survey article, Bloch & Maître (1995:1344) denote it by $D6_\sigma$ and explore its properties. The case already considered is when $c(a,b) = c_A(a,b) = ab$; we may instead take $c(a,b) = c_M(a,b) = a \wedge b$ to get the dilation introduced by Kaufmann & Gupta (1988) and denoted by $D2_\sigma$ by Bloch & Maître (1995:1344); it can be expressed using a restricted infimal convolution.

Now fix an element $a \in [0,1]$ and define a mapping $g_a(x) = \varphi(a,x)$ for a given function $\varphi$ of two variables. It is easy to see that if $\varphi$ is a conjunctor, then the lower inverse $(g_a)_{[-1]}$ defines an implicator $i(a,b) = (g_a)_{[-1]}(b)$. Several such implicators have been studied by Deng & Heijmans (2000:5) and Deng (2000:3). They are used in fuzzy logic.

If $\varphi = c_M = \min$, the implicator $i_M(a,b) = (g_a)_{[-1]}(b)$ is equal to 1 if $b \geqslant a$ and equal to $b$ when $b < a$ (cf. Example 5.8.2). The upper inverse satisfies $(g_a)^{[-1]}(b) = 1$ for $b > a$ and $(g_a)^{[-1]}(b) = b$ for $b \leqslant a$ (cf. Example 5.8.3). Hence the two inverses are equal except for $b = a$.

Taking instead $\varphi = c_A$, we find that

$$i_A(a,b) = (g_a)_{[-1]}(b) = (g_a)^{[-1]}(b) = (b/a) \wedge 1 \text{ when } a > 0,$$

$$(g_a)_{[-1]}(b) = (g_a)^{[-1]}(b) = 1 \text{ when } a = 0 \text{ and } b > 0, \text{ while}$$

$$(g_a)_{[-1]}(b) = 1, (g_a)^{[-1]}(b) = 0 \text{ when } a = b = 0.$$

Both are discontinuous at the origin.

Finally, taking $\varphi$ as the Łukasiewicz $T$-norm $c_L$, we find that the lower and upper inverses agree except when $b = 0$ and $a < 1$: we get $i_L(a, b) = (g_a)_{[-1]}(b) = (g_a)^{[-1]}(b) = (b - a + 1) \wedge 1$ for $0 < b \leqslant 1$, while $(g_a)_{[-1]}(0) = 1 - a$, $(g_a)^{[-1]}(0) = 0$; the first is continuous, the second discontinuous on $[0, 1[ \times \{0\}$. □

## 5.9 Quotients of mappings

We shall now generalize the definitions of upper and lower inverses.

**Definition 5.9.1.** Let a set $X$, a complete lattice $M$, and a preordered set $Y$, as well as two mappings $f\colon X \to M$ and $g\colon X \to Y$ be given. We define two mappings $f/_{\star}g, f/^{\star}g\colon Y \to M$ by

$$(f/_{\star}g)(y) = \bigvee_{x \in X} (f(x);\ g(x) \leqslant_Y y), \qquad y \in Y;$$

$$(f/^{\star}g)(y) = \bigwedge_{x \in X} (f(x);\ g(x) \geqslant_Y y), \qquad y \in Y.$$

We shall call them the ***lower quotient*** and the ***upper quotient*** of the mappings $f$ and $g$. □

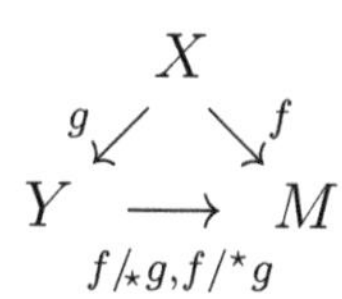

We shall often assume that $X$, $M$ and $Y$ are all complete lattices, but this is not necessary for the definitions to make sense.

We note that $f/_{\star}g = (f^{\mathrm{op}}/^{\star}g^{\mathrm{op}})^{\mathrm{op}}_{\mathrm{op}}$. (For the definition of $_{\mathrm{op}}$ and $^{\mathrm{op}}$, see Definition 2.8.2.)

We may also consider $(f/_{\star}g^{\mathrm{op}})_{\mathrm{op}} = (f^{\mathrm{op}}/^{\star}g)^{\mathrm{op}}$ and $(f/^{\star}g^{\mathrm{op}})_{\mathrm{op}} = (f^{\mathrm{op}}/_{\star}g)^{\mathrm{op}}$; explicitly,

$$((f/_{\star}g^{\mathrm{op}}))_{\mathrm{op}}(y) = ((f^{\mathrm{op}}/^{\star}g))^{\mathrm{op}}(y) = \bigvee_{x \in X} (f(x);\ g(x) \geqslant_Y y); \tag{5.9.1}$$

$$((f/^{\star}g^{\mathrm{op}}))_{\mathrm{op}}(y) = ((f^{\mathrm{op}}/_{\star}g))^{\mathrm{op}}(y) = \bigwedge_{x \in X} (f(x);\ g(x) \leqslant_Y y), \tag{5.9.2}$$

where $y \in Y$.

We refrain from generalizing these definitions to the case when the supremum or infimum does not exist for all choices of $y$.

The mappings $f/_\star g, f/^\star g \in \mathscr{F}(Y, M)$ are always increasing.

The quotients $f/_\star g$ and $f/^\star g$ increase when $f$ increases and they decrease when $g$ increases—just as with division of positive numbers:

$$\text{If } f_1 \leqslant_M f_2 \text{ and } g_1 \geqslant_Y g_2, \text{ then}$$
$$f_1/_\star g_1 \leqslant_M f_2/_\star g_2 \text{ and } f_1/^\star g_1 \leqslant_M f_2/^\star g_2.$$

If $g(x) \leqslant_Y y$, then $f(x) \leqslant_M (f/_\star g)(y)$; if $g(x) \geqslant_Y y$, then $f(x) \geqslant_M (f/^\star g)(y)$. In particular,

$$\text{if } g(x) = y, \text{ then } (f/^\star g)(y) \leqslant_M f(x) \leqslant_M (f/_\star g)(y).$$

We note some special cases.

(1) If we specialize the definitions to the situation when $X = M$ and $f = \mathbf{id}_X$, then $f/_\star g = \mathbf{id}_X/_\star g = g_{[-1]}$ and $f/^\star g = \mathbf{id}_X/^\star g = g^{[-1]}$; cf. Definition 5.2.1. So inverses are quotients.

(2) A second special case is this: Taking $Y = M$ and $g = f$ in the definition we see that, for all mappings $f\colon X \to M$ we have

$$f/_\star f \leqslant \mathbf{id}_M \leqslant f/^\star f; \tag{5.9.3}$$

$$(f/_\star f) \circ f = f = (f/^\star f) \circ f. \tag{5.9.4}$$

(3) If we specialize to the situation where $X$ is an arbitrary subset of $Y$ and $g$ is the inclusion mapping $X \to Y$, then $f/_\star g = f_\vee$ and $f/^\star g = f^\wedge$, where $f_\vee$ and $f^\wedge$ are defined in Definition 4.2.9.

(4) If we specialize further, letting also $f$ be the inclusion mapping $X \to Y = M$, we obtain

$$(f/_\star g)(y) = (f/_\star f)(y) = f_\vee(y) = \bigvee_{x \in X} (x;\ x \leqslant y) = y^\circ \in M. \tag{5.9.5}$$

It is easy to verify that $y \mapsto y^\circ$ is an anoiktomorphism. The elements $y$ such that $y^\circ = y$ are called ***X*-convex**. A well-known situation is described in Example 5.9.2 below.

**Example 5.9.2.** *Convexity theory.*

Let $M$ be the complete lattice $\mathscr{F}(E, \mathbf{R}_!)$ of functions on a vector space $E$ with values in the extended reals, let $F$ be a vector subspace of its algebraic dual $E^\star$; it is allowed to take $F = \{0\}$, and, more interesting, to take the dual space $E'$ of continuous linear forms if $E$ has a topology. We let $W = L \subset M$ be the set of all affine functions with linear part in $F$, i.e., functions of the form $\varphi(x) = \xi(x) + c$ for some linear form $\xi \in F$ and some real constant $c$. A function $f$ such that $f^\circ = f$, where $f^\circ$ is defined by (5.9.5), is called *$W$-convex* by Singer (1997:10).

We see that a function on $E$ is $W$-convex in the sense of Singer if and only if it is equal to the supremum of all its affine minorants belonging to $W$.

We may ask for a characterization of the $W$-convex functions. A generalization of Fenchel's theorem to this setting gives the answer: this happens if and only if the function possesses three properties mentioned in Theorem 9.17.7. □

**Proposition 5.9.3.** (A) *If $f\colon L \to M$ is increasing and $g\colon L \to N$ is coincreasing, then $f/_{\star}g \leqslant f/^{\star}g$.*
(B) *If $f$ is arbitrary and $y \in \mathbf{im}(g)$, say $y = g(a)$, then*

$$(f/^{\star}g)(y) \leqslant f(a) \leqslant (f/_{\star}g)(y).$$

(C) *In particular, if $g$ is surjective and $f$ is arbitrary, then $f/^{\star}g \leqslant f/_{\star}g$.*

**Proof.** (A) We have

$$(f/_{\star}g)(y) = \bigvee_x (f(x);\ g(x) \leqslant y) \quad \text{and} \quad (f/^{\star}g)(y) = \bigwedge_{x'} (f(x');\ g(x') \leqslant y).$$

If $g(x) \leqslant y \leqslant g(x')$, then $x \leqslant x'$ and $f(x) \leqslant f(x')$. (A) follows if we take the supremum over all $x$ and the infimum over all $x'$.

(B) If $g(x) \leqslant g(a) \leqslant g(x')$ we easily get (B).

(C) is just a special case of (B). □

The quotients are the optimal solutions to an inequality:

**Proposition 5.9.4.** *For all mappings $f\colon X \to M$ and $g\colon X \to Y$ we have*

$$(f/^{\star}g) \circ g \leqslant f \leqslant (f/_{\star}g) \circ g \tag{5.9.6}$$

*with equality if $f$ is increasing and $g$ is coincreasing. From this we deduce that $(f/^{\star}g)(y) \leqslant (f/_{\star}g)(y)$ for all $y \in \mathbf{im}g$ as well as for all majorants and minorants of $\mathbf{im}g$. In particular, $f/^{\star}g \leqslant f/_{\star}g$ if $g$ is surjective.*

*Conversely, if $u, v\colon Y \to M$ are two increasing mappings such that*

$$u \circ g \leqslant f \leqslant v \circ g,$$

*then $u \leqslant f/^{\star}g$ and $v \geqslant f/_{\star}g$. Thus $f/^{\star}g$ is the largest increasing mapping $u$ such that $u \circ g \leqslant f$, and $f/_{\star}g$ is the smallest increasing mapping $v$ such that $f \leqslant v \circ g$.*

*In the special case $X = Y$ and $g = \mathbf{id}_X$ we obtain*

$$f/^\star\mathbf{id}_X \leqslant f \leqslant f/_\star\mathbf{id}_X,$$

*where $f/^\star\mathbf{id}_X = f^\wedge$ is the smallest increasing majorant of $f$ and $f_\star/\mathbf{id}_X = f_\vee$ is the largest increasing minorant of $f$; when $f$ itself is increasing we therefore get equality.*

**Proof.** The proof is straightforward. The case of equality in (5.9.6) follows from Proposition 5.9.3. □

We next compare the quotient $f/_\star g$ and the composition $f \circ g_{[-1]}$ (think of $x/y = x \cdot y^{-1}$ for positive numbers):

**Proposition 5.9.5.** *Let $f\colon X \to M$ be an increasing mapping, assuming $X$ to be an ordered set, and $g\colon X \to Y$. We then have $f/_\star g \leqslant f \circ g_{[-1]}$. If we in addition assume that $X$ is a complete lattice and $f$ is a dilation, we have equality here. Similarly, $f/^\star g \geqslant f \circ g_{[-1]}$ with equality if $f$ is an erosion, assuming $X$ to be a complete lattice. If $g$ is coincreasing, then $f/_\star g \leqslant f \circ g_{[-1]} \leqslant f \circ g^{[-1]} \leqslant f/^\star g$.*

**Proof.** The proof is straightforward; see Theorem 5.4.4 for the last statement. □

**Proposition 5.9.6.** *If $P$ is a preordered set and $h\colon M \to P$ is increasing, we have $h \circ (f/_\star g) \geqslant (h \circ f)/_\star g$ with equality if $h$ is a dilation. Similarly, $h \circ (f/^\star g) \leqslant (h \circ f)/^\star g$ with equality if $h$ is an erosion. A special case is Proposition 5.9.5 (take $X = M$ and $f = \mathbf{id}_X$).*

**Proof.** The proof is straightforward. □

## 5.10 Pullbacks and pushforwards

We shall now see how the concepts introduced fit into the study of a very fundamental situation, that of a mapping $f\colon X \to Y$ pulling back mappings $\psi\colon Y \to L$ and pushing forward mappings $\varphi\colon X \to L$.

**Definition 5.10.1.** Let three sets $X$, $Y$ and $L$ be given, as well as a mapping $f\colon X \to Y$. We define the ***pullback*** of $f$, denoted by $f^{\leftarrow}\colon \mathscr{F}(Y, L) \to \mathscr{F}(X, L)$ and having as values mappings $f^{\leftarrow}(\psi)\colon X \to L$, by

$$f^{\leftarrow}(\psi) = \psi \circ f \in \mathscr{F}(X, L), \qquad \psi \in \mathscr{F}(Y, L).$$

If $L$ is a complete lattice, we define the ***lower*** and ***upper pushforwards*** denoted by $f_{\rightarrow}, f^{\rightarrow} \colon \mathscr{F}(X,L) \to \mathscr{F}(Y,L)$ and yielding as values mappings $f_{\rightarrow}(\varphi), f^{\rightarrow}(\varphi) \colon Y \to L$, by

$$(5.10.1) \qquad f_{\rightarrow}(\varphi)(y) = \bigvee_{x \in X} \bigl(\varphi(x);\ f(x) = y\bigr), \quad y \in Y,\ \varphi \in \mathscr{F}(X,L);$$

$$(5.10.2) \qquad f^{\rightarrow}(\varphi)(y) = \bigwedge_{x \in X} \bigl(\varphi(x);\ f(x) = y\bigr)$$

for $y \in Y$, $\varphi \in \mathscr{F}(X,L)$. □

Let us illustrated this with an arrow diagram:

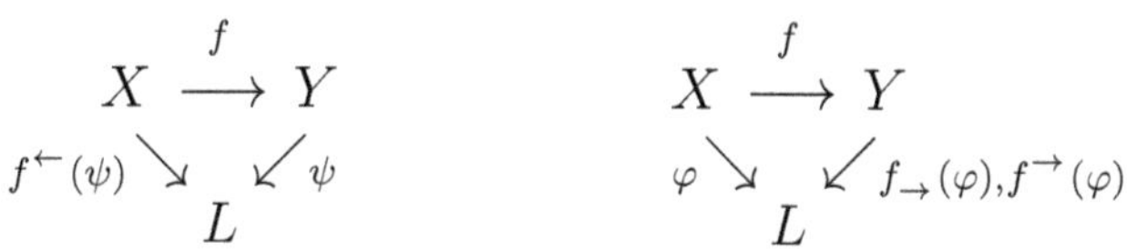

If $y \notin \mathbf{im} f$, we obtain $f_{\rightarrow}(\varphi)(y) = \mathbf{0}_L$; the supremum is not attained. Similarly, $f^{\rightarrow}(\varphi)(y) = \mathbf{1}_L$ for these $y$. Of course there are many other cases when the supremum and infimum in (5.10.1), (5.10.2) are not attained. In fact, $f_{\rightarrow}(\varphi)(y)$ and $f^{\rightarrow}(\varphi)(y)$ determine the smallest interval in $L$ which contains the range of $\varphi$ on the set $\{x;\ f(x) = y\}$, but the range itself need not be an interval (see Exercise 5.12).

If $\psi$ is constant, then $f^{\leftarrow}(\psi)$ is also a constant function, taking the same value.

We note that the lower pushforward defined by (5.10.1) is actually a lower quotient, $f_{\rightarrow}(\varphi) = \varphi/_{\star}f$, and similarly $f^{\rightarrow}(\varphi) = \varphi/^{\star}f$, namely if we provide $Y$ with the discrete order. The results on quotients are therefore available in this setting.

**Proposition 5.10.2.** *Let $f \colon X \to Y$ and a complete lattice $L$ be given. Then the pullback $f^{\leftarrow} \colon \mathscr{F}(Y,L) \to \mathscr{F}(X,L)$ is both a dilation and an erosion.*

**Proof.** We have to prove that $\bigvee f^{\leftarrow}(\psi_j) = f^{\leftarrow}(\bigvee \psi_j)$, equivalently that $\bigvee(\psi_j \circ f) = (\bigvee \psi_j) \circ f$, which is true by the definition of the supremum in $\mathscr{F}(X,L)$ and $\mathscr{F}(Y,L)$. Similarly for the infimum. □

**Proposition 5.10.3.** *If $X$, $Y$ are sets, $f\colon X \to Y$ a mapping, and $L$ a complete lattice, then*

$$\mathbf{epi}\, f_{\to} = \left(\mathbf{hypo}\, f^{\leftarrow}\right)^{\smile} \subset \mathscr{F}(X,L) \times \mathscr{F}(Y,L);$$

$$\mathbf{hypo}\, f^{\to} = \left(\mathbf{epi}\, f^{\leftarrow}\right)^{\smile} \subset \mathscr{F}(X,L) \times \mathscr{F}(Y,L).$$

**Proof.** The statements in the proposition mean that, for all $(\varphi, \psi) \in \mathscr{F}(X,L) \times \mathscr{F}(Y,L)$, we have

$$\psi \geqslant f_{\to}(\varphi) \text{ if and only if } \psi \circ f \geqslant \varphi$$

and

$$\psi \leqslant f^{\to}(\varphi) \text{ if and only if } \psi \circ f \leqslant \varphi.$$

These statements are easy to prove. □

**Corollary 5.10.4.** *Let a mapping $f\colon X \to Y$ and a complete lattice $L$ be given. Then the lower inverse of the lower pushforward is equal to the pullback, $(f_{\to})_{[-1]} = f^{\leftarrow}$, and the upper inverse of the pullback is equal to the lower pushforward, $(f^{\leftarrow})^{[-1]} = f_{\to}$. Also the lower inverse of the pullback is equal to the upper pushforward, $(f^{\leftarrow})_{[-1]} = f^{\to}$, and the upper inverse of the upper pushforward is equal to the pullback, $(f^{\to})^{[-1]} = f^{\leftarrow}$.*

*In all cases, the supremum and infimum defining the lower and upper inverses are attainded.*

**Proof.** The first statement follows from Corollary 5.5.3 applied to (5.9.1); the second from the same corollary applied to (5.9.2). The statement about the suprema and infima being attained is easy to check. Consider for example

$$(f_{\to})^{[-1]}(\psi) = \bigvee_{\varphi} (\varphi;\ f_{\to}(\varphi) \leqslant \psi) = \bigvee_{\varphi} (\varphi;\ \varphi \leqslant f^{\leftarrow}(\psi)) = f^{\leftarrow}(\psi),$$

where it is obvious that the supremum is attained. □

**Corollary 5.10.5.** *The lower pushforward mapping $f_{\to}$ is a dilation, and the upper pushforward $f^{\to}$ is an erosion.*

**Proof.** By Theorem 5.5.5, the upper inverse of an erosion is a dilation; hence $f_{\to} = (f^{\leftarrow})^{[-1]}$ is a dilation. The lower inverse of a dilation is an erosion; hence $f^{\to} = (f^{\leftarrow})^{[-1]}$ is an erosion. □

Summing up, we have two pairs of mutually inverse mappings:

$(f_{\to})_{[-1]} = f^{\leftarrow}$, both a dilation and an erosion, whose upper inverse is

$(f^{\leftarrow})^{[-1]} = f_{\to}$, a dilation; and

$(f^{\leftarrow})_{[-1]} = f^{\to}$, an erosion, whose upper inverse is

$(f^{\to})^{[-1]} = f^{\leftarrow}$, both a dilation and an erosion.

The other inverses $(f^{\to})_{[-1]}$ and $(f_{\to})^{[-1]}$ are not so easy to characterize. Let us note the following.

**Proposition 5.10.6.** *If there is an element $y \in Y \smallsetminus \mathbf{im} f$ such that $\psi(y) < \mathbf{1}_L$, then $(f^{\to})_{[-1]}(\psi)(x) = \mathbf{0}_L$ for all $x$. If on the other hand $\psi$ is identically equal to $\mathbf{1}_L$ in the complement of $\mathbf{im} f$ and $x$ the only point which is mapped to $f(x)$, then, $(f^{\to})_{[-1]}(\psi)(x) = f^{\leftarrow}(\psi)(x) = \psi(f(x))$, and if $x$ is such that more than one point is mapped to $f(x)$, then $(f^{\to})_{[-1]}(\psi)(x) = \mathbf{1}_L$. Similarly, if there is a $y \in Y \smallsetminus \mathbf{im} f$ such that $\psi(y) > \mathbf{0}_L$, then $(f_{\to})^{[-1]}(\psi)(x) = \mathbf{1}_L$ for all $x$. If on the other hand $\psi$ is identically equal to $\mathbf{0}_L$ in the complement of $\mathbf{im} f$ and $x$ the only point which is mapped to $f(x)$, then, $(f_{\to})^{[-1]}(\psi)(x) = f^{\leftarrow}(\psi)(x) = \psi(f(x))$, and if $x$ is such that more than one point is mapped to $f(x)$, then $(f_{\to})^{[-1]}(\psi)(x) = \mathbf{0}_L$.* □

**Example 5.10.7.** We defined two related mappings $f^*$ and $f_*$ to a given mapping $f\colon X \to Y$ in Definition 2.6.1.

We note that $f^{\leftarrow}(\mathbf{indf}_B) = f^*(B)$. The mapping $f^*$ is both a dilation and an erosion; while $f_*$ is in general only a dilation, defined by $f_*(\{x\}) = \{f(x)\}$. □

We consider now the special case when $L = \{0, 1\}$ and $\psi$ is a characteristic function, $\psi = \chi_B \in \mathscr{F}(Y, \{0, 1\})$ for some subset $B$ of $Y$. Then

$$f^{\leftarrow}(\chi_B) = \chi_A, \text{ where } A = f^*(B) = \{x \in X;\ f(x) \in B\},$$

the preimage of $B$.

If $\varphi$ is the characteristic function $\chi_A$ of an arbitrary subset $A$ of $X$, then

$$f_{\to}(\chi_A) = \chi_B, \text{ where } B = f_*(A) = \{f(x);\ x \in A\},$$

the direct image of $A$ under $f$, and

$$f^{\to}(\chi_A) = \chi_C, \text{ where } C = Y \smallsetminus \{f(x);\ x \notin A\},$$

the complement of the direct image of the complement of $A$.

Because of this it is natural to define also a pullback $f^{\leftarrow} : \mathscr{P}(Y) \to \mathscr{P}(X)$ and pushforwards $f_{\to}, f^{\to} : \mathscr{P}(X) \to \mathscr{P}(Y)$, simply by identifying the power set $\mathscr{P}(X)$ with the set $\mathscr{F}(X, \{0, 1\})$ of characteristic functions.

There is in general no inclusion relation between the two sets $B$ and $C$ defined by $f_{\rightarrow}(\chi_A)$ and $f^{\rightarrow}(\chi_A)$, thus between $f_{\rightarrow}(A)$ and $f^{\rightarrow}(A) = \complement f_{\rightarrow}(\complement A)$. In fact, given any subdivision of $Y$ into four disjoint subsets $Y_0 \cup Y_1 \cup Y_2 \cup Y_3$, we can easily find a mapping $f$ and a subset $A$ of $X$ such that $f_{\rightarrow}(A) = Y_0 \cup Y_1$ and $f^{\rightarrow}(A) = Y_0 \cup Y_2$.

## 5.11 Notes

### 5.11.1 *Inverses and their relatives*

Lower and upper inverses were defined by Heijmans & Ronse (1990:265), and some of their properties were developed there. The lower and upper inverses as defined here were presented in my paper (2007a). One can give a more general definition of these inverses by assuming only that $L$ and $M$ are preordered; cf. my paper (2010b). Then the inverses are not necessarily defined at all points in $M$; they may have a smaller domain of definition to keep track of. In this book we refrain from going into this more general situation.

The notion of a Galois connection between ordered sets goes back to Évariste Galois' work on the automorphism groups of a field. For general properties of Galois connections, see (Kuroš 1962) and (Cohn 1981).

Garret Birkhoff (1940:24, 1995:122) defined the concept of *polarity* between power sets; Ore (1944:495) introduced Galois connections in the general setting of ordered sets as well as the terms *Galois correspondence* (for each of the mappings) and *Galois connexion* (for the pair of mappings). Grätzer (1998:69), Gierz et al. (2003:22), Davey & Priestely (2003:155), and Blyth (2005:14) all use the term *Galois connection.*

Quite generally, most problems concerning inverses are ill-posed, meaning that small variations in the input data lead to large variations in the reconstructed situation. (These are more interesting than well-posed problems, by the way.) For an overview of inverse problems in imaging, see (Talbot 2016).

Garret Birkhoff (1995:124) attributes Proposition 5.7.2 to J. Schmidt.

### 5.11.2 *Infimal convolution*

For a survey of the properties and many applications of infimal convolution we refer to (Moreau 1963, 1970), (Rockafellar 1970, 1997), and (Strömberg 1996). For applications see also (Heijmans & Ronse 1990), (Heijmans & Maragos 1997), and (Heijmans & Molchanov 1998).

### 5.11.3 *Fuzzy sets*

For studies of fuzzy sets, see (Kaufmann & Gupta 1988) and (Sladoje 2005).

## 5.12 Exercises

5.1. Let $g$ be a constant mapping $L \to M$, $g(x) = b$, where $L$ and $M$ are complete lattices and $b$ a fixed element of $M$. Determine the lower and upper inverses of $g$.

5.2. Define $g(A) = A^{\perp} = \{x \in \mathbf{R}^n;\ x \cdot a = 0 \text{ for all } a \in A\}$. Determine $g_{[-1]}$ and $g^{[-1]}$.

5.3. We know that $g_{[-1]} \leqslant g^{[-1]}$ if $g$ is coincreasing (Proposition 5.4.1). On the other hand $g^{[-1]} \leqslant g_{[-1]}$ when $g$ is surjective (Corollary 5.4.2). Find lattices $L$ and $M$ and a mapping $g\colon L \to M$ such that $g^{[-1]}$ and $g_{[-1]}$ are not comparable (even with $g$ strictly increasing and injective).

5.4. In Section 5.10 it was mentioned that $f_{\to}(\varphi)(y)$ and $f^{\to}(\varphi)(y)$ determine the smallest interval in $L$ which contains the range of $\varphi$ on the set $\{x;\ f(x) = y\}$, but the range itself need not be an interval. Give an example that this can actually be so.

## Chapter 6

# Structure theorems for mappings

STRUCTURE THEOREMS are the topic of the present chapter. The goal is to state and prove four results with the general character of representing a mapping in terms of simpler mappings. The first gives a set-theoretical representation of dilations, erosions, cleistomorphisms, and anoiktomorphisms; the second of anoiktomorphisms as quotients; the third a represenation of erosions as suprema of elementary erosions; the fourth a represantation of anoiktomorphisms as suprema of elementary anoiktomorphisms. We then give some results on inf-ethmomorphisms and sup-ethmomorphisms.

### 6.1 Set-theoretical representation of dilations, erosions, cleistomorphisms, and anoiktomorphisms

We first present an easy result in the special case of translation-invariant operators on the family of subsets of an abelian group, putting several operations under a common roof.

**Proposition 6.1.1.** *Let $S$ be a subset of an abelian group $G$. Then the dilation, erosion, cleistomorphism and anoiktomorphism with structuring element $S$ can all be written in the form*

$$\varphi_{S,T,U}(A) = \bigcup_{x \in G} (x + S;\ x + T \subset A + U), \qquad A \in \mathscr{P}(G), \tag{6.1.1}$$

*for special choices of the structuring elements $S$, $T$, $U$, viz.*

$$\delta_S = \varphi_{S,\{0\},\{0\}},\ \varepsilon_S = \varphi_{\{0\},S,\{0\}},\ \alpha_S = \varphi_{S,S,\{0\}},\ \kappa_S = \varphi_{\{0\},S,S}. \tag{6.1.2}$$

*We can also write the mappings as*

$$\varphi = (f /_{\star} g) \circ h,$$

*where*

$$f(B) = B \text{ or } B + S,\ g(B) = B + S,\ h(A) = A \text{ or } A + S,$$

*for* $A, B \in \mathscr{P}(G)$.

**Proof.** The dilation $\delta = \delta_S$, the erosion $\varepsilon = \delta_{[-1]}$, the cleistomorphism $\kappa = \varepsilon \circ \delta$, and the anoiktomorphism $\alpha = \delta \circ \varepsilon$ can be written

$$\delta(A) = \bigcup_{B \in \mathscr{P}(G)} (B + S;\ B + S \subset A + S),$$

$$\varepsilon(A) = \bigcup_{B \in \mathscr{P}(G)} (B;\ B + S \subset A),$$

$$\kappa(A) = \bigcup_{B \in \mathscr{P}(G)} (B;\ B + S \subset A + S),$$

$$\alpha(A) = \bigcup_{B \in \mathscr{P}(G)} (B + S;\ B + S \subset A).$$

We now let

$f(B) = B + S$, $h(A) = A + S$ in the first case;

$f(B) = B$, $h(A) = A$ in the second case;

$f(B) = B$, $h(A) = A + S$ in the third case; and

$f(B) = B + S$, $h(A) = A$ in the fourth case;

while $g(B) = B + S$ in all four cases. □

We can think of the points as atoms and the sets $x + S$ as molecules. Then $\delta(A)$ and $\alpha(A)$ consists of molecules, the latter of those that are contained in $A$; whereas $\varepsilon(A)$ and $\kappa(A)$ consists of centers of molecules (which makes sense if the structuring set $S$ is symmetric).

## 6.2 Anoiktomorphisms as quotients

**Theorem 6.2.1.** *Let $f\colon X \to M$ be any mapping from a set $X$ into a complete lattice $M$. Then $\alpha = f/_{\star}f\colon M \to M$ is an anoiktomorphism. Conversely, any anoiktomorphism in $M$ is of this form for some mapping $f\colon X \to M$ with $X = M$. By duality we get analogous statements for the upper quotient and cleistomorphisms.*

**Proof.** It is clear that $\alpha(y) = \bigvee(f(x);\ f(x) \leqslant y)$ defines a mapping $M \to M$ which is increasing and antiextensive. Idempotency remains to be proved. To do so we note that $f(x) \leqslant y$ if and only if $f(x) \leqslant \alpha(y)$. Therefore

$$\alpha(y) = \bigvee_x (f(x);\ f(x) \leqslant y) = \bigvee_x (f(x);\ f(x) \leqslant \alpha(y)) = \alpha(\alpha(y)).$$

Note that, by (5.9.4), $\alpha \circ f = f$, proving that the image of $f$ is invariant under $\alpha$; in other words, all elements in $\mathbf{im} f$ are $\alpha$-open.

The converse follows from the formula $\alpha /_{\star} \alpha = \alpha$, which holds for any anoiktomorphism $\alpha \colon M \to M$. Indeed,

$$(\alpha /_{\star} \alpha)\,(y) = \bigvee_x \left(\alpha(x);\ \alpha(x) \leqslant y\right) = \bigvee_x \left(\alpha(x);\ \alpha(x) \leqslant \alpha(y)\right) = \alpha(y).$$

(Writing out the formula in full if $\alpha = f /_{\star} f$, we obtain $(f /_{\star} f) /_{\star} (f /_{\star} f) = f /_{\star} f = (f /_{\star} f) \circ (f /_{\star} f)$.) □

The set-theoretical anoiktomorphisms were defined as compositions $\delta \circ \varepsilon$. The theorem has a certain formal resemblance to this result. This can be done more generally with $\alpha \colon L \to L$, for a suitable erosion $\varepsilon \colon L \to M$ and a dilation $\delta \colon M \to L$ for some complete lattice $M$. One can take $M$ here as the set of all $\alpha$-open elements of $L$, a complete lattice, albeit in general not a sublattice of $L$ (cf. Gierz et al. 1980:23:3.13, 2003:29:O-3.13).

**Remark 6.2.2.** We note that the choice of the mapping $f$ in Proposition 6.1.1, taking $f(B) = B + S = \delta_S(B)$, is not available in the converse part of Theorem 6.2.1; instead, the mapping $\alpha$ itself can serve as $f$ as we have seen. □

## 6.3 Increasing mappings as suprema of elementary erosions

**Lemma 6.3.1.** *Let $L$ and $M$ be two complete lattices and let $a \in L$ and $b, c \in M$ be three fixed elements. Define a mapping $\varepsilon = \varepsilon_a^{b,c} \colon L \to M$ by*

$$\varepsilon_a^{b,c}(x) = \begin{cases} \mathbf{1}_M, & x = \mathbf{1}_L; \\ c, & x \neq \mathbf{1}_L,\ x \geqslant a; \\ b, & x \ngeqslant a. \end{cases}$$

*If $b \leqslant c$, then $\varepsilon_a^{b,c}$ is an erosion.*

**Proof.** We have to prove that $\varepsilon\big(\bigwedge_{j\in J} x_j\big)$ and $\bigwedge_{j \in J} \varepsilon(x_j)$ are equal. If $x_j = \mathbf{1}_L$ for all $j \in J$ (thus in particular if $J$ is empty), then both expressions take the value $\mathbf{1}_M$; if for all $j$ we have $x_j \geqslant a$ and there is a $j$ such that $x_j \neq \mathbf{1}_L$, then both expressions take the value $c$; finally, if there is $j$

such that $x_j \not\geqslant a$, then $\varepsilon(\bigwedge x_j) = b$ and $\varepsilon(x_j)$ may be either $b$ or $c$ or $\mathbf{1}_M$, but $\varepsilon(x_j) = b$ for at least one $j$, so if $b \leqslant c$ we have $\bigwedge \varepsilon(x_j) = b$ and equality holds. □

We can now prove a structure theorem for increasing mappings which is analogous to Matheron's first structural theorem, Theorem 3.11.1; cf. (Serra 1988a:20, 2001:29), where the result is given for $L = M$.

**Theorem 6.3.2.** *Let $f\colon L \to M$ be an increasing mapping from a complete lattice $L$ into another, $M$. Then $f$ is the supremum of a family of elementary erosions as defined in Lemma 6.3.1; more precisely,*

$$f = \bigvee_{a \in L} \varepsilon_a^{0, f(a)}.$$

**Proof.** On the one hand we have $\varepsilon_a^{0,f(a)}(x) \leqslant f(x)$ for all $a, x \in L$. On the other hand $\varepsilon_a^{0,f(a)}(a) = f(a)$ for every $a \in L$. Hence equality. □

## 6.4 Anoiktomorphisms as suprema of elementary anoiktomorphisms

**Definition 6.4.1.** Given two complete lattices $L$ and $M$ and two elements $a \in L$, $b \in M$ we define mappings $\varepsilon_a^b\colon L \to M$, $\delta_b^a\colon M \to L$, $\alpha_a\colon L \to L$, and $\kappa_b\colon M \to M$ by

$$\varepsilon_a^b(x) = \begin{cases} \mathbf{1}_M, & x \geqslant a, \\ b, & x \not\geqslant a; \end{cases} \qquad \delta_b^a(y) = \begin{cases} \mathbf{0}_L, & y \leqslant b, \\ a, & y \not\leqslant b; \end{cases}$$

$$\alpha_a(x) = \begin{cases} a, & x \geqslant a, \\ \mathbf{0}_L, & x \not\geqslant a; \end{cases} \qquad \kappa_b(y) = \begin{cases} b, & y \leqslant b, \\ \mathbf{1}_M, & y \not\leqslant b. \end{cases}$$

□

Note that $\varepsilon_a^b = \varepsilon_a^{b,\mathbf{1}}$ in the notation of Lemma 6.3.1.

**Lemma 6.4.2.** $(\varepsilon_a^b)^{[-1]} = \delta_b^a$.

**Proof.** Let us recall that the upper inverse is defined by $(\varepsilon_a^b)^{[-1]}(y) = \bigwedge_x (x;\ \varepsilon_a^b(x) \geqslant y)$. If $b \geqslant y$, then $\varepsilon_a^b(x) \geqslant y$ for all $x \in L$, so the upper inverse of $\varepsilon_a^b$ takes the value $\mathbf{0}_L$ at $y$, just as does $\delta_b^a(y)$. If on the other hand $b \not\geqslant y$, then $\varepsilon_a^b(x) \geqslant y$ if and only if $x \geqslant a$, so the upper inverse of $\varepsilon_a^b$ takes the value $a$ at $y$; so does $\delta_b^a$. □

**Lemma 6.4.3.** *For $a \neq \mathbf{0}_L$ and all $b$ we have $\varepsilon_a^b \circ \delta_b^a = \kappa_b$. For $a = \mathbf{0}_L$ and all $b$ we have $\varepsilon_a^b \circ \delta_b^a = \varepsilon_{\mathbf{0}}^b \circ \delta_b^{\mathbf{0}_L} = \kappa_{\mathbf{1}_M}$.*

*For all $a$ and all $b \neq \mathbf{1}_M$ we have $\delta_b^a \circ \varepsilon_a^b = \alpha_a$. For all $a$ and for $b = \mathbf{1}_M$ we have $\delta_b^a \circ \varepsilon_a^b = \delta_{\mathbf{1}_M}^a \circ \varepsilon_a^{\mathbf{1}_M} = \alpha_{\mathbf{0}_L}$.*

**Proof.** If $a \neq \mathbf{0}_L$, then $\varepsilon_a^b(\delta_b^a(y)) = \varepsilon_a^b(\mathbf{0}_L) = b$ if $y \leqslant b$, and $\varepsilon_a^b(\delta_b^a(y)) = \varepsilon_a^b(a) = \mathbf{1}_M$ otherwise. These are the same values as $\kappa_b(y)$. If on the other hand $a = \mathbf{0}_L$ we have $\varepsilon_a^b(\delta_b^a(y)) = \varepsilon_a^b(\mathbf{0}_L) = \mathbf{1}_M = \kappa_{\mathbf{1}_M}(y)$ for all $b$. The proof for the other composition follows by duality. □

We can now prove a structure theorem for anoiktomorphisms which is analogous to Matheron's second structural theorem, Theorem 3.11.7; cf. (Serra 1988a:22, 2001:49).

**Theorem 6.4.4.** *If $\alpha\colon L \to L$ is an anoiktomorphism in a complete lattice $L$, then it is the supremum of a family of elementary anoiktomorphisms $\alpha_a$ as defined in Definition 6.4.1:*

$$\alpha = \bigvee_{a \in A} \alpha_a,$$

*where $A = \mathbf{invar}(\alpha)$ is the invariance set of $\alpha$, the set of all $\alpha$-open elements.*

*Conversely, if $A$ is an arbitrary subset of $L$, then this formula defines an anoiktomorphism. Its invariance set is the set generated by $A$ under the formation of suprema.*

**Proof.** We know from Proposition 3.11.6 that

$$\alpha(x) = \big(\mathbf{id}_{\mathbf{invar}(\alpha)}\big)_\vee(x) = \bigvee_{a \in A} \big(a;\ a \leqslant x\big). \tag{6.4.1}$$

We claim that the last element is equal to $\bigvee_{a \in A} \alpha_a(x)$. Indeed, since $a = \alpha_a(a) = \alpha_a(x)$ for all $a \leqslant x$, we get

$$\bigvee_{a \in A} \big(a;\ a \leqslant x\big) = \bigvee_{a \in A} \big(\alpha_a(a);\ a \leqslant x\big) = \bigvee_{a \in A} \big(\alpha_a(x);\ a \leqslant x\big) = \bigvee_{a \in A} \alpha_a(x),$$

where the last equality follows from the fact that $\alpha_a(x) = \mathbf{0}$ when $a \nleqslant x$ so that these $a$ do not contribute to the supremum. □

## 6.5 Strong ethmomorphisms

If $L$ and $M$ are complete lattices, then the family of increasing mappings $\mathbf{incr}(L, M)$ is in a natural way also a complete lattice. If $L$, $M$ and $N$ are complete lattices and $f\colon L \to M$, $g_j\colon M \to N$ are increasing mappings, then

$$(6.5.1) \qquad (\bigwedge g_j) \circ f = \bigwedge(g_j \circ f) \text{ as well as } (\bigvee g_j) \circ f = \bigvee(g_j \circ f).$$

Here $\bigwedge g_j$ and $\bigvee g_j$ are formed in $\mathbf{incr}(M,N)$, while $\bigwedge(g_j \circ f)$ and $\bigvee(g_j \circ f)$ are formed in $\mathbf{incr}(L,N)$.

However, in contrast to (6.5.1), we have, for increasing mappings $f_j \in \mathbf{incr}(L,M)$ and $g \in \mathbf{incr}(M,N)$, only

$$(6.5.2) \qquad g \circ (\bigwedge f_j) \leqslant \bigwedge(g \circ f_j) \text{ and } g \circ (\bigwedge f_j) \geqslant \bigwedge(g \circ f_j),$$

where the inequalities can be strict (see Exercise 6.7).

In particular, if $L = M = N$, we can take $g_1 = f$ and $g_2 = \mathbf{id}$ in (6.5.1), and note that

$$(6.5.3) \qquad (f \wedge \mathbf{id}) \circ f = (f \circ f) \wedge f \text{ and } (f \vee \mathbf{id}) \circ f = (f \circ f) \vee f,$$

while, with $f_1 = g = f$ and $f_2 = \mathbf{id}$, we get from (6.5.2) only inequalities:

$$(6.5.4) \qquad f \circ (f \wedge \mathbf{id}) \leqslant (f \circ f) \wedge f \text{ and } f \circ (f \vee \mathbf{id}) \geqslant (f \circ f) \vee f.$$

We now define

$$f_{(1)} = f \wedge \mathbf{id} \text{ and } f^{(1)} = f \vee \mathbf{id}, \qquad f \in \mathbf{incr}(L,L),$$

so that (6.5.3) can be written

$$f_{(1)} \circ f = (f \circ f) \wedge f \text{ and } f_{(1)} \circ f = (f \circ f) \vee f.$$

For any ethmomorphism $\eta$ we thus have

$$(6.5.5) \quad \eta \circ \eta_{(1)} \leqslant \eta_{(1)} \circ \eta = \eta = \eta^{(1)} \circ \eta \leqslant \eta \circ \eta^{(1)}, \qquad \eta \in \mathbf{ethmo}(L).$$

**Definition 6.5.1.** Let $\eta \in \mathbf{ethmo}(L)$. Then $\eta$ is called an ***inf-ethmomorphism*** if $\eta \circ \eta_{(1)} = \eta$, and it is called a ***sup-ethmomorphism*** if $\eta \circ \eta^{(1)} = \eta$. If $\eta$ is both an inf-ethmomorphism and a sup-ethmomorphism, we shall say that $\eta$ is a ***strong ethmomorphism***. □

For a strong ethmomorphism $\eta$ we thus have equality throughout in (6.5.5), and we see that a strong ethmomorphism $\eta$ commutes with $\eta_{(1)}$ and $\eta^{(1)}$. Actually also $\eta^{(1)}$ and $\eta_{(1)}$ commute if $\eta$ is a strong ethmomorphism (see Exercise 6.7).

An idempotent mapping $f$ is constant on $\{x, f(x)\}$ for every $x$. An ethmomorphism is therefore constant on the segment $[x, \eta(x)]$ if $x \leqslant \eta(x)$ and on the segment $[\eta(x), x]$ if $\eta(x) \leqslant x$. An inf-ethmomorphism is constant on $[\eta_{(1)}(x), x]$, and a sup-ethmomorphism is constant on the segment

$[y, \eta^{(1)}(y)]$. A strong ethmomorphism is constant on $[\eta_{(1)}(x), \eta^{(1)}(x)]$ for every $x$.

An ethmomorphism $\eta$ is an anoiktomorphism if and only if $\eta_{(1)} = \eta$ (and if and only if $\eta^{(1)} = \mathbf{id}$); it is a cleistomorphism if and only if $\eta^{(1)} = \eta$ (and if and only if $\eta_{(1)} = \mathbf{id}$). Every cleistomorphism and every anoiktomorphism are strong ethmomorphisms. However, there are ethmomorphisms that are not strong:

**Example 6.5.2.** (Matheron 1988b:116) Let us define $\eta\colon L \to L$ by

$$\eta(x) = \begin{cases} x & \text{if } x \geqslant p \text{ or } x \leqslant p, \\ p & \text{otherwise,} \end{cases}$$

where $p$ is any given element of $L$. We can easily prove that $\eta$ is increasing and idempotent, thus an ethmomorphism. Also $\eta_{(1)}$ and $\eta^{(1)}$ are ethmomorphisms.

But $\eta$ is not a sup-ethmomorphism, neither an inf-ethmomorphism if there exists a $q \in L$ which is not comparable with $p$ (implying that $p$, $q$, $p \wedge q$ and $p \vee q$ are all different and that $\eta(q) = p \neq q$). In fact, $\eta(x \vee \eta(x)) = x$ if $x$ is comparable to $p$ and equal to $x \vee p$ otherwise, while $\eta(x \wedge \eta(x)) = x$ if $x$ is comparable to $p$ and equal to $x \wedge p$ otherwise. Thus $(\eta \circ \eta^{(1)})(q) = p \vee q \neq p = \eta(q)$ and $(\eta \circ \eta_{(1)})(q) = p \wedge q \neq p = \eta(q)$. Cf. Exercise 6.7. □

**Lemma 6.5.3.** *If $\eta$ is an ethmomorphism, then $\kappa$, defined as*

$$\kappa = \sup_{k \in \mathbf{N}} \left(\eta^{(1)}\right)^{\circ k},$$

*is a cleistomorphism, in fact equal to $\eta^{\kappa}$, the smallest cleistomorphism majorizing $\eta$. In particular, if $\eta^{(1)}$ is idempotent, then $\eta^{(1)} = \eta^{\kappa}$.*

**Proof.** Clearly $\kappa$ is increasing and larger than the identity. To see that it is idempotent, we write

$$\left(\eta^{(1)}\right)^{\circ k} \circ \left(\eta^{(1)}\right)^{\circ m} = \left(\eta^{(1)}\right)^{\circ (k+m)} \leqslant \kappa,$$

hence $\kappa \circ \kappa \leqslant \kappa$, and

$$\kappa \circ \kappa \geqslant \left(\eta^{(1)}\right)^{\circ k} \circ \left(\eta^{(1)}\right)^{\circ m} = \left(\eta^{(1)}\right)^{\circ (k+m)}.$$

Taking the supremum over all $k$ and $m$ we get $\kappa \circ \kappa \geqslant \kappa$. So $\kappa$ is a cleistomorphism, obviously majorizing $\eta$. Since $\eta^{\kappa}$ is the smallest cleistomorphism majorizing $\eta$, we obtain $\eta^{\kappa} \leqslant \kappa$, hence equality. □

**Theorem 6.5.4.** *If $\alpha\colon L \to L$ is an anoiktomorphism and $\kappa$ a cleistomorphism, then $\eta = \alpha \circ \kappa$ is a sup-ethmomorphism and $\theta = \kappa \circ \alpha$ is an inf-ethmomorphism.*

*Conversely, every sup-ethmomorphism is of the form $\alpha \circ \kappa$ for some anoiktomorphism $\alpha$ and some cleistomorphism $\kappa$.*

**Proof.** The fact that $\eta$ and $\theta$ are ethmomorphisms was proved in Proposition 2.11.6.

That $\eta$ is a sup-ethmomorphism follows from the general inequality $\eta \circ \eta^{(1)} \geqslant \eta$ and the opposite inequality

$$\eta \circ \eta^{(1)} = (\alpha \circ \kappa) \circ ((\alpha \circ \kappa) \vee \mathbf{id}) \leqslant (\alpha \circ \kappa) \circ \kappa = \alpha \circ \kappa = \eta.$$

For the converse, assume that $\eta$ is a sup-ethmomorphism. Then (6.5.5) takes the form

$$\eta \circ \eta_\alpha \leqslant \eta_\alpha \circ \eta = \eta = \eta^\kappa \circ \eta = \eta \circ \eta^\kappa.$$

We have, by (6.5.1), $\eta^{(1)} \circ \eta^{(1)} = (\mathbf{id} \vee \eta) \circ \eta^{(1)} = \eta^{(1)} \vee (\eta \circ \eta^{(1)})$. Since $\eta$ is now assumed to be a sup-ethmomorphism, the latter quantity is equal to $\eta^{(1)} \vee \eta = \eta^{(1)}$, proving that $\eta^{(1)}$ is idempotent, thus, according to Lemma 6.5.3, equal to $\eta^\kappa$.

I claim that $\eta_\alpha \circ \eta^\kappa = \eta \circ \eta^\kappa$. If $x \geqslant \eta(x)$, then $\eta(x) = \eta_\alpha(x)$. We can now take $x = \eta^\kappa(y) = \eta^\kappa(\eta^\kappa(y)) = \eta^\kappa(x)$ for any $y$. Then $x \geqslant \eta(x)$ and we obtain $\eta(\eta^\kappa(y)) = \eta_\alpha(\eta^\kappa(y))$, thus $\eta_\alpha \circ \eta^\kappa = \eta \circ \eta^\kappa$.

We can now finish: $\eta_\alpha \circ \eta^\kappa = \eta \circ \eta^\kappa = \eta \circ \eta^{(1)} = \eta$, proving that $\alpha = \eta_\alpha$ and $\kappa = \eta^\kappa$ yields a solution. □

The factorization $\eta = \alpha \circ \kappa$ in Theorem 6.5.4 is not always unique. This is hardly surprising. If we have three mappings $f, g, h = g \circ f\colon L \to L$ and $f$ is not surjective, we can change the definition of $g$ at will in the complement of $\mathbf{im}(f)$ without changing the composition $g \circ f$. Similarly, if $g|_{\mathbf{im}(f)}$ is not injective, say with $g(f(a)) = g(f(b))$ for some $a$, $b$ with $f(a) \neq f(b)$, then we can replace $f$ by $f_1$ defined by taking $f_1(a) = f(b)$ and $f_1(b) = f(a)$ without changing the values at the other points, and find that $g \circ f = g \circ f_1$. Nevertheless, it could be of interest to note the following result.

**Proposition 6.5.5.** *For any ethmomorphism $\eta$ in a complete lattice $L$ we have $\eta_\alpha \circ \eta^\kappa = \eta_\diamond \circ \eta^\kappa = \eta \circ \eta^\kappa \geqslant \eta$. Dually, $\eta^\kappa \circ \eta_\alpha = \eta^\diamond \circ \eta_\alpha = \eta \circ \eta_\alpha \leqslant \eta$. Here $\eta_\diamond$ and $\eta^\diamond$ are defined in (4.3.1).*

**Lemma 6.5.6.** *With $\eta$ as in Proposition 6.5.5 we have for any $x \in L$: $x \geqslant \eta(x)$ if and only if $\eta_\diamond(x) = \eta(x)$.*

**Proof.** If $x \geqslant \eta(x)$, then $\eta_\diamond(x) \geqslant \eta_\diamond(\eta(x)) = \eta(x)$, since $\eta(x)$ belongs to the invariance set of $\eta$, which is equal to the invariance set of $\eta_\diamond$.

Conversely, we always have $x \geqslant \eta_\diamond(x)$, so if $\eta_\diamond(x) = \eta(x)$, then $x \geqslant \eta(x)$. □

**Proof of Proposition 6.5.5.** We always have $\eta \geqslant \eta_\alpha \geqslant \eta_\diamond$. If $x \geqslant \eta(x)$ this implies that $\eta(x) \geqslant \eta_\alpha(x) \geqslant \eta_\diamond(x) = \eta(x)$, where the last equality follows from Lemma 6.5.6, hence equality $\eta(x) = \eta_\alpha(x) = \eta_\diamond(x)$ when $x \geqslant \eta(x)$. We now take $x = \eta^\kappa(y)$ for any $y$ and note that $x = \eta^\kappa(\eta^\kappa(y)) = \eta^\kappa(x) \geqslant \eta(x)$. This choice of $x$ yields $\eta(\eta^\kappa(y)) = \eta_\alpha(\eta^\kappa(y) = \eta_\diamond(\eta^\kappa(y))$ and completes the proof. □

We know from Example 4.3.11 that it may happen that $\eta_\alpha$ differs from $\eta_\diamond$. We even have $\eta_\alpha(c) \neq \eta_\diamond(c) \neq \eta(c)$ in that example, showing that in three representations of a sup-ethmomorphism,

$$\eta = \eta_\alpha \circ \eta^\kappa = \eta_\diamond \circ \eta^\kappa = \eta \circ \eta^\kappa,$$

any of three different factors, $\eta_\alpha$, $\eta_\diamond$ and $\eta$, can serve.

## 6.6 Notes

For strong ethmomorphisms, see (Matheron & Serra 1988a).

Theorem 6.5.4 is due to Matheron (1988b:130, Corollary 2).

Proposition 6.5.5 follows on combining Matheron's Corollary 2 in (1988b:130) with his Theorem 6.13 on the following page.

## 6.7 Exercises

6.1. Prove that $\alpha/_{\star}\alpha = \alpha$ for all anoiktomorphisms $\alpha$, and hence, in view of Theorem 6.2.1, that $(f/_{\star}f)/_{\star}(f/_{\star}f) = f/_{\star}f$ for any mapping $f \colon X \to M$, where $X$ is a set and $M$ a complete lattice.

6.2. Formulate and prove the statement dual to Theorem 6.2.1.

6.3. Prove that if $L$ is totally ordered, then every ethmomorphism is a strong ethmomorphism.

6.4. Prove that $\eta^{(1)}$ and $\eta_{(1)}$ commute if $\eta$ is a strong ethmomorphism.

6.5. Give examples where the inequalities in (6.5.2) are strict.

6.6. Determine $\eta_{(1)}$, $\eta^{(1)}$, $\eta \circ \eta_{(1)}$, $\eta \circ \eta^{(1)}$ in Example 6.5.2.

6.7. Prove that $\eta_{(1)}$ and $\eta^{(1)}$ in Example 6.5.2 are idempotent.

6.8. Find a complete lattice $L$ and an ethmomorphism $\eta$ in $L$ such that such that $\eta^{(1)}$ is not idempotent. (Consequently $\eta^{(1)} \neq \eta^{\kappa}$.)

# Chapter 7

# Digitization

## 7.1 Introduction

DIGITAL GEOMETRY is about geometry in digital spaces—however, we shall not at this point give a formal definition. Suffice it to say that *digital* is used here as opposed to models based on real or complex numbers. Sometimes the latter are said to be "continuous," but even discrete models can carry a topology other than the discrete one, and therefore allow for non-trivial continuous mappings. A most important example is the Khalimsky topology to be presented in Chapter 15.

So we start our study of digital geometry here and will continue in Chapters 8 and 10, after a survey of classical convexity in Chapter 9.

The space $\mathbf{R}^n$ is a space where we do Euclidean geometry; the space $\mathbf{Z}^n$ is an example of a space where we do digital geometry. However, we shall take a more general approach.

In this chapter we shall first discuss what a good digitization should mean, and then study the concept of a digital line.

Let $X$ be a set and $P$ an arbitary subset of $X$. (Think of $X$ as $\mathbf{R}^2$ and $P$ as $\mathbf{Z}^2$ if you like.) If we want to digitize $X$ we may start with a mapping $f\colon X \to P$ and then define the digitization of a set $A$ as $f_*(A) = \{f(x);\ x \in A\}$ (for the definition of $f_*$, see Definition 2.6.1).

However, this approach is too narrow; it is often not possible to start with a pointwise mapping. Instead, we shall define here a digitization of $X$ into a subset $P$ of $X$ as a mapping $F\colon \mathscr{P}(X) \to \mathscr{P}(P)$ with certain desirable properites. We shall think of $F(A)$ as a digital representation of $A$. A very simple such representation is $F(A) = A \cap P$, but it is not very faithful, since many sets are mapped to the empty set, for example $A = X \setminus P$. (However, it works for sufficiently fat sets.) One desirable condition is therefore that $F(A)$ be empty only if $A$ is empty. We also remark that the mapping $F(A) = A \cap P$ is not of the form $F = f_*$ if $P \neq X$ (Exercise 7.11).

We recall that the mappings $f_*$ are dilations. It seems desirable to require in general that a digitization $F\colon \mathscr{P}(X) \to \mathscr{P}(P)$ be a dilation. In particular this means that it is determined by its images on points, i.e., $F(A) = \bigcup_{x \in A} F(\{x\})$. So it is enough to know the digitization of an arbitrary point in $X$; however, nothing requires the $F(\{x\})$ to be singleton sets.

The following setup seems to be sufficiently flexible.

**Definition 7.1.1.** Let two sets $X$ and $P$ be given, $P$ being a subset of $X$. Let there be given, for every $p \in P$, a subset $Q(p)$ of $X$, called the ***cell with nucleus*** $p$. Then the ***digitization*** determined by these cells is the mapping $F\colon \mathscr{P}(X) \to \mathscr{P}(P)$ defined by

$$F(\{x\}) = \{p \in P;\ x \in Q(p)\}, \qquad x \in X,$$

and

$$F(A) = \bigcup_{x \in A} F(\{x\}) = \{p \in P;\ A \text{ meets } Q(p)\},$$

for $A \in \mathscr{P}(X)$. □

We may think of the cell $Q(p)$ as a pixel, voxel or spel, and of $p$ as its address. If we think of $Q$ as a mapping $Q\colon P \to \mathscr{P}(X)$, then $F(A) = Q^*(\mathscr{Q}_A)$, where $Q^*\colon \mathscr{P}(\mathscr{P}(X)) \to \mathscr{P}(P)$ is defined in Definition 2.6.1 and $\mathscr{Q}_A \in \mathscr{P}(\mathscr{P}(X))$ is the family of all cells which meet $A$.

It is clear that a digitization in this sense is always a dilation. As already pointed out, it is desirable that a nonempty set have a nonempty digitization; this is true if and only if the union of all cells is equal to the whole space $X$.

If $X$ is an abelian group and $P$ a subgroup, it is desirable that the digitization commute with translations, which means that $Q(p) = Q(0) + p$ for all $p \in P$. Indeed, if $Q(p) - p$ varies too much, it is easy to construct strange examples.

**Example 7.1.2.** A simple choice is $Q(p) = \{p\}$. This yields the digitization $F(A) = P \cap A$ already mentioned. If the set is fat, this digitization may work out well. In an abelian group with a metric we can even fatten the set using a dilation with a structuring element $S$, defining $Q(p)$ to be equal to $p + S$; this yields $F(A) = P \cap (A + S)$. □

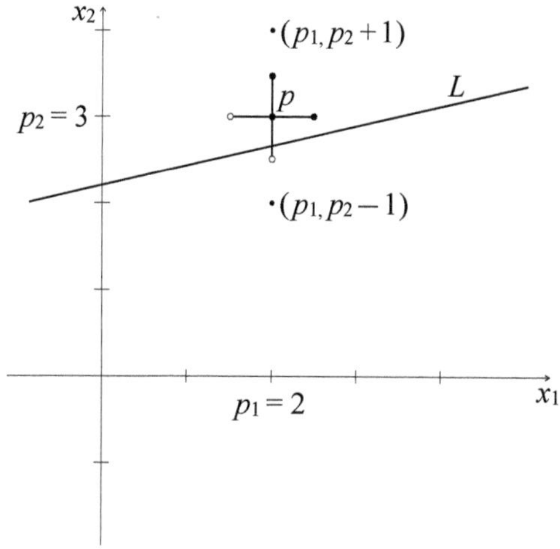


Figure 7.1. Digitization of a Euclidean straight line according to Azriel Rosenfeld. The cross $R(p)$ is defined in formula (7.1.1). (Courtesy Hania Uscka-Wehlou)

**Example 7.1.3.** If $X = \mathbf{R}$ and $P = \mathbf{Z}$ we may choose

$$Q(p) = [p - \tfrac{1}{2}, p + \tfrac{1}{2}].$$

Then every set has a nonempty digitization, but the half-integers have a digitization consisting of two points. If we choose instead $Q(p) = \left]p - \tfrac{1}{2}, p + \tfrac{1}{2}\right[$, then the digitization of a half-integer is empty. As a compromise we may choose $Q(p) = \left]p - \tfrac{1}{2}, p + \tfrac{1}{2}\right]$; the digitization of a point is then always a point: $F(\{x\}) = \left\{\left\lceil x - \tfrac{1}{2}\right\rceil\right\}$. But then a new inconvenience appears: this digitization does not commute with the reflection $x \mapsto -x$.

□

**Example 7.1.4.** If $X = \mathbf{R}^2$ and $P = \mathbf{Z}^2$ we may construct digitizations from what we have already done on the real axis. We may take

$$Q(p) = \left[p_1 - \tfrac{1}{2}, p_1 + \tfrac{1}{2}\right] \times \left[p_2 - \tfrac{1}{2}, p_2 + \tfrac{1}{2}\right], \qquad p \in \mathbf{Z}^2,$$

and similarly for the open and half-open intervals. Another choice is not to take the Cartesian product but to define the cell with nucleus $p$ as

$$(7.1.1)\qquad \begin{aligned} R(p) &= \{x;\ x_1 = p_1 \text{ and } p_2 - \tfrac{1}{2} < x_2 \leqslant p_2 + \tfrac{1}{2}\} \\ &\cup \{x;\ p_1 - \tfrac{1}{2} < x_1 \leqslant p_1 + \tfrac{1}{2} \text{ and } x_2 = p_2\}. \end{aligned}$$

Thus $R(p)$ is a cross with center at $p$. This is the digitization used by Rosenfeld (1974). It is based on the mapping $\mathbf{R} \ni x \mapsto \lceil x - \frac{1}{2} \rceil \in \mathbf{Z}$, already mentioned, a digitization of $\mathbf{R}$ which takes a non-half-integer to the closest integer and moves down by one half in the case of half-integers. Let us call it the ***Rosenfeld digitization***of $\mathbf{R}^2$.

It is clear that in this case the union of the cells is very small compared with $\mathbf{R}^2$, so that many sets have empty digitization. However, the union of all cells is equal to all grid lines $(\mathbf{R} \times \mathbf{Z}) \cup (\mathbf{Z} \times \mathbf{R})$, so that every straight line has a nonempty digitization. So does every rectilinear segment which is sufficiently long. Thus this digitization can be used in the study of digital straight lines—and it was. Note that the family of all cells is disjoint, which implies that the digitization of a point is either empty or a singleton set.

□

The definition of $Q(p)$ says nothing about how close a digitization of a point is to the point. To achieve this we must of course add some requirement that points in the cell $Q(p)$ shall be reasonably close to $p$. This leads us to the topic of Voronoi cells; see Section 7.7. Taking $Q(p)$ equal to $V(p)$ or a subset of $V(p)$ is a reasonable choice. In the special case described in Example 7.1.2, we have $Q(p) = p + S$, so that $Q(p) - p$ is constant.

## 7.2 Serra's four principles on gathering information

Jean Serra (1982:6–15) lists "four principles of quantification." These are about our ways to gather information about the external world. They apply also, but not exclusively, to image analysis. Let us discuss them briefly in mathematical terms.

Serra's first principle is "compatibility under translation." Mathematically speaking, this means that if we translate the object (i.e., move it some distance without rotating it) and then do something to it, the result should be the same as if we perform the two operations in the other order. For a mapping, this simply means that $f(A + b) = f(A) + b$, which we may express as $f \circ T_b = T_b \circ f$, writing $T_b$ for the translation $T_b(A) = A + b$, thus a kind of commutativity. We say that $f$ ***commutes with translations***. However, when it comes to applications of this idea to images on the computer screen, we are in trouble. On a finite screen like

$$\{x \in \mathbf{Z}^2;\ 0 \leqslant x_1 < 1{,}024,\ 0 \leqslant x_2 < 768\}$$

almost nothing can commute with translations. To escape from this difficulty we introduce the ideal, infinite, computer screen with sets of addresses equal to $\mathbf{Z}^2$. The principle is equally useful in $\mathbf{R}^n$ and $\mathbf{Z}^n$.

Serra's second principle is "compatibility under change of scale." For a mapping this means that it commutes with homotheties . An ***homothety*** is a mapping of the form $x \mapsto \lambda x$. So $f(\lambda A)$ should be equal to $\lambda f(A)$ for any positive number $\lambda$. This is not problematic if we are in $\mathbf{R}^n$, but it certainly is if we are in $\mathbf{Z}^n$. This principle will therefore need to be suitably interpreted in $\mathbf{Z}^n$.

The third principle is that of "local knowledge." This principle says that in order to know some bounded part of $f(A)$, we shall not need to know all of $A$, only some bounded part of $A$. Mathematically speaking: for every bounded set $Y$, there exists a bounded set $Z$ such that

$$f(A \cap Z) \cap Y = f(A) \cap Y.$$

To know the result $f(A)$ in an arbitrary bounded set $Y$, we need not know all of $A$; it is enough to know how $A$ looks in some bounded set $Z$, maybe a little larger than $Y$. The principle of local knowledge shall therefore not be understood in the topological sense: the key notion is boundedness.

Serra's fourth principle of quantification is about monotone convergence. It means that if a decreasing sequence $(A_j)$ of closed sets tends to a limit $A$, thus if $A = \bigcap A_j$, then also $f(A_j)$ shall tend to $f(A)$. So if $A_j$ is close to $A$ in some sense and $A_j$ contains $A$, then $f(A_j)$ should be close to $f(A)$.

## 7.3 Distances and metric spaces

Distances in $\mathbf{R}^n$ can be measured in different ways, e.g., by the ***$l^2$-distance*** in the two-dimensional image, the $l^2$-distance between two points $x = (x_1, x_2)$ and $y = (y_1, y_2)$ being

$$d^2(x, y) = \|x - y\|_2 = \sqrt{(x_1 - y_1)^2 + (x_2 - y_2)^2}.$$

Generally speaking, a ***Euclidean distance*** is one that is defined from an inner product, so that the norm is equal to $\|x\| = \sqrt{x \cdot x}$, $x \in \mathbf{R}^n$. After a change of coordinates it can be identified with the $l^2$-distance.

Other distances that have been used are the ***city-block distance*** or ***$l^1$-distance***, ***Manhattan distance*** or ***taxicab distance***

$$d^1(x, y) = \|x - y\|_1 = |x_1 - y_1| + |x_2 - y_2|$$

and the ***chessboard distance***[1] or $l^\infty$***-distance***

$$d^\infty(x, y) = \|x - y\|_\infty = \max(|x_1 - y_1|, |x_2 - y_2|).$$

We shall define more distances on $\mathbf{Z}^n$ in Section 16.5.

**Definition 7.3.1.** Let $X$ be any nonempty set. Let us agree to call a function $d\colon X \times X \to \mathbf{R}$ a ***metric*** or a ***distance function***, if $d$ is ***positive definite***:

(7.3.1) $d(x, y) \geqslant 0$ for all $x, y \in X$ with equality if and only if $x = y$;

***symmetric***:

(7.3.2) $$d(x, y) = d(y, x), \qquad x, y \in X;$$

and, finally, satisfies the ***triangle inequality***:

(7.3.3) $$d(x, z) \leqslant d(x, y) + d(y, z)$$

for $x, y, z \in X$. □

As remarked already in the introduction, every nonempty set can be equipped with a metric, viz. the discrete metric $d_{\text{discr}}$ defined as

(7.3.4) $$d_{\text{discr}}(x, x) = 0, \qquad d_{\text{discr}}(x, y) = 1 \text{ if } x \neq y, \qquad x, y \in X.$$

A ***metric space*** is simply a set provided with a metric. If we have two metric spaces $X_1$ and $X_2$ provided with metrics $d_1$ and $d_2$, then a mapping $f\colon X_1 \to X_2$ is said to be ***Lipschitz*** with ***Lipschitz constant*** $C$ if we have

(7.3.5) $$d_2(f(x), f(y)) \leqslant C d_1(x, y) \text{ for all } x, y \in X_1.$$

In any metric space we define balls as follows: the ***non-strict ball of center at*** $c$ ***and radius*** $r$ is the set of all points $x$ satisfying $d(x, c) \leqslant r$ and will be denoted by $B_\leqslant(c, r)$; the ***strict ball of center*** $c$ ***and radius*** $r$ is the set of all points satisfying the strict inequality $d(x, c) < r$; it will be denoted by $B_<(c, r)$. The set-theoretical difference between the two, $B_\leqslant(c, r) \smallsetminus B_<(c, r) = \{x \in X;\ d(x, c) = r\}$, is the ***sphere of center*** $c$ ***and radius*** $r$.

In the Euclidean case these concepts are well known, and the balls are called ***closed*** and ***open***, respectively. But in general we must be a bit careful: the closure of $B_<(c, r)$ with respect to the topology defined by

[1]This term obviously refers to the king's movements on a chessboard. Since the other pieces can move in different ways, the term should perhaps be avoided.

$d$ is not necessarily equal to $B_{\leqslant}(c, r)$, and the interior of $B_{\leqslant}(c, r)$ is not necessarily equal to $B_<(c, r)$ (see Section 16.7). Also note that if two balls $B_<(c_1, r_1)$ and $B_<(c_2, r_2)$ with $r_1, r_2 > 0$ are disjoint, then we can only conclude that $\max(r_1, r_2) \leqslant d(c_1, c_2)$, whereas in a normed space over the reals a stronger inequality, $\max(r_1, r_2) \leqslant r_1 + r_2 \leqslant \|c_1 - c_2\|$, holds.

Every metric defines a topology: a set is declared to be open if and only if it is a union of open balls. However, we shall often use another topology on $X$ than that defined by $d$.

**Example 7.3.2.** Given a metric space $X$ with distance function $d$, we can define a metric on the power set $\mathscr{P}(X)$ provided we assume $X$ to be bounded for $d$. This metric, called the ***Hausdorff metric*** and named for Felix Hausdorff (1914), is defined by

$$d_H(A, B) = \max\Big[\sup_{x \in A} \inf_{y \in B} d(x, y), \sup_{y \in B} \inf_{x \in A} d(x, y)\Big], \qquad A, B \in \mathscr{P}(X).$$

To allow for the empty set we should make the convention that

$$\inf_{y \in \varnothing} d(x, y) = \sup_{p,q \in X} d(p, q) < +\infty, \qquad x \in X.$$

It is not difficult to see that $d_H$ is actually a metric. □

## 7.4 Discrete sets and uniformly discrete sets

Following (Bourbaki 1961:16) we shall say that a subset $A$ of a metric space $X$ with metric $d$ is ***discrete*** if for each point $a \in A$, $a$ is the only point in $B_<(a, r)$ for some positive $r$. It is called ***uniformly discrete*** if $r$ can be chosen independently of $a$.

For any subset $A$ of $X$ we define its ***separation function***

$$\mathbf{sep}_A \colon A \to \mathbf{R}_!, \qquad \mathbf{sep}_A(a) = \inf_{x \in A} \big(d(x, a);\ x \neq a\big), \qquad a \in A.$$

Equivalently, $\mathbf{sep}_A(a) = \sup\big(r;\ B_<(a, r) = \{a\}\big)$. This function is of importance for $A = X$ as well as when $\varnothing \neq A \neq X$.

Thus $A$ is discrete iff $\mathbf{sep}_A$ is positive everywhere, and $A$ is uniformly discrete iff it is bounded from below by a positive constant.

A ***Delone set*** is a uniformly discrete set such that its separation function is bounded from above. Equivalently, there are numbers $r_0$ and $r_1$ such that

$$0 < r_0 \leqslant \mathbf{sep}_A(a) \leqslant r_1 < +\infty \text{ for all } a \in A.$$

It is also equivalent to state the definition in terms of balls with centers in $A$: There are two positive numbers $r_0$ and $r_1$ such that, for all $a \in A$, $B_{\leqslant}(a, r_0) = \{a\}$ and $B_{\leqslant}(a, r_1) \neq \{a\}$.

## 7.5 To discretize a function

We can discretize a function $f\colon \mathbf{R} \to \mathbf{R}$ partially or completely: we may just take its restriction to the integers, i.e., $f|_{\mathbf{Z}}$, and keep its values in $\mathbf{R}$, thus performing a partial discretization. Or we may discretize also its values, e.g., by taking

$$g(x) = \lfloor f(x) \rfloor, \qquad x \in \mathbf{Z}.$$

Here $\lfloor t \rfloor$ denotes the floor function, defined by $\mathbf{R} \ni t \mapsto \lfloor t \rfloor \in \mathbf{Z}$ and $t \leqslant \lfloor t \rfloor < t + 1$ (see (2.5.1)).

This topic will be treated in more detail in Section 11.14.

## 7.6 Discretization by balayage

Balayage is a method developed in classical potential theory for measures (masses or electric charges).

In this context a common notion is that of ***balayage of a function*** $f$ ***defined in*** $\mathbf{R}^n$ ***onto a discrete set*** $P$ as the function

$$g(p) = \sum_{a \in V(p)} f(a), \qquad p \in P,$$

where the sets $V(p)$, $p \in P$, form a tessellation of $\mathbf{R}^n$, meaning that

$$V(p) \cap V(q) = \emptyset \text{ when } p \neq q, \text{ and } \bigcup_{p \in P} V(p) = \mathbf{R}^n.$$

We may for example take $P = \mathbf{Z}^n \subset \mathbf{R}^n$ and

$$V(p) = \{a \in \mathbf{R}^n;\ \lfloor a_j \rfloor = p_j,\ j = 1, \dots, n\}, \qquad p \in \mathbf{Z}^n,$$

or, in order to get cells closer to $p$, lift it a little:

$$V(p) = \{a \in \mathbf{R}^n;\ \lfloor a_j + \tfrac{1}{2} \rfloor = p_j,\ j = 1, \dots, n\}, \qquad p \in \mathbf{Z}^n.$$

We shall now introduce the Voronoi cells.

## 7.7 Voronoi cells

Let a metric space $X$ be given as well as a subset $P$. The metric of $X$ shall be denoted by $d$. For a point $x \in X$ we view the points in $P$ close to $x$ as approximants; there may be a best approximant. Given $p \in P$ we shall give a name to the set of all $x$ for which this particular $p$ is a (not necessarily unique) best approximant: the ***Voronoi cell with nucleus*** $p$ is

$$\mathbf{vor}(p) = \{x \in X;\ \forall q \in P, d(x,p) \leqslant d(x,q)\}, \qquad p \in P.$$

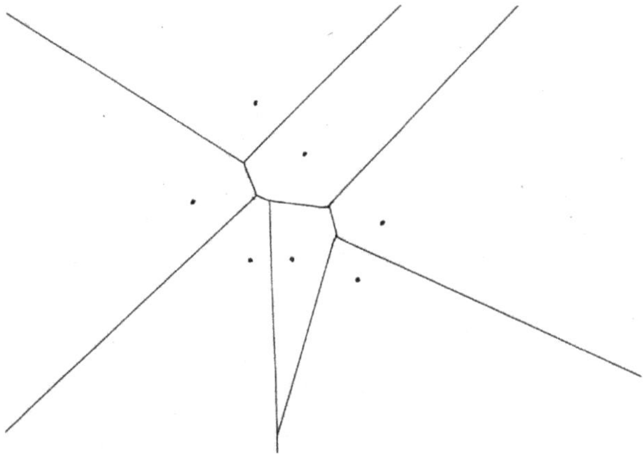

Figure 7.2. Seven Voronoi cells in the plane.

Thus $x \in \mathbf{vor}(p)$ if and only if $p$ is a best approximant of $x$. The Voronoi cells cover the whole space, but they are usually not disjoint, meaning they do not form a partition as described in Example 4.2.8.

We also define the ***strict Voronoi cell*** as

$$\mathbf{vor}_s(p) = \{x \in X;\ \forall q \in P \smallsetminus \{p\}, d(x,p) < d(x,q)\} \qquad p \in P.$$

Thus $x \in \mathbf{vor}_s(p)$ if and only if $p$ is the unique best approximant of $x$. Finally, one might define the ***very strict Voronoi cell*** as

$$\mathbf{vor}_{vs}(p) = \left\{x \in X;\ d(x,p) < \inf_{q \in P \smallsetminus \{p\}} d(x,q)\right\}, \qquad p \in P.$$

It is easy to construct examples where the very strict Voronoi cell is different from the strict Voronoi cell, but in all applications we are interested in they are equal.

Two different strict Voronoi cells are disjoint. Even more can be said: a (non-strict) Voronoi cell is disjoint from every strict Voronoi cell with a different nucleus. The union of all strict Voronoi cells is almost equal to the whole space $X$; there is only some garbage left out: these are the points which have at least two best approximants in $P$. However, since we are mathematicians, we do not have the right to throw away that garbage; we must be careful and consider both the strict and the non-strict Voronoi cells.

We now return to the topic of digitization. It seems reasonable that the digitization of a point should be contained in the set of all nuclei of Voronoi cells which contain that point. After all, these nuclei are the best approximants in $P$ of the point. This argument leads us to the following definition.

**Definition 7.7.1.** Let $X$ be a metric space and $P$ a subset of $X$ such that $P \cap B_<(c,r)$ is finite for all $c \in X$ and all $r \in \mathbf{R}$. If a mapping $\mathbf{dig}\colon \mathscr{P}(X) \to \mathscr{P}(P)$ satisfies

$$\mathbf{dig}(\{x\}) \subset \{p \in P;\ x \in \mathbf{vor}(p)\}, \qquad x \in X, \tag{7.7.1}$$

we shall say that it is a ***Voronoi digitization of*** $X$ ***into*** $P$. □

Note that if $x$ belongs to some strict Voronoi cell $\mathbf{vor}_s(p)$, then it can belong to only one Voronoi cell, viz. the non-strict cell $\mathbf{vor}(p)$ with the same nucleus, so that the right-hand side in (7.7.1) is a singleton set. Hence $\mathbf{dig}(\{x\})$ is either empty or equal to the singleton set $\{p\}$. But if $x$ belongs to, say, two Voronoi cells, the right-hand side in (7.7.1) consists of a set $\{p,q\}$ with $p \neq q$, and there is a choice: $\mathbf{dig}(\{x\})$ may be equal to Ø, $\{p\}$, $\{q\}$, or $\{p,q\}$. And if $x$ belongs to $m$ Voronoi cells, the value can be any of $2^m$ subsets of $P$.

Thus $\mathbf{dig}(\{x\})$ is either empty or a singleton set whenever $x$ belongs to the union of all strict Voronoi cells, but in the complement of that union, the value of the function may be a set with several elements. In some situations we do make a choice and define $\mathbf{dig}(\{x\})$ to be a singleton set by introducing a new criterion. In fact, we shall do so when we come to the Khalimsky topology.

If $X = \mathbf{R}$ and $P = \mathbf{Z}$, then the Voronoi cells are the closed intervals $\left[n - \frac{1}{2}, n + \frac{1}{2}\right]$ and the strict cells are the open intervals $\left]n - \frac{1}{2}, n + \frac{1}{2}\right[$, $n \in \mathbf{Z}$. It is clear that, under a Voronoi digitization, the digitization of a real number which is not of the form $n + \frac{1}{2}$ is the empty set or $\left\{\lfloor x + \frac{1}{2} \rfloor\right\}$. When $x = n + \frac{1}{2}$, we may choose $F(\{x\})$ to be Ø, $\{n\}$, $\{n+1\}$, or $\{n, n+1\}$. When we define the Khalimsky topology in Chapter 14, we shall choose $\{n\}$ for $n$ even and $\{n+1\}$ for $n$ odd. But this is of course only one of many admissible choices.

**Example 7.7.2.** We get examples of Voronoi digitizations by taking $C(p) = \mathbf{vor}(p)$ or $C(p) = \mathbf{vor}_s(p)$. Sometimes it is possible to choose a cell in between these two so that the space is covered exactly once by the different cells; an example was already mentioned: if $X = \mathbf{R}^n$ and $P = \mathbf{Z}^n$ we may choose $C(p) = \prod \left]p_j - \frac{1}{2}, p_j + \frac{1}{2}\right]$ (Example 7.1.4). □

**Example 7.7.3.** The digitization used by Rosenfeld (1974) is a Voronoi digitization, since the cell $C(p)$ defined in (7.1.1) is contained in the Voronoi cell with nucleus $p$, which is $\mathbf{vor}(p) = \left\{x \in \mathbf{R}^2;\ \|x - p\|_\infty \leqslant \frac{1}{2}\right\}$. □

## 7.8 Difficulties in logic

Small children start to count with natural numbers. Real numbers are much more difficult to define.

So one might think that discrete sets, like the set of natural numbers $\mathbf{N} = \{0, 1, 2, 3, \dots\}$, are easier to handle than the set $\mathbf{R}$ of real numbers, with numbers such as $\sqrt{5}$, $\pi$ and e etc.

However, all such assertions need to be considered carefully, in particular in relation to difficulities in logic. To be more precise, there are three levels of difficulty in logical theories.

**(1)** The easiest ones are the decidable theories, i.e., theories which allow for a decision method, an algorithm, by which one can decide in a predictable time whether a statement is true or not. It may take long, but it must be emphasized that an upper bound for the time needed for completion must be known in advance.
*Example:* Propositional calculus with connectives *and:* $p \wedge q$; *or:* $p \vee q$; and *negation:* $\neg p$.

**(2)** Next in line are the theories without a decision method, but which are complete, thus where every true statement but no false statement has a proof. The proof of a true statement can be found by a computer (simply since all proofs consist of finitely many symbols and can therefore be listed, and since we can distinguish a proof from a non-proof), but we cannot tell beforehand when the computer will arrive at a proof.
*Example:* Predicate calculus: Propositional calculus plus the quantifiers *for all:* $\forall x$ and *there exists:* $\exists x$.

Hilbert & Ackermann (1928) asked the question whether certain mathematical theories are decidable and also whether they are complete.

**(3)** Finally, we have the incomplete theories, where there are true statements for which there is no proof.
*Example:* Peano arithmetic: Predicate calculus applied to $\mathbf{N}$ and operations $x \mapsto x + 1$; $(x, y) \mapsto x + y$; $(x, y) \mapsto xy$.

Alfred Tarski showed that the first-order theory of real numbers under addition and multiplication is decidable.[2] Via Descartes, this result applies to elementary Euclidean geometry. Tarski's result is most remarkable, be-

[2]This result was published in full only in his paper (1948) (see also his paper (1951)) but it dates back to 1929 and was briefly mentioned in (Tarski 1931). See the translation in (Tarski 1956:134), where his decision method is called "a mechanical method."

cause Alonzo Church (1936) proved in 1935 that Peano arithmetic is not decidable: there is no algorithm. Alan Turing (1937) independently proved the same result soon after Church.[3]

Unlike the first-order theory of real numbers, Peano arithmetic is not only undecidable but also incomplete by Gödel's first incompleteness theorem (1931).

For these essential distinctions in logic see, e.g., (Ciesielski 1997:5–6) for incompleteness, and (Franzén 2004, 2005) and (Smith 2013) for comprehensive presentations of Gödel's incompleteness theorems. The first of the two theorems was just mentioned above. The second states that in a system of a certain type, the statement that the system is consistent cannot be proved within the system.

Torkel Franzén also published a short article (2006) where he dispels some misinterpretations of the incompleteness theorems. Rebecca Goldstein has published a superbly rich and lively account (2005) on Kurt Gödel's life and his incompleteness theorems.

So, all this points to the fact that, from the point of view of logic, natural numbers are much more difficult to treat than real numbers.

If these difficulties appear to be overwhelming, try to value them just like Piet Hein does, providing us with a better understanding:

Problems worthy
of attack
prove their worth
by hitting back.

But not only in logic!

## 7.9 Comparing differences and derivatives

As a first observation, let us note that Leibniz' rule for the derivative of a product, viz.

$$(fg)' = f'g + fg'$$

takes the form

$$D_a(fg) = (D_af)g + f(D_ag) + (D_af)(D_ag)$$

[3]It follows that the set of integers cannot be defined as a subset of the set of real numbers using only first-order logic—second-order logic, thus with quantifiers ranging over sets of real numbers, is required; cf. (Franzén 2004:114; 2005:26).

for differences of functions $f, g\colon \mathbf{R} \to \mathbf{R}$, with an extra term.

Similarly, the formula for the derivative of a quotient,

$$(f/g)' = \frac{gf' - fg'}{g^2},$$

takes the form

$$(D_a(f/g))(x) = \frac{g(x)(D_af)(x) - f(x)(D_ag)(x)}{g(x)g(x+a)}, \qquad x \in \mathbf{R},$$

where the denominator now is not the square of $g(x)$.

We note in passing that the formula for the derivative of a convolution product,

$$(f * g)' = f' * g = f * g',$$

which looks a little like Leibniz' formula with an equality sign substituted for the plus sign, fares better when we go to differences:

$$D_a(f * g) = (D_af) * g = f * (D_ag).$$

Let us now look at some more concrete examples.

The derivative of $f(x) = x^5$, $x \in \mathbf{R}$, is $f'(x) = 5x^4$.

But the difference quotient is

$$\frac{f(x+h) - f(x)}{h} = 5x^4 + 10x^3h + 10x^2h^2 + 5xh^3 + h^4.$$

And coming to integrals we compare the integral

$$\int_0^a f(x)\mathrm{d}x = \frac{1}{6}a^6$$

with the Riemann sum

$$\frac{1}{n}\sum_{j=0}^{an} f(j/n) = \frac{1}{6}a^6 + \frac{1}{2}\frac{a^5}{n} + \frac{5}{12}\frac{a^4}{n^2} - \frac{1}{12}\frac{a^2}{n^4}.$$

This may explain some of the successes of differential and integral calculus.

We can compare $f(x) = x^m$, $x \in \mathbf{R}$, and its derivative $f'(x) = mx^{m-1}$, with $f(x) = x^m$, $x \in \mathbf{Z}$, and its differences

$$f(x+1) - f(x) = mx^{m-1} + \frac{m(m-1)}{2}x^{m-2} + \cdots + mx + 1.$$

Let us also take a look at some integrals and compare them with the corresponding sums.

Take as before $f(x) = x^m$, $x \in \mathbf{R}$, yielding

$$\int_0^x t^m \, \mathrm{d}t = \frac{1}{m+1} x^{m+1}, \qquad x \geqslant 0,$$

and compare with

$$f(x) = x^m, \qquad x \in \mathbf{Z};$$

$$\sum_{t=0}^{x} t^m = \frac{x^{m+1}}{m+1} + \frac{x^m}{2} + \sum_{j=2}^{m} \frac{m!}{j!(m-j-1)!} B_j x^{m-j+1}, \quad x \in \mathbf{N}, \ m \geqslant 1.$$

Here $B_j$ are the Bernoulli numbers, named for Jakob Bernoulli, of which the first few are (partly according to Gradšteĭn & Ryžik 1962):

$$B_1 = \tfrac{1}{2}, \ B_2 = \tfrac{1}{6}, \ B_3 = 0, \ B_4 = -\tfrac{1}{30}, \ B_5 = 0, \ B_6 = \tfrac{1}{42}, \ B_7 = 0, \ldots;$$

all with odd indices $\geqslant 3$ are zero.

$$m = 0: \quad \sum_{t=0}^{x} t^0 = x + 1;$$

$$m = 1: \quad \sum_{t=0}^{x} t = \frac{1}{2} x(x+1);$$

$$m = 2: \quad \sum_{t=0}^{x} t^2 = \frac{1}{6} x(x+1)(2x+1);$$

$$m = 3: \quad \sum_{t=0}^{x} t^3 = \frac{1}{4} x^2(x+1)^2;$$

$$m = 4: \quad \sum_{t=0}^{x} t^4 = \frac{1}{30} x(x+1)(2x+1)(3x^2+3x-1);$$

$$m = 5: \quad \sum_{t=0}^{x} t^5 = \frac{1}{12} x^2(x+1)^2(2x^2+2x-1);$$

$$m = 6: \quad \sum_{t=0}^{x} t^6 = \frac{1}{42} x(x+1)(2x+1)(3x^4+6x^3-3x+1);$$

$$m = 7: \quad \sum_{t=0}^{x} t^7 = \frac{1}{24} x^2(x+1)^2(3x^4+6x^3-x^2-4x+2).$$

For $m = 2, 4, 6$, the sum contains the factor $x(x+1)(2x+1)$; for $m = 3, 5, 7$, the factor $x^2(x+1)^2$. We may guess that this is so also for even $m \geqslant 8$ and odd $m \geqslant 9$, respectively. That $x$ is a factor for $m \geqslant 1$, and that $x^2$ is a factor for odd $m \geqslant 3$ is easy to see from the general formula.

Comparing with rectangles, we arrive at easy inequalities:

$$\int_1^{x+1} (t-1)^m \, \mathrm{d}t = \frac{x^{m+1}}{m+1} \leqslant \sum_1^x t^m \leqslant \frac{(x+1)^{m+1}-1}{m+1} = \int_1^{x+1} t^m \, \mathrm{d}t$$

for $m$ real and positive.

A more precise inequality for $m \geqslant 1$ is

$$\sum_1^x t^m \leqslant \int_1^{x+1} (t - \tfrac{1}{2})^m \, \mathrm{d}t$$

$$= \frac{(x+\frac{1}{2})^{m+1} - (\frac{1}{2})^{m+1}}{m+1} = \frac{x^{m+1}}{m+1} + \frac{x^m}{2} + \cdots + \frac{x}{2^m}.$$

Partial differential equations—or rather difference equations—are used can serve as a tool in image analysis; see, e.g., the book by Guillermo Sapiro (2001, reprinted 2006).

## 7.10 Notes

The term *balayage* is of French origin and means 'sweeping (using a broom)'; *balayer* 'to sweep'.

The Voronoi cells are named for Georgiĭ Feodosevič Voronoĭ, Georges Voronoï.

The Educational Broadcasting Service (EBS), a public TV channel in South Korea, sent a lecture on discrete mathematics on 2006 March 15 at 06:30 local time. Suddenly the formula

$$\sum_1^n k^3 = \left(\frac{n(n+1)}{2}\right)^2$$

appeared on my screen. This is the formula for $m = 3$ on page 138.

So much for public understanding of summation formulas ... at least in some countries ...

For later developments, see (Hübler et al. 1981), with an algorithm for the calculation of convex hulls; (Bruckstein 1991), where continued fractions are briefly mentioned; (Voss 1991); (Debled 1995); and (Uscka-Wehlou 2009d), which contains a survey of chain codes and continued fractions. See also (Uscka-Wehlou 2009b, 2010) on continued fractions.

So much for the many studies of discrete analogues of the equation $F'' = 0$. Discrete analogues of the inequality $F'' \geqslant 0$, on the other hand, are not so well known. We will develop integer analogues of this inequality, which then yield a new way of looking at convexity for functions $f\colon \mathbf{Z} \to \mathbf{Z}$; cf. (Eckhardt 2001), (Murota 2003)[1] and (Kiselman 2004).

The focus of interest in the present chapter is the space $\mathscr{F}(\mathbf{Z}, \mathbf{Z})$ of functions with discrete values. However, as a preparation for future work, the basic definitions in Section 8.3 are given for $\mathscr{F}(\mathbf{Z}^n, \mathbf{Z})$ and even more generally—that does not cost more.

## 8.2 Digital lines and Rosenfeld's chord property

In $\mathbf{R}^2$ we know what a ***straight line*** is: it is a set of the form

$$\{(1-t)a + tb;\ t \in \mathbf{R}\},$$

where $a$ and $b$ are two distinct points in the plane. And a ***straight-line segment***, or ***rectilinear segment***, is a connected subset of that line. We shall consider here closed segments of finite length, and may then write them as $\{(1-t)a + tb;\ 0 \leqslant t \leqslant 1\}$, where $a$ and $b$ are the endpoints. We shall denote this segment by $[a, b]$.

We shall take the set $P$, introduced in Section 7.7 to define Voronoi cells, as $\mathbf{Z}^2$ in the discussion which follows. The digitization of a straight line segment is the image under **dig** of $[a, b]$, thus

$$\mathbf{dig}([a,b]) = \bigcup_{t\in[0,1]} F(\{(1-t)a + tb\}) \subset \mathbf{Z}^2,$$

where $F$ is a mapping $\mathscr{P}(\mathbf{R}^2) \to \mathscr{P}(\mathbf{Z}^2)$.

Suppose we are dealing with a Voronoi digitization (see Definition 7.7.1). When

$$x = (1-t)a + tb$$

[1] Admittedly, Murota's book is mainly about functions with extended real values, but since he considers convexity of sets as well, it also concerns integer-valued functions via their finite epigraphs.

belongs to a strict Voronoi cell, which in this case is

$$\mathbf{vor}_s(p) = \{x;\ \|x-p\|_\infty < \tfrac{1}{2}\}, \qquad p \in \mathbf{Z}^2,$$

then

$$F(\{x\}) = \{(\lfloor x_1 + \tfrac{1}{2}\rfloor, \lfloor x_1 + \tfrac{1}{2}\rfloor)\},$$

the unique point in $\mathbf{Z}^2$ closest to $x$. However, when $x_1$ is a half-integer, and $x_2$ is not, the digitization may be empty or consist of one or two points; when both coordinates are half-integers, the value may be a set of zero, one, two, three or four points.

In his famous paper (1974), Azriel Rosenfeld defined the digitization as in (7.1.1). In particular, a point is mapped to a point or to the empty set. For straight lines with slope less than 45°, he considered the intersections of its line segments with the vertical grid lines only. However, a line segment may intersect a horizontal grid line but no vertical grid line at all. In this case the cell is just the first segment in the union in (7.1.1), but this does not matter so much, since the result will be trivially true for empty digitizations and the digitization is nonempty anyway for sufficiently long line segments.

**Definition 8.2.1.** We shall say with Rosenfeld (1974:1264) that a subset $A$ of $\mathbf{R}^2$ has ***the chord property in the sense of Rosenfeld*** if for all points $a, b \in A$ the segment $[a, b]$ is contained in $A + B^{\infty}_{<}(0,1)$, the dilation of $A$ by the open unit ball (a square in this case) for the $l^\infty$ metric. □

Theorems 8.2.2 and 8.2.4 below give together a characterization of the digitization of a straight line segment.

**Theorem 8.2.2.** *The Rosenfeld digitization of a straight line segment has the chord property.*

**Proof.** If the digitization of a line $L$ has the chord property, so does the digitization of every segment of $L$. We may therefore restrict attention to the case of a whole line $L$. Let $L$ be a straight line and $D \subset \mathbf{Z}^2$ its digitization. Let $p, q$ be two points in $D$, and $r$ an arbitrary point on the segment $[p, q]$. We shall prove that there exists a point $d \in D$ such that $\|d - r\|_\infty < 1$.

First we reduce to the case when the slope of $L$ is between 0 and 1—note that the hypothesis and the conclusion are invariant under reflection and permutation of the coordinates. When it is exactly 0 or 1 the result is easy.

Instead of the digitization defined by (7.1.1) we shall now use only the vertical part of the cell,

$$R_v(p) = \{x;\ x_1 = p_1 \text{ and } p_2 - \tfrac{1}{2} < x_2 \leqslant p_2 + \tfrac{1}{2}\}.$$

When the slope of a line is strictly between 0 and 1, $R$ and $R_v$ yield the same result.

When the slope of $L$ is strictly between 0 and 1 we consider first the case when $r_1 \in \mathbf{Z}$. In this case we define $s \in \mathbf{R}^2$ as the point in $L$ with $s_1 = r_1$. The digitization $d = \mathbf{dig}(\{s\}) \in D$ of $s$ satisfies $\|d - s\|_\infty \leqslant \frac{1}{2}$, thus $\|d - r\|_\infty \leqslant \|d - s\|_\infty + \|s - r\|_\infty \leqslant 1$. But can equality occur here? No. If we analyze the definition of the digitization we find that

$$r_2 - \tfrac{1}{2} < s_2 \leqslant r_2 + \tfrac{1}{2} \tag{8.2.1}$$

because of corresponding inequalities for $p$ and $q$ with respect to points on $L$, and that $d_2 - \frac{1}{2} < s_2 \leqslant d_2 + \frac{1}{2}$. Combining the two inequalities we see that

$$d_2 - 1 < s_2 - \tfrac{1}{2} \leqslant r_2 < s_2 + \tfrac{1}{2} \leqslant d_2 + 1,$$

so that actually $|d_2 - r_2| < 1$, while $|d_1 - r_1| = 0$, thus $r \in B^\infty_<(d, 1)$.

Next we consider the case $r_1 \notin \mathbf{Z}$, implying that $m < r_1 < m + 1$ for some integer $m$. We now define $s$, $s'$ and $s''$ as the points on $L$ such that $s_1 = r_1$, $s'_1 = m$ and $s''_1 = m + 1$, and let $d'$ and $d''$ be the digitizations of $s'$ and $s''$. Concerning $d'$ and $d''$ we must have $d'_2 \leqslant d''_2 \leqslant d'_2 + 1$. We shall therefore look separately at the two cases $d''_2 = d'_2$ and $d''_2 = d'_2 + 1$.

In case $d''_2 = d'_2$ we have $d'_2 - \frac{1}{2} < s'_2, s''_2 \leqslant d'_2 + \frac{1}{2}$, so the same inequality follows also for $s_2$, since $s_2$ is between $s'_2$ and $s''_2$. Combining with (8.2.1) we see that

$$d'_2 - 1 < s_2 - \tfrac{1}{2} \leqslant r_2 < s_2 + \tfrac{1}{2} \leqslant d'_2 + 1,$$

and conclude that $r \in B^\infty_<(d', 1) \cap B^\infty_<(d'', 1)$.

In case $d''_2 = d'_2 + 1$ we must have $r_2 > d'_2 - 1$ and $r_2 < d''_2 + 1$ so that $r \in B^\infty_<(d', 1) \cup B^\infty_<(d'', 1)$. The theorem is now completely proved. □

To prove a converse we shall need the concept of digital arc. Let us say that two points in $\mathbf{Z}^2$ are ***eight-neighbors*** if their $l^\infty$ distance is 1.[2] Then a ***digital arc*** is a mapping $\gamma$ from a finite integer interval $[a, b]_\mathbf{Z}$ into the plane $\mathbf{Z}^2$ which is Lipschitz-1, i.e., with Lipschitz constant as defined by (7.3.5) equal to 1 for the $l^\infty$ norm and such that $\gamma(a)$ and $\gamma(b)$ have one eight-neighbor in $\gamma_*([a, b]_\mathbf{Z})$ and $\gamma(x)$ has two eight-neighbors for $x = a + 1, \dots, b - 1$.

[2]Two points are ***four-neighbors*** if their $l^1$ distance is 1.

Figure 8.1. A set $S$ with five elements having the chord property and another set $T$ not having the property: a chord of $T$ leaves $T + B^{\infty}_{<}(0,1)$ at its midpoint.

**Example 8.2.3.** The set $S = \{(0,0), (1,0), (2,0), (3,0), (4,1)\}$ to the left in Figure 8.1 has the chord property, while the set $T$, consisting of the five points $(0,0)$, $(1,0)$, $(2,0)$, $(3,1)$, $(4,2)$ does not. Indeed, the point $(2,1)$ belongs to the chord $[(0,0),(4,2)]$, but it does not belong to the dilated set $T + B^{\infty}_{<}(0,1)$, although it does belong to the closed set $T + B^{\infty}_{\leqslant}(0,1)$. Thus, in view of Theorem 8.2.2, it cannot be the Rosenfeld digitization of a straight line segment. □

We may define a Voronoi digitization by declaring the digitization of $\left(0,-\frac{1}{2}\right)$ to be $(0,0)$, that of $\left(2,\frac{1}{2}\right)$ to be $(2,0)$, and that of $\left(4,1\frac{1}{2}\right)$ to be $(4,2)$. Then $T$ is the digitization of the straight line segment $\left[\left(0,-\frac{1}{2}\right),\left(4,1\frac{1}{2}\right)\right]$. This procedure does not commute with translations, which offers a kind of explanation—of course it should not be allowed to move up by one half from $\left(0,-\frac{1}{2}\right)$ and $\left(4,1\frac{1}{2}\right)$ and down by one half from $\left(2,\frac{1}{2}\right)$. Rosenfeld avoided this by always moving down in the case of half-integers.

**Theorem 8.2.4.** *If a digital arc $D$ in $\mathbf{Z}^2$ has the chord property, then it is the Rosenfeld digitization of some straight line segment in $\mathbf{R}^2$.*

**Lemma 8.2.5.** *Denote by $\pi_j\colon \mathbf{Z}^2 \to \mathbf{Z}$ the projection $(x_1,x_2) \mapsto x_j$, $j = 1,2$. If a digital arc $D$ has the chord property, then one of the restrictions $\pi_j\big|_D\colon D \to \mathbf{Z}$, $j = 1,2$, is injective.*

**Proof.** Since $D$ is a finite set it is contained in a minimal rectangle $[p_1,q_1] \times [p_2,q_2]$, where $p = (p_1,p_2)$ and $q = (q_1,q_2)$ are two vertices, the other two being $(p_1,q_2)$ and $(q_1,p_2)$. If $p_1 = q_1$ or $p_2 = q_2$ we are done, so assume that $p_1 < q_1$ and $p_2 < q_2$. We claim that $\pi_1\big|_D$ is injective if $q_1 - p_1 \geqslant p_2 - q_2$; otherwise that $\pi_2\big|_D$ is injective. So assume that $q_1 - p_1 \geqslant q_2 - p_2 > 0$. Each side of the rectangle must contain an endpoint of the arc; otherwise it cannot have the chord property. Since there are only two endpoints, they must be mapped to the vertices of the rectangle. After a possible reflection of the coordinates we may assume that the endpoints are $\gamma(a) = (p_1, p_2) = p$ and $\gamma(b) = (q_1, q_2) = q$. We claim that there are no two points on the arc with the same abscissa. If this were so, there would exist two such points with distance 1: $s = (s_1, s_2)$ and $t = (t_1, t_2)$ with $t_1 = s_1$ and $t_2 = s_2 + 1$. The point $t$ cannot be an endpoint—that would violate the chord property for the segment $[p, t]$ and the point $r \in [p, t]$ with $r_1 = t_1 - 1$. Therefore $t$ has a second neighbor in addition to $s$. But then this other neighbor must be $t' = (t_1 + 1, t_2 + 1)$, which violates the chord property for the segment $[p, t']$ and the point $r' \in [p, t']$ with $r'_1 = t_1 - 1$. This contradiction proves the lemma. □

**Proof of Theorem 8.2.4.** Let $D$ be a digital arc with the chord property. In view of the lemma and the symmetry of the digitization procedure, we may assume that there are no pairs of points $a, b$ in $D$ with $a_1 = b_1$, $a_2 \neq b_2$. Given three real numbers $\alpha$, $\beta$, $\gamma$ we define a strip in the plane by

$$S(\alpha, \beta, \gamma) = \{x \in \mathbf{R}^2;\ \alpha x_1 + \beta \leqslant x_2 \leqslant \alpha x_1 + \gamma\}.$$

Let us define the ***height*** of the strip as $\gamma - \beta$. The boundary $\partial S(\alpha, \beta, \gamma)$ of the strip has two components, given by the straight lines $x_2 = \alpha x_1 + \beta$ and $x_2 = \alpha x_1 + \gamma$. A finite set $D$ of integer points is a subset of the digitization of a non-vertical straight line segment if and only if $D$ is contained in a strip of height strictly less than 1.

For every given $\alpha$ there is a smallest strip $S(\alpha, \beta, \gamma)$ containing $D$. Moreover, varying also $\alpha$, there is a strip $S_0 = S(\alpha_0, \beta_0, \gamma_0)$ of smallest height. If $D$ consists of only one or two points, the conclusion follows easily, so let us assume that $D$ has at least three points. Clearly there must be at least one point of $D$ in each component of the boundary of $S_0$; otherwise we could increase $\beta$ or decrease $\gamma$ to obtain a narrower strip. And one of these lines must contain a second point of $D$; otherwise we could rotate the line slightly to obtain a strip of smaller height. For definiteness we may and shall assume that the three points on the boundary of the strip are $p$, $s$, $q$ with $p_1 < s_1 < q_1$ and where $p$ and $q$ are on the lower boundary and $s$ on the upper boundary. (See Figure 8.2 on page 149, where $p = (-1, 0)$, $s = (0, 1)$ and $q = (1, 1)$ when the slope is $1/2$, and $p = (-1, 0)$, $s = (0, 1)$ and $q = (6, 1)$ when the slope is $1/7$.) Let $r$ be the point on $[p, q]$ with abscissa equal to that of $s$. (We note that $p, s, q$ belong to $\mathbf{Z}^2$, while $r$ need not do so.)

Now assume that $D$ is not a subset of the digitization of a straight line. Then the height of this smallest strip is at least 1, so that $s_2 \geqslant r_2 + 1$,

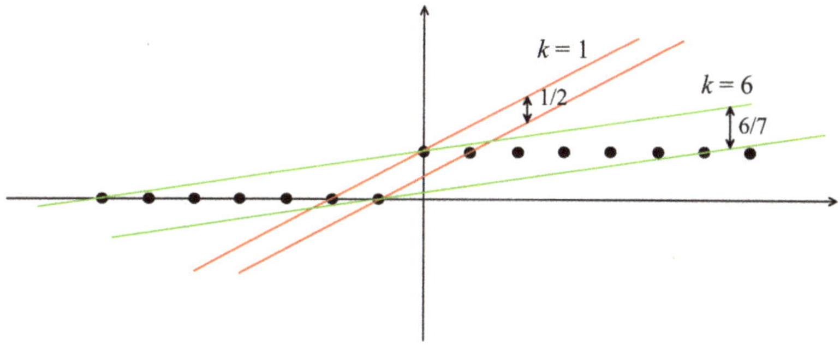


Figure 8.2. Minimal strips which contain $2k + 1$ points for $k = 1$ and $k = 6$, thus three points $(-1, 0), (0, 1), (1, 1)$ (and in addition a fourth point $(-2, 0)$), and thirteen points $(-6, 0)$, $(-5, 0)$, $(-4, 0)$, $(-3, 0)$, $(-2, 0)$, $(-1, 0)$, $(0, 1)$, $(1, 1)$, $(2, 1)$, $(3, 1)$, $(4, 1)$, $(5, 1)$, $(6, 1)$ (and in addition a fourteenth point $(-7, 0)$). (Courtesy Hania Uscka-Wehlou)

showing that $r$ does not belong to $B^{\infty}_{<}(s, 1)$. To see that $D$ does not satisfy the chord property we must however show that there is no $d \in D$ such that $r \in B^{\infty}_{<}(d, 1)$. So far we only know that $r$ does not belong to $B^{\infty}_{<}(s, 1)$. However, $s$ is the only point in $D$ on the vertical line $x_1 = s_1$ and all other points $d \in D$ satisfy $|d_1 - r_1| = |d_1 - s_1| \geqslant 1$, so that $\|r - d\|_\infty \geqslant |r_1 - d_1| \geqslant 1$. Therefore $D$ does not satisfy the chord property.

We have thus proved that a digital arc $D$ having the chord property is a subset of the digitization of some straight line $L$. However, since $D$ is a digital arc, it is the digitization of a connected subset of $L$. Obviously this subset can be taken to be compact, i.e., a straight line segment. □

## 8.3 Difference operators

**Definition 8.3.1.** Given abelian groups $G$ and $H$, and any element $a \in G$, we define a difference operator $D_a \colon \mathscr{F}(G, H) \to \mathscr{F}(G, H)$ by

(8.3.1) $$(D_a F)(x) = F(x + a) - F(x), \qquad x \in G,\ F \in \mathscr{F}(G, H).$$

We note that $D_a F = (\delta_{-a} - \delta_0) * F$. Here the asterisk means convolution, defined in (3.3.3). □

Most of the time we will take $G = \mathbf{R}^n$, $H = \mathbf{R}^m$. If $n = 1$ and $a \in \mathbf{N}$, then $D_a$ operates also from $\mathscr{F}(\mathbf{Z}, \mathbf{R})$ to $\mathscr{F}(\mathbf{Z}, \mathbf{R})$ and from $\mathscr{F}(\mathbf{Z}, \mathbf{Z})$ to $\mathscr{F}(\mathbf{Z}, \mathbf{Z})$; we shall use the same symbol for its restrictions to $\mathscr{F}(\mathbf{Z}, \mathbf{R})$ and $\mathscr{F}(\mathbf{Z}, \mathbf{Z})$.

**Definition 8.3.2.** Given two vector spaces $E$ and $H$ over $\mathbf{Q}$, we combine two difference operators to obtain the ***Jensen operator***

$$J_{a,b}\colon \mathscr{F}(E,H) \to \mathscr{F}(E,H)$$

for two points $a, b \in E$ such that there are two numbers $s, t \geqslant 0$ satisfying $s+t=1$ and $sa+tb=0$:

$$(8.3.2)\qquad \begin{aligned}(J_{a,b}F)(x) &= s(D_aF)(x) - t(D_{-b}F)(x+b)\\ &= sF(x+a) - F(x) + tF(x+b), \quad x \in G.\end{aligned}$$

We define

$$(8.3.3)\qquad \mu_{a,b} = s\delta_{-a} - \delta_0 + t\delta_{-b},$$

so that $J_{a,b}(F) = \mu_{a,b} * F$. □

When $a \neq b$, the numbers $s$ and $t$, if they exist, are uniquely determined.

For a more general Jensen operator, see condition (D) in Theorem 8.5.1 and Definitions 9.9.2 and 9.9.3.

A function $F \in \mathscr{F}(\mathbf{R},\mathbf{R})$ is convex if and only if $J_{a,b}F \geqslant 0$ for all positive real numbers $a, b$.

When $a$ and $b$ are integers, the Jensen operator maps $\mathscr{F}(\mathbf{Z},\mathbf{Z})$ into $\mathscr{F}(\mathbf{Z},\mathbf{Q})$.

We shall use the fact that $(J_{a,b}F)(0) = F(0) - H(0)$, where $H$ is the affine function which takes the same values as $F$ at the points $a$ and $b$, and thus measures the deviation from being affine.

Another second-order difference operator is $D_bD_a$, given by

$$(8.3.4)\qquad (D_bD_aF)(x) = F(x+a+b) - F(x+a) - F(x+b) + F(x).$$

It is well known that a continuous function $F\colon \mathbf{R} \to \mathbf{R}$ is convex if and only if $D_aD_aF \geqslant 0$ for all real $a > 0$; equivalently $D_bD_aF \geqslant 0$ for all $a, b > 0$. We note that $(J_{-b,b}F)(x+b) = \frac{1}{2}(D_bD_bF)(x)$.

For functions in $\mathscr{F}(\mathbf{Z},\mathbf{R})$, the conditions $D_1D_1f = 0$ and $D_1D_1f \geqslant 0$ give easy and satisfying results; see Theorem 8.5.1 below. For functions in $\mathscr{F}(\mathbf{Z},\mathbf{Z})$, on the other hand, these conditions yield very narrow classes of functions (Exercise 8.12). But if we relax them to $|D_1D_1f| \leqslant 1$ and $D_1D_1f \geqslant -1$, we get classes of functions which are much too wide to be of interest (Exercise 8.12). It turns out that a simple compromise, intermediate between the two conditions, viz.

$$(8.3.5)\qquad |D_bD_af| \leqslant 1 \qquad a, b \in \dot{\mathbf{N}},$$

for straight lines, and

$$D_b D_a f \geqslant -1, \qquad a, b \in \dot{\mathbf{N}}, \tag{8.3.6}$$

for convexity, yields classes with good properties. These inequalities are equivalent to $|J_{a,b}f| < 1$ and $J_{a,b}f > -1$ for all $a, b \in \mathbf{Z}$, respectively; see Theorem 8.6.1.

Defining $T$ as the translation operator $(Tf)(x) = f(x+1)$ and $I$ as the identity operator, we can write $D_1 = T - I$ and

$$D_a = T^a - I = P_a \circ (T - I) = P_a \circ D_1 = D_1 \circ P_a, \qquad a \in \dot{\mathbf{N}},$$

a telescoping series, where

$$P_a = T^{a-1} + T^{a-2} + \cdots + T^2 + T + I, \qquad a \in \dot{\mathbf{N}}. \tag{8.3.7}$$

The composition $D_b D_a = D_b \circ D_a$, $a, b \in \dot{\mathbf{N}}$, can be factorized as

$$D_b D_a = P_b P_a D_1 D_1. \tag{8.3.8}$$

(Here and in the sequel we write just $D_b D_a$ for the composition, omitting the circle.) Therefore the condition $D_b D_a f \geqslant -1$ can be expressed using instead $g = D_1 f$ or $h = D_1 D_1 f$: it is equivalent to $P_b P_a D_1 g \geqslant -1$ as well as to $P_b P_a h \geqslant -1$.

Since $P_a$ and $P_b$ have positive coefficients, the condition $h = D_1 D_1 f \geqslant 0$ implies that $D_b D_a f = P_b P_a h \geqslant 0$ for all $a, b \in \dot{\mathbf{N}}$. But if $h \geqslant -1$, we can conclude only that $D_b D_a f \geqslant -ab$. (The function $u(x) = -\frac{1}{2}x(x-1)$ with $D_1 D_1 u = -1$ and $D_b D_a u = -ab$ shows that the conclusion cannot be improved.) This indicates that the condition $D_b D_a f \geqslant -1$ for all $a, b \in \dot{\mathbf{N}}$ is much stronger than requiring it just for $a = b = 1$.

For more on the calculus for the difference operators $D_b D_a$, also in several variables, see Chapter 12.

## 8.4 Real-valued functions on the integers

For functions in $\mathscr{F}(\mathbf{Z}, \mathbf{R})$ questions on convexity can be resolved easily:

**Theorem 8.4.1.** *A function $f \colon \mathbf{Z} \to \mathbf{R}$ satisfies the equation*

$$D_1 D_1 f = 0, \tag{8.4.1}$$

*equivalently*

$$D_b D_a f = 0, \qquad a, b \in \dot{\mathbf{N}}, \tag{8.4.2}$$

*if and only if there are real constants $A$ and $B$ such that $f(x) = Ax + B$.*

*It satisfies the inequality*

$$(8.4.3) \qquad D_1 D_1 f \geqslant 0,$$

*equivalently*

$$(8.4.4) \qquad D_b D_a f \geqslant 0, \qquad a, b \in \dot{\mathbf{N}},$$

*if and only if the points with integer abscissae in* $\mathbf{cvxh}(\mathbf{epi}^{\text{finite}}(f))$ *all belong to* $\mathbf{epi}^{\text{finite}}(f)$.

*Equivalent conditions are* $J_{-1,1}f = 0$ *and* $J_{-1,1}f \geqslant 0$, *respectively.*

**Proof.** The equivalence of (8.4.1) and the seemingly stronger condition (8.4.2) follows from the factorization $D_b D_a = P_b P_a D_1 D_1$ (see (8.3.8)). Similarly, (8.4.3) is equivalent to (8.4.4).[3]

If $(x, y) \in \mathbf{Z} \times \mathbf{R}$ belongs to the convex hull of $\mathbf{epi}^{\text{finite}}(f)$, then $y \geqslant (1-t)f(a) + tf(b)$ for some points $a, b \in \mathbf{Z}$ with $a \leqslant x \leqslant b$ and with $t = (x-a)/(b-a) \in [0,1]$ (if $a = b$ there is nothing to prove). We shall prove that $(x, y)$ then belongs to $\mathbf{epi}^{\text{finite}}(f)$, i.e., that $y \geqslant f(x)$. By the lemma below we know that $\mu_{-1,1} * \nu_{a,b} = \mu_{a,b}$ for some function $\nu_{a,b}$ with only nonnegative values. Here $\mu_{a,b}$ is defined in (8.3.3). It follows that $\mu_{a,b} * f \geqslant 0$, i.e., that $f(x) \leqslant (1-t)f(a) + tf(b) \leqslant y$.

Conversely, if $\mathbf{cvxh}(\mathbf{epi}^{\text{finite}}(f))$ is contained in $\mathbf{epi}^{\text{finite}}(f)$, we see that we have in particular $f(x) \leqslant \frac{1}{2}f(x-1) + \frac{1}{2}f(x+1)$, i.e., that $D_1 D_1 f \geqslant 0$.

The last statement, about $J_{-1,1}$, follows from the identity $(J_{-1,1}F)(x+1) = \frac{1}{2}(D_1 D_1 F)(x)$. □

In the course of the proof we used the following result.

**Lemma 8.4.2.** *For every* $a, b \in \mathbf{Z}$ *there exists a function* $\nu_{a,b} \geqslant 0$ *such that* $\mu_{-1,1} * \nu_{a,b} = \mu_{a,b}$. *(Here* $\mu_{a,b}$ *is defined in formula* (8.3.3).*)*

**Proof.** We define $\nu_{a,b} = \mu_{a,b} * g$, where the function $g$ is defined by $g(x) = |x|$, $x \in \mathbf{Z}$, and satisfies $\mu_{-1,1} * g = \delta_0$. It follows that $\mu_{-1,1} * \nu_{a,b} = \mu_{a,b} * \delta_0 = \mu_{a,b}$ just as we wanted.

That $\nu_{a,b} \geqslant 0$ follows from the fact that, writing $g_{-1}(x) = -x$, $g_1 = x$, $g = g_{-1} \vee g_1$, we have $\mu_{a,b} * g_j = 0$, $j = -1, 1$, and therefore $\nu_{a,b} = \mu_{a,b} * (g_{-1} \vee g_1) \geqslant 0$. □

[3]The latter inequality should be compared with inequality (1.39) in Murota (2003:25), which is his starting point for the introduction of M-convex functions.

## 8.5 Functions taking real values on $\mathbf{Z}^n$

For functions defined in $\mathbf{Z}^n$ and with values in $\mathbf{R}_!$ we can characterize the convex extensible functions on $\mathbf{Z}^n$ with values in the extended reals in several ways:

**Theorem 8.5.1.** *For any function $f\colon \mathbf{Z}^n \to \mathbf{R}_!$ the following four conditions are equivalent.*

(A) *There exists a convex subset $C$ of $\mathbf{R}^n \times \mathbf{R}$ such that $C \cap (\mathbf{Z}^n \times \mathbf{R}) = \mathbf{epi}^{\mathrm{finite}}(f)$;*

(B) $\mathbf{cvxh}(\mathbf{epi}^{\mathrm{finite}}(f)) \cap (\mathbf{Z}^n \times \mathbf{R}) \subset \mathbf{epi}^{\mathrm{finite}}(f)$;

(C) *The function $f$ is convex extensible;*

(D) *The function $f$ satisfies Jensen's inequality*

$$f(x) \leqslant \sum_{j=0}^{n} \lambda_j f(p^{(j)})$$

*for all $\lambda_j \geqslant 0$ with $\sum \lambda_j = 1$ and all points $p^{(0)}, \ldots, p^{(n)}, x \in \mathbf{Z}^n$ such that $x = \sum_0^n \lambda_j p^{(j)}$.*

**Proof.** (A) $\Rightarrow$ (B). If (A) holds, we have $\mathbf{cvxh}(\mathbf{epi}^{\mathrm{finite}}(f)) \cap (\mathbf{Z}^n \times \mathbf{R}) \subset C \cap (\mathbf{Z}^n \times \mathbf{R}) = \mathbf{epi}^{\mathrm{finite}}(f)$.

(B) $\Rightarrow$ (C). If (B) is satisfied, we define

$$F(x) = \inf_y \big(y \in \mathbf{R};\ (x, y) \in \mathbf{cvxh}(\mathbf{epi}^{\mathrm{finite}}(f))\big), \qquad x \in \mathbf{Z}^n.$$

Then $F$ is convex and $F$ restricted to $\mathbf{Z}^n$ gives us $f$ back.

(C) $\Rightarrow$ (D). If $F$ is a convex extension of $f$, it satisfies Jensen's inequality (8.5.1)

$$F((1-\lambda)x + \lambda x') \leqslant (1-\lambda)F(x) \dotplus \lambda F(x'), \qquad x, x' \in \mathbf{R}^n,\ 0 \leqslant \lambda \leqslant 1,$$

which, as is well-known, implies the more general inequality in (D).

(D) $\Rightarrow$ (A). If (D) holds we define a set $C$ consisting of all points $(x, y) \in \mathbf{R}^n \times \mathbf{R}$ such that $\sum \lambda_j p^{(j)} = x$ and $\sum \lambda_j f(p^{(j)}) \leqslant y$ for some points $p^{(1)}, \ldots, p^{(n)} \in \mathbf{Z}^n$ and some real numbers $\lambda_j \geqslant 0$ with $\sum \lambda_j = 1$. It can be shown that $C$ is convex. It is clear that $C \cap (\mathbf{Z}^n \times \mathbf{R}) \subset \mathbf{epi}^{\mathrm{finite}}(f)$. Thus (A) holds. □

For functions in $\mathscr{F}(\mathbf{Z}, \mathbf{Z})$ there is no simple characterization like the equivalence of (A) and (C) in the theorem above. Condition (A) is quite natural in this context, whereas only very special functions in $\mathscr{F}(\mathbf{Z}, \mathbf{Z})$ are convex extensible.

## 8.6 Characterization of straightness: The chord property

In order to characterize straightness of finite subsets of $\mathbf{Z}^2$, Azriel Rosenfeld (1974) introduced the chord property; see Definition 8.2.1.

We may define the $U$-digitization of a subset $A$ of $\mathbf{R}^n$ as the set

$$\mathbf{dig}_U(A) = (A+U) \cap \mathbf{Z}^n, \qquad A \in \mathscr{P}(\mathbf{R}^n).$$

Here $U$ is a set of pixels or voxels located at the origin—it may in fact be any subset of $\mathbf{R}^n$. We may take $U = \{0\}$, but then many sets will have empty digitization; the role of $U$ is to fatten $A$ before intersecting it with the grid $\mathbf{Z}^n$.

We note that $\mathbf{dig}_U$ is a dilation $\mathscr{P}(\mathbf{R}^n) \to \mathscr{P}(\mathbf{Z}^n)$ for any $U$, i.e.,

$$\mathbf{dig}_U(\bigcup A_j) = \bigcup \mathbf{dig}_U(A_j)$$

for any family $(A_j)$ of subsets of $\mathbf{Z}^n$. We even have

$$\mathbf{dig}_{\bigcup U_k}(\bigcup A_j) = \textstyle\bigcup_k \bigcup_j \mathbf{dig}_{U_k}(A_j).$$

We also note that the operation commutes with translations by an integer vector:

$$\mathbf{dig}_U(c+A) = c + \mathbf{dig}_U(A), \qquad c \in \mathbf{Z}^n,\ A \in \mathscr{P}(\mathbf{R}^n),$$

as well as the symmetry $\mathbf{dig}_U(A) = \mathbf{dig}_A(U)$.

Rosenfeld took $U$ as the cross $R$ defined in (7.1.1); see Figure 7.1 on page 127. Then the straight line $L$ in $\mathbf{R}^2$ defined by an equation $x_2 = F(x_1) = \alpha x_1 + \beta$ with $|\alpha| < 1$ gives rise to a function $f\colon \mathbf{Z} \to \mathbf{Z}$. Indeed, given $p_1 \in \mathbf{Z}$, there is one and only one $p_2$ such that $(p_1, p_2)$ belongs to $L+R$. Actually $p_2 = f(p_1) = \lceil \alpha p_1 + \beta - \frac{1}{2} \rceil$, so that this digitization of the real line with equation $x_2 = F(x_1)$ has the equation $p_2 = \lceil \alpha p_1 + \beta - \frac{1}{2} \rceil$. For each $p_1$ one chooses the integer closest to $\alpha p_1 + \beta$ if there is a unique closest integer, and, by convention, $\alpha p_1 + \beta - \frac{1}{2}$ if $\alpha p_1 + \beta$ is a half-integer (the choice between $\alpha p_1 + \beta - \frac{1}{2}$ and $\alpha p_1 + \beta + \frac{1}{2}$, made to obtain uniqueness, introduces, as already noted in Example 7.1.3, a certain asymmetry). The differences $(D_1 f)(p) = f(p_1+1) - f(p_1)$ form an ***upper mechanical word***, also called a $\beta$-sequence (see, e.g., Uscka-Wehlou 2009d), to be compared with the constant $F(x_1+1) - F(x_1) = \alpha$ for the original function. As we have seen in Section 8.2, Rosenfeld (1974) characterized the digitizations of a straight line segment when the Rosenfeld cross $R(p)$ is used for the digitization (this cross is defined in formula (7.1.1)). We shall now determine when the graph of a function $f\colon [c,d]_{\mathbf{Z}} \to \mathbf{Z}$ with $|D_1 f| \leqslant 1$, has the chord property.

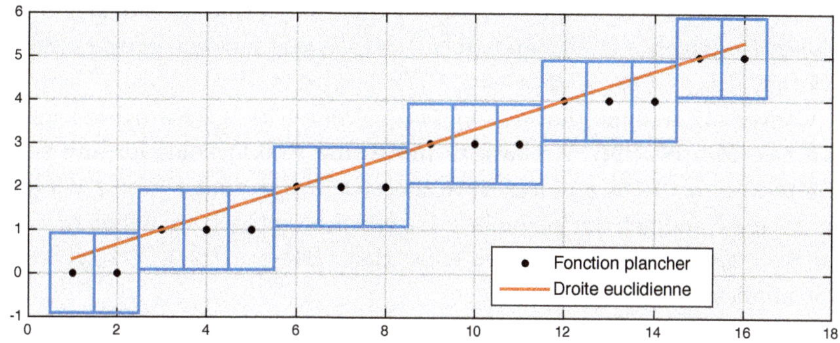


Figure 8.3. Covering the Euclidean straight line of equation $y = \frac{1}{3}x$ by a dilation of its discretization using the floor function and with structuring element equal to the rectangle $[-\frac{1}{2}, \frac{1}{2}] \times [-\frac{5}{6}, \frac{5}{6}]$. Jean-Marc Chassery & Isabelle Sivignon (2013) started this approach but used only squares as structuring elements. (Courtesy Adama Arouna Koné; see his paper 2017)

When $A$ is the graph of a function $f \in \mathscr{F}(\mathbf{Z}, \mathbf{Z})$ the chord property can be formulated as follows. Given $p < t < q$ with integers $p$ and $q$ and a real number $t$, we let $H\colon \mathbf{R} \to \mathbf{R}$ be the affine function which takes the values of $f$ at $p$ and $q$. Then the chord property says that

$$(8.6.1) \qquad |H(t) - f(\lfloor t \rfloor)| < 1 \text{ or } |H(t) - f(\lceil t \rceil)| < 1.$$

If $t$ happens to be an integer, this simplifes to

$$(8.6.2) \qquad |H(t) - f(t)| < 1.$$

**Theorem 8.6.1.** *Let $f\colon \mathbf{Z} \to \mathbf{Z}$ be a function with integer values. Then its graph has the chord property if and only if $|D_1 f(x)| \leqslant 1$ and $|J_{a,b} f(x)| < 1$ for all $(x, a, b) \in \mathbf{Z} \times \dot{\mathbf{N}} \times \dot{\mathbf{N}}$. The corresponding result holds also for a function defined on an interval $[c, d]_{\mathbf{Z}}$ or $[c, +\infty[_{\mathbf{Z}}$ or $]-\infty, d]_{\mathbf{Z}}$ of $\mathbf{Z}$.*

**Proof.** Assume first that $\mathbf{graph}(f)$ has the chord property. Then $|D_1 f(x)| \leqslant 1$, for otherwise the midpoint $(x + \frac{1}{2}, \frac{1}{2} f(x) + \frac{1}{2} f(x+1))$ of the segment

$$[(x, f(x)), (x+1, f(x+1))]$$

would not belong to $\mathbf{graph}(f) + U$.

Let $x < x + a < x + a + b$ be given with $x \in \mathbf{Z}$, $a, b \in \dot{\mathbf{N}}$. Then $(J_{a,b} f)(x) = H(x+a) - f(x+a)$, where $H$ is the affine function which takes the values $f(x)$ at $x$ and $f(x+a+b)$ at $x+a+b$. In the chord property we take $p = x$, $t = x + a$, $q = x + a + b$. It follows that $(x + a, H(x + a))$ belongs to $\mathbf{graph}(f) + U$, and since $t = x + a$ is now an integer, (8.6.2) says that $|(J_{a,b}f)(x)| = |H(x + a) - f(x + a)| < 1$.

Conversely, assume that $|(J_{a,b}f)(x)| = |H(x+a) - f(x+a)| < 1$ for all $(x, a, b) \in \mathbf{Z} \times \dot{\mathbf{N}} \times \dot{\mathbf{N}}$. We have to prove that (8.6.1) holds for any real $t$ with $p < t < q$. If $t$ is an integer we choose $x, a, b$ so that $p = x$, $t = x + a$, $q = x+a+b$ and get the inequality. If $t$ is not an integer, we define $t_0 = \lfloor t \rfloor$ and $t_1 = t_0 + 1 = \lceil t \rceil$. Then we know that $|H(t_j) - f(t_j)| < 1$, $j = 0, 1$. This implies that, for $j = 0, 1$,

$$v_0 = \min(f(t_0), f(t_1)) - 1 \leqslant f(t_j) - 1 < H(t_j)$$
$$< f(t_j) + 1 \leqslant \max(f(t_0), f(t_1)) + 1 = v_1,$$

where $v_0$ and $v_1$ are defined by the equalities. Thus $H(t_0)$ and $H(t_1)$ both belong to the open interval $]v_0, v_1[$, which implies that $H(t)$, which is obtained by interpolation between $H(t_0)$ and $H(t_1)$, is also in this interval, in other words that the point $(t, H(t))$ belongs to the open rectangle $\Omega = ]t_0, t_1[ \times ]v_0, v_1[$.

We now invoke the other hypothesis, viz. that $|D_1 f| \leqslant 1$, which implies that $v_1 - v_0 \leqslant 3$ and hence that $\Omega$ is a subset of the dilation $\mathbf{graph}(f) + U$. Thus finally $(t, H(t)) \in \mathbf{graph}(f) + U$. We are done. □

## 8.7 Characterization of straightness: Balanced words

**Theorem 8.7.1.** *A function $f \in \mathscr{F}(\mathbf{Z}, \mathbf{Z})$ with $0 \leqslant D_1 f \leqslant 1$ satisfies*

(8.7.1) $$|D_b D_a f(x)| \leqslant 1, \qquad x \in \mathbf{Z}, \quad a, b \in \dot{\mathbf{N}},$$

*if and only if the binary word $D_1 f$ is balanced.*

For the proof we recall some concepts from word theory. By a ***word*** we understand here a doubly infinite sequence $(w_j)_{j\in\mathbf{Z}}$ of letters $w_j$; it is ***binary*** if there are only two letters; we shall then take them as 0 and 1. (Often one studies words that are infinite in only one direction, $(w_j)_{j\in\mathbf{N}}$.)

A ***factor*** $w' = (w_j)_{j=p}^{q}$ of a word $w$ is said to have ***length*** $q - p + 1$:

$$\text{length}(w') = q - p + 1.$$

The empty word $\varepsilon = (w_j)_{j=p}^{p-1}$ has length 0.

If $w$ is binary, the number of ones in a factor $w' = (w_j)_{j=p}^{q}$ is called its ***height***:

$$\mathbf{height}(w') = \sum_{j=p}^{q} w_j.$$

A function $f \in \mathscr{F}(\mathbf{Z}, \mathbf{Z})$ is said to have ***chain code*** $c = c(f) = (c_j)_{j\in\mathbf{Z}}$, where

$$c_j = f(j+1) - f(j) = D_1 f(j), \qquad j \in \mathbf{Z}.$$

Conversely, every sequence $(c_j)_{j\in\mathbf{Z}}$ determines a family of functions having this chain code; we take

$$f(x) = C + \sum_{j=0}^{x-1} c_j \text{ for } x \geqslant 0 \text{ and } f(x) = C - \sum_{j=x}^{-1} c_j \text{ for } x < 0,$$

where $C$ is an arbitrary constant (equal to $f(0)$).

A binary word $w$ is said to be ***balanced*** if for any two factors $w'$ and $w''$ of $w$ we have

(8.7.2) $\text{length}(w') = \text{length}(w'')$ implies $|\,\mathbf{height}(w') - \mathbf{height}(w'')| \leqslant 1.$

Let now $w' = (w_j)_{j=p'}^{q'}$, $w'' = (w_j)_{j=p''}^{q''}$ be two factors of the same binary word $w$. That they have the same length means that $q'-p'+1 = q''-p''+1$. Their heights are

$$\mathbf{height}(w') = \sum_{j=p'}^{q'} w_j, \quad \mathbf{height}(w'') = \sum_{j=p''}^{q''} w_j.$$

Now, writing $w_j = D_1 f(j)$, we obtain

$$\mathbf{height}(w') = \sum_{j=p'}^{q'} D_1 f(j) = D_a f(p'), \qquad \text{where } a = q' - p' + 1.$$

**Proof of Theorem 8.7.1.** Given $f$, let $w' = (w_j)_{j=p'}^{q'}$ and $w'' = (w_j)_{j=p''}^{q''}$ be two factors of the same length of the binary word $w = D_1 f$. For reasons of symmetry we may assume that $p' \leqslant p''$. Define $x = p'$, $a = q' - p' + 1 = q'' - p'' + 1$, the common length of the intervals, and $b = p'' - p' = q'' - q'$, the distance between their left endpoints. Then $x + a = q' + 1$, $x + b = p''$, and $x + a + b = q'' + 1$, so that

$$\mathbf{height}(w'') - \mathbf{height}(w') = D_a f(p'') - D_a f(p') = D_b D_a f(p') = D_a D_b f(p').$$

We see that condition (8.7.1) translates directly to condition (8.7.2). □

Thus the equality $D_1 D_1 f = 0$ or $D_b D_a f = 0$ for all $a, b \in \dot{\mathbf{N}}$ for functions in $\mathscr{F}(\mathbf{Z}, \mathbf{R})$ is replaced by the inequality (8.7.1) for functions in $\mathscr{F}(\mathbf{Z}, \mathbf{Z})$, which we can understand as a kind of approximate equality. Note that we require this inequality for $(x, a, b) \in \mathbf{Z} \times \dot{\mathbf{N}} \times \dot{\mathbf{N}}$; it seems that this requirement cannot be considerably weakened (except of course that we may restrict attention to $0 < a \leqslant b$).

## 8.8 Hyperplanes in the sense of Reveillès

Jean-Pierre Reveillès (1991:45) introduced digital lines in the digital plane as solutions to a double Diophantine inequality: he considered sets of the form

$$\{x \in \mathbf{Z}^2; \beta \leqslant \alpha_1 x_1 + \alpha_2 x_2 < \gamma\}, \tag{8.8.1}$$

where $\alpha_1$ and $\alpha_2$ are real numbers, not both of them zero, and $\beta$ and $\gamma$ are real numbers. We shall refer to such a set as a ***digital straight line in the sense of Reveillès***. He considers in particular the case when $\alpha_1$ and $\alpha_2$ are integers; then he calls the digital line *rational*—indeed, if $\alpha_2 \neq 0$, its slope $-\alpha_1/\alpha_2$ is a rational number. It is obvious how to generalize this definition to higher dimensions: we then speak about a ***digital hyperplane in the sense of Reveillès***.

## 8.9 Refined digital hyperplanes

Let us define slabs in $\mathbf{R}^n$:

$$\begin{aligned} T &= T(\alpha,\beta,\gamma) = \{x \in \mathbf{R}^n; \beta \leqslant \alpha \cdot x \leqslant \gamma\}, \\ T^* &= T^*(\alpha,\beta,\gamma) = \{x \in \mathbf{R}^n; \beta \leqslant \alpha \cdot x < \gamma\}, \\ T_* &= T_*(\alpha,\beta,\gamma) = \{x \in \mathbf{R}^n; \beta < \alpha \cdot x \leqslant \gamma\}, \\ T_*^* &= T_*^*(\alpha,\beta,\gamma) = \{x \in \mathbf{R}^n; \beta < \alpha \cdot x < \gamma\}. \end{aligned} \tag{8.9.1}$$

We shall also need to talk about the real hyperplanes

$$T^0 = \{x \in \mathbf{R}^n; \alpha \cdot x = \beta\}, \qquad T^1 = \{x \in \mathbf{R}^n; \alpha \cdot x = \gamma\}.$$

If $\beta \leqslant \gamma$, we have $T = T_*^* \cup T^0 \cup T^1$.

A digital hyperplane $D$ in the sense of Reveillès is of the form $T^*(\alpha,\beta,\gamma) \cap \mathbf{Z}^n$ and therefore satisfies

$$(T \cap \mathbf{Z}^n) \smallsetminus D \subset T^0,$$

i.e., the points in $T \cap \mathbf{Z}^n$ not in $D$ all belong to a single real hyperplane in $\mathbf{R}^n$.

In my paper (2004:456) this was generalized to the following.

**Definition 8.9.1.** Let us denote by $\pi_k \colon \mathbf{Z}^n \to \mathbf{Z}^{n-1}$ the projection which forgets the coordinate $x_k$, $k = 1, \dots, n$. A set $D$ is a ***refined digital hyperplane*** if

(1) $D$ is such that all points with integer coordinates in $\mathbf{cvxh}(D)$ belong to $D$;

(2) $T^*_* \cap \mathbf{Z}^n \subset D \subset T \cap \mathbf{Z}^n$ for some choice of $\alpha \in \mathbf{R}^n \smallsetminus \{0\}$ and $\beta, \gamma \in \mathbf{R}$, and, finally,

(3) For some $k$, the sets $\pi_k(D \cap T^0)$ and $\pi_k(D \cap T^1)$ are disjoint and together fill all of $\pi_k((T^0 \cup T^1) \cap \mathbf{Z}^2)$. □

In two dimensions this definition can be expressed in a simple way. We take $n = 2$, $(\alpha_1, \alpha_2) = (-\alpha, 1)$ and define strips in $\mathbf{R}^2$ as follows.

$$
\begin{aligned}
S(\alpha,\beta,\gamma) &= T = \{x \in \mathbf{R}^2; \beta \leqslant x_2 - \alpha x_1 \leqslant \gamma\},\\
S^*(\alpha,\beta,\gamma) &= T^* = \{x \in \mathbf{R}^2; \beta \leqslant x_2 - \alpha x_1 < \gamma\},\\
S_*(\alpha,\beta,\gamma) &= T_* = \{x \in \mathbf{R}^2; \beta < x_2 - \alpha x_1 \leqslant \gamma\},\\
S^*_*(\alpha,\beta,\gamma) &= T^*_* = \{x \in \mathbf{R}^2; \beta < x_2 - \alpha x_1 < \gamma\}.
\end{aligned}
\tag{8.9.2}
$$

Then a straight line in $\mathbf{Z}^2$ in the sense of Reveillès is, possibly after a permutation and a sign change of the coordinates, equal to the intersection $S^*(\alpha,\beta,\gamma) \cap \mathbf{Z}^2$, for some $\alpha, \beta, \gamma$, $|\alpha| \leqslant 1$. (We could as well have used $S_*(\alpha,\beta,\gamma)$ here, for $S_*(\alpha,\beta,\gamma) = -S^*(\alpha,-\gamma,-\beta)$.)

A refined digital line with $|\alpha| \leqslant 1$ and $\gamma = \beta + 1$ is either a digital line in the sense of Reveillès or, possibly after a reflection, of the form

$$
\begin{aligned}
D(\alpha,\beta,p) = \{&x \in \mathbf{Z}^2 \cap S^*(\alpha,\beta,\beta+1);\ x_1 < p\}\\
&\cup \{x \in \mathbf{Z}^2 \cap S_*(\alpha,\beta,\beta+1);\ x_1 \geqslant p\}
\end{aligned}
\tag{8.9.3}
$$

for some $p \in \mathbf{Z}$. This is because the only pairs of the form $(A, \complement A)$ with both $A$ and $\complement A$ with the property that all points with integer coordinates in $\mathbf{cvxh}(A)$ belong to $A$ and all points with integer coordinates in $\mathbf{cvxh}(\complement A)$ belong to $\complement A$ are the pairs $(\mathbf{Z}, \varnothing)$, $(\varnothing, \mathbf{Z})$ and $(]-\infty, p[_{\mathbf{Z}}, [p, +\infty[_{\mathbf{Z}})$, $p \in \mathbf{Z}$.

**Theorem 8.9.2.** *Every digital line in the sense of Reveillès is a refined digital line.*

*Conversely, given $|\alpha| \leqslant 1$ and $\beta$ real, we consider four cases for the set*

$$D = S(\alpha,\beta,\beta+1) \cap \mathbf{Z}^2,$$

*defining*

$$D^j = \{x \in D; x_2 - \alpha x_1 = \beta + j\}, \qquad j = 0, 1:$$

($\alpha'$) *The slope $\alpha$ is rational and $\beta \in \mathbf{Z} + \alpha\mathbf{Z}$. Then $D^0$ and $D^1$ contain infinitely many points and $D$ is not a refined digital line. For any integer $p$, the set $D(\alpha,\beta,p)$, obtained by removing from $D$ certain points in $D^0 \cup D^1$ (see* (8.9.3)*), is a refined digital line. The sets $D \smallsetminus D^0$ and $D \smallsetminus D^1$ are digital lines in the sense of Reveillès.*

($\beta'$) *The slope $\alpha$ is rational and $\beta \notin \mathbf{Z} + \alpha\mathbf{Z}$ (for instance when $\beta$ is irrational). Then $D^0$ and $D^1$ are empty, so that $D = S_*^*(\alpha, \beta, \beta + 1) \cap \mathbf{Z}^2$ and $D$ is a digital straight line in the sense of Reveillès.*

($\gamma'$) *The slope $\alpha$ is irrational and $D^0$ is empty. Then*

$$D = S(\alpha, \beta, \beta + 1) \cap \mathbf{Z}^2 = S_*^*(\alpha, \beta, \beta + 1) \cap \mathbf{Z}^2$$

*is a digital straight line in the sense of Reveillès.*

($\delta'$) *The slope $\alpha$ is irrational and $D^0$ is a singleton set. Then $D^1$ is also a singleton set, and $D$ is not a refined digital line. But $D \smallsetminus D^0$ and $D \smallsetminus D^1$ are digital straight lines in the sense of Reveillès.*

Thus in cases ($\beta'$), ($\gamma'$) and ($\delta'$) the two concepts coincide; in case ($\alpha'$) they are different. In cases ($\alpha'$) and ($\delta'$) we have to remove certain points in the bounding lines $D^0$, $D^1$ to obtain what we want, while this is not necessary in cases ($\beta'$) and ($\gamma'$).

For case ($\alpha'$), cf. Example 11.14.4; for case ($\gamma'$) take $\alpha = 1/\sqrt{2}$ and $\beta = \frac{1}{2}$; for case ($\delta'$), cf. Example 11.14.5.

**Proof.** We note that generally $D^1 = D^0 + (0, 1)$, which means that, in order to get a digital line in the sense of Reveillès, we always have to remove one of $D^0$ and $D^1$ unless they are empty. Cases ($\alpha'$) and ($\beta'$) are then straightforward.

For cases ($\gamma'$) and ($\delta'$) we note that, since $\alpha$ is irrational, we cannot have two points $p$ and $q$ with $p_1 \neq q_1$ in $D^0$; otherwise the slope of the line, which is $\alpha = (q_2 - p_2)/(q_1 - p_1)$ would be rational. Thus $D^0$ and $D^1$ are either empty or singleton sets. □

## 8.10 Extending rectilinear segments

Let us consider functions defined on an interval: let $c$ and $d$ be two integers and consider functions $f\colon [c, d]_{\mathbf{Z}} \to \mathbf{Z}$. We can then form $D_b D_a f(x)$ only for $c \leqslant x \leqslant d - a - b$, $a, b \in \dot{\mathbf{N}}$. A natural question is whether the conditions $|D_b D_a f(x)| \leqslant 1$ for these finitely many $a, b, x$ are sufficient to ensure that $f$ represents a straight line segment; in other words, whether we can find an extension $g$ to all of $\mathbf{Z}$ of the function $f$ which satisfies the conditions everywhere. The answer is in the affirmative, but the extension is never unique.

We recall the definitions in (8.9.2) of various strips $S(\alpha, \beta, \gamma)$ etc. in $\mathbf{R}^2$. All these strips have height $\gamma - \beta$.

**Theorem 8.10.1.** *If $f\colon [c,d]_{\mathbf{Z}} \to \mathbf{Z}$ satisfies $|D_bD_af(x)| \leqslant 1$ for all $x, a, b$ for which the expression is defined, then its graph is contained in an open strip $S_*^*(\alpha, \beta, \gamma)$ with rational $\alpha$ and of height $\gamma - \beta < 1$. If a function $f\colon \mathbf{Z} \to \mathbf{Z}$ defined on the whole integer axis satisfies $|D_bD_af| \leqslant 1$, its graph is contained in a closed strip $S(\alpha, \beta, \beta + 1)$ of height* 1.

**Proof.** See Figure 8.2 on page 149. Given $f\colon [c,d]_{\mathbf{Z}} \to \mathbf{Z}$ with $c \leqslant d$ and a real number $\alpha$, there exist real numbers $\beta$ and $\gamma$ such that the strip $S(\alpha, \beta, \gamma)$ contains the graph of $f$. For every $\alpha$ we choose $\beta = \beta(\alpha)$ maximal and $\gamma = \gamma(\alpha)$ minimal. Then there is at least one $p$ and one $q$ such that $f(p) = \alpha p + \beta$ and $f(q) = \alpha q + \gamma$.

Next we vary $\alpha$ to minimize the height $\gamma(\alpha) - \beta(\alpha)$: we obtain a strip with smallest height which contains the graph of $f$. If $d - c \geqslant 1$ it is unique. Unless $d - c \leqslant 1$ (a trivial case), the graph has at least three points and it is clear that there must be either at least two points on the lower boundary and one on the upper boundary of the strip, or vice versa. For reasons of symmetry, we may assume that we have $p < s < q$ with $(p, f(p))$ and $(q, f(q)$ on the lower boundary, $(s, f(s))$ on the upper boundary, i.e., $f(p) = \alpha p + \beta$ and $f(q) = \alpha q + \beta$, while $f(s) = \alpha s + \gamma$. If for instance $(p, f(p))$ and $(s, f(s))$ are on the lower boundary and $(q, f(q))$ on the upper boundary, and there is no point on the upper boundary to the left of $q$ and no point on the lower boundary to the right of $s$, it is easy to see that there are strips with a smaller height. (Cf. the discussion on minimal strips in Section 8.2 above.)

The condition $|J_{a,b}f(x)| < 1$ can now be written, taking $x = p$, $a = s - p$, and $b = q - s$,

$$-1 < \frac{b}{a+b} f(p) - f(s) + \frac{a}{a+b} f(q) \leqslant 0,$$

which, if we insert the values of $a$, $b$, $f(p)$, $f(s)$, and $f(q)$, just says that $\beta \leqslant \gamma < \beta + 1$.

Therefore all points in the graph of $f$ lie in the closed strip $S(\alpha, \beta, \gamma)$, which is contained in the half-open strip $S^*(\alpha, \beta, \beta + 1)$. It is of course also contained in the open strip $S_*^*(\alpha, \beta - t, \gamma + t)$ of height $\gamma + t - (\beta - t) < 1$ for any small positive number $t$.

This means that $f(x) = \lceil \alpha x + \beta \rceil$ for all $x$ in the domain of definition of $f$.

We can choose $\alpha = (f(q) - f(p))/(q - p)$, a rational number. However, since $\gamma - \beta < 1$, we can also vary $\alpha$ in some interval and choose infinitely many rational or irrational values for the slope of the line.

Finally, if $f\colon \mathbf{Z} \to \mathbf{Z}$, we consider its restriction $f_k$ to the interval $[-k, k]_{\mathbf{Z}}$, $k \in \mathbf{N}$, and apply what we know about $f_k$: there is a strip $S(\alpha_k, \beta_k, \gamma_k)$ of height $\gamma_k - \beta_k < 1$. It is not difficult to show that, as $k \to +\infty$, the sequences $(\alpha_k)$, $(\beta_k)$ and $(\gamma_k)$ tend to limits $\alpha$, $\beta$ and $\gamma$ and that the graph of $f$ is contained in the closed strip $S(\alpha, \beta, \gamma)$. But we can only say that $\gamma - \beta \leqslant 1$; cf. Example 11.14.4. □

**Theorem 8.10.2.** *If the graph of a function* $f\colon \mathbf{Z} \to \mathbf{Z}$ *or* $f\colon [c, d]_{\mathbf{Z}} \to \mathbf{Z}$ *is contained in a half-open strip* $S^*(\alpha, \beta, \beta + 1)$ *or* $S_*(\alpha, \beta, \beta + 1)$, *then* $|(D_b D_a f)(x)| \leqslant 1$ *for all* $x$ *and* $a, b \in \dot{\mathbf{N}}$ *for which the expression is defined.*

**Proof.** If $\alpha x + \beta \leqslant f(x) < \alpha x + \beta + 1$ or $\alpha x + \beta < f(x) \leqslant \alpha x + \beta + 1$ we get

$$(D_b D_a f)(x) = f(x + a + b) - f(x + a) - f(x + b) + f(x)$$

$$< \alpha(x + a + b) + \beta + 1 - \alpha(x + a) - \beta - \alpha(x + b) - \beta + \alpha x + \beta + 1 = 2.$$

Since $(D_b D_a f)(x)$ is an integer for functions with integer values, we must have $(D_b D_a f)(x) \leqslant 1$. By symmetry, $(D_b D_a f)(x) \geqslant -1$. □

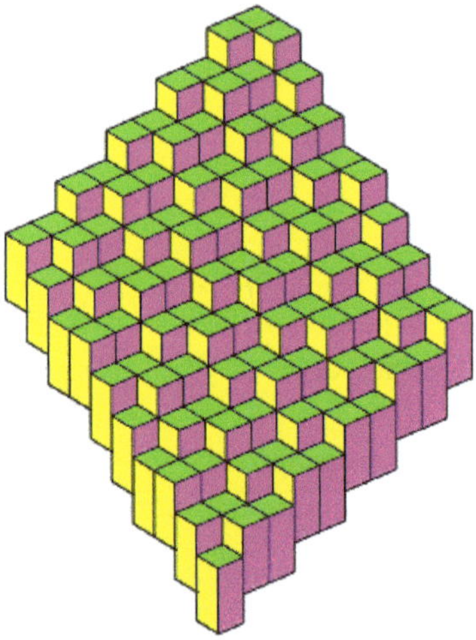

Figure 8.4. Covering a Euclidean plane by a dilation of its dicretization using the floor function and with structuring element equal to the box $[-\frac{1}{2}, \frac{1}{2}] \times [-\frac{1}{2}, \frac{1}{2}] \times [-\frac{9}{8}, \frac{9}{8}]$. (Courtesy Adama Arouna Koné)

**Theorem 8.10.3.** *Let $f\colon [c,d]_{\mathbf{Z}} \to \mathbf{Z}$ be given such that $|D_bD_af(x)| \leqslant 1$ for all $a$, $b$, $x$ for which the expression is defined, i.e., for $c \leqslant x \leqslant d-a-b$, $a,b \in \dot{\mathbf{N}}$. Then $f$ can be extended to a function $g\colon \mathbf{Z} \to \mathbf{Z}$ such that $|D_bD_ag(x)| \leqslant 1$ for all $x \in \mathbf{Z}$ and all $a,b \in \dot{\mathbf{N}}$.*

If we look at this as a combinatorial problem for chain codes, i.e., for binary words, the theorem says, in case $0 \leqslant D_1f \leqslant 1$, that a balanced finite binary word can be extended to a periodic balanced infinite word, moreover to infinitely many words with different periods—and also to infinitely many balanced nonperiodic infinite words.

**Proof.** Applying Theorem 8.10.1, we find a half-open strip $S^*(\alpha,\beta,\beta+1)$ or $S_*(\alpha,\beta,\beta+1)$ which contains the graph of $f$. In the first case $f(x) = \lceil \alpha x+\beta\rceil$, in the second $f(x) = \lfloor \alpha x+\beta+1\rfloor$ for all $x \in [c,d]_{\mathbf{Z}}$. We can now define the extension $g(x) = \lceil \alpha x+\beta\rceil$ or $g(x) = \lfloor \alpha x+\beta+1\rfloor$ for all $x \in \mathbf{Z}$. According to Theorem 8.10.2 we now have $|D_bD_ag| \leqslant 1$. □

**Example 8.10.4.** Let $f_k$ be the restriction to $[-k,k]_{\mathbf{Z}}$ of the Heaviside function $Y_0$ with any $k \in \dot{\mathbf{N}}$. (The Heaviside function is defined in (3.3.5).) Then the construction gives a slope $\beta_k = 1/(k+1) > 0$ and height $\gamma_k - \beta_k = k/(k+1) < 1$. Actually we may choose any $\beta_k$ with $0 < \beta_k \leqslant 1/(k+1)$ and still get $\gamma_k - \beta_k < 1$. But it is not possible to choose $\beta_k = 0$, for then $\gamma_k - \beta_k = 1$.

This means that the straight line constructed from a restriction of $f$ to a finite interval $[c,d]$ containing $-1$ and $0$ must always have a positive slope, although $f$ itself represents a line with slope zero. If we choose a rational slope, the chain code $D_1g$ of the extension will be periodic, while $D_1f$ is not. The function $Y_0$ can never appear as an extension in the construction in the proof of Theorem 8.10.3. See also Example 11.14.4 □

## 8.11 Notes

### 8.11.1 *Difference operators*

A pioneer in the theory of linear difference equation was George David Birkhoff (father of Garret Birkhoff) with his paper (1911).

An early, systematic study of difference operators, still of more than historical interest, is Niels Erik Nörlund's book (1924).

The characerizations of straightness using difference operators presented here I published in my paper (2011), with a short preview in (2010a), and followed up with estimates in (2015b).

With the advent of computers, numerical analysis has been enormously developed. In particular, difference operators are used to approximate differential operators. Often, however, researchers are content to develop an algorithm which approximates the solution to a differential equation with an error that is $O(h^2)$ of the step length $h$, not caring too much about exact solutions. However, in some publications, like that by Esbelin & Malgouyres (2016), a careful comparison between derivatives and differences are performed.

Quantum calculus, also known as $q$-calculus, is about a discretization where the derivative is replaced by the so called "$q$-derivative," which is the quotient

$$\frac{f(qx) - f(x)}{qx - x}, \qquad x \in \mathbf{R} \smallsetminus \{0\}, \tag{8.11.1}$$

for a given constant $q > 1$. Quantum calculus is used in theoretical physics.

### 8.11.2 *Sources*

Theorems 8.2.2 and 8.2.4 are due to Rosenfeld (1974). The proof of Theorem 8.2.4 given here is shorter than the original proof.

Theorem 8.6.1 is due to Samieinia (2010a: Theorem 3.1). She measures the deviation from a chord $[(p, f(p)), (q, f(q))]$ of the graph of $f$, where $p < s < q$, by what she calls a *boomerang* rather than by the Jensen operator. The boomerang in question is the union of the two segments $[(p, f(p)), (s, f(s))]$ and $[(s, f(s)), (q, f(q))]$ with $(s, f(s))$ as the vertex, and the vertical distance between the vertex of the boomerang and the chord is equal to $|J_{p-s,q-s}f(s)|$, which establishes the equivalence.

For digitizations of convex sets, see Chapters 10 and 11.

Rosenfeld & Kim (1982) explain the chord property and prove that it is equivalent to a collinearity property.

Dorst & Duin (1984) establish a link between digitized straight lines and the theory of numbers (Farey series, continued fractions).

### 8.11.3 *Etymology*

We have in Latin the noun *differentia*$_{(f)}$ 'difference' from the verb *differo* 'I carry in different directions', with present participle *differens, differentis*.

The Latin verb *derivare* means 'to divert', 'to turn away into another river', from which we have *derivative* 'diverted'; *rivus* means 'a (small) river'.

## 8.12 Exercises

8.1. Consider the sets

$$A = \{(0,0),(1,0),(2,0),(3,0),(4,1),(5,2)\}$$

and

$$B = \{(0,0),(1,0),(2,0),(3,1),(4,2),(5,3)\},$$

each with six points (cf. the sets $S$ and $T$, each with five points, in Figure 7.1 and the discussion in Example 8.2.3). Prove that these sets do not have the chord property.

8.2. Show that the assertion on page 150 that the condition $D_1D_1f = 0$ for functions in $\mathscr{F}(\mathbf{Z},\mathbf{Z})$ gives a class which is too narrow to be an acceptable class of straight lines; the equation has very few solutions.

8.3. Show that the assertion on page 150 that the condition $|D_1D_1f| \leqslant 1$ for functions in $\mathscr{F}(\mathbf{Z},\mathbf{Z})$ gives a class which is too wide to be of interest as an acceptable class of straight lines; many funny functions satisfy the inequality.

8.4. As a first step in quantum calculus, determine all functions $f\colon \mathbf{R} \to \mathbf{R}$ with $q$-derivative equal to zero for all $x \neq 0$; see (8.11.1). What about functions $f\colon \mathbf{Z} \to \mathbf{R}$?

# Chapter 9

# Convexity in vector spaces

## 9.1 Introduction

THE THEORY OF CONVEXITY of sets and functions in vector spaces is a highly developed and very rich theory.

Before we go on further with the discretization in Chapters 7 and 8, it seems that a survey of convexity in vector spaces could be useful.

Therefore, the purpose of the present chapter is to collect definitions and basic properties of convex sets and functions in vector spaces over the field of real numbers. On the one hand, this is a source of methods to be used also in discrete environments; on the other hand, it serves as a model for desirable results in discrete settings.

Here we present merely a tiny fraction of the classical theory, which, however, might be enough as a background for later chapters. Readers who are already familiar with convexity in vector spaces can skip the present chapter.

## 9.2 Defining convex sets

It is most convenient to define convex functions with the help of convex sets. This also has the advantage that we can treat functions with infinite values without difficulty. That is why we start now with convex sets.

**Definition 9.2.1.** A ***rectilinear segment*** in a vector space $E$ is the set

$$[a,b] = \{(1-t)a + tb;\ t \in \mathbf{R},\ 0 \leqslant t \leqslant 1\}, \tag{9.2.1}$$

where $a$ and $b$ are its ***endpoints***.

A subset $A$ of a vector space is said to be ***convex*** if

$$\{a,b\} \subset A \text{ implies } [a,b] \subset A. \tag{9.2.2}$$

A segment $[a, b]$ with endpoints $a$, $b$ in a given set will be called a ***chord*** of that set, and we define the ***chord set*** of any set $A$ as

$$\mathbf{chord}(A) = \bigcup_{a,b \in A} [a, b] \subset E \tag{9.2.3}$$

for $A \subset E$. □

Thus a set $A$ is convex if and only if

$$\mathbf{chord}(A) \subset A. \tag{9.2.4}$$

It is justified, I think, to call this property the ***chord property in the sense of Euclid.*** Indeed, Definition 4 in his first book of the Στοιχεῖα (Stoikheĩa) 'Elements' reads according to Heath: "A straight line is a line which lies evenly with the points on itself." (Euclid 1956:165). This can arguably be interpreted as $\mathbf{chord}(A) \subset A$, which together with the property of a line being a "breadthless length" (Definition 2; Euclid 1956:158) implies that the set is an εὐθεῖα (eutheĩa), a rectilinear segment or an infinite straight line.[1]

Every segment, every straight line, and all affine subspaces are convex sets.

It is worth noticing that to check convexity of a set, only its intersections with one-dimensional subspaces need to be considered. In other words, a set $A \subset E$ is convex if and only if the inverse image $f^*(A)$ is an interval in $\mathbf{R}$ for every mappling $f \colon \mathbf{R} \to E$ of the form $f(t) = ta + b$, $t \in \mathbf{R}$, $a, b \in E$.

A convex set $A$ in $\mathbf{R}^n$ need not have interior points, but it does if we consider it as a subset of the smallest affine space which contains it. We define the ***relative interior*** of $A$, denoted by $\mathbf{relint}(A)$, as the interior taken with respect to the topology in this affine subspace which is induced by the usual topology on $\mathbf{R}^n$. In this way every convex set, even if only a singleton set, has a nonempty relative interior. The set $\overline{A} \smallsetminus \mathbf{relint}(A)$ is the boundary of $A$ taken in the smallest affine space containing $A$ and is more interesting than the boundary of $A$ if the space mentioned is not all of $\mathbf{R}^n$.

## 9.3 Properties of convex sets

**Proposition 9.3.1.** *If $A$ and $B$ are convex subsets of a vector space, then also $A \cap B$, $A + B$ and $sA$ for $s \in \mathbf{R}$ are convex.*

---

[1]What Euclid actually meant with the word *eutheĩa* is the theme in my paper (2015a).

**Proof.** If $c, c'$ are two points in $A+B$, then $c=a+b$ and $c'=a'+b'$ for some $a, a' \in A$ and $b, b' \in B$. For $t \in [0,1]$ we have

$$(1-t)c+tc=\big((1-t)a+ta'\big)+\big((1-t)b+tb'\big) \in A+B,$$

proving that $A+B$ is convex. The other properties have even simpler proofs. □

**Proposition 9.3.2.** *If $(A_j)_{j\in J}$ is an increasing family of convex subsets of a vector space indexed by a directed ordered family $J$, then also their union $A=\bigcup_{j\in J} A_j$ is convex.*

**Proof.** If $a, b$ are two points in $A$, then $a \in A_i$ and $b \in A_j$ for some $i, j \in J$. By hypothesis there is an index $k \in J$ such that and $i, j \leqslant k$. So $a \in A_i \subset A_k$ and $b \in A_j \subset A_k$; therefore also $(1-t)a+tb \in A_k \subset A$ for $0 \leqslant t \leqslant 1$. □

## 9.4 The convex hull

**Definition 9.4.1.** Given a subset $A$ of a vector space $E$, we define the ***convex hull*** of $A$ as the smallest convex set containing $A$. It will be denoted by $\mathbf{cvxh}(A)$. □

The convex hull is well defined since any intersection of convex sets is convex.

The operation **cvxh** is increasing, idempotent, and extensive, in other words, a cleistomorphism in $\mathscr{P}(E)$. If we instead regard it as a mapping from the complete lattice $\mathscr{P}(E)$ to the complete lattice of all convex subsets of $E$, it is a dilation. The operation **chord**, on the other hand, is increasing and extensive, but not idempotent in dimension $n \geqslant 2$.

In one dimension we have **chord** = **cvxh**; in $\mathbf{R}^2$ we have, in view of Carathéodory's theorem (Theorem 9.7.1),

$$A \subset \mathbf{chord}(A) \subset \mathbf{chord}(\mathbf{chord}(A)) = \mathbf{cvxh}(A).$$

In $\mathbf{R}^n$ we need to take the operation **chord** $n$ times to arrive at $\mathbf{cvxh}(A)$.

We also note that

$$\mathbf{chord}(A) \subset A \;\Leftrightarrow\; \mathbf{cvxh}(A) \subset A \;\Leftrightarrow\; \mathbf{cvxh}(A) = A.$$

## 9.5 The Hahn–Banach theorem

Among the affine subspaces we will pay attention to the ***hyperplanes***, those of codimension 1, which means that they are defined by a single equation $\xi(x) = c$, where $\xi$ is a nonzero linear form on the vector space. Other important sets are the ***half-spaces***, which are defined by an inequality $\xi(x) \geqslant c$ or $\xi(x) > c$.

A ***topological vector space*** is a vector space $E$ equipped with a topology such that both ***addition***

$$E \times E \ni (x, y) \mapsto x + y \in E$$

and ***multiplication by scalars***

$$\mathbf{R} \times E \ni (t, x) \mapsto tx \in E$$

are continuous.

After having restricted some variables we see that all translations $x \mapsto x + a$ are continuous as well as all mappings $t \mapsto ta$ for a fixed $a$. The latter property implies that the inverse image of an open set in $E$ under the mapping $t \mapsto ta$ is open for the usual topology in $\mathbf{R}$.

As already mentioned in Section 2.7, an ordered set $X$ is said to be totally ordered if for any two elements $x, y \in X$ we have $x \leqslant y$ or $y \leqslant x$. An ordered set $X$ is said to be ***inductive*** or ***inductively ordered*** (Bourbaki 1963:34) if every totally ordered subset of $X$ possesses a majorant in $X$.[2] This means that for every $Y \subset X$ which is totally ordered, there exists an element $b \in X$ such that $y \leqslant b$ for all $y \in Y$. This concept is of interest because it is used as an hypothesis in the Kuratowski–Zorn lemma, which guarantees the existence of maximal elements.

**Theorem 9.5.1** (The Kuratowski–Zorn lemma). *Every inductively ordered set possesses a maximal element.* □

We shall accept this result here. One can prove it using the Axiom of Choice; conversely, the latter can be proved from the Kuratowski–Zorn lemma (see, e.g., Ciesielski 1997: Theorem 4.3.4 and Section 4.4).

To establish the existence of a maximal element we shall have to prove that a certain order is inductive.

The example of interest here is the family of all vector subspaces of a vector space, with the order given by inclusion: $L \leqslant M$ if and only if

---

[2]The empty set is totally ordered, so an inductively ordered set must be nonempty.

$L \subset M$. It is clearly an inductive order: If $\mathscr{L}$ is a totally ordered family of vector subspaces, then the union of all members of $\mathscr{L}$ is a maximal element.

**Theorem 9.5.2** (The Hahn–Banach theorem). *Every open convex set in a topological vector space $E$ is the intersection of a family of open half-spaces.*

*Every closed convex set in a topological vector space is the intersection of a family of closed half-spaces—and also the intersection of a family of open half-spaces.*

This theorem gives us a duality, or polarity, between, on the one hand, convex sets which are either open or closed, and, on the other hand, open half-spaces being given by nonzero continuous linear forms, elements of the dual space. Actually the class of sets which are intersections of a family of open half-spaces is larger than those mentioned: Werner Fenchel (1952) called this class the *evenly convex sets* and proved that the duality works for them, just as in the form of the Hahn–Banach Theorem stated here. All sets which are strictly convex in the sense of Definition 9.11.1 are evenly convex. So are sets like a triangle in $\mathbf{R}^2$ with one vertex removed, for example the set defined by $0 < x_1 + x_2 \leqslant 1$, $x_1 \geqslant 0$, $x_2 \geqslant 0$. An open triangle with one boundary point added is not evenly convex. This observation leads us to define the refined half-spaces below; see Definition 9.5.3.

**Proof of Theorem 9.5.2.** Given a nonempty open and convex set $A$ and a point $c \notin A$ we shall prove that there is an open half-space which contains $A$ but not $c$. We may take $c = 0$. Look at the family of all vector subspaces which do not meet $A$. This is an inductively ordered family, and it has a maximal element $M$. We claim that $M$ is of codimension 1. We construct the quotient space $E/M$ and denote the quotient mapping $E \to E/M$ by $\psi$. We shall thus prove that $E/M$ is of dimension 1. If $E/M$ were of dimension 2 or higher, we could find a line $L$ passing through the origin and not meeting the image $\psi_*(A)$ of $A$ in $E/M$.

To see this, we take any two-dimensional subspace $E_2$ of $E/M$ and look at the rays $R(\theta) = \{(r\cos\theta, r\sin\theta);\ r \geqslant 0\}$, $\theta \in \mathbf{R}$, for some coordinate system in $E_2$. Let $U$ be the open set of all $\theta \in \mathbf{R}$ such that $R(\theta)$ cuts the image $\psi_*(A)$ of $A$ in $E/M$. Because of the convexity of $\psi_*(A)$ we cannot have $\theta \in U$ and $\theta + \pi \in U$ at the same time. Hence both $U$ and its complement must be nonempty; pick any $\theta_0 \notin U$ and define $\theta_1 = \inf_{\theta \geqslant \theta_0}(\theta;\ \theta \in U)$. Then we must have $\theta_1 \notin U$ and $\theta_1 + \pi \notin U$, so that the whole line $L = R(\theta_1) \cup R(\theta_1 + \pi)$ is disjoint from $\psi_*(A)$. The inverse image $\psi^*(L)$ of $L$ is a vector subspace of $E$ strictly containing $M$, contradicting its maximality.

If $A$ is closed and nonempty and $p \notin A$, we take a point $a \in A$ and then a point $c \neq p$ on the segment $[p, a]$ which does not belong to $A$. We then proceed as above for an open set $A$; we find that $c$ does not belong to an open half-space containing $A$. The closure of this half-space does not contain $p$. □

**Definition 9.5.3.** Let us say that a subset $Y$ of a vector space $E$ is a ***refined half-space*** if it is convex and satisfies

$$\{x \in E;\ \xi(x) < c\} \subset Y \subset \{x \in E;\ \xi(x) \leqslant c\}$$

for some nonzero $\xi$ in the algebraic dual to $E$. Let us denote by $\mathscr{Y}$ the family of all such sets $Y$. □

We can now define a cleistomorphism $\kappa_{\mathscr{Y}}$ by defining $\kappa_{\mathscr{Y}}(A)$ to be the intersection of all refined half-spaces containing $A$. It is easy to see that $\kappa_{\mathscr{Y}} = \mathbf{cvxh}$.

## 9.6 Supporting hyperplanes

**Definition 9.6.1.** Given any set $A$ in a topological vector space $E$, a ***supporting hyperplane*** is a hyperplane $Y$ such that $A$ is contained in one of the closed half-spaces defined by $Y$ and such that the closure of $A$ meets $Y$. □

**Theorem 9.6.2.** *Let $A$ be a convex subset of a topological vector space $E$ with $\varnothing \neq A \neq E$ and let $a$ be any boundary point of $A$. Then there exists a supporting hyperplane of $A$ passing through $a$.*

**Proof.** In view of the Hahn–Banach theorem there is a closed hyperplane $Y$ which passes through $a$ and such that $A$ is in one of the closed half-spaces defined by $Y$. □

A set $A$ in a topological vector space is said to be ***bounded*** if for any neighborhood $U$ of the origin there exists a number $\lambda_0$ such that $\lambda U$ contains $A$ for all $\lambda$ with $|\lambda| \geqslant \lambda_0$.

**Corollary 9.6.3.** *Let $A$ be a bounded nonempty subset of a topological vector space and let $a$ be any boundary point of* $\mathbf{cvxh}(A)$. *Then there exists a supporting hyperplane of $A$ passing through $a$.*

**Proof.** We apply the theorem to $\mathbf{cvxh}(A)$. A supporting hyperplane of $\mathbf{cvxh}(A)$ must also be a supporting hyperplane of $A$. □

## 9.7 Carathéodory's theorem

It is easy to see that the convex hull of a set $A$ is given as the set of all finite sums

$$\sum_{j=1}^{N} \lambda_j a^{(j)}, \tag{9.7.1}$$

where the $a^{(j)}$ belong to $A$, the scalars $\lambda_j$ satisfy $\lambda_j > 0$, $\sum \lambda_j = 1$, and where $N$ can be any positive integer. In a space of infinite dimension, the number $N$ must sometimes be taken arbitrarily large, but in $\mathbf{R}^n$ it is enough to take $N = n+1$. This is the result of Carathéodory's theorem:

**Theorem 9.7.1** (Carathéodory's theorem). *For any subset $A$ of $\mathbf{R}^n$, its convex hull is equal to the set of all points*

$$x = \sum_{j=1}^{n+1} \lambda_j a^{(j)}, \tag{9.7.2}$$

*where the $a^{(j)}$ belong to $A$, and the scalars $\lambda_j$ satisfy $\lambda_j > 0$ and $\sum \lambda_j = 1$.*

**Proof.** The theorem is visibly true for $n = 0$ and $n = 1$. We shall assume that it is true for $n-1$ and deduce that it must be true for $n$. Denote by $A_N$ the set of all points which can be represented as in (9.7.1) for a certain $N$. So $\mathbf{cvxh}(A) = \bigcup_N A_N$ and we shall prove that $\mathbf{cvxh}(A)$ is equal to $A_{n+1}$.

Let now $x$ be an arbitrary point in $A_N$. If $x = a^{(1)} \in A_1$, we are through; if not, we consider the straight line passing through $a^{(1)}$ and $x$. The set $A_N$ is closed and bounded and therefore there must exist a point $b$ belonging to the boundary of $A_N$ such that $x$ lies between $a^{(1)}$ and $b$. Thus $x$ belongs to the convex hull of $\{a^{(1)}, b\}$. Now $b$ lies in a supporting hyperplane $Y$ of $A_N$ (see Theorem 9.6.2 above), where all the points $a^{(j)}$ must lie in $Y$. Since $Y$ is of dimension $n-1$, $b$ belongs to $A_n$ by the induction hypothesis. It follows that $x$ belongs to $A_{n+1}$. □

## 9.8 Approximation of the convex hull

By taking iterated Minkowski sums of a set we can approximate the convex hull:

**Theorem 9.8.1** (Rådström's Approximation Theorem). *Let $A$ be any set in a vector space $E$ and define for $k \in \dot{\mathbf{N}}$,*

$$A_{[k]} = \frac{1}{k}(A + A + \cdots + A) \ (k \text{ terms}).$$

*Then* $A \subset A_{[k]} \subset \mathbf{cvxh}(A)$ *with equality if* $A$ *is convex.*

*If* $E = \mathbf{R}^n$ *and* $A$ *is bounded, then* $A_{[k]}$ *converges to* $\mathbf{cvxh}(A)$ *in the sense that, for every positive number* $r$*, there exists an index* $k_r$ *such that*

$$\mathbf{cvxh}(A) \subset A_{[k]} + B_{\leqslant}(0,r) \subset \mathbf{cvxh}(A) + B_{\leqslant}(0,r), \qquad k \geqslant k_r.$$

We note that the sets $A$, $A_{[k]}$, and $\mathbf{cvxh}(A)$ all have the same support function (Definition 9.17.1 below).

**Proof.** The set $kA$ is always a subset of the sum of $k$ sets $A$, and equality holds if $A$ is convex.

For the second part of the proposition we fix a positive integer $k$ and consider an arbitrary point $p$ in the convex hull of $A$. We have, in view of Carathédory's theorem, Theorem 9.7.1, $p = \sum_1^{n+1} \lambda_j a^{(j)}$ for some points $a^{(j)}$ in $A$ and some numbers $\lambda_j > 0$ with $\sum \lambda_j = 1$. (If some of the $\lambda_j$ are irrational, we have to work a little more.) Define integers

$$s_j = \lfloor k\lambda_j \rfloor \text{ and } u_j = \lceil k\lambda_j \rceil .$$

Then

$$k\lambda_j - 1 < s_j \leqslant k\lambda_j \leqslant u_j < k\lambda_j + 1,$$

from which we conclude that

$$k - n - 1 < \sum s_j \leqslant k \leqslant \sum u_j < k + n + 1.$$

We have either $u_j = s_j$ or $u_j = s_j + 1$, implying that, with $k_s = \sum s_j$ and $k_u = \sum u_j$,

$$k - n \leqslant k_s \leqslant k \leqslant k_u \leqslant k + n.$$

If $k_s$ happens to be equal to $k$, we are finished, since then the point $b = (1/k)\sum s_j a^{(j)}$ belongs to $A_{[k]}$ (and also to $\mathbf{cvxh}(A)$), and the distance $\|p - b\|$ between $b$ and $p$ is not larger than

$$\left\|\sum (\lambda_j - s_j/k) a^{(j)}\right\| \leqslant \sum (\lambda_j - s_j/k)\|a^{(j)}\| \leqslant \frac{n+1}{k} \sup_{a\in A} \|a\|,$$

which is smaller than $r$ for $k$ large enough.

If it so happens that $k_u = k$, we can conclude similarly with $d = (1/k)\sum u_j a^{(j)} \in A_{[k]}$ .

The case when $k_s < k < k_u$ remains to be considered. Let us denote by $I$ the set of indices $j$ such that $u_j = s_j + 1$. We have $\mathbf{card}(I) = k_u - k_s$. We choose a subset $J$ of $I$ of cardinality $k - k_s \leqslant k_u - k_s$. Then we define for $j \in J$, a new integer $t_j = u_j = s_j + 1$. For all other values of $j$ we define $t_j = s_j$. Then $\sum t_j = k$ and we can proceed with $c = (1/k)\sum t_j a^{(j)} \in A_{[k]}$.

□

## 9.9 Defining convex functions

**Definition 9.9.1.** A function $f\colon E \to \mathbf{R}_!$ defined in a vector space $E$ is said to be ***convex*** if its finite epigraph is convex as a subset of $E \times \mathbf{R}$. We shall write $CVX(E, \mathbf{R}_!)$ for the set of these functions. □

The functions that take one of the values $-\infty$, $+\infty$ identically are convex, since their finite epigraphs are, respectively, the whole space $E \times \mathbf{R}$ and the empty set. An affine function is convex, since its finite epigraph is a closed half-space.

It is easy to see that a function $f\colon E \to \mathbf{R}_!$ is convex if and only if it satisfies ***Jensen's inequality***

$$f((1-t)x+ty) \leqslant (1-t)f(x) \dotplus tf(y), \quad x, y \in E,\ 0 \leqslant t \leqslant 1. \tag{9.9.1}$$

Here we define $0 \cdot (+\infty) = 0 \cdot (-\infty) = 0$ (or else consider only $0 < t < 1$). There are more general inequalites to be described now.

**Definition 9.9.2.** A ***Jensen function*** on a real vector space $E$ is a function $\mu\colon E \to \mathbf{R}$ of the form

$$\mu = \sum_{j=1}^{N} \lambda_j \delta_{a^{(j)}} - \delta_0, \tag{9.9.2}$$

where the scalars $\lambda_j$ and the points $a^{(j)}$ satisfy the conditions

$$\lambda_j > 0, \quad j = 1, \dots, N, \qquad \sum_{j=1}^{N} \lambda_j = 1, \text{ and } \sum_{j=1}^{N} \lambda_j a^{(j)} = 0. \tag{9.9.3}$$

The set of all Jensen functions on $E$ will be denoted by $\mathscr{J}(E)$. □

A function is convex if and only if it satisfies the more general inequality, also called Jensen's inequality,

$$\mu * f \geqslant 0 \tag{9.9.4}$$

for all Jensen functions $\mu$. We define here the convolution $\mu * f$ using upper addition, meaning that the sum should take the value $+\infty$ as soon as one of the terms has this value.

**Definition 9.9.3.** We shall say that $\mu\colon E \to \mathbf{R}$ is a ***generalized Jensen function*** if

($\alpha'$) $\mu$ is nonzero at finitely many points only;

($\beta'$) $\sum_{x \in E} \mu(x) = 0$;

($\gamma'$) $\sum_{x \in E} \mu(x)x = 0$;

($\delta'$) $\sum_{x \in E} \mu(x)g^+(x) \geqslant 0$ for all affine functions $g$.

The set of all generalized Jensen functions will be denoted by $\mathscr{J}_{\text{gen}}(E)$. □

Condition ($\beta'$) means that the positive part $\mu^+$ and the negative part $\mu^-$ of $\mu$ have the same mass ($l^1$ norm), while condition ($\gamma'$) means that $\mu^+$ and $\mu^-$ have the same barycenter. We define here the ***barycenter***, to be denoted by $\mathbf{bary}(f)$, of a nonzero function $f \geqslant 0$ of finite support as

$$\mathbf{bary}(f) = \frac{\sum_{x \in E} f(x)x}{\sum_{x \in E} f(x)}. \tag{9.9.5}$$

Obviously every Jensen function is a generalized Jensen function.

**Definition 9.9.4.** Given a function $f\colon A \to \mathbf{R}_!$, where $A$ is a subset of a vector space $E$, we shall say that an element $\alpha$ of the algebraic dual of $E$ is a ***subgradient of*** $f$ ***at a point*** $a \in A$ if $f(x) \geqslant f(a) + \alpha(x - a)$ for all $x \in A$. The set of all subgradients at $a$ is denoted by $\partial f(a)$ and called the ***subdifferential of*** $f$ ***at*** $a$; the union of the subdifferentials at all points $a \in A$ is called the ***subdifferential of*** $f$, denoted by $\partial f$. □

## 9.10 Properties of convex functions

**Proposition 9.10.1.** *If $f$ and $g$ are convex functions defined on a vector space, then also $f \vee g$, $f + g$, $f \sqcap g$ and $sf$ for $s > 0$ are convex.*

**Proof.** That $f \sqcap g$ is convex when $f$ and $g$ are convex follows from the equality (3.4.7). The other properties are even more simply proved. □

**Proposition 9.10.2.** *If $(f_j)_{j \in J}$ is a decreasing family of convex functions defined on a vector space indexed by a directed ordered family $J$, then also their infimum $f = \inf_{j \in J} f_j$ is convex.*

**Proof.** This follows from Proposition 9.3.2 with the translation offered by the equality (3.4.7). □

**Proposition 9.10.3.** *Form the infimal convolution*

$$f_t(x) = \inf_{y\in\mathbf{R}^n} \left(f(y) + t\|x-y\|_2^2\right), \qquad x \in \mathbf{R}^n, \quad t > 0,$$

*for a function* $f\colon \mathbf{R}^n \to \mathbf{R}$, *and assume that* $f(x) \geqslant -C\|x\|_2^2$ *for some constant* $C > 0$. *If* $f$ *is lower semicontinuous, then* $f_t \nearrow f$ *as* $t \nearrow +\infty$. *If* $f$ *is convex, then also* $f_t$ *is convex and of class* $C^1$.

**Proof.** The proof of convergence is quite similar to that of Proposition 3.5.1. That $f_t$ is convex follows from Proposition 9.10.1 and (3.4.7).

To prove that $f_t$ is of class $C^1$, we argue as follows. Given any point $a \in \mathbf{R}^n$ and a positive number $t$, there is a (unique) point $b$ such that $f_t(a) = f(b) + t\|a-b\|_2^2$. It follows that

$$f(b) + t\|a-b\|_2^2 \geqslant f_t(x) \geqslant f(b) + \alpha(x-b),$$

for all $x \in \mathbf{R}^n$, where $\alpha$ is a suitable subgradient. Thus the graph of the convex function $f_t$ is squeezed between a paraboloid and a hyperplane, which implies that its partial derivatives exist at $b$; the gradient at $b$ is equal to $\alpha$.

In general, a function can have a gradient at every point without being of class $C^1$, but if it is convex, existence of a gradient at every point implies that the gradient is continuous. □

## 9.11 Strict and strong convexity

**Definition 9.11.1.** A subset $A$ of a topological vector space $E$ is said to be ***strictly convex*** if every supporting hyperplane contains but one point of the closure of $E$.

A function $f\colon E \to \mathbf{R}_!$ is said to be ***strictly convex*** if its finite epigraph is strictly convex in $E \times \mathbf{R}$. □

For functions it is equivalent to require that the graph does not contain any interval of nonzero length.

**Definition 9.11.2.** A function $f\colon \mathbf{R}^n \to \mathbf{R}_!$ is said to be ***strongly convex*** if, for every point $a \in \mathbf{R}^n$, there is a positive number $s$ such that $x \mapsto f(x) - s\|x\|_2^2$ is convex in a neighborhood of $a$.

A subset $A$ of $\mathbf{R}^n$ is said to be ***strongly convex*** if, for every point $a$ in the boundary of $A$, there is a strongly convex function $f$ of $n-1$ variables

$x' = (x_1, \ldots, x_{n-1})$ defined near $a'$, and we can choose coordinates in $\mathbf{R}^n$ such that we have

$$\{x \in V;\ x_n > f(x')\} \subset A \cap V \subset \{x \in V;\ x_n \geqslant f(x')\}$$

for some neighborhood $V$ of $a$. □

The function $\mathbf{R} \ni x \mapsto \sqrt{1+x^2} \in \mathbf{R}$ is strongly convex and illustrates the fact that $s$ cannot be chosen independently of $a$.

The function $\mathbf{R} \ni x \mapsto x^4 \in \mathbf{R}$ is strictly convex but not strongly convex.

## 9.12 Functions taking the value minus infinity

**Proposition 9.12.1.** *For any convex function defined in $\mathbf{R}^n$, the set $M_f$ where it takes the value $-\infty$ is convex. If $M_f$ is nonempty and $f$ takes finite value at some point, then it must be equal to $+\infty$ in the complement of the closure of $M_f$.*

**Proof.** Take any two points $a$ and $b$ such that $f(a) = f(b) = -\infty$. Then the points $(a, s)$ and $(b, s)$ for any real number $s$ must belong to the convex set $\mathbf{epi}^{\text{finite}}(f)$. Therefore this must be true also for the points

$$((1-t)a + tb, s) \text{ with } 0 \leqslant t \leqslant 1,$$

which implies that $f((1-t)a + tb)$ is equal to minus infinity.

Now, if $f(a) = -\infty$ and $f(c)$ is finite, the value of $f((1-t)a + tc)$ must be equal to minus infinity for $0 \leqslant t < 1$ and equal to plus infinity when $t > 1$. So, on each ray $\{(1-t)a + tc;\ t \geqslant 0\}$ emanating from $a$ and containing $c$ there can be at most one point where $f$ is finite. □

It is clear from the proof that the result holds for any vector space, even without a topology, using a convenient definition of the closure.

## 9.13 The convex envelope

**Definition 9.13.1.** Given a function $f\colon A \to \mathbf{R}_!$, where $A$ is a subset of a vector space $E$, the largest convex function $F\colon E \to \mathbf{R}_!$ such that $F|_A \leqslant f$ is called the ***convex envelope*** of $f$ and will be denoted by $\mathbf{cvxe}(f)$. □

**Remark 9.13.2.** In general we have

$$(9.13.1) \qquad \big(\mathbf{cvxe}(f)\big)(x) = \inf_{t \in \mathbf{R}} \big[t;\ (x,t) \in \mathbf{cvxh}\big(\mathbf{epi}^{\text{finite}}(f)\big)\big].$$

The convex envelope of $f\colon E \to \mathbf{R}_!$ evaluated at a point $x$ is equal to the infimum of all expressions

$$\sum_{j=1}^{N} \lambda_j f(x + a^{(j)}),$$

where the $\lambda_j$ and the $a^{(j)}$ are subject to the conditions (9.9.3). We note that the points $(x + a^{(j)}, f(x + a^{(j)}))$ belong to $\mathbf{epi}^{\text{finite}}(f)$, implying that the point

$$\big(\sum \lambda_j (x + a^{(j)}), \sum \lambda_j f(x + a^{(j)})\big)$$

belongs to $\mathbf{cvxh}(\mathbf{epi}^{\text{finite}}(f))$ and that therefore $(x, (\mathbf{cvxe}(f))(x) + t)$ belongs to $\mathbf{cvxh}(\mathbf{epi}^{\text{finite}}(f))$ for every positive number $t$. □

**Remark 9.13.3.** We have inclusions

$$\begin{aligned} \mathbf{epi}^{\text{finite}}_{\text{strict}}(\mathbf{cvxe}(f)) &= \mathbf{cvxh}(\mathbf{epi}^{\text{finite}}_{\text{strict}}(f)) \\ &\subset \mathbf{cvxh}(\mathbf{epi}^{\text{finite}}(f)) \subset \mathbf{epi}^{\text{finite}}(\mathbf{cvxe}(f)). \end{aligned} \tag{9.13.2}$$

The two inclusion relations here can be strict (Exercise 9.23). □

**Definition 9.13.4.** Let $A$ be any subset of a vector space $E$ and let $f\colon A \to \mathbf{R}_!$ be any function defined on $A$. We shall say that $f$ is ***convex extensible*** if it is the restriction to $A$ of a convex function defined on all of $E$, i.e., if there is a convex function $F\colon E \to \mathbf{R}_!$ such that $F|_A = f$. □

There may exist more than one convex extension of a given function $f$. For example $F(x) = |x| - 1$ and $F^+(x) = \max(|x| - 1, 0)$, $x \in \mathbf{R}$, have the same restriction to $A = \mathbf{Z} \smallsetminus \{0\}$. If there exists a convex extension, then $\mathbf{cvxe}(f)$ is the largest one.

The convex envelope $\mathbf{cvxe}(f)$ of a function $f$ defined on $\mathbf{Z}^n$ is piece-wise affine in $\mathbf{R}^n$. It is convenient to have a term for subsets where the envelope is not affine:

**Definition 9.13.5.** Given a function $f\colon \mathbf{Z}^2 \to \mathbf{R}$ such that $\mathbf{cvxe}(f)$ is an extension of $f$, we shall say that a rectilinear segment $[p, q]$ in $\mathbf{R}^2$ is a ***fold line for*** $\mathbf{cvxe}(f)$ if $\mathbf{cvxe}(f)$ is not affine in any neighborhood of $[p, q]$ and $[p, q]$ is maximal with this property. We shall also say that a ray or a straight line is a ***fold line*** if it has the same property. In three dimensions we have similarly ***fold planes*** and in general ***fold hyperplanes***. □

## 9.14 Approximation of the convex envelope

In analogy with Proposition 9.8.1 we can approximate the convex envelope of a function by iterated infimal convolutions:

**Theorem 9.14.1.** *Let $f\colon E \to \mathbf{R}_!$ be any function on a vector space $E$ and define for $k \in \dot{\mathbf{N}}$,*

$$f_{[k]}(x) = \frac{1}{k}(f \sqcap f \sqcap \cdots \sqcap f)(kx) \ (k \text{ terms}), \qquad x \in E.$$

*Then $\mathbf{cvxe}(f) \leqslant f_{[k]} \leqslant f$ with equality if $f$ is convex.*

*If $E = \mathbf{R}^n$ and $\mathbf{cvxe}(f)$ is realvalued and bounded in some neighborhood of a given bounded set $A$, then $f_{[k]}$ converges to $\mathbf{cvxe}(f)$ in $A$ in the sense that, for every positive $r$, there exists an index $k_r$ such that, for $k \geqslant k_r$ and $x \in A$ we have,*

$$(\mathbf{cvxe}(f))(x) \leqslant \sup_{\|y\| \leqslant r} f_{[k]}(x+y) + r \leqslant \sup_{\|y\| \leqslant r} (\mathbf{cvxe}(f))(x+y) + r.$$

We note that the functions $\mathbf{cvxe}(f)$, $f_{[k]}$ and $f$ all have the same Fenchel transform.

**Proof.** That $\mathbf{cvxe}(f) \leqslant f_{[k]} \leqslant f$ can be seen as in Proposition 9.8.1 by looking at the strict finite epigraphs: $k\,\mathbf{epi}^{\text{finite}}_{\text{strict}}(f_{[k]})$ is the sum of $k$ copies of $\mathbf{epi}^{\text{finite}}_{\text{strict}}(f)$, which contains the multiple $k\,\mathbf{epi}^{\text{finite}}_{\text{strict}}(f)$, so $\mathbf{epi}^{\text{finite}}_{\text{strict}}(f_{[k]}) \supset \mathbf{epi}^{\text{finite}}_{\text{strict}}(f)$, equivalently, $f_{[k]} \leqslant f$. If $\mathbf{cvxe}(f) = f$ we obtain equality. (These inequalities can also be proved by a direct calculation of the $f_{[k]}$.)

To prove the second part of the proposition we argue as in the proof of Proposition 9.8.1, using the fact that the Minkowski sum of $k$ copies of $\mathbf{epi}^{\text{finite}}_{\text{strict}}(f)$ is the finite strict epigraph of $f \sqcap f \sqcap \cdots \sqcap f$ ($k$ terms) (see (3.4.7)). A difference is that the strict epigraphs are unbounded, but if we restrict attention to what happens below a certain ceiling, everything works.

In the approximation we use balls of radius $t$ in $\mathbf{R}^{n+1}$, and then replace them by a margin of $t$ in the vertical direction and a supremum over a neighborhood in the horizontal direction. To be precise, let us define for any function $f$ and any $r > 0$,

$$f_r(x) = \sup_{\|y\| \leqslant r} f(x+y) + r \text{ and } f_{-r}(x) = \inf_{\|y\| \leqslant r} f(x+y) - r.$$

Then

$$\mathbf{epi}^{\text{finite}}_{\text{strict}}(f_r) \subset \mathbf{epi}^{\text{finite}}_{\text{strict}}(f) \subset \mathbf{epi}^{\text{finite}}_{\text{strict}}(f) + B_{\leqslant}(0,r) \subset \mathbf{epi}^{\text{finite}}_{\text{strict}}(f_{-r}),$$

implying that the two ways of approximating functions are equivalent. Here we can let the norm defining the ball be the $l^2$ norm $\|\cdot\|_2$ or, perhaps more elegantly, the norm defined as $\|(x,t)\|_{\mathbf{R}^{n+1}} = \max(\|x\|_{\mathbf{R}^n}, |t|)$ for a norm $\|\cdot\|_{\mathbf{R}^n}$ in $\mathbf{R}^n$. □

We now present a somewhat complicated example to illustrate the action of taking the upper and lower inverses of simple mappings in various contexts:

**Example 9.14.2.** Let $L$ be a complete lattice and $f, g$ the mappings $L \to L$ given by $f(x) = c \wedge x$, $g(x) = c \vee x$, where $c$ is a fixed element of $L$. We shall study the four inverses $f^{[-1]}$, $g_{[-1]}$, $f_{[-1]}$ and $g^{[-1]}$. (The first two are easier than the last two.) We shall do so in four different contexts.

(A1), (A2) For a general complete lattice $L$.
(B1), (B2) When $L = \mathscr{P}(W)$, the power set of a set $W$.
(C1), (C2) When $L = \mathscr{F}(W, \mathbf{R}_!)$, the complete lattice of all functions on a set $W$ with values in $\mathbf{R}_!$.
(D1), (D2) When $L$ is the complete lattice $CVX(\mathbf{R}^n, \mathbf{R}_!)$ of all convex functions on $\mathbf{R}^n$ with values in $\mathbf{R}_!$.

(A1) For a general lattice $L$. With $f(x) = c \wedge x$ we get $f^{[-1]}(y) = y$ if $y \leqslant c$ and equal to $\mathbf{1}$ otherwise. We can write this briefly as $f^{[-1]} = \mathbf{id} \vee h_c$, where $h_c$ is the function which takes the value $h_c(y) = \mathbf{0}$ when $y \leqslant c$ and $\mathbf{1}$ otherwise.

With $g(x) = c \vee x$, we have $g_{[-1]}(y) = y$ when $y \geqslant c$ and equal to $\mathbf{0}$ otherwise. Briefly $g_{[-1]} = \mathbf{id} \wedge k_c$, where $k_c(y) = \mathbf{1}$ for $y \geqslant c$ and equal to $\mathbf{0}$ otherwise.

(A2) We have $f_{[-1]}(y) = \mathbf{1}$ if $y \geqslant c$ and $\geqslant y$ otherwise, but strict inequality can occur.

Similarly, $g^{[-1]}(y) = \mathbf{0}$ when $y \leqslant c$ and $\leqslant y$ otherwise, but also here the inequality can be strict. In special cases we can say more; see (B2), (C2), (D2) below.

(B1) With $L = \mathscr{P}(W)$ we obtain from (A1) for $f(A) = C \cap A$, $A \subset W$: $f^{[-1]} = \mathbf{id} \cup h_C$, where $h_C(B) = \emptyset$ when $B \subset C$ and is equal to the whole set $W$ otherwise.

For $g(A) = C \cup A$, $A \subset W$, we get from (A1) $g_{[-1]}(B) = B \cap k_C(B)$, where $k_C(B) = W$ when $B \supset C$ and equal to $\emptyset$ otherwise.

(B2) The lower inverse of $f$ is given by

$$f_{[-1]}(B) = \bigcup_A (A;\ C \cap A \subset B) = B \cup \complement C.$$

The latter set can obviously be strictly larger than $B$.

The upper inverse of $g$ is $g^{[-1]}(B) = \bigcap_A (A;\ C \cup A \supset B) = B \smallsetminus C$. And $B \smallsetminus C$ can be strictly smaller than $B$.

(C1) With $f(\varphi) = \gamma \wedge \varphi$ for a given function $\gamma \colon \mathbf{R}^n \to \mathbf{R}_!$, we study its upper inverse, given by

$$f^{[-1]}(\psi) = \bigwedge(\varphi;\ \gamma \wedge \varphi \geqslant \psi), \qquad \psi \in \mathscr{F}(W, \mathbf{R}_!).$$

We can use (A1) to get $f^{[-1]}(\psi) = \psi$ if $\gamma \leqslant \psi$ and $f^{[-1]}(\psi) = +\infty$ otherwise. Briefly $f^{[-1]} = \mathbf{id} \vee \theta_\gamma$, where $\theta_\gamma(\psi) = -\infty$ if $\psi \leqslant \gamma$ and $\theta_\psi) = +\infty$ if not.

Similarly we get $g_{[-1]}$ from (A1) and find that

$$g_{[-1]}(\psi) = \bigvee(\varphi \in \mathscr{F}(W, \mathbf{R}_!);\ \gamma \vee \varphi \leqslant \psi) = \psi$$

if $\psi \geqslant \gamma$ and equal to $-\infty$ when there is a point $a$ such that $\psi(a) < \gamma(a)$. This can be written $g_{[-1]} = \mathbf{id} \wedge \tau_\gamma$, where $\tau_\gamma(\psi) = +\infty$ when $\psi \geqslant \gamma$ and equal to $-\infty$ otherwise.

(C2) We find that $\big(f_{[-1]}(\psi)\big)\big)(x) = \psi(x)$ if $\psi(x) < \gamma(x)$ and equal to $+\infty$ if $\psi(x) \geqslant \gamma(x)$. Indeed, in the definition of the lower inverse,

$$f_{[-1]}(\psi) = \bigvee(\varphi;\ \gamma \wedge \varphi \leqslant \psi), \qquad \psi \in \mathscr{F}(W, \mathbf{R}_!),$$

we can take $\varphi(x)$ equal to $+\infty$ for $x$ such that $\psi(x) \geqslant \gamma(x)$ and equal to $\psi(x)$ if $\psi(x) < \gamma(x)$. Hence we have a strict inequality $f_{[-1]}(\psi)(x) > \psi(x)$ at all points $x$ such that $+\infty > \psi(x) \geqslant \gamma(x)$ and not only when $\psi \geqslant \gamma$ everywhere.

Similary as for $f_{[-1]}$ we can in the definition of the upper inverse,

$$g^{[-1]}(\psi) = \bigwedge(\varphi \in \mathscr{F}(W, \mathbf{R}_!);\ \gamma \vee \varphi \geqslant \psi), \qquad \psi \in \mathscr{F}(W, \mathbf{R}_!),$$

take $\varphi$ with the value $-\infty$ for $x$ such that $\psi(x) \leqslant \gamma(x)$ and equal to $\psi(x)$ if $\psi(x) > \gamma(x)$. This is because we get an admissible function $\varphi$ by taking $\varphi(a) = -\infty$ and equal to $+\infty$ at all other points. Hence we have a strict inequality $\big(g^{[-1]}(\psi)\big)(x) < \psi(x)$ at all points $x$ such that $-\infty < \psi(x) \leqslant \gamma(x)$ and not only when $\psi \leqslant \gamma$ everywhere.

(D1) With $f(\varphi) = \gamma \wedge \varphi$ for a given convex function $\gamma \colon \mathbf{R}^n \to \mathbf{R}_!$ we find that

$$f^{[-1]}(\psi) = \bigwedge_{cvx} \big(\varphi \in CVX(\mathbf{R}^n, \mathbf{R}_!);\ \gamma \wedge_{cvx} \varphi \geqslant \psi\big), \qquad \psi \in CVX(\mathbf{R}^n, \mathbf{R}_!).$$

This yields the value $\psi$ if $\psi \leqslant \gamma$ and equal to $+\infty$ if there exists a point $a$ such that $\psi(a) > \gamma(a)$.

As to the lower inverse of $g$, defined by

$$g_{[-1]}(\psi) = \bigvee(\varphi \in CVX(\mathbf{R}^n, \mathbf{R}_!);\ \gamma \vee \varphi \leqslant \psi), \qquad \psi \in CVX(\mathbf{R}^n, \mathbf{R}_!),$$

we see that here $\vee_{cvx}$ agrees with the pointwise supremum $\vee$, so the result from (C1) can be applied directly. We obtain $g_{[-1]}(\psi) = \psi$ when $\psi \geqslant \gamma$ and equal to $-\infty$ if there exists a point $a$ such that $\psi(a) < \gamma(a)$.

(D2) Now about $f_{[-1]}$ in the convex case, defined by

$$f_{[-1]}(\psi) = \bigvee(\varphi \in CVX(\mathbf{R}^n, \mathbf{R}_!);\ \gamma \wedge_{cvx} \varphi \leqslant \psi), \qquad \psi \in CVX(\mathbf{R}^n, \mathbf{R}_!).$$

Define $A^\gamma_\psi$ as the set of all points $a$ such hat $\gamma(a) \geqslant \psi(a)$. For every $a \in A^\gamma_\psi$ we can choose $\varphi \in CVX(\mathbf{R}^n, \mathbf{R}_!)$ taking the value $\psi(x)$ if $x$ belongs to $\mathbf{cvxh}(A^\gamma_\psi)$ and taking the value $+\infty$ in its complement. Then

$$(\gamma \wedge_{cvx} \varphi)(x) \leqslant \gamma(x) \wedge \varphi(x) = \gamma(x) \wedge \psi(x) \leqslant \psi(x)$$

for $x \in \mathbf{cvxh}(A^\gamma_\psi)$, and $(\gamma \wedge_{cvx} \varphi)(x) = -\infty \leqslant \psi(x)$ for $x \notin \mathbf{cvxh}(A^\gamma_\psi)$, so that $\gamma \wedge_{[cvx} \varphi \leqslant \psi$ everywhere. It follows that $\big(f_{[-1]}(\psi)\big)(x) \geqslant \varphi(x) = +\infty$ if $x \notin \mathbf{cvxh}(A^\gamma_\psi)$ and $\geqslant \psi(x)$ if $x \in \mathbf{cvxh}(A^\gamma_\psi)$.

Here it can happen that we have strict inequality, for instance if we take $\gamma(x) = x^2$, $\psi(x) = (x-1)^2$, and $\varphi(x) = (x-2)^2$ for $x \in \mathbf{R}$, so that $\gamma \wedge_{cvx} \varphi \leqslant \psi$, thus that $\varphi$ is admissible, and therefore $\big(f_{[-1]}(\psi)(1) \geqslant \varphi(1) = 1 > \psi(1)$.

Finally, for $g^{[-1]}$, defined by

$$g^{[-1]}(\psi) = \bigwedge_{cvx}(\varphi \in CVX(\mathbf{R}^n, \mathbf{R}_!);\ \gamma \vee \varphi \geqslant \psi), \qquad \psi \in CVX(\mathbf{R}^n, \mathbf{R}_!).$$

If $A^\gamma_\psi$ is equal to the whole space $\mathbf{R}^n$, we have $\psi \leqslant \gamma$ everywhere, so every convex function $\varphi$ is admissible in the formation of the infimumum, making its value equal to $-\infty$.

If $A^\gamma_\psi$ is empty, we have $\psi > \gamma$ everywhere and we see that $g^{[-1]} = \psi$.

If on the other hand $A^\gamma_\psi$ is nonempty, we can, for any point $a$ in $A^\gamma_\psi$ define an admissible function $\varphi$ by taking $\varphi(a) = -\infty$ and $\varphi(x) = +\infty$ for $x \neq a$. This shows that $g^{[-1]}(\psi) \leqslant \varphi(a) = -\infty$. Since $g^{[-1]}$ is convex it must be equal to minus infinity also in $\mathbf{cvxh}(A^\gamma_\psi)$, and equal to plus infinity in the complement of $\overline{\mathbf{cvxh}(A)}$ (Proposition 9.12.1).

If a point $b$ belongs to the boundary of $A^\gamma_\psi$ and also to $A^\gamma_\psi$, we have of course $\big(g^{[-1]}(\psi)\big)(b) = -\infty$. If $b$ belongs to the boundary of $A^\gamma_\psi$ but not to $A^\gamma_\psi$, i.e., if $\psi(b) > \gamma(b)$, we have $\big(g^{[-1]}(\psi)\big)(b) = +\infty$.

An example is with $\gamma(x) = x^2 - 2/9$, $x \in \mathbf{R}$, and $\psi(x) = x^4$ when $x < 1$ and $\psi(x) = +\infty$ when $x \geqslant 1$. Then $A^\gamma_\psi = [-q, -p] \cup [p, q[$, where $p = \sqrt{1/3}$ and $q = \sqrt{2/3}$, and $\mathbf{cvxh}(A^\gamma_\psi) = [-q, q[$. Here

$$-q \in A^\gamma_\psi \cap \partial A^\gamma_\psi \text{ and consequently } \big(g^{[-1]}(\psi)\big)(-q) = -\infty,$$

while $q \in \overline{\mathbf{cvxh}(A^\gamma_\psi)} \smallsetminus A^\gamma_\psi$ and $\big(g^{[-1]}(\psi)\big)(q) = +\infty$. □

## 9.15 Separating hyperplanes

Let $Y$ be a hyperplane in a vector space $E$. It is the intersection of two half-spaces $T_1$ and $T_2$ defined by non-strict inequalities. We shall say that two subsets $A$ and $B$ of $E$ are ***separated by*** $Y$ if $A$ is a subset of $T_1$ and $B$ is a subsets of $T_2$ or the other way around. (We thus speak of separation in a non-strict sense: if $A = B = Y$, $A$ and $B$ are separated by $Y$.) Since the $T_j$ are convex, $Y$ also separates their convex hulls $\mathbf{cvxh}(A)$ and $\mathbf{cvxh}(B)$. If in addition $E$ has a topology and $Y$ is closed, then the $T_j$ are closed and convex, so that $Y$ also separates their closed convex hulls $\overline{\mathbf{cvxh}(A)}$ and $\overline{\mathbf{cvxh}(B)}$.

If $A$ and $B$ are empty, they are separated by any hyperplane. If $A$ is empty it can be separated from $B$ if and and only if $B$ is contained in a half-space. Only the case when both $A$ and $B$ are nonempty is therefore of interest.

A special case is when $A = \{a\}$ is a singleton set. But we should note also that we can separate two sets $A$ and $B$ if and only the origin can be separated from the set $A - B = \{x - y;\ x \in A,\ y \in B\}$.

If $E = \mathbf{R}^{n+1}$, $Y$ can be given by an equation $\alpha_1 x_1 + \cdots + \alpha_{n+1}x_{n+1} + c = 0$ and we may, possibly after a permutation of the coordinates, assume that $\alpha_{n+1}$ is nonzero, and even that it is equal to $-1$. Thus $Y$ is given by the equation $t = \alpha \cdot x + c$, and $(x,t) \in A$ implies $t \leqslant \alpha \cdot x + c$; $(x,t) \in B$ implies $t \geqslant \alpha \cdot x + c$, or the other way around, where we write $t$ for $x_{n+1}$, $x$ for $(x_1, \dots, x_n)$, and $\alpha$ for $(\alpha_1, \dots, \alpha_n)$.

Defining $F(x) = -\sup(t;\ (x,t) \in A)$ and $G(x) = \inf(t;\ (x,t) \in B)$ we see that also the sets $\mathbf{hypo}^{\text{finite}}(-F)$ and $\mathbf{epi}^{\text{finite}}(G)$, which contain $A$ and $B$, respectively, are separated by $Y$.

We establish criteria first for two sets to be disjoint:

**Proposition 9.15.1.** *Let $F, G\colon X \to \mathbf{R}_!$ be two given functions, where $X$ is any subset of $\mathbf{R}^n$, and define*

$$A = A^{\mathbf{R}}_X = \mathbf{hypo}^{\text{finite}}(-F) = \{(x,t) \in X \times \mathbf{R};\ t \leqslant -F(x)\},$$
$$(9.15.1)\qquad B = B^{\mathbf{R}}_X = \mathbf{epi}^{\text{finite}}(G) = \{(x,t) \in X \times \mathbf{R};\ G(x) \leqslant t\}.$$

*Then $A$ and $B$ are disjoint if and only if $0 < F \dotplus G$.*

**Proof.** We observe that for two arbitrary elements $a, b$ of $\mathbf{R}_!$ we have $0 < a \dotplus b$ if and only if the interval $[b, -a] = \{x \in \mathbf{R}_!;\ b \leqslant x \leqslant -a\}$ (formed in $\mathbf{R}_!$) is disjoint from $\mathbf{R}$. Thus if $A$ and $B$ are disjoint, the intersection $[G(x), -F(x)] \cap \mathbf{R} = \emptyset$ for every $x \in X$, so that $0 < F \dotplus G$.

Conversely, if $0 < F(x) \dotplus G(x)$, there is no real number $t$ satisfying $G(x) \leqslant t \leqslant -F(x)$, so that, if this holds for every $x \in X$, $A \cap B$ must be empty. □

**Theorem 9.15.2.** *Any two disjoint open convex sets in a topological vector space $E$ can be separated by a closed hyperplane.*

There are topological vector spaces with dual equal to zero. The separation mentioned in the theorem can then seem like a paradox. It is resolved by the fact that in these vector spaces, the only open convex sets are the empty set and the whole space.

**Proof.** If $A$ and $B$ are two convex subsets of $E$ we form $C = A - B$, a convex set not containing the origin if $A$ and $B$ are disjoint. It is open if one of $A$ and $B$ is open. According to the Hahn–Banach Theorem, Theorem 9.5.2, there is a continuous linear form $\xi$ on $E$ such that $\xi(c) \geqslant 0$ for all $c \in C$, implying that $\xi(a) \geqslant \xi(b)$ for all $a \in A$, $b \in B$. □

**Theorem 9.15.3.** *Let $F, G \colon X \to \mathbf{R}_!$ be two given functions defined on a subset $X$ of $\mathbf{R}^n$, and define $A$ and $B$ by (9.15.1). Consider the following conditions.*

(A) *There is a non-vertical hyperplane separating $A$ and $B$.*

(B) *There exists a vector $\xi \in \mathbf{R}^n$ such that $\widetilde{F}(-\xi), \widetilde{G}(\xi) < +\infty$ and $\widetilde{F}(-\xi) + \widetilde{G}(\xi) \leqslant 0$, where $\widetilde{F}$ is the Fenchel transform of $F$.*

(C) $0 \leqslant \widetilde{\widetilde{F}} \dotplus \widetilde{\widetilde{G}}$, *where $\widetilde{\widetilde{F}}$ is the second Fenchel transform of $F$.*

(D) $0 \leqslant \mathbf{cvxe}(F) \dotplus \mathbf{cvxe}(G)$.

(E) *There is a hyperplane separating $A$ and $B$.*

*Assume that both $A$ and $B$ are nonempty. Then* (A) $\Leftrightarrow$ (B) $\Rightarrow$ (C) $\Rightarrow$ (D) $\Rightarrow$ (E) *but, if $X = \mathbf{R}^n$, $n \geqslant 1$, we have* (B) $\not\Leftarrow$ (C) $\not\Leftarrow$ (D) $\not\Leftarrow$ (E).

For the definition of the Fenchel transform see Subsection 9.17.2 below.

**Proof.** (A) $\Leftrightarrow$ (B). A non-vertical separating hyperplane must be given by an affine function $H$ such that $-F \leqslant H \leqslant G$. This affine function in turn has an equation $H(x) = \xi \cdot x + c$ for some vector $\xi$ and some real number $c$. Hence $(-\xi) \cdot x - F(x) \leqslant c$ for all $x$, which implies $\widetilde{F}(-\xi) \leqslant c$. Similarly

$\xi \cdot x - G(x) \leqslant -c$ for all $x$, which implies $\widetilde{G}(\xi) \leqslant -c$. Adding we get (B). This argument can be reversed to show also that (B) implies (A).

(A) $\Rightarrow$ (C). If we have $-F \leqslant H \leqslant G$ we apply the Fenchel transformation twice to deduce that $-H = -\widetilde{\widetilde{H}} \leqslant \widetilde{\widetilde{F}}$ and $H = \widetilde{\widetilde{H}} \leqslant \widetilde{\widetilde{G}}$, so that (C) follows.

(C) $\Rightarrow$ (D). This is easy since $\widetilde{\widetilde{F}} \leqslant \mathbf{cvxe}(F)$ and $\widetilde{\widetilde{G}} \leqslant \mathbf{cvxe}(G)$.

(D) $\Rightarrow$ (E). If we have (D), the two convex sets

$$\mathbf{hypo}^{\text{finite}}_{\text{strict}}(-\,\mathbf{cvxe}(F)) \text{ and } \mathbf{epi}^{\text{finite}}_{\text{strict}}(\mathbf{cvxe}(G))$$

are disjoint. Hence there exists, by the Hahn–Banach theorem, a hyperplane $Y$ separating them. Since the closure of $\mathbf{hypo}^{\text{finite}}_{\text{strict}}(-\,\mathbf{cvxe}(F))$ contains $\mathbf{hypo}^{\text{finite}}(-\,\mathbf{cvxe}(F))$ and similarly for $\mathbf{epi}^{\text{finite}}_{\text{strict}}(\mathbf{cvxe}(G))$, this hyperplane also serves to separate $\mathbf{hypo}^{\text{finite}}(-\,\mathbf{cvxe}(F))$ and $\mathbf{epi}^{\text{finite}}(\mathbf{cvxe}(G))$.

The three non-implications will be proved by the following three examples. Note that we only use functions which do not take the value $-\infty$.

**Example 9.15.4.** (A) $\not\Leftarrow$ (C). We take $X = \mathbf{R}$ and define $F(x) = +\infty$ for $x < 0$, $F(x) = -\sqrt{x}$ for $x \geqslant 0$; $G(x) = -\sqrt{-x}$ for $x \leqslant 0$, $G(x) = +\infty$ for $x > 0$. Then $F + G = 0$ at the origin and $F + G = +\infty$ elsewhere, and $\widetilde{\widetilde{F}} = F$ and $\widetilde{\widetilde{G}} = G$ so that (C) holds, but (A) does not hold—the only separating line is the vertical line $x = 0$. □

**Example 9.15.5.** (C) $\not\Leftarrow$ (D). Let $n = 1$ and define $F(x) = +\infty$ for $x \leqslant 0$, $F(x) = 0$ for $x > 0$; $G(x) = -x^2$ for $x < 0$, $G(x) = +\infty$ for $x \geqslant 0$. Then $\mathbf{cvxe}(F) \dotplus \mathbf{cvxe}(G)$ is $+\infty$ everywhere, so (D) holds, but $\widetilde{\widetilde{F}}(0) = 0$ and $\widetilde{\widetilde{G}} = -\infty$ everywhere, so that $\widetilde{\widetilde{F}}(0) \dotplus \widetilde{\widetilde{G}}(0) = -\infty$, and (C) does not hold. □

**Example 9.15.6.** (D) $\not\Leftarrow$ (E). Let $n = 1$ and define $F(x) = +\infty$ for $x < 0$, $F(x) = -1$ for $x \geqslant 0$; $G(x) = 0$ for $x \leqslant 0$, $G(x) = +\infty$ for $x > 0$. Then $\mathbf{cvxe}(F) = F$, $\mathbf{cvxe}(G) = G$, and $F(0) + G(0) = -1 < 0$, so (D) does not hold, but (E) holds: $x = 0$ is a separating line. □

□

## 9.16 Normed spaces

To any vector space $E$ over the field of real numbers we associate its ***algebraic dual*** $E^\star$, the vector space of all linear forms on $E$, i.e., the functions $\xi\colon E \to \mathbf{R}$ satisfying $\xi(x + ty) = \xi(x) + t\xi(y)$ for all $x, y \in E$ and all $t \in \mathbf{R}$.

We say that a function $E \ni x \mapsto \|x\| \in \mathbf{R}$ is a ***norm*** if $\|x\| \geqslant 0$ with equality if and only if $x = 0$; $\|x + y\| \leqslant \|x\| + \|y\|$ for all $x$ and $y$; and finally $\|tx\| = t\|x\|$ for all $x \in E$ and all positive numbers $t$. This means that the function $E^2 \ni (x, y) \mapsto \|x - y\|$ is a metric with the extra property of being positively homogeneous. The subadditivity and the homegeneity together imply that $x \mapsto \|x\|$ is convex.

The space $\mathbf{R}^n$ of all $n$-tuples can be normed by the ***$l^p$-norm*** $\|\cdot\|_p$, $1 \leqslant p \leqslant +\infty$, which is defined for $1 \leqslant p < +\infty$ by

$$\|x\|_p = \Big(\sum |x_j|^p\Big)^{1/p}, \qquad x = (x_1, \ldots, x_n) \in \mathbf{R}^n.$$

When $p = +\infty$ this has to be interpreted as a limit. More explicitly one defines

$$\|x\|_\infty = \max_j |x_j|, \qquad x \in \mathbf{R}^n.$$

In addition to the algebraic dual $E^\star$, which is defined for any vector space, we consider for any normed vector space $E$ also its ***dual space***, denoted by $E'$ and consisting of all continuous linear forms on $E$. These are the linear mappings $\xi \colon E \to \mathbf{R}$ such that $|\xi(x)| \leqslant C\|x\|$ for some constant $C$. On the dual we define the norm ***dual*** to $\|\cdot\|$ by

$$\|\xi\|' = \sup_{\|x\| \leqslant 1} |\xi(x)|, \qquad \xi \in E'. \tag{9.16.1}$$

It follows that $|\xi(x)| \leqslant \|\xi\|' \cdot \|x\|$ for all $x \in E$ and all $\xi \in E'$.

When $E = \mathbf{R}^n$, we may identify both $E^\star$ and $E'$ with $\mathbf{R}^n$, and the evaluation of $\xi$ at the point $x$, i.e., the number $\xi(x)$, is then the ***inner product***, defined by $\xi \cdot x = \xi_1 x_1 + \cdots + \xi_n x_n$. The Euclidean norm $\|\cdot\|_2$, defined by $\|x\|_2^2 = x \cdot x$, is dual to itself:

$$\|\xi\|_2' = \sup_{\|x\|_2 \leqslant 1} \xi(x) = \|\xi\|_2 = \sqrt{\textstyle\sum \xi_j^2}.$$

It is not difficult to prove that the norm dual to $\|\cdot\|_1$ is $\|\cdot\|_\infty$ and vice versa. More generally, one can prove that the norm dual to $\|\cdot\|_p$ is $\|\cdot\|_q$, where $q = p/(p-1)$, $1 < p < +\infty$, with a natural interpretation also when $p = 1, +\infty$. This statement follows from Hölder's inequality and its converse.

A vector space provided with a Euclidean norm is called a ***Euclidean space***.

## 9.17 Duality in convex analysis

By the term *duality* we aim at properties and results which involve a vector space and its dual. The most important examples are the support function and the Fenchel transformation, to be defined now.

### 9.17.1 *The support function*

**Definition 9.17.1.** Given any subset $A$ of a vector space $E$ we define its ***support function*** $H_A$ by

$$H_A(\xi) = \sup_{x \in A} \xi(x), \qquad \xi \in E^\star.$$

Here $E^\star$ is the algebraic dual of $E$. □

**Example 9.17.2.** If $A$ is a ball, $A = B_{\leqslant}(c, r)$ with $r \geqslant 0$, or $B_<(c, r)$ with $r > 0$, then $H_A(\xi) = \xi(c) + r\|\xi\|'$, $\xi \in E^\star$, where $\|\cdot\|$ is an arbitrary norm in $E$ and $\|\cdot\|'$ its dual norm, defined on $E^\star$ in (9.16.1) above. □

We list a few properties of the support function.

**Proposition 9.17.3.** *In any real vector space $E$, the support function of the empty set is the constant $-\infty$: $H_\varnothing = -\infty$.*

*The support function of a singleton set $\{p\}$ is the linear form*

$$E^\star \ni \xi \mapsto \xi(p) \in \mathbf{R}.$$

*We have*

$$H_{A+B} = H_A \dotplus H_B \text{ and } H_{tA} = tH_A, \qquad A, B \subset E, \quad t > 0.$$

The proof is easy.

### 9.17.2 *The Fenchel transformation*

**Definition 9.17.4.** To any function $\varphi\colon E \to \mathbf{R}_!$ we define its ***Fenchel transform*** $\widetilde{\varphi}\colon E^\star \to \mathbf{R}_!$ by

$$\widetilde{\varphi}(\xi) = \sup_{x \in E} \big(\xi(x) - \varphi(x)\big), \qquad \xi \in E^\star,$$

where $E^\star$ is the algebraic dual of $E$. □

It follows that

$$(9.17.1) \qquad \xi(x) \leqslant \varphi(x) \dot{+} \widetilde{\varphi}(\xi), \quad x \in E, \quad \xi \in E^\star,$$

called ***Fenchel's inequality***. If $\varphi(a)$ is finite and $\alpha$ is a subgradient of $\varphi$ at $a$, then we have equality in (9.17.1): $\alpha(a) = \varphi(a) + \widetilde{\varphi}(\alpha)$.

It is evident that $\widetilde{\varphi}$ is the smallest function $g$ such that $\xi(x) \leqslant \varphi(x) \dot{+} g(\xi)$ holds.

Obviously the points where $\varphi(x) = +\infty$ play no role in the definition of the Fenchel transform of $\varphi$: we have

$$(9.17.2) \qquad \widetilde{\varphi}(\xi) = \sup_{x \in \mathbf{dom}(\varphi)} \big(\xi(x) - \varphi(x)\big), \qquad \xi \in E^\star,$$

where **dom** denotes the effective domain, defined in formula (3.4.4).

For any family of functions $(\varphi_j)_{j\in J}$, $\varphi_j \in \mathscr{F}(E, \mathbf{R}_!)$, we clearly have

$$(9.17.3) \qquad \sup_{j\in J} \widetilde{\varphi} = \left(\inf\nolimits_{j\in J} \varphi\right)^{\sim}.$$

Therefore the transformation $\mathbf{w}\colon \varphi \mapsto -\widetilde{\varphi}$ is an erosion and so satisfies $(\mathbf{w}^{[-1]})_{[-1]} = \mathbf{w}$; cf. Proposition 5.6.1.

We see that the support function of a set is the Fenchel transform of its indicator function: $H_A = \widetilde{\mathbf{indf}_A}$.

**Example 9.17.5.** If the graph of a function $\varphi$ is a paraboloid,

$$\varphi(x) = a + \beta \cdot x + \tfrac{1}{2}c\|x\|_2^2, \qquad x \in \mathbf{R}^n,$$

where $a \in \mathbf{R}$, $\beta \in \mathbf{R}^n$ and $c > 0$, then the same is true of the graph of its transform: $\widetilde{\varphi}(\xi) = -a + \frac{1}{2}c^{-1}\|\xi - \beta\|_2^2$. □

We list some properties of the transformation.

**Proposition 9.17.6.** *In any vector space $E$ the Fenchel transform of the constant $+\infty$ is the constant $-\infty$, and the transform of any function which takes the value $-\infty$ at some point is the constant $+\infty$.*

*The transform of the indicator function $\mathbf{indf}_{\{p\}}$ of a singleton set $\{p\}$ is the linear form*

$$E^\star \ni \xi \mapsto \xi(p) \in \mathbf{R},$$

*and the transform of a linear form $\alpha \in E^\star$ is the indicator function $\mathbf{indf}_{\{\alpha\}}\colon E^\star \to \mathbf{R}$ of the set $\{\alpha\}$.*

*Moreover*

$$(tf)^{\sim}(\xi) = t\widetilde{f}(\xi/t), \qquad t > 0, \quad f\colon E \to \mathbf{R}_!.$$

The proof is easy. Cf. Proposition 9.17.3. See also Theorem 9.18.1.

We now ask what happens if we apply the transformation again. We let $\Xi$ be any subset of $E^\star$, and define the Fenchel transform of any function $f$ defined on $\Xi$ by

$$\widetilde{f}(x) = \sup_{\xi\in\Xi} \big(\xi(x) - f(\xi)\big), \qquad x \in E.$$

Here we can take $\Xi = \{0\}$ as well as $\Xi = E^\star$. In a normed space it is customary to take $\Xi = E'$, the dual of $E$.

The equivalence

$$(9.17.4) \quad \widetilde{\varphi} \leqslant f \text{ if and only if } \widetilde{f} \leqslant \varphi, \qquad \varphi \in \mathscr{F}(E, \mathbf{R}_!), \ f \in \mathscr{F}(\Xi, \mathbf{R}_!),$$

follows easily from the definition.

We may form the second transform $\widetilde{\widetilde{\varphi}}$ of a function defined on $E$. The main result in the theory of the Fenchel transformation is the following.

**Theorem 9.17.7.** *Let $\varphi$ be a function defined on a vector space $E$ and let $\Xi$ be any subset of its algebraic dual $E^\star$. Then we always have $\widetilde{\widetilde{\varphi}} \leqslant \varphi$. Equality holds if and only if*

(C) *$\varphi$ is convex;*
(L) *$\varphi$ is lower semicontinuous for the weakest topology for which all linear forms in $\Xi$ are continuous; and*
(M) *$\varphi$ does not take the value $-\infty$ unless it is identically equal to $-\infty$.*

Property (L) here means that if $\varphi(a) > s$, then there are linear forms $\xi_1, \dots, \xi_m \in \Xi$ and a number $\theta > 0$ such that $\varphi(x) > s$ when $|\xi_j(x-a)| \leqslant \theta$, $j = 1, \dots, m$. We shall denote this topology by $\sigma(E, \Xi)$. (We get the chaotic topology when $\Xi$ is empty or if $\Xi = \{0\}$, implying that the only lower semicontinuous functions are the constants.)

In $\mathbf{R}^n$ we usually choose $\Xi = \mathbf{R}^n$; the semicontinuity is then semicontinuity with respect to the usual topology of $\mathbf{R}^n$.

Before we go on, let us consider other expressions of this semicontinuity. The topology $\sigma(E, \Xi)$ on $E$ gives rise to a topology in $E \times \mathbf{R}$, viz. the product topology, for which a basis for the neighborhoods of a point $(a, s) \in E \times \mathbf{R}$ are given by the sets

$$\{(x,t);\ |\xi_j(x-a)| < \theta,\ j = 1, \dots, N,\ |t-s| < \theta\}, \qquad \xi_j \in \Xi,\ \theta > 0.$$

Let us call this topology $\tau(E \times \mathbf{R}, \Xi)$.

**Lemma 9.17.8.** *Given a vector space $E$ and a subset $\Xi$ of its algebraic dual, the following three conditions on a function $f \colon E \to \mathbf{R}_!$ are equivalent.*

(a) *The function $f$ is lower semicontinuous for $\sigma(E, \Xi)$, i.e.,*

$$\liminf_{y \to x} f(y) = f(x) \text{ for every } x \in E;$$

(b) *The finite epigraph of $f$ is a closed subset of $E \times \mathbf{R}$ for the topology $\tau(E \times \mathbf{R}, \Xi)$;*

(c) *For every real number $s$ the sublevel set $\{x \in E;\ f(x) \leqslant s\}$ is closed for $\sigma(E, \Xi)$.*

**Proof.** All implications here have straightforward proofs. Let us just note that, for the proof of the implication (b) $\Rightarrow$ (c), the sublevel set

$$\{x \in E;\ f(x) \leqslant s\} \subset E \text{ is equal to } \mathbf{epi}^{\text{finite}}(\varphi) \cap \{(x,t) \in E \times \mathbf{R};\ t \leqslant s\},$$

which is the intersection of two $\tau(E \times \mathbf{R}, \Xi)$-closed sets in $E \times \mathbf{R}$. □

**Proof of Theorem 9.17.7.** The necessity of the three properties is easy to prove. In particular we note that every affine function $x \mapsto \xi(x) + c$ with $\xi \in \Xi$ is continuous, in particular lower semicontinuous for the topology $\sigma(E, \Xi)$. The supremum of any family of lower semicontinuous functions is lower semicontinuous.

The converse implication is based on the Hahn–Banach theorem (Theorem 9.5.2), where every closed convex set is an intersection of closed half-spaces. Here the term *closed* shall be understood as closed for the topology $\tau(E \times \mathbf{R}, \Xi)$ defined before Lemma 9.17.8. We note that both $E$ and $E \times \mathbf{R}$ are topological vector spaces under the topologies defined.

So let now $\varphi$ be a convex function with closed finite epigraph. If $\varphi$ takes the value $-\infty$, the result is true, for $(-\infty)\tilde{} = +\infty$ and $(+\infty)\tilde{} = -\infty$. Therefore we can from now on assume that $-\infty < \varphi \leqslant +\infty$. Moreover the conclusion is true $\varphi$ is equal to $+\infty$ identically. Let $(a, r)$ be an arbitrary point not belonging to the finite epigraph, i.e., with $r < \varphi(a)$. It can happen that $\varphi(a) < +\infty$ (an easy case) but also that $\varphi(a) = +\infty$ (a more difficult case).

Since the finite epigraph is closed, there exists, according to Theorem 9.15.2, a hyperplane in $E \times \mathbf{R}$ which separates $\mathbf{epi}^{\text{finite}}(\varphi)$ and $(a, r)$. Such a hyperplane has the equation $\xi(x) + bt = c$ for some $\xi \in \Xi$ and some $b, c \in \mathbf{R}$. The half-space defined by $\xi(x) + bt \leqslant c$ is either the finite epigraph of an affine function (when $b < 0$) or vertical (when $b = 0$), or, finally, the finite hypograph of an affine function (when $b > 0$). The last type can contain only the empty finite epigraph and so appears only when $\varphi$ is $+\infty$ identically; then the result is valid as we already observed.

The problem is to prove that the vertical half-spaces, which actually can occur, do not influence the intersection.

We observe that it is possible to take the hyperplane as strictly separating, which means that there is a number $\theta$ such that

$$\xi(x) + bt \leqslant c < c + \theta \leqslant \xi(a) + br \tag{9.17.5}$$

for every $(x, t) \in \mathbf{epi}^{\text{finite}}(\varphi)$. If there is a point where $\varphi < +\infty$ (and if this is not true, $\varphi$ is identically plus infinity and the result is already proved), then we see that we must have $b \leqslant 0$ (let $t \geqslant \varphi(x)$ tend to plus infinity). If $b < 0$, it is easy to finish the proof: we just divide by $-b > 0$ and obtain

$$-\frac{\xi(x)}{b} - \varphi(x) \leqslant -\frac{c}{b} < -\frac{c+\theta}{b} \leqslant -\frac{\xi(a)}{b} - r.$$

Hence $\widetilde{\varphi}(-\xi/b) \leqslant -\xi(a)/b - r$ and therefore

$$\widetilde{\widetilde{\varphi}}(a) \geqslant -\frac{\xi(a)}{b} - \widetilde{\varphi}(-\xi/b) \geqslant r.$$

This means that $\widetilde{\widetilde{\varphi}}(a)$ is at least as large as $r$, which can be as close as we like to $\varphi(a)$. Hence $\widetilde{\widetilde{\varphi}}(a) \geqslant \varphi(a)$, implying equality. In particular we see that if $\varphi(a) < +\infty$, then $b$ must be negative and the reasoning already done is valid.

But it can happen that $b = 0$. This situation corresponds to the fact that the hyperplane is vertical: it is then not the graph of any affine function on $E$. This is the most difficult part of the proof.

In this situation $\varphi(a) = +\infty$ and we shall use a special trick to tilt the vertical hyperplane. When $b = 0$, (9.17.5) says that

$$\xi(x) \leqslant c < c + t \leqslant \xi(a)$$

for every $x$ such that there is a $t$ with $(x, t)$ in the finite epigraph, i.e., for every $x$ with $\varphi(x) < +\infty$. This yields, according to the definition of the Fenchel transform that

$$\begin{aligned}\widetilde{\varphi}(\theta + s\xi) &= \sup_{x \in \mathbf{dom}(\varphi)} \big(\theta(x) + s\xi(x) - \varphi(x)\big)\\ &\leqslant \widetilde{\varphi}(\theta) + \sup_{x \in \mathbf{dom}(\varphi)} s\xi(x) \leqslant \widetilde{\varphi}(\theta) + sc,\end{aligned}$$

for every $s > 0$, for in the definition of $\widetilde{\varphi}$ it is clear that only points $x$ with $\varphi(x) < +\infty$ contribute: see (9.17.2). Therefore we obtain

$$\begin{aligned}\widetilde{\widetilde{\varphi}}(a) &\geqslant (\theta + s\xi)(a) - \widetilde{\varphi}(\theta + s\xi) \geqslant (\theta + s\xi)(a) - \widetilde{\varphi}(\theta) - sc\\ &\geqslant \theta(a) + st - \widetilde{\varphi}(\theta).\end{aligned} \tag{9.17.6}$$

Here we now choose $\theta \in \Xi$ such that $\widetilde{\varphi}(\theta) < +\infty$; that this is possible we saw in the first part of the proof where we assumed that $\varphi(a) < +\infty$. If we now let $s$ tend to $+\infty$ in (9.17.6), we see that $\widetilde{\widetilde{\varphi}}(a) = +\infty$, which is what we wanted. □

So a function $\varphi$ which satisfies the three conditions can be represented as the supremum of a family of affine functions $\varphi(x) = \sup_\xi \big(\xi(x) - \widetilde{\varphi}(\xi)\big)$. This can be most helpful in proving that certain functions are convex: see Exercises 9.6–9.8.

It is now clear that Fenchel's inequality can be improved to

$$\xi(x) \leqslant \widetilde{\widetilde{\varphi}}(x) \dotplus \widetilde{\varphi}(\xi).$$

The topology defined by $\Xi$ is relevant also for the closure of convex hulls of sets:

**Corollary 9.17.9** (to Theorem 9.17.7). *For any given subset $\Xi$ of the algebraic dual $E^\star$ of a vector space $E$, we can retrieve the closure for the topology $\sigma(E,\Xi)$ defined by $\Xi$ of the convex hull of a set $A \subset E$ from its support function:*

$$\overline{\mathbf{cvxh}(A)} = \bigcap_{\xi\in\Xi} \{x \in E;\ \xi(x) \leqslant H_A(\xi)\}.$$

*As a consequence, the four sets $A$, $\overline{A}$, $\mathbf{cvxh}(A)$ and $\overline{\mathbf{cvxh}(A)}$ all have the same support function.*

**Proof.** We note that the indicator function of a set is lower semicontinuous for the topology defined by $\Xi$ if and only if the set is closed for this topology. Therefore the second Fenchel transform of an indicator function is the indicator function of the closure of the convex hull of the set. So this is just a special case of Fenchel's main theorem. □

Let us denote by $CVX_0(E, \mathbf{R}_!)$ the family of all functions with the three properties mentioned in the theorem. The restriction of the Fenchel transformation to $CVX_0(E, \mathbf{R}_!)$ is thus an involution, i.e., it is equal to its own inverse. Artstein-Avidan & Milman (2009:662) have proved that any decreasing involution $CVX_0(\mathbf{R}^n, \mathbf{R}_!) \to CVX_0(\mathbf{R}^n, \mathbf{R}_!)$ is, up to some obvious modifications, the restriction to $CVX_0(\mathbf{R}^n, \mathbf{R}_!)$ of the Fenchel transformation. Thus the Fenchel transformation is, essentially, unique among all transformations with the indicated properties. While this is (intuitively) hardly surprising, it is a very nice property and a convincing witness to the significance of Fenchel's work.

We shall also need a partial transform, defined as follows.

**Definition 9.17.10.** To any function $\varphi\colon X \times E \to \mathbf{R}_!$, where $X$ is any set and $E$ a vector space, we define its ***partial Fenchel transform*** $\varphi^\diamond\colon X \times E^\star \to \mathbf{R}_!$ on the Cartesian product of $X$ and the algebraic dual $E^\star$ of $E$ by

$$\varphi^\diamond(x, \eta) = \sup_{y\in E} \big(\eta(y) - \varphi(x, y)\big)$$

for $(x, \eta) \in X \times E^\star$. □

Usually we have $X = \mathbf{R}^n$ and $E = \mathbf{R}^m$.

The equivalence noted in (9.17.4) can be expressed in terms of epigraphs: $\mathbf{epi}\, \widetilde{\varphi} \supset \mathbf{epi}\, f$ if and only if $\mathbf{epi}\, \widetilde{f} \supset \mathbf{epi}\, \varphi$. This means that the lower inverse of the Fenchel transformation is equal to the Fenchel transformation if we order functions in the usual way. If we order functions instead by the inclusion relation between their epigraphs, it is about the upper inverse of the Fenchel transformation.

After a change of order on one of the sides it satisfies the second condition in Proposition 5.7.2, so that we have a Galois connection (cf. also Theorem 5.5.1 and Corollary 5.5.4). The property (9.17.3) means that we have an antierosion or a duality in the sense of Singer; i.e., after a change of the order relation we have a dilation or erosion (cf. condition (E) in Theorem 5.5.1). Singer (1997) studies several other dualities in convexity theory.

## 9.18 Duality of infimal convolution and addition

Infimal convolution is dual to addition under the Fenchel transformation:

**Theorem 9.18.1.** *Let $f, g\colon E \to \mathbf{R}_!$ be arbitrary functions on a vector space $E$. Then $\widetilde{f \sqcap g} = \tilde{f} \mathbin{\underset{\cdot}{+}} \tilde{g}$.*

**Proof.** The proof is an easy calculation, where Lemma 2.4.1 plays an essential role. If $\varphi$ and $\psi$ are convex, then their upper sum $\varphi \mathbin{\dot{+}} \psi$ is convex, but in general not their lower sum. However, when $\varphi = \tilde{f}$ and $\psi = \tilde{g}$, then also the lower sum is convex (since it is realized as a Fenchel transform). Moreover the lower sum is equal to the upper sum except when $f$ or $g$ is identically $+\infty$. □

This duality enables us to compare the two convolution operations we have studied, which we do in the next section.

## 9.19 Comparing two convolution operations

Let us compare classical convolution by means of sums or integrals on the one hand, and infimal convolution on the other hand. For definitions, see Definitions 3.3.1, 3.3.2 and 3.4.1, respectively.

There is a far-reaching analogy between the two—which should not come as a surprise, since infimal convolution is a tropicalization of classical convolution and also a limiting case of the latter.

### *Neutral elements*

Ordinary convolution has a neutral element $\delta_0$: $f*\delta_0 = f$, where $\delta_0$ denotes Kronecker delta placed at the origin in the discrete case and the Dirac measure when we work with integrals or distributions. For infimal convolution, $\mathbf{indf}_{\{0\}}$ is the neutral element as noted on page 49.

### *Loss of information*

When we take the Fourier transform of a convolution product we get, under suitable hypotheses, the multiplicative product of the Fourier transforms of the two factors:

$$\widehat{f*g} = \hat{f}\cdot\hat{g}.$$

For infimal convolution, the Fenchel transformation serves the same purpose, being the tropical counterpart of the Laplace or Fourier transformations (see Theorem 9.18.1 above):

$$\widetilde{f \sqcap g} = \tilde{f} \dotplus \tilde{g}.$$

In both cases these formula can yield a kind of inverse: for $h = f * g$ and $h = f \sqcap g$ the formulas

$$\hat{f} = \hat{h}/\hat{g} \text{ and } \tilde{f} = \tilde{h} - \tilde{g}$$

can sometimes be given a meaning. They enable us to reconstruct an unknown function $f$ if $g$ and $h = f*g$, or $g$ and $h = f \sqcap g$, respectively, are known.

It can happen that there is no loss of information at all: $f * \delta_0 = f$ and $f \sqcap \mathbf{indf}_{\{0\}} = f$. Also a total loss of information can occur: $f * 0 = 0$ and $f \sqcap (-\infty) = -\infty$. Typically, for both operations, there is some loss of information concerning $f$ when a convolution with $g$ is performed, but the loss is not total.

### *Approximation*

With $\varphi \in C^\infty(\mathbf{R}^n)$ with compact support, satisfying $\varphi \geqslant 0$ and of integral 1, and defining $\varphi_k(x) = k^n\varphi(kx)$, we have a well-known convergence

$f * \varphi_k \to f$ as $k \to +\infty$ (depending of the regularity of $f$; also for distributions). Similarly, with $V_k(x) = k\|x\|$, $f \sqcap V_k$ converges to $f$ under certain conditions; see Proposition 3.5.1. Here $f * \varphi_k$ is a weighted mean over a moving interval, and $f \sqcap V_k$ the infimum of $f$ over a moving interval, both containing useful information, sometimes in fact more useful than that of $f$ itself, since small variations have been eliminated—think of raw data in the form of temperatures registered every third hour when you want to study the climate over decades or centuries.

## 9.20 Coppel's axioms

William Andrew Coppel presented in his book (1998:6) axioms for generalized types of convexity. They are the following.

**Definition 9.20.1.** Given any set $X$, an ***alignment*** on $X$ is a family $\mathscr{A}$ of subsets of $X$ which satisfies the following two conditions.

(A1) $\mathscr{A}$ is a Moore family (see Definition 4.3.7).
(A2) The union of any nonempty family of sets in $\mathscr{A}$ which is totally ordered by inclusion belongs to $\mathscr{A}$. □

It is equivalent to take in Axiom (A2) a subfamily which is directed by inclusion (see Definition 2.7.2 and the discussion in Coppel 1998:24–25).

The family of closed sets in $\mathbf{R}^n$, $n \geqslant 1$, satisfies (A1) but not (A2).

It is easy to see that the convex sets form an alignment:

**Proposition 9.20.2.** *In any vector space $E$ over the reals, the family of all convex sets form an anlignment.* □

When we come to various forms of discrete convexity in the next chapter it is essential to check whether the new classes also form alignments.

Of relevance in this context is the concept of an algebraic cleistomorphism, which is defined in Definition 2.11.3.

In relation to Coppel's axioms in a discrete setting, see the study by Roy & Stell (2003).

## 9.21 Three fundamental properties

There are three fundamental properties in convex analysis in vector spaces over the field of real numbers that could be desirable to have also in digital convexity. In this section we present these properties. In the next chapter we shall show that they do not hold in a naive approach to digital convexity, and then, in a following chapter, indicate some of the attempts at saving them. Most definitions and some of the results are given also in higher dimension.

The classical results in convexity theory to be discussed are:

(1) *The image of a convex set under an affine mapping is convex. Equivalently, the marginal function of a convex function is convex;*
(2) *A local minimum of a convex function is global;*
(3) *There exists a separating hyperplane between two given disjoint convex sets.*

The first two results are very easily seen to hold; the third has been considered in Section 9.15.

### 9.21.1 *Property 1: Images of convex sets*

It is an everyday experience that the shadow of a convex set is convex. Expressed mathematically, we say that the image of a convex set in any dimension under an affine mapping is convex. This result is so easily proved that it often is mentioned just in passing and not as a theorem. To be explicit, if $A$ is a convex set in some $\mathbf{R}^n$ and $f\colon \mathbf{R}^n \to \mathbf{R}^m$ is affine, $a = f(x)$ and $b = f(y)$ for some $x, y \in A$, then the whole segment $[a, b]$ is in the range of $f$, since we have $(1-t)a + tb = (1-t)f(x) + tf(y) = f((1-t)x + ty)$.

It is enough here to consider a projection of the form

$$\mathbf{R}^n \ni (x_1, \ldots, x_n) \mapsto \pi(x_1, \ldots, x_n) = (x_1, \ldots, x_k) \in \mathbf{R}^k, \qquad k < n,$$

i.e., the mapping which forgets the $n-k$ variables $(x_{k+1}, \ldots, x_n)$. Indeed, any affine mapping $f\colon \mathbf{R}^n \to \mathbf{R}^m$ can be written as a composition

$$f = i \circ \tau \circ g \circ \pi,$$

where $\pi$ is the projection $\pi\colon \mathbf{R}^n \to \mathbf{R}^k$, $g\colon \mathbf{R}^k \to \mathbf{R}^k$ a linear bijection, $\tau$ a translation, and finally $i$ is the inclusion mapping of $\mathbf{R}^k$ into $\mathbf{R}^m$. The three mappings $g$, $\tau$ and $i$ all obviously preserve convexity, so it suffices to consider the projection $\pi$.

A special case is the result that the marginal function of a convex function is convex. Let us first define the marginal function in general.

**Definition 9.21.1.** Given any function $f\colon X \times Y \to \mathbf{R}_!$, we define its ***marginal function*** $f_{\mathrm{marg}}\colon X \to \mathbf{R}_!$ as

$$f_{\mathrm{marg}}(x) = \inf_{y \in Y} f(x, y) \tag{9.21.1}$$

for $x \in X$. □

Now, given any convex function $F$ on $\mathbf{R}^n \times \mathbf{R}^m$, its marginal function $F_{\mathrm{marg}}$, defined on $\mathbf{R}^n$, is convex.

This follows from the statement on projections since the image of the strict finite epigraph of $f$ under the projection which forgets the variable $y$ is equal to the strict finite epigraph of $f_{\mathrm{marg}}$:

$$\mathbf{epi}^{\mathrm{finite}}_{\mathrm{strict}}(f_{\mathrm{marg}}) = \pi_*(\mathbf{epi}^{\mathrm{finite}}_{\mathrm{strict}}(f)),$$

where $\pi\colon X \times Y \to X$ is the projection defined by $\pi(x, y) = x$, $x \in X$, $y \in Y$. Expressed otherwise, we have $t > f(x, y)$ for some $y$ if and only if $t > \inf_y f(x, y)$.

However, this special case implies the general case with the help of the indicator functions: for any subset $A$ of $X \times Y$ we have

$$\inf_{y \in Y} (\mathbf{indf}_A)(x, y) = 0 \text{ iff } x \in \pi_*(A).$$

So if the marginal function of $\mathbf{indf}_A$ is convex, then the projection $\pi_*(A)$ is convex, and from this we conclude that any affine mapping preserves convexity.

Perhaps more practical is the observation that if two convex functions $f$, $g$ are given, then the set $A = \mathbf{hypo}^{\mathrm{finite}}(-f) \cap (\mathbf{epi}^{\mathrm{finite}}(g)$ is convex, and the marginal functions of $f$ and $g$ define the projection of $A$. So the boundary of $A$ consists of a lower boundary and an upper boundary; each is given by a function which can then be studied using derivates or differences.

This property of marginal functions is of great importance in many results in convex optimization.

**Remark 9.21.2.** The operation of taking the marginal function is a special case of infimal convolution. Indeed, if we define $g(x)$, $x \in \mathbf{R}^n$, to be zero when $x_1 = x_2 = \cdots = x_m = 0$ and $+\infty$ elsewhere, then $h = f \sqcap g$, defined by

$$h(x_1, \ldots, x_m) = \inf_{x_{m+1}, \ldots, x_n} f(x_1, \ldots, x_m, x_{m+1}, \ldots, x_n),$$

is the marginal function $f_{\mathrm{marg}}$ of $f$. □

### 9.21.2 *Property 2: Local minima are global*

A fundamental property of convex functions of real variables is that a local minimum of a convex function is automatically global. In other words, if $F$ is convex on $\mathbf{R}^n$ and $F \geqslant F(0)$ in a neighborhood of the origin, then

$F \geqslant F(0)$ everywhere. Actually we do not even need a neighborhood: it is enough that $F \geqslant F(0)$ in a set such that, for every $x \in \mathbf{R}^n$, there exists a positive number $s_x$ such that $F(tx) \geqslant F(0)$ for all real numbers $t$ with $|t| \leqslant s_x$. So the result is valid in any dimension; no topology is needed.

This quite easy result of course facilitates everything in convex optimization.

### 9.21.3 *Property 3: Separating hyperplanes*

In $\mathbf{R}^n$ one can always separate two disjoint convex sets by a hyperplane. See Theorem 9.15.3.

## 9.22 Notes

### 9.22.1 *Sources for convexity theory*

Sources of historical interest are the publications by Hermann Minkowski (1903, 1910, 1911b) and Mandelbrojt (1939)—the latter with a first form of the Fenchel transform on the real line.

For the results touched upon here, the classical book by Rockafellar (1970, 1997) is still an important source. Another classic is that by Bonnesen & Fenchel (1934).

Important later works are those by Fenchel (1949, 1953, 1983), Moreau (1966, 1966–1967, 1967), Ioffe & Tikhomirov (1968), Hiriart-Urruty & Lemaréchal (1993, 2001), Gruber & Wills (1993) with several chapters, like those by Gruber (1993a, 1993b), Hörmander (1994), Coppel (1998), Boyd & Vandenberghe (2009), Barbu & Precupanu (2012), Simon (2014), Borwein & Giladi (2016), and Eckhoff (2016).

For a survey of the pioneering work of Jean-Jacques Moreau, see the article by Lionel Thibault (2016). Werner Fenchel's work and life is described in my paper (2019a).

The French term *sousgradient* and its English equivalent *subgradient* (Definition 9.9.4) was introduced by Jean-Jacques Moreau according to a letter from him to Jean-Baptiste Hiriart-Urruty dated 1993 March 26.

Zong (1996) discusses "strange" examples in convex and discrete geometry.

Rådström's Approximation Theorem 9.8.1, which says that iterated Minkowski sums of a set, properly scaled, converge to its convex hull, was published in his article (1952:115) and is actually more general than the theorem stated here. His result is often quoted as the Shapley–Folkman–

Starr theorem, also called the Emerson–Folkman–Greenleaf–Shapley–Starr theorem; see (Fradelizi et al. 2018:3, footnote 2). The lemma of Lloyd Shapley and Jon Folkman was first published by Ross M. Starr (1969). William R. Emerson and Frederick P. Greenleaf (1969) proved the result independently. For a survey and a detailed discussion on how fast the sets $A_{[k]}$ approach the convex hull of $A$, see (Fradelizi et al. 2018). This article does not mention Rådström's results in (1952).

For a survey of phenomena related to the support function, see (Ghosh & Kumar 1998).

In the family of all functions on a topological vector space with values in $\mathbf{R}_!$ we can define three operators related to the conditions (C), (L) and (M) in Theorem 9.17.7: that of taking the convex envelope **cvxe**, now to be denoted by $c$; that of taking the largest lower semicontinuous minorant, denoted by $l$; and that of mapping the function to the constant $-\infty$ if the function takes that value at some point and leaving it unchanged if not, here denoted by $m$. These three operators are idempotent and satisfy in addition the rules $cmc = mcm = mc$; $lml = mlm = ml$; $clc = lcl = lc$, where for brevity composition is written as juxtaposition. They generate a semigroup with at most 18 elements and was introduced and studied in my paper (2002a); it was later generalized to more than three generators by several persons, for instance by Kudryavtseva & Mazorchuk (2009). In a normed space of infinite dimension, the semigroup is realized with all the 18 elements possible; in some topological vector spaces the number is 17; in $\mathbf{R}^n$ with $n \geqslant 2$ three elements lead to the same operator so that the number is 16; in $\mathbf{R}$ four elements define the same operator, so that the semigroup has only 15 elements.

### 9.22.2 *The Kuratowski–Zorn lemma*

For the history of the Kuratowski–Zorn lemma (Theorem 9.5.1) and its other (more frequent) name Zorn's lemma, see (Campbell 1978).

### 9.22.3 *Terms used*

The term *supporting hyperplane* (Definition 9.6.1) is the usual one for the concept defined here, to be found in (Rockafellar 1970:100, 1997:100), (Hiriart-Urruty & Lemaréchal 1993:113, 2001:44), (Simon 2014:290), (Coppel 2014:169) and other books and articles. More unusual are *tangent hyperplane* (Rockafellar 1970:169, 1997:169; Simon 2014:122) and *support hyperplane* (Simon 2014:122).

The support function (Definition 9.17.1) is so called in many sources, for instance in (Rockafellar 1970:112, 1997:112), (Fenchel 1983:127), (Hiriart-Urruty & Lemaréchal 1993:208, 2001:134). (Fenchel 1949:73) has the *function of support.* Hörmander (1983:105, 1994:69) calls it the *supporting function.*

For the Fenchel transformation (Definition 9.17.4) many different terms have been used. It is related to the Legendre transformation, and therefore sometimes called just the Legendre transformation, but its character is quite different, since the Legendre transformation belongs to differential geometry, whereas the Fenchel transformation is a tropical counterpart to the Laplace transformation and belongs to the now rapidly growing field of tropical geometry as well as to the theory of complete lattices. For the plethora of terms, see my paper (2019a).

Werner Fenchel, when asked which term to prefer for the transformation, wrote in a private letter to the author dated 1977 March 07: "I do not want to add a new name, but nevertheless say, that if I now would give the transformation one, I would let me be guided by the analogy and the connection with polarity between convex sets (in dual vector spaces) and for instance call it parabolic polarity." (My translation from Danish; see my paper 2019a.)

## 9.23 Exercises

9.1. In any vector space we have $2A \subset A + A$ with equality if $A$ is convex. Show by example that the equality $2A = A + A$ is not sufficient for $A$ to be convex. Prove that, in $\mathbf{R}^n$, equality implies that the closure $\overline{A}$ is convex.

9.2. Prove that $\mathbf{cvxh}(A + B) \subset \mathbf{cvxh}(A) + \mathbf{cvxh}(B)$ for all subsets $A$ and $B$ of $\mathbf{R}^n$. Can the inclusion relation be strict?

9.3. Find examples to show that the inclusion relations in (9.13.2) can be strict.

9.4. Prove that for bounded closed convex sets $A$ and $B$ in $\mathbf{R}^n$, assuming $B$ to be nonempty, we have $\kappa_B(A) = A$, i.e., $A$ is invariant (closed) for $\kappa_B$. *Hint:* Start with $n = 1$.

9.5. Let $A$ be the filled square $\{x \in \mathbf{R}^2;\ \|x\|_\infty \leqslant 1\}$ and $B$ a Euclidean closed disk of radius $r$, viz.

$$B = B_{\leqslant}(0, r) = \{x \in \mathbf{R}^2;\ \|x\|_2 \leqslant r\}.$$

Describe for all positive $r$ the five sets $A+B = \delta_B(A) = \delta_A(B)$, $\varepsilon_B(A)$, $\varepsilon_A(B)$, $\alpha_B(A)$, $\alpha_A(B)$ (the sets $\kappa_B(A)$ and $\kappa_A(B)$ are taken care of in the previous exercise).

9.6. Let $f$ be a convex function of one variable for $0 < x < 1$ and such that $f(x) < 0$ for these $x$. Prove that

$$g(x, y) = -\frac{y^2}{f(x)}$$

is convex as a function of $(x, y)$ in the strip $0 < x < 1$, $y \in \mathbf{R}$.

9.7. Suppose that $f$ is a convex function of one variable $y$ for $a < y < b$ and that $f(y) < 0$ there. Prove that

$$g(x, y) = x \log \left( \frac{-x}{f(y)} \right)$$

is convex as a function of $(x, y)$ for $x > 0$, $a < y < b$. *Hint:*

$$x \log x = \sup_{\xi \in \mathbf{R}} \left( x\xi - e^{\xi - 1} \right), \qquad x > 0.$$

9.8. Take an affine function $f$ in $\mathbf{R}^3$ and a convex function $g$ which is negative in the cube $|x| < 1$, $|y| < 1$, $|z| < 1$ in $\mathbf{R}^3$. Prove that

$$h(x, y, z) = -\frac{f(x, y, z)^4}{g(x, y, z)^3}$$

is convex in the cube. *Hint:* Try to prove this using differential calculus if $g$ is of class $C^2$ ... no, this is a bad piece of advice.

9.9. Let $f(x) = \|x\|_2$, $x \in \mathbf{R}^n$. Determine the function $f_t$, $t > 0$, in Proposition 9.10.3. Note that the gradient of $f_t$ is actually continuous.

# Chapter 10

# Discrete convexity

## 10.1 Introduction

DISCRETE CONVEXITY is a term used to describe, in a discrete setting, many efforts to build up theories and construct algorithms with properties which are analogous to the ones in classical convexity theory. For a background to the latter, we refer to the preceeding chapter.

In this chapter we study these questions in any finite dimension; in the next, a more detailed study of geometry in two dimensions, including functions of one variable, will be presented.

## 10.2 Restrictions and extensions of functions

Since we have a well-established theory of convexity in $\mathbf{R}^n$ it is only natural to utilize that knowledge by considering extensions to $\mathbf{R}^n$ of functions on $\mathbf{Z}^n$.

Given a function $f\colon \mathbf{Z}^n \to \mathbf{R}_!$ we extend it in two ways to a function defined on $\mathbf{R}^n$ depending on the situation. We define $f_0 = 0$ in $\mathbf{R}^n \smallsetminus \mathbf{Z}^n$, and $f_\infty = +\infty$ in $\mathbf{R}^n \smallsetminus \mathbf{Z}^n$. In the first case, zero is used since that value does not influence sums; in the second, $+\infty$ since that value does not influence infima.

The extension $f_0$ is suitable when studying convolutions and $f_\infty$ when studying Fenchel transforms. Indeed

$$
\begin{aligned}
(f_0 * g)(x) &= \sum_{y\in\mathbf{Z}^n} f(y)g(x-y), \qquad x \in \mathbf{R}^n; \\
(f_\infty \sqcap g)(x) &= \inf_{y\in\mathbf{Z}^n} (f(y) \dot{+} g(x-y)), \; x \in \mathbf{R}^n; \text{ and} \\
\widetilde{f_\infty}(\xi) &= \sup_{x\in\mathbf{Z}^n} (\xi\cdot x - f(x)), \qquad \xi \in \mathbf{R}^n.
\end{aligned}
$$

## 10.3 Convexity with respect to a subset of a vector space

Already in Chapter 8, for instance in property (B) in Theorem 8.5.1 and in property (1) in Definition 8.9.1, we encountered a kind of discrete convexity, a property which we shall now give a name.

**Definition 10.3.1.** Given any subset $Y$ of a vector space $E$, we define the $Y$-***convex hull*** of a set $A \subset Y$ as $\mathbf{cvxh}(A) \cap Y$. We shall say that a subset $A$ of $Y$ is $Y$-***convex*** if its $Y$-convex hull is equal to $A$. □

When $Y = E$ we get usual convexity. Of interest here are primarily the cases $E = \mathbf{R}^n$, $Y = \mathbf{Z}^n \subset \mathbf{R}^n$ and $E = \mathbf{R}^n \times \mathbf{R}$, $Y = \mathbf{Z}^n \times \mathbf{R} \subset \mathbf{R}^n \times \mathbf{R}$. The $Y$-convex hull is equal to the smallest $Y$-convex set containing $A$; see the next proposition.

Since we always have $A \subset \mathbf{cvxh}(A) \cap Y$ for subsets $A$ of $Y$, $Y$-convexity of $A$ is equivalent to the inclusion

$$\mathbf{cvxh}(A) \cap Y \subset A. \tag{10.3.1}$$

**Proposition 10.3.2.** *In any vector space $E$ and any subset $Y$ of $E$ the $Y$-convex sets form an alignment in the sense of Definition 9.20.1.*

This will allow us to apply the general results on abstract convexity satisfying Coppel's axioms to $Y$-convexity.

**Proof.** (A1) We shall prove that the $Y$-convex sets form a Moore family, i.e., that if $A_j$, $j \in J$, are all $Y$-convex, then so is their intersection $A$. So assume that $\mathbf{cvxh}(A_j) \cap Y \subset A_j$; we shall prove that $\mathbf{cvxh}(A) \cap Y \subset A$. We have for all indices $j$,

$$\mathbf{cvxh}(A) \cap Y \subset \mathbf{cvxh}(A_j) \cap Y \subset A_j.$$

Taking the intersection over all $j$, we get the result.

(A2) Let now a family $(A_j)_{j\in J}$ of $Y$-convex sets $A_j$ be given, where $J$ is a totally ordered index set. Assume that $A_j \subset A_k$ for all $j, k \in J$ with $j \leqslant k$. We shall then prove that the union $A$ of all the $A_j$ is $Y$-convex. Let now $x$ belong to the $Y$-convex hull $\mathbf{cvxh}(A) \cap Y$ of $A$. If $x$ belongs to this set, there is a finite subset $B$ of $\mathbf{cvxh}(A)$ such that $x$ belongs to the convex hull of $B$ (see the discussion in the beginning of Section 9.7). Then there exists an index $k \in J$ such that $B$ is a subset of $A_k$, and we conclude that $x$ belongs to $\mathbf{cvxh}(A_k) \cap Y \subset A_k \subset A$. Since this is true for all $x$ in $\mathbf{cvxh}(A) \cap Y$, we see that $\mathbf{cvxh}(A) \cap Y$ is contained in $A$, which is thus $Y$-convex. □

**Definition 10.3.3.** Given subsets $X$ and $Z \subset X$ of a vector space $E$, and defining $Y = Z \times \mathbf{R} \subset X \times \mathbf{R} \subset E \times \mathbf{R}$, we shall say that a function $f\colon X \to \mathbf{R}_!$ is ***Y-convex*** if its finite epigraph is a $Y$-convex set in the sense of Definition 10.3.1. The largest $Y$-convex minorant of $f$ shall be called the ***Y-convex envelope*** of $f$. □

Thus $f$ is $Y$-convex if and only if

$$\text{(10.3.2)} \qquad \mathbf{cvxh}(\mathbf{epi}^{\text{finite}}(f)) \cap Y \subset \mathbf{epi}^{\text{finite}}(f);$$

cf. (10.3.1).

When $X = Z = \mathbf{R}^n$, hence $Y = \mathbf{R}^n \times \mathbf{R}$, $Y$-convexity is the usual convexity for functions $F \in \mathscr{F}(\mathbf{R}^n, \mathbf{R}_!)$.

When $Y = \mathbf{Z}^n \times \mathbf{R}_!$, $Y$-convexity means that $f$ is the restriction of some convex function defined in $\mathbf{R}^n$. It is then equal to $\mathbf{cvxe}(f)|_{\mathbf{Z}^n}$.

**Remark 10.3.4.** A $\mathbf{Z}^n$-convex set need not be connected in any sense. We can take $A = \{(0,0),(3,2)\} \subset \mathbf{Z}^2$, which is $\mathbf{Z}^2$-convex since it is the intersection of $\mathbf{Z}^2$ and the segment $\{(x, 2x/3);\ 0 \leqslant x \leqslant 3\} \subset \mathbf{R}^2$. More generally, we can take $A = \{(0,0),(p,q)\}$ with $p, q \in \dot{\mathbf{N}}$ and relatively prime. A strengthened definition introduced and studied by Jacques-Olivier Lachaud (2021) eliminates these bad examples in that it requires that $A$ and $\mathbf{cvxh}(A)$ shall yield the same result when $n+1$ naturally defined cleistomorphisms are applied to them, thus satisfying more conditions than those in Definition 10.3.1. □

## 10.4 The integer neighborhood and the canonical extension

**Definition 10.4.1.** We define the ***integer neighborhood*** of a real number $a$, denoted by $N(a)$, as the set $\{\lfloor a \rfloor, \lceil a \rceil\} \subset \mathbf{Z}$. We define the ***integer neighborhood*** of a point $a = (a_1, \dots, a_n) \in \mathbf{R}^n$ as the set

$$N(a) = N(a_1) \times \cdots \times N(a_n),$$

which is a subset of $\mathbf{Z}^n$. □

The integer neighborhood has $2^k$ elements, where $k$ is the number of indices $j$ such that $a_j \in \mathbf{R} \smallsetminus \mathbf{Z}$. Equivalently,

$$N(a) = \big(a + B^{\infty}_{<}(0,1)\big) \cap \mathbf{Z}^n, \qquad a \in \mathbf{R}^n,$$

where $B^{\infty}_{<}(c,r)$ denotes the strict ball for the $l^{\infty}$-norm with center at $c$ and of radius $r$. The mapping

$$\mathbf{R}^n \supset A \mapsto \nu(A) = \bigcup_{a \in A} N(a) \subset \mathbf{Z}^n$$

is one of many digitizations of $\mathbf{R}^n$ and commutes with the formation of arbitrary unions, thus is a dilation, i.e.,

$$\nu\big(\bigcup_{j\in J} A_j\big) = \bigcup_{j\in J} \nu(A_j), \qquad A_j \subset \mathbf{R}^n.$$

**Definition 10.4.2.** The ***canonical extension*** of a function $f\colon \mathbf{Z}^n \to \mathbf{R}_!$ is defined, for every $a \in \mathbf{R}^n$, as the value at $a$ of the convex envelope of $f|_{N(a)}$, the restriction of $f$ to the integer neighborhood of $a$. We shall denote it by $\mathbf{can}(f)\colon \mathbf{R}^n \to \mathbf{R}_!$. □

The canonical extension is actually an extension, since $N(a) = \{a\}$ for every $a \in \mathbf{Z}^n$.

It is useful to have another characterization of the canonical extension:

**Proposition 10.4.3.** *For any function $f\colon \mathbf{Z}^n \to \mathbf{R}_!$, any point $a \in \mathbf{R}^n$, and any $p \in \mathbf{Z}^n$ such that the cube $p + [0,1]^n$ contains $a$, the value of the canonical extension at $a$ is equal to the value at $a$ of the convex envelope of $f|_{p+\{0,1\}^n}$.*

**Proof.** For brevity, let us denote by $C(p)$ the cube $p+[0,1]^n$ and by $V(p)$ its set of vertices, $p + \{0,1\}^n$, for any point $p$ with integer coordinates.

If a point $a$ belongs to only one cube $C(p)$, then $N(a) = V(p)$ and there is nothing to prove.

However, a point $a$ may belong to two different cubes $C(p)$ and $C(q)$, $p, q \in \mathbf{Z}^n$, $p \neq q$. Then $N(a)$ is a subset of $V(p) \cap V(q)$. Since $N(a)$ is a subset of $V(p)$ if $a \in V(p)$, we have $\mathbf{cvxe}\,\big(f|_{N(a)}\big) \geqslant \mathbf{cvxe}\,\big(f|_{V(p)}\big)$.

To prove the converse inequality, we define, given $a \in \mathbf{R}^n$ and $p \in \mathbf{Z}^n$ such that $a \in C(p)$, two sets of indices

$$J_k = \{j \in [1,n]_{\mathbf{Z}};\ a_j = p_j + k\}, \qquad k = 0, 1,$$

and an affine function

$$G(x) = \sum_{j\in J_0} (x_j - p_j) + \sum_{j\in J_1} (p_j + 1 - x_j), \qquad x \in \mathbf{R}^n.$$

If both $J_0$ and $J_1$ are empty, then $G$ is identically zero and $C(p)$ is the only cube to which $a$ belongs. We now assume that this is not the case. Then the zero set of $G$ is a hyperplane $Y_{a,p}$ in $\mathbf{R}^n$. Obviously $G$ is nonnegative in the cube $C(p)$, and $Y_{a,p}$ is a supporting hyperplane of this cube. In general a supporting hyperplane intersects $V(p)$ in a set which contains other vertices than those in $N(a)$, but in view of our construction, the hyperplane has the important property that

$$Y_{a,p} \cap V(p) = N(a).$$

This implies that any convex combination of points in $V(p)$ yielding a point in $Y_{a,p}$ is already a convex combination of points in $N(a)$. This proves that the convex envelope of $f|_{V(p)}$ is equal to the convex envelope of $f|_{N(a)}$ at $a$. We are done. □

## 10.5 Properties of the canonical extension

If a function $f\colon \mathbf{R}^n \to \mathbf{R}_!$ does not take the value $-\infty$, then

$$\mathbf{can}(f) = \inf_{p\in\mathbf{Z}^2} \mathbf{cvxe}\left(f|_{p+\{0,1\}^2}\right) = \inf_{p\in\mathbf{Z}^2} \left(f|_{p+\{0,1\}^2}\right)^{\approx}.$$

For the next theorem we shall need a function $\psi\colon \mathbf{R}^2 \to \mathbf{R}$ defined as follows

$$\psi(s,t) = \begin{cases} 0 & \text{if } st \geqslant 0 \\ \min(|s|,|t|) & \text{if } st < 0. \end{cases} \tag{10.5.1}$$

This function is thus zero if $s$ and $t$ are of the same sign, and positive if $st < 0$. It mesures to what extent the two numbers are of different sign. Indeed, its value at a point $(s,t)$ is the $l^\infty$-distance from $(s,t)$ to the union of the first and third quadrants.

In the following theorem we shall use the difference operators $D_a$ defined in Definition 8.3.1.

**Theorem 10.5.1.** *The operation $f \mapsto \mathbf{can}(f)$ is increasing, positively homogenous, i.e.,*

$$\mathbf{can}(\lambda f) = \lambda\mathbf{can}(f), \qquad \lambda \geqslant 0, \quad f \in \mathscr{F}(\mathbf{Z}^2,\mathbf{R}),$$

*and superadditive in the sense that, for all $x \in p + [0,1]^2$,*

$$\begin{aligned}(\mathbf{can}(f+g))(x) &\geqslant (\mathbf{can}(f))(x) + (\mathbf{can}(g))(x) \\ &\geqslant (\mathbf{can}(f+g))(x) - \tfrac{1}{2}\psi(D_{(0,1)}D_{(1,0)}f(p), D_{(0,1)}D_{(1,0)}g(p)),\end{aligned} \tag{10.5.2}$$

*where $\psi$ is defined by* (10.5.1). *More precisely, defining a function $\Psi_+(f,g)\colon \mathbf{Z}^2 \to \mathbf{R}$ as*

$$\Psi_+(f,g) = \mathbf{can}(f) + \mathbf{can}(g) - \mathbf{can}(f+g), \quad f,g \in \mathscr{F}(\mathbf{Z}^2,\mathbf{R}), \tag{10.5.3}$$

*we have for $x$ in a square $p + [0,1]^2$ with $p \in \mathbf{Z}^2$,*

$$\begin{aligned}\Psi_+(f,g)(x) &= \psi(D_{(0,1)}D_{(1,0)}f(p), D_{(0,1)}D_{(1,0)}g(p)) \\ &\quad\times \left[\max(|x_1 - p_1 - \tfrac{1}{2}|, |x_2 - p_2 - \tfrac{1}{2}|) - \tfrac{1}{2}\right].\end{aligned} \tag{10.5.4}$$

*This quantity is* $\leqslant 0$ *in the square* $p + [0,1]^2$, *zero when* $x$ *belongs to the sides of the square, and it attains its minimum at the center* $p + (\frac{1}{2}, \frac{1}{2})$ *of the square. In particular, we have*

$$\Psi_+(f, -f)(x) = |D_{(0,1)}D_{(1,0)}f(p)| \times \left[\max(|x_1 - p_1 - \tfrac{1}{2}|, |x_2 - p_2 - \tfrac{1}{2}|) - \tfrac{1}{2}\right].$$

*For every point* $p \in \mathbf{Z}^2$ *we have*

$$\mathbf{can}(f + g) = \mathbf{can}(f) + \mathbf{can}(g)$$

*in the square* $p + [0,1]^2$ *if and only if* $[D_{(0,1)}D_{(1,0)}f(p)] \cdot [D_{(0,1)}D_{(1,0)}g(p)] \geqslant 0$. *In particular, we have*

$$\mathbf{can}(f + H|_{\mathbf{Z}^2}) = \mathbf{can}(f) + H$$

*for all affine functions* $H$. *In other words,* $\Psi_+(f + H|_{\mathbf{Z}^2}, g + K|_{\mathbf{Z}^2}) = \Psi_+(f, g)$ *for all affine functions* $H, K \colon \mathbf{R}^2 \to \mathbf{R}$, *and* $\Psi_+(f,g)$ *vanishes in a square* $p + [0,1]^2$ *if and only if* $[D_{(0,1)}D_{(1,0)}f(p)] \cdot [D_{(0,1)}D_{(1,0)}g(p)] \geqslant 0$.

**Proof.** For the proof of (10.5.4) we first observe that we can reduce the problem to the case when $p = (0,0)$ and $f$ and $g$ vanish at the three points $(0,0)$, $(1,0)$ and $(0,1)$. Indeed, we can add affine functions to $f$ and $g$ to attain this, and $\Psi_+(f,g)$ does not change when we do this. Then, writing $s = f(1,1)$ and $t = g(1,1)$ we see that $D_{(0,1)}D_{(1,0)}f(0,0) = s$ and $D_{(0,1)}D_{(1,0)}g(0,0) = t$.

It is now easy to see that $\mathbf{can}(f + g) = \mathbf{can}(f) + \mathbf{can}(g)$ if $st \geqslant 0$. Indeed, if for instance $s, t \geqslant 0$, then

$$\begin{aligned}(\mathbf{can}(f))(x) &= s(x_1 + x_2 - 1)^+,\\ (\mathbf{can}(g))(x) &= t(x_1 + x_2 - 1)^+,\\ (\mathbf{can}(f+g))(x) &= (s+t)(x_1 + x_2 - 1)^+\end{aligned}$$

in the square.

If $st < 0$, say for definiteness that $s \geqslant -t > 0$, then

$$\begin{aligned}(\mathbf{can}(f))(x) &= s(x_1 + x_2 - 1)^+,\\ (\mathbf{can}(g))(x) &= \max(tx_1, tx_2),\\ (\mathbf{can}(f+g))(x) &= (s+t)(x_1 + x_2 - 1)^+\end{aligned}$$

in the square. This shows that $\Psi_+(f,g)$ equals the convex function

$$\Psi_+(f,g)(x) = |t| \left[\max\left(|x_1 - \tfrac{1}{2}|, |x_2 - \tfrac{1}{2}|\right) - \tfrac{1}{2}\right], \qquad x \in [0,1]^2.$$

We note that, in this case, $|t| = \min(|s|, |t|) = \psi(s,t)$. In view of the symmetry it suffices to look at the two cases already considered. □

## 10.6 Integral convexity

To every function $f\colon \mathbf{Z}^n \to \mathbf{R}_!$ we associate two functions $\mathbf{R}^n \to \mathbf{R}_!$. The first is the convex envelope of $f$, already defined in Definition 9.13.1. The second is the canonical extension of $f$, which was defined in Definition 10.4.2.

**Definition 10.6.1.** We shall say with Favati & Tardella (1990:9) that a function $f\colon \mathbf{Z}^n \to \mathbf{R}_!$ is ***integrally convex*** if $\mathbf{can}(f)$ is convex. □

The canonical extension of a function defined in $\mathbf{Z}^2$ is thus defined in $\mathbf{R}^2$ and convex in each of the squares $p+[0,1]^2$, $p \in \mathbf{Z}^2$, but it is not necessarily convex on all of $\mathbf{R}^2$: there is no reason why it should be convex across the vertical and horizontal lines.

The canonical extension is always an extension of $f$ but not necessarily convex; the convex envelope is always convex but not necessarily an extension of $f$.

We always have $\mathbf{cvxe}(f) \leqslant \mathbf{can}(f)$ with equality if and only if $f$ is integrally convex. Every integrally convex function is convex extensible, since for such a function, $\mathbf{can}(f)$ is a convex extension.

**Remark 10.6.2.** To determine $(\mathbf{can}(f))(x)$ for a function of two variables, we need to know the values of $f$ at no more than four points, viz. on the set

$$N(x) = \{(\lfloor x_1 \rfloor, \lfloor x_2 \rfloor), (\lceil x_1 \rceil, \lfloor x_2 \rfloor), (\lceil x_1 \rceil, \lceil x_2 \rceil), (\lfloor x_1 \rfloor, \lceil x_2 \rceil)\}.$$

To determine $\mathbf{cvxe}(f)$ for a function $f\colon \mathbf{Z}^n \to \mathbf{R}_!$ on the other hand, it may well be that no finite amount of information on $f$ is sufficient (see Exercise 10.15). Therefore there does not exist any algorithm to compute $\mathbf{cvxe}(f)$. □

## 10.7 Lateral convexity: Definitions

The second-order operators $D_b D_a$, given by

$$(D_b D_a F)(x) = F(x+a+b) - F(x+b) - F(x+a) + F(x) \tag{10.7.1}$$

will play an important role in the sequel.

We can extend $D_a$ to an operator $(D_a)_!\colon \mathscr{F}(\mathbf{R}^n, \mathbf{R}_!) \to \mathscr{F}(\mathbf{R}^n, \mathbf{R}_!)$ operating on functions with infinite values:

$$((D_a)_! F)(x) = F(x+a) \dotplus (-F(x)) \tag{10.7.2}$$

for $x, a \in \mathbf{R}^n$ and $F \in \mathscr{F}(\mathbf{R}^n, \mathbf{R}_!)$, and where $\dot{+}$ denotes upper addition, defined in (2.4.1).

We extend also the operator $D_a D_b$ to functions taking infinite values by defining

$$((D_b D_a)_! F)(x) = F(x+a+b) \dot{+} [(-F(x+a)) \underset{\cdot}{+} (-F(x+b))] \dot{+} F(x).$$

In particular the definition implies that $(D_b D_a)_! F \geqslant 0$ if and only if

$$F(x) \dot{+} F(x+a+b) \geqslant F(x+a) \dot{+} F(x+b), \qquad x \in \mathbf{Z}^n.$$

Of course $(D_b D_a)_! = (D_a D_b)_!$, but in general $(D_b D_a)_!$ is different from $(D_b)_!(D_a)_!$ as shown by a simple example:

**Example 10.7.1.** Define $f(0,0) = f(1,0) = +\infty$ and $f(1,1) = f(2,1) = 0$. Then for $a = (1,0)$ and $b = (1,1)$ we have

$$((D_b D_a)_! f)(0,0) = ((D_a D_b)_! f)(0,0) = ((D_a)_!(D_b)_! f)(0,0) = +\infty,$$

while $((D_b)_!(D_a)_! f)(0,0) = -\infty$. □

**Definition 10.7.2.** Given any subset $A$ of $\mathbf{Z}^n \times \mathbf{Z}^n$, we shall say that a function $f\colon \mathbf{Z}^n \to \mathbf{R}$ is ***A*-*laterally convex*** if $D_b D_a f \geqslant 0$ for all $(a,b) \in A$. If the function has infinite values, thus $f\colon \mathbf{Z}^n \to \mathbf{R}_!$ we shall say that it is ***A*-*laterally convex in the generalized sense*** if $(D_b D_a)_! f \geqslant 0$ for all $(a,b) \in A$. □

**Definition 10.7.3.** We denote by $\Phi(A)$ the set of all $A$-laterally convex functions. In the other direction, given any subset $\mathscr{G}$ of $\mathscr{F}(\mathbf{Z}^n, \mathbf{R})$, we define $\Psi(\mathscr{G})$ as the set of all pairs $(a,b) \in \mathbf{Z}^n \times \mathbf{Z}^n$ such that $D_b D_a f \geqslant 0$ for all $f \in \mathscr{G}$. We also define $\Lambda$ as the composition $\Psi \circ \Phi\colon \mathscr{P}(\mathbf{Z}^n \times \mathbf{Z}^n) \to \mathscr{P}(\mathbf{Z}^n \times \mathbf{Z}^n)$. □

The compositions $\Phi \circ \Psi$ and $\Psi \circ \Phi$ are cleistomorphisms, to be studied in Chapters 12 and 13.

We note what happens under a change of variables. If $f\colon \mathbf{Z}^n \to \mathbf{R}$ and $g = f \circ \varphi$, where $\varphi$ is any mapping $\mathbf{Z}^s \to \mathbf{Z}^n$, then

$$D_p g(x) = f(\varphi(x+p)) - f(\varphi(x)), \qquad x \in \mathbf{Z}^s. \tag{10.7.3}$$

In particular, if $\varphi$ is affine, then

$$\varphi(x+p) = \varphi(x) + Mp = \varphi(x) + \sum_{k=1}^{s} m_{jk} p_k, \qquad x, p \in \mathbf{Z}^s, \tag{10.7.4}$$

where $M$ is the constant matrix $M = (m_{jk})_{j=1}^{n}{}_{k=1}^{s}$ with $m_{j,k} = D_{e^{(k)}}\varphi_j$, defining $e^{(k)}$ to be the unit vector with 1 at place $k$ and 0 at the other places. Hence

$$D_p g = (D_{Mp} f) \circ \varphi, \qquad p \in \mathbf{Z}^s.$$

Continuing, we get

$$D_q D_p g = (D_{Mq} D_{Mp} f) \circ \varphi \in \mathscr{F}(\mathbf{Z}^s, \mathbf{R}), \qquad p, q \in \mathbf{Z}^s. \tag{10.7.5}$$

## 10.8 Lateral convexity: Morphological aspects

The notions of mathematical morphology are very helpful when it comes to understanding lateral convexity.

The mappings $\Phi$ and $\Psi$ defined in Definition 10.7.3 are decreasing and $\Psi \circ \Phi$ and $\Phi \circ \Psi$ are larger than the respective identity mappings. One expresses this fact by saying that the pair of mappings $(\Phi, \Psi)$ forms a Galois connection, defined in Subsection 5.7.1. This fact can also be expressed using the concept of the lower inverse of a mapping between ordered sets; see Chapter 5.

We defined in Definition 10.7.3 also

$$\Lambda(A) = \Psi(\Phi(A)), \qquad A \in \mathscr{P}(\mathbf{Z}^n \times \mathbf{Z}^n).$$

It is well known from Galois theory and easy to see that the operation $A \mapsto \Lambda(A)$ is increasing and idempotent, thus an ethmomorphism. It is also larger than the identity, and so a cleistomorphism.

If a function is $A$-laterally convex, it is automatically $\Lambda(A)$-laterally convex; any set $B$ satisfying $A \subset B \subset \Lambda(A)$ defines the same class of functions.

From the definition it is obvious that the class $\Phi(A)$ of $A$-laterally convex functions is closed under addition and multiplication by a nonnegative scalar. From the formulas

$$(D_{-a}f)(x) = -(D_a f)(x-a), \qquad (D_{-b}D_{-a}f)(x) = (D_b D_a f)(x-a-b)$$

it follows that $-A = \{(-a,-b);\ (a,b) \in A\}$ is contained in $\Lambda(A)$. The same is true of $A^{\smile} = \{(b,a);\ (a,b) \in A\}$. We define

$$A^{\text{sym}} = A \cup (-A) \cup A^{\smile} \cup (-A)^{\smile},$$

which may have up to four times as many elements as $A$ but still defines the same class, i.e., $\Phi(A^{\text{sym}}) = \Phi(A)$.

The formula

$$D_b D_{-a} f(x) = -D_b D_a f(x-a)$$

shows that $f$ is $\{(-a,b)\}$-laterally convex if and only if $-f$ is $\{(a,b)\}$-laterally convex. So the concepts introduced will enable us to study also $A$-laterally concave functions and $A$-laterally affine functions.

The formula

$$(D_b f)(x) + (D_c f)(x+b) = (D_{b+c} f)(x)$$

applied to $D_a f$ yields

$$(D_b D_a f)(x) + (D_c D_a f)(x+b) = (D_{b+c} D_a f)(x), \tag{10.8.1}$$

which implies that if $D_b D_a f \geqslant 0$ and $D_c D_a f \geqslant 0$, then we also have $D_{b+c} D_a f \geqslant 0$. This means that the set of pairs $\{(a,b) \in \mathbf{Z}^n \times \mathbf{Z}^n\}$ such that the inequality holds is closed under partial addition:

$$(a,b) +_{\{1\}} (a,c) = (a, b+c), \tag{10.8.2}$$

i.e., if the first elements agree, we may add the two pairs. For sets we define

$$B +_{\{1\}} C = \{(a, b+c);\ (a,b) \in B, (a,c) \in C\}.$$

Similarly we can define of course

$$(a,b) +_{\{2\}} (c,b) = (a+c, b) \tag{10.8.3}$$

and

$$B +_{\{2\}} C = \{(a+c, b);\ (a,b) \in B, (c,b) \in C\}$$

when the two second elements are the same (see the definition of restricted Minkowski addition defined in Section 3.8).

By repeated use of these formulas we see that $\Lambda(A) = \Psi \circ \Phi$, defined in Definition 10.7.3, contains the sets

$$A^{\text{sym}} +_{\{1\}} A^{\text{sym}}, \quad A^{\text{sym}} +_{\{1\}} (A^{\text{sym}} +_{\{2\}} A^{\text{sym}}) \text{ and so on.}$$

We sum up the discussion on $\Lambda(A)$ in the following lemma.

**Lemma 10.8.1.** *Let $A$ be any subset of $\mathbf{Z}^n \times \mathbf{Z}^n$.*

(1) *For any $a \in \mathbf{Z}^n$, $(a, \mathbf{0})$ and $(\mathbf{0}, a)$ belong to $\Lambda(A)$.*
(2) *If $(a,b) \in \Lambda(A)$, then $(b,a), (-a,-b), (-b,-a)$ all belong to $\Lambda(A)$.*
(3) *If $(a,b), (a,c) \in \Lambda(A)$, then $(a,b) +_{\{1\}} (a,c) = (a, b+c)$ belongs to $\Lambda(A)$.*
(4) *If $(a,c), (b,c) \in \Lambda(A)$, then $(a,c) +_{\{2\}} (b,c) = (a+b, c)$ belongs to $\Lambda(A)$.*
(5) *For any given family $\mathscr{G}$ of functions $\mathbf{Z}^n \to \mathbf{R}$, if $\Psi(\mathscr{G})$ contains a set $A$, it also contains $\Lambda(A)$.* □

When $n = 1$ and $A = \{(1,1)\}$, $f$ is $A$-laterally convex if and only if it is convex extensible. As already mentioned, this is the only reasonable definition of convexity in one integer variable. We note that it is equivalent to $B$-lateral convexity for any $B$ such that

$$(1,1) \in B \subset \Lambda(A) \text{ or } (-1,-1) \in B \subset \Lambda(A).$$

In this case, $\Lambda(A)$ is easy to determine: It is equal to

$$\{(s,t) \in \mathbf{Z} \times \mathbf{Z};\ st \geqslant 0\}.$$

More generally, for any $n$ and any $j \in [1,n]_{\mathbf{Z}}$, if $A = \{(e^{(j)}, e^{(j)})\}$, then a function is $A$-laterally convex if and only if it is convex extensible in the variable $x_j$ when the others are kept fixed. Since this is a convenient property, we shall normally require that

$$(e^{(j)}, e^{(j)}) \in A, \qquad j = 1, \dots, n. \tag{10.8.4}$$

If this is so, all $A$-laterally convex functions are $\{(1,1)\}$-laterally convex in each variable when the others are kept fixed.

## 10.9 Lateral convexity: Examples

**Example 10.9.1.** If $f$ is the restriction to $\mathbf{Z}^n$ of a polynomial of degree at most two,

$$f(x) = \alpha + \sum_{j=1}^{n} \beta_j x_j + \sum_{j=1}^{n}\sum_{k=1}^{n} \gamma_{jk} x_j x_k, \qquad x \in \mathbf{Z}^n,$$

with $\gamma_{jk} = \gamma_{kj}$, we see that

$$(D_b D_a f)(x) = 2\sum_{j=1}^{n}\sum_{k=1}^{n} \gamma_{jk} a_j b_k,$$

so that $f$ is $A$-laterally convex if and only if the last expression is nonnegative for all $(a,b) \in A$. When $A$ is the full set $\mathbf{Z}^n \times \mathbf{Z}^n$, this means that the matrix $(\gamma_{jk})$ is positive semidefinite. In particular the restriction to $\mathbf{Z}^n$ of an arbitrary affine function is $A$-laterally convex (the matrix $(\gamma_{jk})$ is zero).

We also see that the special polynomial $f(x) = x_j^2$ is $A$-laterally convex if and only if $a_j b_j \geqslant 0$ for all $(a,b) \in A$. Conversely, if $a_j b_j \geqslant 0$ and $g$ is any convex-extensible function of one variable, then the function $x \mapsto g(x_j)$ is $\{(a,b)\}$-laterally convex. □

In view of this example we shall normally require that

$$(a,b) \in A \text{ implies } a_j b_j \geqslant 0, \qquad j = 1, \dots, n. \tag{10.9.1}$$

**Example 10.9.2.** A special kind of laterally convex functions are the L♮-convex functions, which are defined by Murota in (2003: 1.33) by the property

$$f(\lfloor \tfrac{1}{2}x + \tfrac{1}{2}y \rfloor) + f(\lceil \tfrac{1}{2}x + \tfrac{1}{2}y \rceil) \leqslant f(x) + f(y), \qquad x, y \in \mathbf{Z}^n. \tag{10.9.2}$$

A function $f\colon \mathbf{Z}^n \to \mathbf{R}$ is L♮-convex if and only if it is $Q$-laterally convex with

$$Q = \{(a,b) \in \mathbf{Z}^n \times \mathbf{Z}^n;\ b - a \in \{0,1\}^n \cup \{-1,0\}^n\}.$$

So, in all dimensions, L♮-convexity is a special case of lateral convexity.

We shall prove first that $Q$-lateral convexity implies L♮-convexity. Let $x$ and $y$ be given and define

$$a = \lfloor \tfrac{1}{2}x + \tfrac{1}{2}y \rfloor - x \text{ and } b = \lceil \tfrac{1}{2}x + \tfrac{1}{2}y \rceil - x.$$

Note that $b-a \in \{0,1\}^n \subset Q$. Then $x+a+b = y$, so that, if $f$ is $Q$-laterally convex, we obtain $f(\lfloor \frac{1}{2}x + \frac{1}{2}y \rfloor) + f(\lceil \frac{1}{2}x + \frac{1}{2}y \rceil) = f(x+a) + f(x+b) \leqslant f(x) + f(x+a+b) = f(x) + f(y)$, proving (10.9.2).

Next we shall see that L♮ convexity implies $Q$-lateral convexity. If $x$, $a$ and $b$ are given with $b - a \in \{0,1\}^n \subset Q$, we define $y = x + a + b$. Then $\lfloor \frac{1}{2}x + \frac{1}{2}y \rfloor = x + a$ and $\lceil \frac{1}{2}x + \frac{1}{2}y \rceil = x + b$ so that, if $f$ is L♮-convex, we get $f(x+a) + f(x+b) = f(\lfloor \frac{1}{2}x + \frac{1}{2}y \rfloor) + f(\lceil \frac{1}{2}x + \frac{1}{2}y \rceil) \leqslant f(x) + f(y) = f(x) + f(x+a+b)$. If instead $b - a \in \{-1,0\}^n$, we interchange $a$ and $b$. This shows the implication. □

## 10.10 Representations in terms of elementary convex functions

The most elementary convex functions are the affine ones, and the Fenchel transformation shows that certain convex functions can be represented as a supremum of affine functions, viz.,

$$\tilde{\tilde{\varphi}}(x) = \sup_{\xi \in \mathbf{R}^n} \big(\xi \cdot x - \tilde{\varphi}(\xi)\big), \qquad x \in \mathbf{R}^n.$$

So the affine functions $x \mapsto h(x) = \xi \cdot x + c$, where $\xi$ is a vector in $\mathbf{R}^n$ and $c$ a real constant, are the building blocks of the convex function $\tilde{\tilde{\varphi}}$; we know the exact conditions under which the latter equals $\varphi$. This is in analogy with Fourier synthesis, where a function is represented as a sum or integral of pure oscillations $x \mapsto \mathrm{e}^{\mathrm{i}\xi \cdot x}$, $\xi \in \mathbf{R}^n$.

However, in relation to convolution it would be of interest to build up a convex function as a sum rather than a supremum of elementary convex functions. Of course a sum of affine functions is itself affine, so we must go one step further:

**Definition 10.10.1.** Let us say that a function $g$ on $\mathbf{R}^n$ is an ***elementary convex function*** if it has the form $g = (h_1 \vee h_2)|_{\mathbf{Z}^n}$ for some affine functions $h_1, h_2 \colon \mathbf{R}^n \to \mathbf{R}$. □

The elementary convex functions are thus those which have at most one fold hyperplane. Since $h_1 \vee h_2 = h_1 + (h_2 - h_1)^+$, it is equivalent to use affine functions and their positive part.

**Example 10.10.2.** For functions of one real variable we know that $f \in C^2(\mathbf{R})$ is convex if and only if $f'' \geqslant 0$. Suppose for simplicity that $f(x)$ vanishes for $x \leqslant 0$, and define $h(x) = x$, $x \in \mathbf{R}$. Then the convolution

$$(f'' * h^+)(x) = \int_{\mathbf{R}} f''(y)h^+(x-y)\mathrm{d}y, \qquad x \in \mathbf{R},$$

has a good sense; in fact the integral goes only over $0 \leqslant y \leqslant x$. We see that $F = f'' * h^+$ has the second derivative $F'' = f''$, so we must have $F(x) = f(x) + Ax + B$ for some constants $A$ and $B$. However, since both $F$ and $f$ vanish on the negative axis, they are equal. So $f$ is a linear combination (in the form of an integral) of the elementary convex functions $y \mapsto h^+(x-y)$. A similar formula for the negative axis completes the representation.

Now this can be done for any convex function on $\mathbf{R}$: $f''$ has a sense as a distribution, and so does the convolution $f'' * h^+$. So even if $f$ is only continuous, it is a linear combination of elementary convex functions using well-known concepts from distribution theory. □

We avoid here the more complicated situation when we have convex functions of several real variables, and instead consider functions defined on $\mathbf{Z}^n$. The question is then whether all convex extensible functions on $\mathbf{Z}^n$ can be written as the restriction of a sum of elementary convex functions.

**Example 10.10.3.** The function given by $f(x) = |x_1 + x_2| \vee |x_1 - x_2|$, $x \in \mathbf{Z}^2$, can be written as a sum $g$ of four elementary convex functions: $g(x) = x_1^+ + x_1^- + x_2^+ + x_2^-$. □

**Example 10.10.4.** Let now $f(x) = \|x\|_2^2$, $x \in \mathbf{Z}^2$. The affine function $h_a$, $a \in \mathbf{Z}^2$, defined by

$$h_a(x) = \|a\|_2^2 + (2a_1 + 1)(x_1 - a_1) + (2a_2 + 1)(x_2 - a_2), \qquad x \in \mathbf{R}^2,$$

agrees with $\mathbf{cvxe}(f)$ in the square

$$P_a = [a_1, a_1 + 1] \times [a_2, a_2 + 1] \subset \mathbf{R}^2, \qquad a \in \mathbf{Z}^2.$$

We note that $h(x) = \lambda_1 x_1 + \lambda_2 x_2$ is a maximal affine minorant of $\mathbf{cvxe}(f)$ for all $\lambda_j$ satisfying $|\lambda_j| \leqslant 1$, but not all of these agree with $\mathbf{cvxe}(f)$ in a maximal set.

We now measure the distance from the origin to a point $a$ in $\mathbf{Z}^2$ by the infinity norm $\|a\|_\infty = \max |a_j|$ and define

$$(10.10.1) \quad g(x) = h_0 + \sum_{k=1}^{\infty} \sum_{\substack{a \in \mathbf{Z}^2 \\ \|a\|_\infty = k-1}} \sum_{\substack{b \in \mathbf{Z}^2 \\ \|b\|_\infty = k}} (h_b - h_a)^+, \qquad x \in \mathbf{Z}^2,$$

a denumerable sum of elementary convex functions. We have $g|_{\mathbf{Z}^2} = f$. □

Let us say that a function $f\colon \mathbf{Z}^n \to \mathbf{R}_!$ is of ***fast growth*** if

$$\liminf_{\|x\|_2 \to +\infty} \frac{f(x)}{\|x\|_2} = +\infty.$$

This property implies that the graph of the convex envelope of $f$ consists of bounded polytopes in $\mathbf{R}^n \times \mathbf{R}$.

**Theorem 10.10.5.** *Given a convex extensible function $f\colon \mathbf{Z}^n \to \mathbf{R}$ of fast growth, there are denumerably many elementary convex functions $g_j\colon \mathbf{R}^n \to \mathbf{R}$, $j \in \mathbf{Z}$, such that $f(x) = \sum_{j\in\mathbf{Z}} g_j(x)$, $x \in \mathbf{Z}^n$.*

**Proof.** This result is easy to prove for $n = 1$ and less so for $n \geqslant 2$. Let us therefore take the case of one dimension first.

Given $f\colon \mathbf{Z} \to \mathbf{R}$ we define $h_j(x) = f(j) + \lambda_j(x - j)$, $x \in \mathbf{R}$, where $\lambda_j = f(j+1) - f(j)$. Thus $h_j$ is an affine function taking the same value as $f$ at $x = j$ and $x = j + 1$. We note that $\lambda_j \leqslant \lambda_{j+1}$ and that

$$(h_{j+1}(x) - h_j(x))^+ = 0 \text{ for } j, x \in \mathbf{Z} \text{ satisfying } j \geqslant x - 1.$$

Then

$$g = \sum_{j=-\infty}^{-1} (h_j - h_{j+1})^+ + h_0 + \sum_{j=1}^{\infty} (h_j - h_{j-1})^+$$

is a sum of elementary convex functions taking the same values as $f$ at all integer points. Indeed, for a given $x \in \mathbf{N}$, the first sum is zero and the last sum goes only over $j = 0, \ldots, x - 1$, yielding a telescoping series with sum equal to $h_x(x) - h_0(x)$, implying that $g(x) = h_x(x) = f(x)$ since we add

$h_0(x)$. Similarly, for $x \in \mathbf{Z}$, $x \leqslant 0$, the last sum is zero and the first goes only over $j = x, \ldots, -1$, resulting again in $g(x) = f(x)$.

For dimensions higher than 2, we have to construct a distance like we did in Example 10.10.4. This will be done by means of a neighborhood relation.

We note that the convex envelope of $f$ is a supremum of denumerably many maximal affine functions $h_j$, $j \in \mathbf{N}$. If $f$ is not the restriction of an affine function, it has in fact non-denumerably many maximal affine minorants, but we shall restrict attention to those that are maximal not only as to their values, but also in the sense that the set of all $x \in \mathbf{R}^n$ where the affine minorant agrees with $\mathbf{cvxe}(f)$ is maximal. It cannot then be contained in a hyperplane and must have a nonempty interior.

We define a polytope

$$P_j = \{x \in \mathbf{R}^n;\ h_j(x) = \sup_{i \in \mathbf{N}} h_i(x)\}, \qquad j \in \mathbf{N},$$

thus a convex set of nonempty interior. Then we define a neighborhood relation $\diamond$ in $\mathbf{N}$ by declaring that $j \diamond k$ if and only if $j \neq k$ and $P_j \cap P_k \neq \emptyset$. We start with an arbitrary polytope $P_0$ and define a distance from 0 by saying that the distance between $a \in \mathbf{N}$ and 0 is equal to $k$ if there are indices $t_0, t_1, \ldots, t_k$ such that $t_0 = 0$ and $t_k = a$ and $t_i \diamond t_{i+1}$, $i = 0, \ldots, k-1$, but there is no shorter chain. Then the index set $M_k$, $k \in \mathbf{N}$, is defined to be the set of indices with distance $k$ to 0. So $\mathbf{N}$ is a disjoint union of all the index sets $M_k$. With this notation we define

$$g = h_0 + \sum_{k \in \mathbf{N}} \sum_{j \in M_k} \sum_{i \in M_{k-1}} (h_j - h_i)^+,$$

where we define $M_{-1} = \emptyset$. It follows that $g = \sup_{j \in \mathbf{N}} h_j = \mathbf{cvxe}(f)$. Thus $g|_{\mathbf{Z}^n} = f$. □

## 10.11 Separating partially discretized sets

In Section 9.15 we looked at the separation of sets by hyperplanes in a vector space $E$, in particular in $\mathbf{R}^n$. We now restrict the functions to $\mathbf{Z}^n$ and note what happens. We recall Theorem 9.15.3, where we can take $X = \mathbf{Z}^n$.

**Proposition 10.11.1.** *Let $F, G \colon \mathbf{R}^n \to \mathbf{R}_!$ be two given functions, define $f = F|_{\mathbf{Z}^n}$ and $g = G|_{\mathbf{Z}^n}$. Define two sets $A = A_{\mathbf{Z}}^{\mathbf{R}}$ and $B = B_{\mathbf{Z}}^{\mathbf{R}}$ by (9.15.1) with $X = \mathbf{Z}^n$. Then $A$ and $B$ are disjoint if and only if $0 < f \dotplus g$.*

**Proof.** We note that $[b, -a]_{\mathbf{Z}} \cap \mathbf{Z} = [b, -a] \cap \mathbf{Z}$ and proceed as in the proof of Proposition 9.15.1. $\square$

**Theorem 10.11.2.** *Let the functions $f$ and $g$ and the sets $A = A^{\mathbf{R}}_{\mathbf{Z}}$ and $B = B^{\mathbf{R}}_{\mathbf{Z}}$ be as in Proposition 10.11.1. Consider the following conditions.*

(a) *There is a non-vertical hyperplane separating $A$ and $B$.*
(b) *There exists a vector $\xi \in \mathbf{R}^n$ such that $0 \geqslant \widetilde{f}(-\xi) + \widetilde{g}(\xi)$, where $\widetilde{f}$ is the Fenchel transform of $f$.*
(c) $0 \leqslant \widetilde{\widetilde{f}} \dotplus \widetilde{\widetilde{g}}$, *where $\widetilde{\widetilde{f}}$ is the second Fenchel transform of $f$.*
(d) $0 \leqslant \mathbf{cvxe}(f) \dotplus \mathbf{cvxe}(g)$.
(e) *There is a hyperplane separating $A$ and $B$.*

*Assume that both $A$ and $B$ are nonempty.*

*If $n = 1$, then* (a) $\Leftrightarrow$ (b) $\Leftrightarrow$ (c) $\Rightarrow$ (d) $\Rightarrow$ (e) *but* (c) $\not\Leftarrow$ (d) $\not\Leftarrow$ (e).

*If $n \geqslant 2$, then* (a) $\Leftrightarrow$ (b) $\Rightarrow$ (c) $\Rightarrow$ (d) $\Rightarrow$ (e) *but* (b) $\not\Leftarrow$ (c) $\not\Leftarrow$ (d) $\not\Leftarrow$ (e).

**Proof.** The implications in Theorem 9.15.3 remain true in this setting and with the same proofs.

(c) $\Rightarrow$ (a), $n = 1$. If $n = 1$ and (c) holds, we know in view of the Hahn–Banach theorem that there is a separating line. If it happens to be vertical, we shall prove that we can tilt it. Suppose for definiteness that $x = 0$ is a separating line, and that $A$ is to the right of it, $B$ to the left of it. This implies that $f(x) = +\infty$ for $x \leqslant -1$ and that $g(x) = +\infty$ for $x \geqslant 1$. However, since (c) holds, we must have $0 \leqslant f(0) \dotplus g(0)$. I claim that the line $L_\alpha$ with equation $y = \alpha x$ is a separating line when $\alpha$ is large enough. If $L_\alpha$ is not separating, there must either exist a point $(x, g(x))$ with $x \leqslant -1$ and $g(x) < \alpha x$, or a point $(x, -f(x))$ with $x \geqslant 1$ and $-f(x) > \alpha x$. If the first possibility is realized for arbitrarily large $\alpha$, it follows that $\widetilde{g}(-1) = -\infty$ and so that $\widetilde{g} = -\infty$ everywhere. Then, in view of (c) we must have $\widetilde{\widetilde{f}}$ equal to $+\infty$ everywhere, i.e., $A$ must be empty, contrary to the hypothesis. If the second possibility is realized, a similar arguments works.

Two of the counterexamples have to be reconsidered.

**Example 10.11.3.** (a) $\not\Leftarrow$ (c), $n \geqslant 2$. Example 9.15.4 cannot serve in the discrete case. Instead we let $n = 2$ and define $f(x, y) = +\infty$ for $x \leqslant 0$, $f(x, y) = y$ for $x \geqslant 1$; $g(x, y) = 0$ for $x \leqslant 0$, $g(x, y) = +\infty$ for $x \geqslant 1$. Then $f$ and $g$ are convex extensible and $f \dotplus g = +\infty$ everywhere, so (d) holds. And every plane $x = c$ with $0 \leqslant c \leqslant 1$ separates $A$ and $B$, so (e) holds. But a non-vertical plane can never separate $A$ and $B$. Indeed, let $z = \alpha x + \beta y + c$ be the equation of a non-vertical plane. Then, taking $x = 0$, we must have $z = \beta y + c \leqslant 0$ and, taking $x = 1$, $z = \alpha + \beta y + c \geqslant -y$. Since this must hold for all choices of $y \in \mathbf{Z}$, we get a contradiction: the first inequality can hold only for $\beta = 0$, the second only for $\beta = -1$. □

**Example 10.11.4.** (c) $\not\Leftarrow$ (d). Example 9.15.5 works. To be precise, define $f(x) = +\infty$ for $x \leqslant -1$, $f(x) = 0$ for $x \geqslant 0$; $g(x) = -x^2$ for $x \leqslant -1$, $g(x) = +\infty$ for $x \geqslant 0$. Then $\mathbf{cvxe}(f) \dotplus \mathbf{cvxe}(g) = f \dotplus g = +\infty$ everywhere, so (d) holds, but $\widetilde{\widetilde{f}}(0) \dotplus \widetilde{\widetilde{g}}(0) = -\infty$; (c) does not hold. □

**Example 10.11.5.** (d) $\not\Leftarrow$ (e), $n = 1$. We need a new example. Define $f(x) = +\infty$ for $x \leqslant -1$, $f(x) = -1$ for $x \geqslant 0$; $g(x) = -\infty$ for $x \leqslant 0$, $g(x) = +\infty$ for $x \geqslant 1$. Then both $f$ and $g$ are convex, and (e) holds but not (d). □

With these examples we have finished the proof of Theorem 10.11.2. □

## 10.12 Separating completely discretized sets

We now discretize also the values of the functions.

**Proposition 10.12.1.** *Let two functions $f, g\colon \mathbf{Z}^n \to \mathbf{Z}$ be given, and define now $A$ and $B$ by*

$$A = A_{\mathbf{Z}}^{\mathbf{Z}} = \mathbf{hypo}^{\text{finite}}(-f) \cap (\mathbf{Z}^n \times \mathbf{Z}),$$
$$B = B_{\mathbf{Z}}^{\mathbf{Z}} = \mathbf{epi}^{\text{finite}}(g) \cap (\mathbf{Z}^n \times \mathbf{Z}). \tag{10.12.1}$$

*Then $A$ and $B$ are disjoint if and only if $0 < f \dotplus g$.*

**Proof.** The proof of Proposition 10.11.1 works also here. □

**Theorem 10.12.2.** *Let $f$, $g$, $A_{\mathbf{Z}}^{\mathbf{Z}}$ and $B_{\mathbf{Z}}^{\mathbf{Z}}$ be as in Proposition 10.12.1 above. Consider the following conditions.*

(a) *There is a non-vertical hyperplane separating $A$ and $B$.*
(b) *There exists a vector $\xi \in \mathbf{R}^n$ such that $0 \geqslant \widetilde{f}(-\xi) + \widetilde{g}(\xi)$, where $\widetilde{f}$ is the Fenchel transform of $f$.*
(c) *$0 \leqslant \widetilde{\widetilde{f}} \dotplus \widetilde{\widetilde{g}}$, where $\widetilde{\widetilde{f}}$ is the second Fenchel transform of $f$.*
(d) *$0 \leqslant \mathbf{cvxe}(f) \dotplus \mathbf{cvxe}(g)$.*
(e) *There is a hyperplane separating $A$ and $B$.*

*Assume that both $A$ and $B$ are nonempty.*

*If* $n = 1$, *then* (a) $\Leftrightarrow$ (b) $\Leftrightarrow$ (c) $\Rightarrow$ (d) $\Rightarrow$ (e) *but* (c) $\not\Leftarrow$ (d) $\not\Leftarrow$ (e).

*If* $n \geqslant 2$, *then* (a) $\Leftrightarrow$ (b) $\Rightarrow$ (c) $\Rightarrow$ (d) $\Rightarrow$ (e) *but* (b) $\not\Leftarrow$ (c) $\not\Leftarrow$ (d) $\not\Leftarrow$ (e).

**Proof.** The implications in Theorem 9.15.3 remain true in this setting and with the same proofs.

(c) $\Rightarrow$ (a), $n = 1$. This can be done just as we did in the proof of Theorem 10.11.2.

One of the counterexamples have to be reconsidered.

**Example 10.12.3.** (a) $\not\Leftarrow$ (c), $n \geqslant 2$. We do like in Example 10.11.3. □

**Example 10.12.4.** (c) $\not\Leftarrow$ (d). Example 9.15.5 works. □

**Example 10.12.5.** (d) $\not\Leftarrow$ (e), $n = 1$. We need a new example. Define $f(x) = +\infty$ for $x \leqslant -1$, $f(x) = 0$ for $x \geqslant 0$; $g(x) = -\infty$ for $x \leqslant 0$, $g(x) = +\infty$ for $x \geqslant 1$. Then $x = 0$ separates $A$ and $B$, so (e) holds, but $f(0) \dot{+} g(0) = 0 \dot{+} (-\infty) = -\infty$, so (d) does not hold. □

These examples complete the proof of Theorem 10.12.2. □

## 10.13 Possible future studies

Given a subset $A$ of $\mathbf{Z}^n \times \mathbf{Z}^n$, can we characterize the class of Fenchel transforms of $A$-laterally convex functions? The simplest example is the following.

**Example 10.13.1.** For $n = 1$, take $A = (1, 1) \in \mathbf{Z} \times \mathbf{Z}$. Then $f \colon \mathbf{Z} \to \mathbf{R}_!$ is $A$-laterally convex if and only if it is convex extensible. We note that $\tilde{\tilde{f}}$ is convex and its finite graph is affine in each interval $[a, a+1]$, $a \in \mathbf{Z}$.

The Fenchel transform $\tilde{f} \colon \mathbf{R} \to \mathbf{R}_!$ of an $A$-laterally convex function is convex and its finite graph is piecewise affine with slopes that are integers. This answers the question in this case. □

## 10.14 Notes

### 10.14.1 *Other convexity properties*

Several kinds of discrete convexity have been studied.

Miller (1971:168) introduced a class of discretely convex functions (see (12.4.1) above). For these functions local minima are global. They are not convex extensible—nor is the class closed under addition; see (Murota & Shioura 2001:156 and 2001:161, respectively).

The canonical extension is related to the extension studies by Singer (1984/85).

Two other concepts of convexity were introduced by Kazuo Murota (1996, 1998). They are called M-convexity and L-convexity, respectively. For functions with either of these two properties, local minima are global. Two other classes of functions are obtained by a special restriction of M- and L-convex functions to a space of one dimension less. These functions are called $M^\natural$-convex and $L^\natural$-convex.[1] They were introduced by Murota & Shioura (1999:96) and Fujishige & Murota (2000:135), respectively. The class of $M^\natural$-convex ($L^\natural$-convex) functions properly contains the class of M-convex (L-convex) functions. These classes of functions have been studied with respect to some operations such as infimal convolution, addition, and addition by an affine function; see (Murota & Shioura 2001). However, these classes are quite small (see Example 11.7.4).

For a survey of L-convex and M-convex functions, see (Murota 2009).

### 10.14.2 *Difference operators and discrete convexity*

As we have seen in Sections 10.7, 10.8, 10.9, difference operators play an important role in defining convexity properties. Among the studies is a chapter in Samieinia's thesis (2010), published in (2014).

### 10.14.3 *Invariance properties*

The class of integrally convex functions has the property of being invariant under simple coordinate transformations: If we put $g(x, y) = f(x, -y)$, $(x, y) \in \mathbf{Z}^2$, then $f$ and $g$ are integrally convex at the same time, and $f$ and $g$ have the same marginal function:

$$\inf_{y \in \mathbf{Z}} f(x, y) = \inf_{y \in \mathbf{Z}} g(x, y), \qquad x \in \mathbf{Z}.$$

Several of the classes mentioned in Subsection 10.14.1 do not have this property, which implies that they are not suited for the study of marginal functions, and indeed provide poor analogues of convex functions of real variables, which are invariant under such simple coordinate transformations.

[1] These expressions should be read, respectively, as "M-natural-convex" (Murota 2003:27, footnote 23), and "L-natural-convex" (Murota 2003:23, footnote 18). Here M stands for *matroid* and L for *lattice* (Murota 2003:xxi).

The definition of integral convexity (Definition 10.6.1) extends that for two variables in (Kiselman 2011).

For comparisons of discrete convexity and classical convexity, see, e.g., (Yüceer 2002, 2006).

Representations in terms of elementary convex functions (Section 10.10) were studied in my paper (2019b).

For connectedness of digital hyperplanes see, e.g., (Jamet & Toutant 2006, 2009) and (Domenjoud et al. 2009, 2019).

### 10.14.4 *Discrete convexity, symmetry, and asymmetry*

Jourlin & Laget (1988) and Matheron & Serra (1988b) have studied symmetry and asymmetry related to discrete convexity in $\mathbf{Z}^2$ and $\mathbf{Z}^n$.

### 10.14.5 *Terms used*

Favati & Tardella (1990:6) call $\mathbf{can}(f)$ the *extension of* $f$. Murota (2003:93) calls it the *local convex extension of* $f$. See also (Murota 2003:99) for notes on the history and terminology on the topics discussed here.

In my note (2008a) I called the integrally convex functions *fortement convexes.*

## 10.15 Exercises

10.1. Consider the three properties

$$\text{(A)}\ \ \mathbf{cvxh}\left(\mathbf{epi}^{\text{finite}}_{\text{strict}}(f)\right) \cap Y = \mathbf{epi}^{\text{finite}}_{\text{strict}}\left(\mathbf{cvxe}(f)\right) \cap Y \subset \mathbf{epi}^{\text{finite}}_{\text{strict}}\left(f\right),$$

$$\text{(B)}\ \ \mathbf{cvxh}\left(\mathbf{epi}^{\text{finite}}(f)\right) \cap Y \subset \mathbf{epi}^{\text{finite}}(f),$$

$$\text{(C)}\ \ \mathbf{epi}^{\text{finite}}\left(\mathbf{cvxe}(f)\right) \cap Y \subset \mathbf{epi}^{\text{finite}}(f),$$

of which (A) = (10.3.2) was used in Definition 10.3.3 to define $Y$-convexity of functions $f\colon \mathbf{R}^n \to \mathbf{R}_!$. In view of the inclusions (9.13.2) on page 179 we have (C) $\Rightarrow$ (B) and (B) $\Rightarrow$ (A).

In spite of the fact that both inclusions in (9.13.2) can be strict, prove that (A) $\Rightarrow$ (C), so that all three (A), (B) and (C) are equivalent.

10.2. Find an example of a subset $A$ of $\mathbf{Z}^2$ such that knowledge of any finite subset of $A$ never suffices to decide whether a given point belongs to its closed convex hull.

10.3. Find an example of a function $f\colon \mathbf{Z}^2 \to \mathbf{R}_!$ such that knowledge of its values at finitely many points never suffices to calculate its convex envelope at a given point.

# Chapter 11

# Discrete convexity in two dimensions

DISCRETE PHENOMENA IN TWO DIMESIONS is the topic of the present chapter. The graphs of functions of one variable are of course typical candidates for this study.

We sum up the results on straightness of digital lines and segments in Section 11.12.

## 11.1 Jensen's inequality in the discrete case

The difference operator $D_bD_a$ (see Definition 8.3.1) has the advantage that it is symmetric in $a$ and $b$ and that it has entire coefficents. A drawback is that it involves four points if $0 \neq a \neq b \neq 0$. On the other hand, the Jensen operator $J_{a,b}$ (see Definition 8.3.2) involves only three points, but has the drawback that it does not have integer coefficients. In the proofs here either one can be used—it is mostly a matter of taste which to choose. To pass from one to the other, we note the following result.

**Theorem 11.1.1.** *A function $f\colon \mathbf{Z} \to \mathbf{Z}$ satisfies the condition*

$$D_bD_af \geqslant -1 \quad \textit{for all } a,b \in \dot{\mathbf{N}}$$

*if and only it satisfies*

$$J_{a,b}f > -1 \quad \textit{for all } a,b \in \dot{\mathbf{N}}.$$

*Similarly, $|D_bD_af| \leqslant 1$ for all $a,b \in \dot{\mathbf{N}}$ is equivalent to $|J_{a,b}f| < 1$ for all $a,b \in \dot{\mathbf{N}}$. The corresponding results hold for functions which are defined on an interval $[c,d]_{\mathbf{Z}} = [c,d] \cap \mathbf{Z}$; in this case $(D_bD_af)(x)$ and $(J_{a,b}f)(x)$ can only be defined for $c \leqslant x < x+a < x+a+b \leqslant d$.*

**Proof.** The equality $D_bD_af = J_{a,b} + J_{b,a}$ shows that $J_{a,b}f, J_{b,a}f > -1$ implies $D_bD_af > -2$; hence, since $D_bD_af$ has integer values, that $D_bD_af \geqslant -1$.

Conversely, we shall prove that if there exists points $x, a, b$ such that $J_{a,b}f(x) \geqslant 1$, then there exists points $x', a', b'$ such that $D_{b'}D_{a'}f(x') > 1$. (Actually these points may be chosen so that $x' + a' = x + a$.) So suppose that $J_{a,b}f(x) \geqslant 1$. We may then assume that $a$ and $b$ are minimal with this property, for otherwise we can either replace $x$ by a larger value and $a$ by a smaller value or $b$ by a smaller value, in both cases keeping $x + a$ fixed and keeping $J_{a,b}f(x)$ or making it even larger. If $a = b$, then $D_bD_af(x) = 2J_{a,b} \geqslant 2 > 1$ and we have obtained what we want. If $a \neq b$, say $a < b$, then the minimality implies that $J_{b,a}f(x) > 0$, for otherwise the points $x$, $x + a$ and $x + b = x + a + b'$ would be new points with $0 < b' = b - a < b$ such that $(J_{a,b'}f)(x) \geqslant (J_{a,b}f)(x) \geqslant 1$. If on the other hand $a > b$, then similarly $J_{b,a}f(x) > 0$, for otherwise the points $x' = x + b$, $x' + a' = x + a$ and $x + a + b$ would be new points with $a' = a - b < a$ such that $(J_{a',b}f)(x') \geqslant (J_{a,b}f)(x) \geqslant 1$. Hence in all cases $D_bD_af(x) = J_{a,b}f(x) + J_{b,a}f(x) > J_{a,b}f(x) \geqslant 1$. We are done. $\square$

We extend the Jensen operator to functions with infinite values as

$$(11.1.1) \quad ((J_{a,b})_!F)(x) = \frac{b}{a+b}F(x) \dotplus \big(-F(x+a)\big) \dotplus \frac{a}{a+b}F(x+a+b).$$

Here $\dotplus$ denotes upper addition, defined in (2.4.1).

The following lemma will be used in the next theorem.

**Lemma 11.1.2.** *Let $f \in \mathscr{F}(\mathbf{Z}, \mathbf{Z}_!)$. Every point in $\mathbf{cvxh}(\mathbf{epi}^{\text{finite}}(f))$ is of the form*

$$(x_1, y_2 + z_2) \in \mathbf{R} \times \mathbf{R},$$

*where $z_2 \geqslant 0$ and $(x_1, y_2)$ is on a segment $[p, q]$ with $p, q \in \mathbf{epi}^{\text{finite}}(f)$.*

*If $f > -\infty$, it is enough to take segments $[(p_1, f(p_1)), (q_1, f(q_1))]$ with $p_1 \leqslant q_1$ and such that for all points $s_1$ with $p_1 < s_1 < q_1$, the point $(s_1, f(s_1))$ lies strictly above the segment $[(p_1, f(p_1)), (q_1, f(q_1))]$.*

**Proof.** In view of Carathéodory's theorem, Theorem 9.7.1, every point in $\mathbf{cvxh}(\mathbf{epi}^{\text{finite}}(f))$ is in the convex hull of three points in $\mathbf{epi}^{\text{finite}}(f)$, but in view of the special form of a finite epigraph, we can simplify the description as follows.

Let $x$ be a point in the convex hull of three points $p, q, r$ in $\mathbf{epi}^{\text{finite}}(f)$. Now, any point inside a triangle in $\mathbf{R}^2$ must be on or above one of its three sides. This means that $x$ is on or above one of the three segments $[p,q]$, $[q,r]$, $[r,p]$; let us say the first one. If $p_1 = q_1$, we may assume that $p_2 \leqslant q_2$, and then $x_2 = p_2 + z_2$ with $z_2 \geqslant 0$. If on the other hand $p_1 \neq q_1$, then we define $y_2$ by letting $(x_1, y_2)$ be the point on the segment $[p,q]$, $x_1$ being given. So $(x_1, y_2)$ belongs to a segment with endpoints in $\mathbf{epi}^{\text{finite}}(f)$, and $x_2 \geqslant y_2$, so that $x_2 = y_2 + z_2$ with $z_2 \geqslant 0$.

If $f > -\infty$ we need only points on the graph, thus we may take $p = (p_1, f(p_1))$ etc. If there is a point $(s_1, f(s_1))$ with $p_1 \neq s_1 \neq q_1$ on or below the segment $[(p_1, f(p_1)), (q_1, f(q_1))]$, we can use instead one of the segments

$$[(p_1, f(p_1)), (s_1, f(s_1))], \; [(s_1, f(s_1)), (q_1, f(q_1))]$$

to get a new representation, and then go on until there are no more points $(s_1, f(s_1))$ of the graph with $p_1 \neq s_1 \neq q_1$ on or below the segments used.

□

**Theorem 11.1.3.** *A function* $f\colon \mathbf{Z} \to \mathbf{Z}_!$ *is* $(\mathbf{Z} \times \mathbf{Z})$*-convex if and only if*

$$((J_{a,b})_! f)(x) > -1 \qquad \text{for all } \; (x, a, b) \in \mathbf{Z} \times \dot{\mathbf{N}} \times \dot{\mathbf{N}},$$

*equivalently*

$$\lceil ((J_{a,b})_! f)(x) \rceil \geqslant 0 \qquad \text{for all } \; (x, a, b) \in \mathbf{Z} \times \dot{\mathbf{N}} \times \dot{\mathbf{N}}.$$

*Explicitly, this is the case if and only if, for all points* $p, s, q \in \mathbf{Z}$ *with* $p < s < q$, *we have*

$$f(s) \leqslant \Big\lceil (1-\lambda) f(p) \dotplus \lambda f(q) \Big\rceil, \quad \text{where } \lambda = \frac{s-p}{q-p}. \tag{11.1.2}$$

Thus a suitable weakening of Jensen's inequality (8.1.1) gives the right condition.

**Proof.** For the definition of $(\mathbf{Z} \times \mathbf{Z})$-convexity, see Definition 10.3.1. First assume that $f$ is $(\mathbf{Z} \times \mathbf{Z})$-convex, and take three points $p < s < q$. If one of $f(p)$, $f(q)$ is equal to $+\infty$, then (11.1.2) obviously holds. If one of them, say $f(p)$, is equal to $-\infty$ while $f(q) < +\infty$, then $(p, p')$ belongs to $\mathbf{epi}^{\text{finite}}(f)$ for every $p' \in \mathbf{Z}$, which implies that $(s, s')$ belongs to $\mathbf{cvxh}(\mathbf{epi}^{\text{finite}}(f))$ for negative numbers $s'$ with arbitrarily large absolute values. Hence $f(s) = -\infty$; the inequality holds.

The case when both $f(p)$ and $f(q)$ are finite remains to be considered. Then $(p, f(p))$ and $(q, f(q))$ belong to $\mathbf{epi}^{\text{finite}}(f)$, so the point

$$(s, s') = (1-\lambda)(p, f(p)) + \lambda(q, f(q))$$

belongs to its convex hull, hence also $(s, \lceil s' \rceil)$. Since $f$ is convex extensible and $(s, \lceil s' \rceil)$ has integer coordinates, this point must belong to $\mathbf{epi}^{\text{finite}}(f)$. We are done.

Conversely, if (11.1.2) holds, then we shall prove that every point in

$$\mathbf{cvxh}(\mathbf{epi}^{\text{finite}}(f)) \cap \mathbf{Z}^2$$

belongs to $\mathbf{epi}^{\text{finite}}(f)$. In view of Lemma 11.1.2, every point in $\mathbf{cvxh}(\mathbf{epi}^{\text{finite}}(f))$ is of the form $(x, y' + z')$, where $z' \geqslant 0$ and $(x, y')$ is on some segment $[(p, p'), (q, q')]$ with $(p, p'), (q, q') \in \mathbf{epi}^{\text{finite}}(f)$. But (11.1.2) says that such a point $(x, y' + z')$ with $x = s \in \mathbf{Z}$ and $y' + z' \in \mathbf{Z}$ belongs to $\mathbf{epi}^{\text{finite}}(f)$: $f(s) \leqslant \lceil y' \rceil \leqslant y' + z'$. □

**Corollary 11.1.4.** *The graph of a function $f\colon \mathbf{Z} \to \mathbf{Z}$ satisfies the chord property in the sense of Rosenfeld if and only if, for all $x \in \mathbf{Z}$, $a, b \in \dot{\mathbf{N}}$, the pair* $(\lfloor (J_{a,b}f)(x) \rfloor, \lceil (J_{a,b}f)(x) \rceil)$ *is equal to* $(0,1)$ *or* $(0,0)$ *or* $(-1,0)$.

**Proof.** This follows on applying Theorem 11.1.3 to $f$ and $-f$. □

## 11.2 Discrete convexity and the chord property

Since the chord property of Euclid (9.2.4) is unreasonable in a digital setting, it has been weakened by Azriel Rosenfeld (1974) in a sense which turned out to be successful: We already formulated in Definition 8.2.1 the chord property in the sense of Rosenfeld for a set $A$ in $\mathbf{R}^n$:

(11.2.1) $$\mathbf{chord}(A) \subset A + U,$$

where $U$ is the open unit ball in $\mathbf{R}^2$ for the $l^\infty$-norm $\|x\|_\infty = \max(|x_1|, |x_2|)$,

$$U = \{x \in \mathbf{R}^2,\ \|x\|_\infty < 1\}.$$

We note that

$$(c + U) \cap \mathbf{Z}^2 = \{c\}, \qquad c \in \mathbf{Z}^2,$$

which implies that, for any set $A \subset \mathbf{Z}^2$ having the chord property, we have

(11.2.2) $$\mathbf{chord}(A) \cap \mathbf{Z}^2 = A;$$

cf. (10.3.1) above. In particular, $\{p, q\} \subset A$ implies $[p, q] \cap \mathbf{Z}^2 \subset A$ if $A$ has the chord property and $p_1 = q_1$ or $p_2 = q_2$ (we say that $A$ is vertically or horizontally convex).

Kim & Rosenfeld (1982) established a perfect digital analogue in $\mathbf{Z}^2$ of the Euclidean definition of convexity (9.2.2): they proved that a subset $A$

of $\mathbf{Z}^2$ is $\mathbf{Z}^2$-convex if and only if any two of its points can be connected by a digital straight-line segment in the sense of Rosenfeld contained in $A$.

**Proposition 11.2.1.** *For subsets of* $\mathbf{Z}^2$, *the chord property* (11.2.1) *in the sense of Rosenfeld implies* $\mathbf{Z}^2$*-convexity. The converse implication does not hold.*

**Proof.** Assume that $A \subset \mathbf{Z}^2$ has the chord property, and let $t$ belong to $\mathbf{cvxh}(A) \cap \mathbf{Z}^2$. We have to prove that $t \in A$. By Carathéodory's theorem (Theorem 9.7.1) there are three not necessarily distinct points $a, b, c \in A$ such that $t \in \mathbf{cvxh}(\{a, b, c\})$. The vertical line $\{x;\ x_1 = t_1\}$ cuts two of the segments $[a, b]$, $[b, c]$, $[c, a]$. Denote these two points by $p = (p_1, p_2) = (t_1, p_2)$ and $q = (q_1, q_2) = (t_1, q_2)$. By the chord property there exists a point $r \in A$ with $\|r - p\|_\infty < 1$. Since $t_1$ is an integer, we must have $r_1 = p_1 = t_1$. Similarly there is a point $s \in A$ with $s_1 = q_1 = t_1$ and $|s_2 - q_2| < 1$.

We thus have two points $r, s \in A$ with the same first coordinate, and the vertical segment $[r, s]$ contains the given point $t$. It follows from (11.2.2) that $[r, s]_{\mathbf{Z}^2}$ must be a subset of $A$. In particular $t \in [r, s]_{\mathbf{Z}^2} \subset A$, and we are done.

That the converse implication does not hold is shown by the set $\{(0, 0), (2, 1)\}$, which is $\mathbf{Z}^2$-convex but does not have the chord property. (There is a lack of connectivity here.) □

**Remark 11.2.2.** A set in $\mathbf{Z}^2$ which is $\mathbf{Z}^2$-convex need not be connected in a natural way. We can start with a connected $\mathbf{Z}^2$-convex set and lose connectedness under a simple transformation like shearing,

$$\mathbf{Z}^2 \ni (x, y) \mapsto (x + 2y, y) \in \mathbf{Z}^2,$$

taking a $\mathbf{Z}^2$-convex set $A = \{(0, 0), (1, 0), (1, 1), (2, 1), (1, 2)\}$ with image $\{(0, 0), (1, 0), (3, 1), (4, 1), (5, 2)\}$; see (Crombez et al. 2019:410). □

However, the graph of a function $f\colon \{0, 1, 2\} \to \mathbf{Z}$ with $f(0) = 0$, $f(2) = 1$ is $\mathbf{Z}^2$-convex if and only if it has the chord property (this happens if and only if $f(1) \in \{0, 1\}$). This indicates that it is possible to get a converse for graphs, and indeed this is what we shall see in the following theorem.

**Theorem 11.2.3.** *The graph of an integer-valued function defined on an interval of* $\mathbf{Z}$ *and satisfying* $|D_1 f| \leqslant 1$ *is* $\mathbf{Z}^2$*-convex if and only if it has the chord property.*

**Proof.** One direction has already been proved in Proposition 11.2.1. Let $f\colon [c,d]_{\mathbf{Z}} \to \mathbf{Z}$ be such that its graph $A$ is $\mathbf{Z}^2$-convex. We claim that $|J_{a,b}f| < 1$. To reach a contradiction, we assume that $J_{a,b}f(x) \leqslant -1$; then $H(x+a) - f(x+a) \leqslant -1$, where $H$ is the affine function which takes the same values as $f$ at $x$ and $x+a+b$. Since $(x+a, f(x+a))$ belongs to $A$ and

$$H(x+a) \leqslant f(x+a) - 1 < f(x+a)$$

and since also $(x+a, H(x+a))$ belongs to $\mathbf{cvxh}(A)$, it follows that the point $(x+a, f(x+a)-1)$ belongs to $\mathbf{cvxh}(A)$, hence to $A$ in view of its $\mathbf{Z}^2$-convexity. But this contradicts the fact that $A$ is a graph. The conclusion now follows from Theorem 8.6.1. □

To any function $f \in \mathscr{F}(\mathbf{Z}, \mathbf{Z}_!)$ we associate the function $g \in \mathscr{F}(\mathbf{Z}, \mathbf{R}_!)$ taking the same values. Then $\mathbf{epi}^{\text{finite}}(f) = \mathbf{epi}^{\text{finite}}(g) \cap (\mathbf{Z} \times \mathbf{Z}) \subset \mathbf{epi}^{\text{finite}}(g)$ with a strict inclusion except when both finite epigraphs are empty. However, their convex hulls are the same. This is because, for every $(p,q) \in \mathbf{epi}^{\text{finite}}(f)$, the whole ray $(p,q) + L$, where

$$L = \{(0,z) \in \mathbf{R}^2;\ z \geqslant 0\},$$

is contained in $\mathbf{cvxh}(\mathbf{epi}^{\text{finite}}(f))$, so that both convex hulls can be described as the convex hull of the union of all sets $(p,q)+L$ with $(p,q)$ varying in $\mathbf{epi}^{\text{finite}}(f)$. (When $f(p)$ is finite, we can take $q = f(p)$.)

## 11.3 Submodular and separable functions

The quantity

$$(D_{(0,1)}D_{(1,0)}f)(x,y) = f(x+1,y+1) - f(x+1,y) - f(x,y+1) + f(x,y),$$

defined for $(x,y) \in \mathbf{Z}^2$, which I shall call the ***module*** of $f$, will play an important role in the sequel.

We shall say that $f\colon \mathbf{Z}^2 \to \mathbf{R}_!$ is

***modular*** if $(D_{(0,1)}D_{(1,0)})_!f = 0$ everywhere (then it can have only finite values, so that $D_{(0,1)}D_{(1,0)}f = 0$).
***submodular*** if $(D_{(0,1)}D_{(1,0)})_!f \leqslant 0$ everywhere; and
***supermodular*** if $(D_{(0,1)}D_{(1,0)})_!f \geqslant 0$ everywhere.

The definition of submodular functions goes back to the concept of strongly subadditive functions on the family of subsets of a given set introduced by

Gustave Choquet (1955:132). When the set is finite, their relation with convex functions of real variables is given by the Lovász extension (1983). However, in this correspondence it is necessary that the empty set corresponds to the origin and the full set to the point$(1, 1, \dots, 1)$; no translation or reflection is allowed.

These definitions are all special cases of $A$-lateral convexity. Indeed, $f\colon \mathbf{Z}^2 \to \mathbf{R}$ is submodular iff it is $A$-laterally convex for $A = \{((-1,0),(0,1))\}$ and supermodular iff it is $B$-laterally convex for $B = \{((1,0),(0,1))\}$; finally it is modular iff it is $(A \cup B)$-laterally convex.

A modular function is also called ***separable***. This is because such a function can be written $f(x_1, x_2) = f_1(x_1)+f_2(x_2)$, a sum of two univariate functions. Indeed, a modular function can be written in this way if we take $f_1(x_1) = f(x_1, 0)$ and $f_2(x_2) = f(0, x_2) - f(0, 0)$.

We note that all types of $A$-lateral convexity can be expressed in terms of submodularity of a composition of the function with linear mappings. Indeed (10.7.5) gives in the special case $n = s = 2$, $p = (1, 0)$, $q = (0, 1)$,

$$D_{(0,1)}D_{(1,0)}g = (D_bD_af) \circ \varphi, \tag{11.3.1}$$

where $a = (m_{11}, m_{21})$ and $b = (m_{12}, m_{22})$. Given any $(a, b) \in \mathbf{Z}^2 \times \mathbf{Z}^2$ we can thus express $D_bD_af$ as the modulus of a composition of $f$ with some linear mapping, viz. the mapping $\varphi$ defined by $\varphi(x)_j = a_jx_1 + b_jx_2$, $j = 1, 2$. We need a family of linear mappings, one for each element $(a, b)$ of $A$.

## 11.4 The piecewise separately affine extension

**Definition 11.4.1.** Let $f\colon \mathbf{Z}^2 \to \mathbf{R}$ be any function. Then its ***piecewise separately affine extension***, denoted $\mathbf{psa}(f)\colon \mathbf{R}^2 \to \mathbf{R}$, is the unique extension of $f$ to $\mathbf{R}^2$ which is separately affine in each square $a + [0, 1]^2$, $a \in \mathbf{Z}^2$. □

We can get $\mathbf{psa}(f)$ by first extending $f$ canonically in the first variable to the set of all lines $\mathbf{R} \times \mathbf{Z}$, then by extending the resulting function canonically in the second variable—or in the opposite order; the result is the same.

Explicitly, the extension is given for $(x, y) \in \mathbf{Z}^2$, $(s, t) \in [0, 1]^2$, by

$$\begin{aligned}(\mathbf{psa}(f))(x + s, y + t) &= (1 - s)(1 - t)f(x, y) + s(1 - t)f(x + 1, y)\\ &\quad + (1 - s)tf(x, y + 1) + stf(x + 1, y + 1).\end{aligned}$$

Obviously $\mathbf{psa}(f)$ is separately affine in each square $(x,y)+[0,1]^2$ for $(x,y)\in\mathbf{Z}^2$.

We note that $\mathbf{psa}(f)(x+s,y+t)$ is a weighted mean of the values of $f$ at the four points $(x,y)$, $(x+1,y)$, $(x+1,y+1)$, $(x,y+1)$, the weights being just those that give $(x+s,y+t)$ from these points.

Writing $F=\mathbf{psa}(f)$, we find that, in the open square $(x,y)+\,]0,1[\,^2$, $\partial_1^2F=\partial_2^2F=0$ (the coefficients for $s^2$ and $t^2$ are zero) and that $\partial_2\partial_1F=D_{(0,1)}D_{(1,0)}f(x,y)$ (the coefficient for $st$ is equal to the module of $f$ in the square). So $\mathbf{psa}(f)$ is in each square equal to $D_{(0,1)}D_{(1,0)}f(x,y)st$ plus an affine function. Now the function $(s,t)\mapsto Cst$ is convex only when $C$ is zero. We can conclude that $\mathbf{psa}(f)$ is convex in the square only if it is affine. This happens precisely when the module of $f$ at $(x,y)$ vanishes (see Section 11.3).

## 11.5 Rhomboidal convexity

The concept of $A$-lateral convexity (see Definition 10.7.2) is particularly important for the special choice of $A\subset\mathbf{Z}^2$ to be presented here.

**Definition 11.5.1.** A function $f\colon\mathbf{Z}^2\to\mathbf{R}$ is said to be ***rhomboidally convex*** if it is $R$-laterally convex for

$$R=\{(a,b)\in\mathbf{Z}^2\times\mathbf{Z}^2;\ a=(1,0),\ b=(1,-1),(1,0),(1,1)\}$$
$$\cup\ \{(a,b)\in\mathbf{Z}^2\times\mathbf{Z}^2;\ a=(0,1),\ b=(-1,1),(0,1),(1,1)\}.$$

If $f$ has infinite values, we say that it is ***rhomboidally convex in the generalized sense*** if it is $R$-laterally convex in the generalized sense.

A function $F\colon\mathbf{R}^2\to\mathbf{R}$ is said to be ***rhomboidally convex*** if $\mathbf{Z}^2\ni(x,y)\mapsto F_t(x,y)=F(tx,ty)\in\mathbf{R}$ is rhomboidally convex in the sense just defined for every positive number $t$. □

**Remark 11.5.2.** We note that a rhomboidally convex function also satisfies the diagonal inequality $D_{(1,1)}D_{(1,1)}f\geqslant 0$. Indeed,

$$D_{(1,1)}D_{(1,1)}f(x)=D_{(1,1)}D_{(0,1)}f(x)+D_{(1,1)}D_{(1,0)}f(x+(0,1)),$$

a sum of two second-order difference operators. □

**Remark 11.5.3.** If $f$ and $g$ are rhomboidally convex, then clearly $f+g$ has the same property. But this is not always true for $f\vee g$. Indeed, taking $f(x,y)=y$ and $g(x,y)=x-y$, $(x,y)\in\mathbf{Z}^2$, which are restrictions of affine functions and therefore rhomboidally convex, we see that $h=f\vee g$ does not satisfy $(D_bD_ah)(0,0) \geqslant 0$ with $a = (1,0)$, $b = (1,1)$, although the function in $\mathbf{R}^2$ defined by the same expressions is convex. The canonical extension of $h$ is not convex. □

Let $\mathbf{feb}(\mathbf{R}^2)$ be the set of the eight cardinal compass directions in the Finnish–Estonian–Breton sense, i.e.,

$$\mathbf{feb}(\mathbf{R}^2) = \{a \in \mathbf{R}^2;\ a_1a_2(a_1 - a_2)(a_1 + a_2) = 0\},$$

and

$$\mathbf{feb}(\mathbf{Z}^2) = \mathbf{feb}(\mathbf{R}^2) \cap \mathbf{Z}^2.$$

**Theorem 11.5.4.** *A function $f\colon \mathbf{Z}^2 \to \mathbf{R}$ is rhomboidally convex if and only if it is $\mathbf{feb}(\mathbf{Z}^2)$-laterally convex.*

In Theorem 11.6.2 we shall discuss several equivalent definitions of this concept.

**Proof.** Let $f$ be a rhomboidally convex function. In view of the symmetry it is enough to study elements $a = (a_1, 0)$ in $\mathbf{feb}(\mathbf{Z}^2)$ with $a_1 \geqslant 1$; elements $a = (a_1, a_1)$ with $a_1 \geqslant 1$; and elements $a = (0, a_2)$ with $a_2 \geqslant 1$; and for $a = (-a_2, a_2)$ with $a_2 \geqslant 1$. To obtain the more general inequalities involving $\mathbf{feb}(\mathbf{Z}^2)$, we shall perform an induction. Assume first that $a$ is of the form $(a_1, 0)$ and that $b$ is of the form $(b_1, 0)$ or $(b_1, \pm b_1)$. The formula

$$(D_bD_{a+(1,0)}f)(x) = (D_bD_af)(x) + (D_bD_{(1,0)}f)(x+a)$$

shows that if $D_bD_cf \geqslant 0$ for all $c = (c_1, 0)$ satisfying $1 \leqslant c_1 \leqslant a_1$, then the inequality holds also for $D_bD_{a+(1,0)}f$.

Similar formulas hold for the other cases. □

However, the calculations in the proof above have no sense for functions with infinite values:

**Example 11.5.5.** Let $p$ and $q$ be two points in $\mathbf{Z}^2$, define

$$f(p) = f(q) = 0, \text{ and take } f = +\infty \text{ at all other points.}$$

This function satisfies $(D_bD_a)_!f \geqslant 0$ for all $a$ and $b$ such that $\|a+b\|_\infty < \|q-p\|_\infty$, but not if $a + b = q - p$ and $a, b \neq 0$. In particular, it satisfies $(D_bD_a)_!f \geqslant 0$ for the pairs $(a,b) \in R$ mentioned in Definition 11.5.1 as soon as $\|q-p\|_\infty \geqslant 3$.

This example also shows that it is not sufficient to use a finite number of conditions in Definition 11.5.1 when $+\infty$ is allowed as a value—indeed $\|q-p\|_\infty$ can be arbitrarily large.

The function $f$ is separately $(\mathbf{Z}\times\mathbf{R})$-convex if and only if $\|q-p\|_\infty \leqslant 1$ or $(q_1-p_1)(q_2-p_2) \neq 0$. It is $(\mathbf{Z}^2\times\mathbf{R})$-convex if and only if $[p,q]\cap\mathbf{Z}^2 = \{p,q\}$. Finally, it is integrally convex if and only if $\|p-q\|_\infty \leqslant 1$. □

## 11.6 Conditions for rhomboidal convexity

**Theorem 11.6.1.** *Consider the following four conditions on a function* $f\colon \mathbf{Z}^2 \to \mathbf{R}$.

($\alpha'$) *$f$ is rhomboidally convex (Definition 11.5.1);*
($\beta'$) *$f$ is integrally convex (Definition 10.6.1);*
($\gamma'$) *$f$ is convex extensible (Definition 9.13.4);*
($\delta'$) *The restriction of $f$ to any digital line $\{c+ta;\ t\in\mathbf{Z}\}$, $c,a\in\mathbf{Z}^2$, is convex extensible.*

*Then* $(\alpha') \Leftrightarrow (\beta') \Rightarrow (\gamma') \Rightarrow (\delta')$, *and, in general,* $(\beta') \not\Leftarrow (\gamma') \not\Leftarrow (\delta')$.

**Proof.** $(\alpha') \Rightarrow (\beta')$. We first note:

(11.6.1) *For $a\in\mathbf{Z}^2$ and $x_2$ satisfying $a_2 < x_2 < a_2+1$, the derivative from the right of* $\mathbf{can}(f)$ *at $(a_1,x_2)$ is equal to* $\min\big((D_{(1,0)}f)(a),(D_{(1,0)}f)(a_1,a_2+1)\big)$.

This minimum is $\geqslant (D_{(1,0)}f)(a_1-1,a_2)$ if $(\alpha')$ holds. Hence the bend in $\mathbf{can}(f)$ at the line $x_2=a_2$ goes in the right direction: $\mathbf{can}(f)$ is convex as a function of $x_1$ there. Together with a study accross vertical lines this proves $(\beta')$.

$(\alpha') \Leftarrow (\beta')$. In view of (11.6.1) we see that $(\beta')$ implies that

$$(D_{(1,0)}f)(a_1-1,a_2) \leqslant \min\big((D_{(1,0)}f)(a),(D_{(1,0)}f)(a_a,a_2+1)\big).$$

Together with other similar inequalities this proves that $f$ is rhomboidally convex.

$(\beta') \Rightarrow (\gamma')$. See the comment after Definition 10.6.1.

$(\gamma') \Rightarrow (\delta')$. If $F$ is a convex extension of $f$, then $D_aD_aF \geqslant 0$ for all $a\in\mathbf{R}^2$. In particular $D_aD_af \geqslant 0$ for all $a\in\mathbf{Z}^2$.

$(\beta') \not\Leftarrow (\gamma')$. The function defined in (12.2.1) below with $m=1$ shows this. Here $(\mathbf{can}(f))(x,\frac12)$ takes the values $1,\ \frac12,\ 1,\ \frac12,\ 1$ for $x=0,\ \frac12,\ 1,\ \frac32, 2$, respectively, so $\mathbf{can}(f)$ is not convex.

$(\gamma') \not\Leftarrow (\delta')$. Define

$$G(x,y) = (2y-x-1)^+ \vee (2x-y-1)^+ \vee (-x-y-1)^+, \qquad (x,y)\in\mathbf{R}^2.$$

The function $G$ is certainly convex, so its restriction $g = G|_{\mathbf{Z}^2}$ is convex extensible. Now define $f(x,y) = g(x,y)$ for $(x,y) \neq (0,0)$ and $f(0,0) = g(0,0) + \frac{1}{2} = \frac{1}{2}$. For $f$ to satisfy $D_a D_a f \geqslant 0$ it is enough to consider $D_a D_a f$ on a digital line

$$L = \{ta;\ t \in \mathbf{Z}\} = \{t(p,q);\ t \in \mathbf{Z}\}$$

which passes through the origin, since $f$ and $g$ differ only at the origin. It is sufficient to prove that $\frac{1}{2}f(p,q) + \frac{1}{2}f(-p,-q) \geqslant f(0,0) = \frac{1}{2}$ for two relatively prime integers $p$, $q$, since the points $(p,q)$ and $(-p,-q)$ are the integer points closest to the origin on $L$. We see that

$$f(p,q) \vee f(-p,-q) \leqslant f(p,q) + f(-p,-q) < 1$$

only if

$$|2p - q| < 2,\ |2q - p| < 2,\ |p + q| < 2.$$

This happens only if $(p,q) = (0,0)$. So the restriction of $f$ to $L$ is convex extensible, but $f$ is not convex extensible. Indeed, the origin is the barycenter of the three points $(1,1)$, $(-1,0)$, $(0,-1)$:

$$(0,0) = \tfrac{1}{3}(1,1) + \tfrac{1}{3}(-1,0) + \tfrac{1}{3}(0,-1),$$

but

$$f(0,0) = \tfrac{1}{2} > \tfrac{1}{3}f(1,1) + \tfrac{1}{3}f(-1,0) + \tfrac{1}{3}f(0,-1) = 0,$$

so Jensen's inequality is not satisfied. □

**Theorem 11.6.2.** *Given a function $f\colon \mathbf{Z}^2 \to \mathbf{R}$ the following properties are equivalent.*

(A) *$f$ is rhomboidally convex in the sense of Definition 11.5.1, meaning that*

(11.6.2) $\quad D_b D_a f \geqslant 0$ *in* $\mathbf{Z}^2$, $\quad a = (1,0),\ b = (1,b_2),\ b_2 = -1,0,1;$

(11.6.3) $\quad D_b D_a f \geqslant 0$ *in* $\mathbf{Z}^2$, $\quad a = (0,1),\ b = (b_1,1),\ b_1 = -1,0,1;$

(B) *We let $\varphi$ be an affine mapping,*

$$\varphi(x,y) = \begin{pmatrix} x_0 \\ y_0 \end{pmatrix} + \begin{pmatrix} m_{11} & m_{12} \\ m_{21} & m_{22} \end{pmatrix} \begin{pmatrix} x \\ y \end{pmatrix}, \quad (x,y) \in Z^2,$$

*and consider the composition $f \circ \varphi$. The function $f \circ \varphi$ is supermodular for all $x_0, y_0 \in \mathbf{Z}$ and all $m_{11}$, $m_{12}$, $m_{21}$, $m_{22} \in \mathbf{Z}$ satisfying $m_{11} = 1$, $m_{12} = 1$, $m_{21} = -1,0,1$, $m_{22} = 0$; or $m_{11} = 0$, $m_{12} = -1,0,1$, $m_{21} = 1$, $m_{22} = 1$.*

(C) *With notation as in* (B), *the function* $f \circ \varphi$ *is supermodular for all* $x_0, y_0 \in \mathbf{Z}$ *and all* $m_{11}, m_{12}, m_{21}, m_{22} \in \mathbf{Z}$ *satisfying* $m_{11} > 0$, $m_{12} > 0$, $m_{21} = -m_{11}, 0, m_{11}$, $m_{22} = 0$; *or* $m_{11} = 0$, $m_{12} = -m_{22}, 0, m_{22}$, $m_{21} > 0$, $m_{22} > 0$.

(D) *For every point* $(x, y) \in \mathbf{Z}^2$ *we have*

$$D^2_{(1,0)} f(x,y) \geqslant \max\big(0, D_{(0,1)}D_{(1,0)}f(x,y), -D_{(0,1)}D_{(1,0)}f(x,y-1)\big) \tag{11.6.4}$$

*and*

$$D^2_{(0,1)} f(x,y) \geqslant \max\big(0, D_{(0,1)}D_{(1,0)}f(x,y), -D_{(0,1)}D_{(1,0)}f(x-1,y)\big). \tag{11.6.5}$$

(E) *For all points* $a, b \in \mathbf{Z}^2$ *with* $\|a - b\|_1 = 1$, *the subdifferentials* $\partial f(a)$ *and* $\partial f(b)$ *have a common element.*

(F) $\mathbf{can}(f)^{\widetilde{\ }}\big|_{\mathbf{Z}^2} = f$;

(G) $f$ *is the restriction to* $\mathbf{Z}^2$ *of some convex function on* $\mathbf{R}^2$ *which is affine on all segments* $[p, p + (1,0)]$ *as well as on all segments* $[p, p + (0,1)]$, $p \in \mathbf{Z}^2$.

(H) $f$ *is integrally convex in the sense of Favati & Tardella (1990:9), i.e., the canonical extension* $\mathbf{can}(f)$ *of* $f$ *is convex in* $\mathbf{R}^2$;

(I) *The canonical extension* $\mathbf{can}(f)$ *of* $f$ *is equal to the convex envelope* $\mathbf{cvxe}(f)$ *of* $f$;

(J) $\mathbf{can}(f)$ *satisfies Jensen's inequality*

$$\mathbf{can}(f)(x'', y'') \leqslant (1-\lambda)(\mathbf{can}(f))(x,y) + \lambda(\mathbf{can}(f))(x', y'), \tag{11.6.6}$$

*where* $\lambda \in [0,1]$, $(x,y), (x',y') \in \mathbf{Z}^2$ *and*

$$(x'', y'') = (1-\lambda)(x,y) + \lambda(x',y') \in \mathbf{R}^2.$$

(K) $\mathbf{can}(f)$ *satisfies Jensen's inequality* (11.6.6) *when* $(x,y), (x',y') \in \mathbf{Z}^2$ *and* $x' = x + 2$, $|y' - y| \leqslant 1$ *as well as when* $|x' - x| \leqslant 1$, $y' = y + 2$;

(L) $\mathbf{psa}(f)$, *defined in Definition 11.4.1, satisfies Jensen's inequality as in* (K).

**Remark 11.6.3.** We note that conditions (F)–(L) refer to functions defined on $\mathbf{R}^2$, whereas in conditions (A)–(E) only functions defined on the integer points appear. For epistemological reasons the latter may be preferred. It is however perhaps easier to grasp the significance of property (G). □

In (A) there will be six conditions, one for each of the six elements in the set $R$ in Definition 11.5.1. They will be named as follows.

*the two one-variable conditions:*

(11.6.7) $$D_{(1,0)}D_{(1,0)}f \geqslant 0 \text{ and } D_{(0,1)}D_{(0,1)}f \geqslant 0;$$

*the left horizontal lozenge condition:*

(11.6.8) $$D_{(1,-1)}D_{(1,0)}f \geqslant 0;$$

*the right horizontal lozenge condition:*

(11.6.9) $$D_{(1,1)}D_{(1,0)}f \geqslant 0;$$

*the left vertical lozenge condition:*

(11.6.10) $$D_{(-1,1)}D_{(0,1)}f \geqslant 0;$$

*the right vertical lozenge condition:*

(11.6.11) $$D_{(1,1)}D_{(0,1)}f \geqslant 0.$$

In addition to these, we have
*the two diagonal conditions:*

(11.6.12) $$D_{(1,1)}D_{(1,1)}f \geqslant 0 \text{ and } D_{(-1,1)}D_{(-1,1)}f \geqslant 0,$$

which follow from the others; see Remark 11.5.2.

**Proof of Theorem 11.6.2.** The equivalence of (K) and (L) is due to the fact that the two extensions $\mathbf{can}(f)$ and $\mathbf{psa}(f)$ agree on $(\mathbf{Z}\times\mathbf{R})\cup(\mathbf{R}\times\mathbf{Z})$, thus in particular on all points that can occur in Jensen's inequality in (K) and (L).

The equivalence between (A) and (K) is easy; one is essentially a reformulation of the other.

Let us prove the equivalence between (A) and (H). The derivative to the left with respect to $x$ of $F = \mathbf{can}(f)$ at the point $(x, y+\frac{1}{2})$ is either equal to that at the point $(x, y+1)$ or equal to that at the point $(x, y)$. The first alternative holds when the module at $(x-1, y)$ is nonnegative; the second when it is nonpositive. The largest of the two derivatives from the left is equal to the derivative from the left at $(x, y+\frac{1}{2})$, in other words

$$\partial_x^- F(x, y+\tfrac{1}{2}) = \max\big(D_{(1,0)}f(x-1, y), D_{(1,0)}f(x-1, y+1)\big).$$

Similarly, the derivative to the right of $F$ at the same point is

$$\partial_x^+ F(x, y+\tfrac{1}{2}) = \min\big(D_{(1,0)}f(x, y), D_{(1,0)}f(x, y+1)\big).$$

For $F$ to be convex near the vertical interval $](x, y), (x, y+1)[$ it is necessary and sufficient that its derivative to the right be at least as large as its derivative to the left, thus that

$$D_{(1,0)}f(x-1, y+a) \leqslant D_{(1,0)}f(x, y+b), \qquad a, b = 0, 1.$$

This is equivalent to (11.6.2).

Condition (B) is just a reformulation of (A), and a special case of (C). Condition (C) on the other hand follows from (B) in view of the following lemma, which enables us to add parallelograms. It is easily proved by induction. □

**Lemma 11.6.4.** *For any function $f \in \mathscr{F}(\mathbf{Z}^2, \mathbf{R})$ we have*

$$D_{(0,1)}D_{(1,0)}(f \circ \varphi) + D_{(0,1)}D_{(1,0)}(f \circ \varphi') = D_{(0,1)}D_{(1,0)}(f \circ \varphi''),$$

*where*

$$\varphi(x, y) = M\begin{pmatrix} x \\ y \end{pmatrix}, \; \varphi'(x, y) = \begin{pmatrix} a \\ a \end{pmatrix} + M'\begin{pmatrix} x \\ y \end{pmatrix}, \; \varphi''(x, y) = M''\begin{pmatrix} x \\ y \end{pmatrix},$$

*and*

$$M = \begin{pmatrix} a & b \\ a & 0 \end{pmatrix}, \quad M' = \begin{pmatrix} a' & b \\ a' & 0 \end{pmatrix}, \quad M'' = \begin{pmatrix} a+a' & b \\ a+a' & 0 \end{pmatrix}.$$

*Also,*

$$D_{(0,1)}D_{(1,0)}f \circ \psi + D_{(0,1)}D_{(1,0)}f \circ \psi' = D_{(0,1)}D_{(1,0)}f \circ \psi'',$$

*where*

$$\psi(x, y) = N\begin{pmatrix} x \\ y \end{pmatrix}, \; \psi'(x, y) = \begin{pmatrix} b' \\ 0 \end{pmatrix} + N'\begin{pmatrix} x \\ y \end{pmatrix}, \; \psi''(x, y) = N''\begin{pmatrix} x \\ y \end{pmatrix},$$

*and*

$$N = \begin{pmatrix} a & b \\ a & 0 \end{pmatrix}, \quad N' = \begin{pmatrix} a & b' \\ a & 0 \end{pmatrix}, \quad N'' = \begin{pmatrix} a & b+b' \\ a & 0 \end{pmatrix}.$$

*There are similar addition formulas that can be obtained by interchanging $x$ and $y$.* □

Cf. the proof of Theorem 11.5.4. We see that $M''$ is obtained from $M$ and $M'$ by partial addition.

**Remark 11.6.5.** It might seem paradoxical that, for integrally convex functions $f$, the functions $f \circ \varphi$ are submodular when the linear mapping $\varphi$ is given by one of the four matrices

$$M = \begin{pmatrix} 1 & -1 \\ 1 & 0 \end{pmatrix}, \quad \begin{pmatrix} 1 & -1 \\ -1 & 0 \end{pmatrix}, \quad \begin{pmatrix} 0 & -1 \\ 1 & -1 \end{pmatrix}, \quad \begin{pmatrix} 0 & -1 \\ -1 & 0 \end{pmatrix}$$

but not necessarily for

$$M = \begin{pmatrix} 1 & 0 \\ 0 & 1 \end{pmatrix}.$$

It should be noted that $g(x, y) = f(x, -y)$ is integrally convex if and only if $f$ is, whereas $g$ is submodular if and only if $f$ is supermodular. The invariance properties of the two classes are different. □

## 11.7 Independence of the conditions for rhomboidal convexity

As to the independence of the four lozenge conditions, the following example yields sufficient information.

**Example 11.7.1.** Define for a real number $\alpha$,

$$f(x, y) = (y - \alpha x)^2, \qquad (x, y) \in \mathbf{Z}^2.$$

Then $f$ satisfies the one-variable conditions as well as the diagonal conditions for all $\alpha$. Moreover,

(1) $f$ satisfies the left horizonal lozenge condition if and only if $\alpha(\alpha+1) \geqslant 0$;
(2) $f$ satisfies the right horizonal lozenge condition if and only if $\alpha(\alpha - 1) \geqslant 0$;
(3) $f$ satisfies the left vertical lozenge condition if and only if $\alpha \geqslant -1$;
(4) $f$ satisfies the right vertical lozenge condition if and only if $\alpha \leqslant 1$.

We find that $D^2_{(1,0)} f = \alpha^2$, $D^2_{(0,1)} f = 2$, so that both one-variable conditions are satisfied. Moreover, we get

$$D_{(1,1)} D_{(1,0)} f = 2\alpha(\alpha - 1), \qquad D_{(-1,1)} D_{(0,-1)} f = 2\alpha(\alpha + 1)$$

and

$$D_{(1,1)} D_{(0,1)} f = 2(1 - \alpha), \qquad D_{(-1,1)} D_{(0,1)} f = 2(\alpha + 1).$$

Thus

(1) for $\alpha = -\frac{1}{2}$ all conditions except the left horizontal lozenge condition (11.6.8) are satisfied.
(2) for $\alpha = \frac{1}{2}$ all conditions except the right horizontal lozenge condition (11.6.9) are satisfied.
(3) for $\alpha = -\frac{3}{2}$ all conditions except the left vertical lozenge condition (11.6.10) are satisfied.

(4) for $\alpha = \frac{3}{2}$ all conditions except the right vertical lozenge condition (11.6.11) are satisfied.

For all these values of $\alpha$, the marginal function takes the value zero for even $x$ and the value $\frac{1}{4}$ for $x$ odd; hence it is not $(\mathbf{Z} \times \mathbf{R})$-convex.

This shows that each of the four lozenge conditions is necessary for the marginal function to be $(\mathbf{Z} \times \mathbf{R})$-convex. □

**Remark 11.7.2.** For $a = b$, condition (A) means just that $f$ is separately $(\mathbf{Z} \times \mathbf{R})$-convex (the two one-variable conditions). But we require in addition that $D_{(1,0)}f(x+1,y)$ shall majorize $D_{(1,0)}f(x, y \pm 1)$ (the horizontal lozenge condition) and that $D_{(0,1)}f(x, y+1)$ shall majorize $D_{(0,1)}f(x \pm 1, y)$ (the vertical lozenge condition).

A similar remark holds for the case $m_{21} = 0$ in the first part of (B) and (C), and the case $m_{12} = 0$ in the second part: these cases just mean that the function shall be separately $(\mathbf{Z} \times \mathbf{R})$-convex. If we already know that the function is separately $(\mathbf{Z} \times \mathbf{R})$-convex, we need to take only $m_{11} = |m_{21}| = 1$ in the first part of (B) and only $m_{11} = |m_{21}| > 0$ in the first part of (C). In these cases, it is enough to take $x_0 = y_0 = 0$, which will simplify calculations. □

Two other examples comparing convexity properties may be of interest.

**Example 11.7.3.** When $n = 2$ and $A = \{(e^{(1)}, e^{(2)})\}$, a function is $A$-laterally convex (Definition 10.7.2) if and only if it is supermodular. Note that (10.8.4) is not satisfied in this case. (Cf. Murota 2003:26, 206–207.) □

**Example 11.7.4.** Since, for $n = 2$, $Q \supset R$ (see Example 10.9.2), we have $\Phi(Q) \subset \Phi(R)$, i.e., every L$^\natural$-convex function is rhomboidally convex. In fact, the L$^\natural$-convex functions form a tiny fraction of the rhomboidally convex functions. To illustrate this fact, let us mention that a function $f(x_1, x_2) = g(x_1 + x_2)$, $(x_1, x_2) \in \mathbf{Z}^2$, is L$^\natural$-convex if and only if $g \colon \mathbf{Z} \to \mathbf{R}$ is the restriction to $\mathbf{Z}$ of an affine function defined on $\mathbf{R}$, while it is rhomboidally convex if and only if $g$ is convex extensible. (If $f(x_1, x_2) = h(x_1 - x_2)$, the situation is quite different.) □

## 11.8 The maximal set of pairs which defines rhomboidal convexity

It is natural to ask for the maximal set $\Lambda(R) = \Psi(\Phi(R))$ of pairs which defines the class of rhomboidally convex functions. The answer is the following.

**Proposition 11.8.1.** *Let $S_a = \{ta;\ t \geqslant 0\}$, $a \in \mathbf{R}^2 \smallsetminus \{0\}$, be the ray containing $a$. With $R$ defined in Definition 11.5.1, $\Lambda(R)$ is equal to the set of all pairs $(a, b) \in \mathbf{Z}^2 \times \mathbf{Z}^2$ such that the interior $(S_a + S_b)^\circ$ of the sector $S_a + S_b$ does not intersect the set*

$$E = S^\infty(0,1) \cap \mathbf{Z}^2 = \{(\pm 1, 0), (0, \pm 1), (\pm 1, \pm 1)\}.$$

The set $E$ has eight points and is the set of integer points in the unit sphere for the $l^\infty$-metric, equal to $B^\infty_\leqslant(0,1) \smallsetminus B^\infty_<(0,1)$ and denoted by $S^\infty(0,1)$.

If the angle of the sector $S_a + S_b$ is larger than $\pi/4$, then its interior must contain a point of $E$. If the angle is at most $\pi/4$, then the interior may or may not contain such a point. In particular, $\Delta \subset \Lambda(R)$, since the interior of $S_a + S_a$ is empty. Here $\Delta$ is the diagonal.

**Proof.** Define $T$ as the set of all pairs with the mentioned property, thus

$$T = \{(a, b);\ (S_a + S_b)^\circ \cap E = \varnothing\}.$$

If $a$ and $b$ are contained in one of the sectors defined by $E$, we shall prove that the pair $(a, b) = ((a_1, a_2), (b_1, b_2)) \in \mathbf{Z}^2 \times \mathbf{Z}^2$ belongs to $\Lambda(R)$. It is enough to consider the case $0 \leqslant a_2 \leqslant a_1$, $0 \leqslant b_2 \leqslant b_1$. Then the point $((a_1 - a_2, 0), (b_1 - b_2, 0))$ can be obtained from $((1, 0), (1, 0)) \in R$ by a number of partial additions and so belongs to $\Lambda(R)$. Similarly $((a_1 - a_2, 0), (b_2, b_2))$ can be obtained from $((1, 0), (1, 1)) \in R$. By partial addition we then get

$$((a_1 - a_2, 0), (b_1 - b_2, 0)) +_{\{1\}} ((a_1 - a_2, 0), (b_2, b_2)) = ((a_1 - a_2, 0), b) \in \Lambda(R)$$

in the notation of Section 3.8. Similarly $((a_2, a_2), b) \in \Lambda(R)$. We finally have

$$((a_1 - a_2, 0), b) +_{\{2\}} ((a_2, a_2), b) = (a, b) \in \Lambda(R).$$

We have proved that $T \subset \Lambda(R)$.

Conversely, if the sector $S_a + S_b$ does contain an element of $E$ in its interior, then we can construct a $\Lambda(R)$-laterally convex function which satisfies $(D_b D_a f)(x) < 0$ at some point $x$. Take as a typical example $(1, 0) \in E$.

Then if the interior of $S_a + S_b$ contains $(1,0)$, say with $a_1, b_1 > 0$, $a_2 > 0$ and $b_2 < 0$, we take $f(x) = |x_2|$ and obtain

$$(D_b D_a f)(0) = |a_2 + b_2| - a_2 - |b_2| < 0.$$

If for example $(1,1)$ belongs to the interior of $S_a + S_b$, then we take, for $a_2 > a_1$ and $b_2 < b_1$, $f(x) = |x_1 - x_2|$ and can conclude that $(D_b D_a f)(0) < 0$.

This proves that $\Lambda(R) \subset T$. □

## 11.9 $(\mathbf{Z} \times \mathbf{R})$-convexity and separate $(\mathbf{Z} \times \mathbf{R})$-convexity

A function $f\colon \mathbf{Z} \to \mathbf{R}_!$ is $(\mathbf{Z} \times \mathbf{R})$-convex if and only if $(D_1 D_1)_! f \geqslant 0$. Hence we may define also separate convexity using the operators $D_a D_a$; we say that $f\colon \mathbf{Z}^n \to \mathbf{R}_!$ is ***separately*** $(\mathbf{Z} \times \mathbf{R})$***-convex*** if it is $A$-laterally convex for $A = \{(e^{(1)}, e^{(1)}), \dots, (e^{(n)}, e^{(n)})\}$, where the $e^{(j)}$ denote the unit vectors. For a function $f\colon \mathbf{R}^n \to \mathbf{R}_!$ we say that it is ***separately convex*** if it is $A$-laterally convex for $A = \{(te^{(j)}, te^{(j)});\ t \in \mathbf{R},\ j = 1, \dots, n\}$.

## 11.10 Smooth rhomboidally convex functions

We note the following characterization of laterally convex function of two real variables which are sufficiently smooth.

**Proposition 11.10.1.** *A function $F\colon \mathbf{R}^2 \to \mathbf{R}$ of class $C^2$ is rhomboidally convex if and only if it satisfies*

$$\partial_1^2 F \geqslant |\partial_2 \partial_1 F| \text{ and } \partial_2^2 F \geqslant |\partial_2 \partial_1 F|. \tag{11.10.1}$$

**Proof.** Let $A$ denote the parallelogram

$$\{(x,y) \in \mathbf{R}^2;\ b \leqslant y \leqslant b+1,\ y-b \leqslant x-a \leqslant y-b+1\},$$

the convex hull of $\{(a,b), (a+1,b), (a+1,b+1), (a+2,b+1)\}$. For a function $F \in C^2(\mathbf{R}^2)$ we note that

$$\begin{aligned} &\iint_A \left(\partial_1^2 F + \partial_1 \partial_2 F\right) \mathrm{d}x\, \mathrm{d}y \\ &\quad = F(a+2,b+1) - F(a+1,b) - F(a+1,b+1) + F(a,b). \end{aligned} \tag{11.10.2}$$

If $F$ satisfies (11.10.1), then the integrand is nonnegative; hence also the right-hand side. However, the right-hand side is equal to $D_{(1,1)} D_{(1,0)} F(a,b)$.

With similar formulas for $\partial_1^2 F - \partial_1\partial_2 F$ and $\partial_2^2 F \pm \partial_1\partial_2 F$ we conclude that $F\big|_{\mathbf{Z}^2}$ is rhomboidally convex. The same holds of course for $F_t(x,y) = F(tx,ty)$.

Conversely, assume that $\mathbf{Z}^2 \ni (x,y) \mapsto F(tx,ty)$ is rhomboidally convex for every $t > 0$. Then $\partial_1^2 F(x,y) + \partial_2\partial_1 F(x,y)$ is, for all $(x,y) \in \mathbf{RQ}^2$, the limit of expressions

$D_{(t,t)}D_{(t,0)}F(x,y) = F(x+2t,y+t) - F(x+t,y) - F(x+t,y+t) + F(x,y)$

where $t \to 0$ through values such that $x/t$ and $y/t$ are both integers. Since all such expressions are nonnegative, also all limits must be nonnegative, and $\partial_1^2 F + \partial_2\partial_1 F$ is nonnegative in $\mathbf{RQ}^2$, hence everywhere in $\mathbf{R}^2$. □

## 11.11 Extending functions from integer points

It is natural to ask what the relations are between the functions defined on $\mathbf{R}^2$ and their restrictions to $\mathbf{Z}^2$. It turns out that there is a satisfying relation in the three cases of separately $(\mathbf{Z} \times \mathbf{R})$-convex, $(\mathbf{Z}^2 \times \mathbf{R})$-convex, and rhomboidally convex functions. We consider them in the next three subsections.

### 11.11.1 *Extension of separately* $(\mathbf{Z} \times \mathbf{R})$*-convex functions*

**Theorem 11.11.1.** *The restriction to $\mathbf{Z}^2$ of a separately convex function $F\colon \mathbf{R}^2 \to \mathbf{R}$ is also separately $(\mathbf{Z} \times \mathbf{R})$-convex.*

*Conversely, given a separately $(\mathbf{Z} \times \mathbf{R})$-convex function $f\colon \mathbf{Z}^2 \to \mathbf{R}$, it has a separately convex extension to $\mathbf{R}^2$. More precisely, the extension $F$ can be chosen so that, in each square $(x,y) + ]0,1[^2$, $(x,y) \in \mathbf{Z}^2$, it is equal to a polynomial of degree two and satisfies three partial differential equations: $\partial_1^2 F = \partial_1^2 F = 0$ and $\partial_2\partial_1 F = D_{(0,1)}D_{(1,0)}f(x,y)$.*

**Proof.** If $F\colon \mathbf{R}^2 \to \mathbf{R}$ is separately convex, its restriction to $\mathbf{Z}^2$ obviously is.

Conversely, given $f$, the question is whether $F = \mathbf{psa}(f)$ has a bend in the right direction over vertical and horizontal lines $x = x_0$ and $y = y_0$. This depends on the signs of $D^2_{(1,0)}F(x,y+t)$ and $D^2_{(0,1)}F(x+s,y)$, $(x,y) \in \mathbf{Z}^2$, $(s,t) \in ]0,1[^2$. By symmetry, it is enough to study one of them. We find that, for $0 < t < 1$,

$$\begin{aligned}(D^2_{(1,0)}F)(x,y+t) &= F(x+2,y+t) - 2F(x+1,y+t) + F(x,y+t)\\ &= (1-t)(D^2_{(1,0)}f)(x,y) + t(D^2_{(1,0)}f)(x,y+1) \geqslant 0.\end{aligned}$$

Hence $F$ is separately convex in $\mathbf{R}^2$. □

### 11.11.2 *Extension of* $(\mathbf{Z}^2 \times \mathbf{R})$*-convex functions*

**Theorem 11.11.2.** *The restriction to* $\mathbf{Z}^2$ *of a convex function* $F\colon \mathbf{R}^2 \to \mathbf{R}$ *is* $(\mathbf{Z}^2 \times \mathbf{R})$*-convex.*

*Conversely, every* $(\mathbf{Z}^2 \times \mathbf{R})$*-convex function* $f\colon \mathbf{Z}^2 \to \mathbf{R}$ *is convex extensible. In fact,* $\mathbf{cvxe}(f)$ *is an extension.*

**Proof.** Obviously $F|_{\mathbf{Z}^2}$ is $(\mathbf{Z}^2 \times \mathbf{R})$-convex in the sense of Definition 10.3.3 if $F \in \mathscr{F}(\mathbf{R}^2, \mathbf{R})$ is convex.

In the other direction we note that, if $t > (\mathbf{cvxe}(f))(x)$, then, by (9.13.1), $t \geqslant (\mathbf{cvxh}(\mathbf{epi}^{\text{finite}}(f)))(x)$, which implies that $t \geqslant f(x)$ if $f$ is $(\mathbf{Z}^2 \times \mathbf{R})$-convex. Hence $f \leqslant \mathbf{cvxe}(f)|_{\mathbf{Z}^2}$; the latter is actually an extension of $f$. □

### 11.11.3 *Extension of rhomboidally convex functions*

**Theorem 11.11.3.** *The restriction to* $\mathbf{Z}^2$ *of a rhomboidally convex function* $F\colon \mathbf{R}^2 \to \mathbf{R}_!$ *is rhomboidally convex.*

*Conversely, given a rhomboidally convex function* $f\colon \mathbf{Z}^2 \to \mathbf{R}_!$*, it has a rhomboidally convex extension to* $\mathbf{R}^2$*. In fact,* $\mathbf{can}(f)$ *is such an extension.*

**Proof.** If $F\colon \mathbf{R}^2 \to \mathbf{R}_!$ is rhomboidally convex, $\mathbf{Z}^2 \ni (x, y) \mapsto F(tx, ty)$ is by definition rhomboidally convex for every positive $t$. In particular this is true for $t = 1$.

Conversely, if $f\colon \mathbf{Z}^2 \to \mathbf{R}_!$ is rhomboidally convex, we shall prove that $\mathbf{Z}^2 \ni (x, y) \mapsto (\mathbf{can}(f))(tx, ty) = F_t(x, y)$ is rhomboidally convex for any positive $t$. The proof will proceed in two steps: First we prove that $F_t|_{\mathbf{Z}^2}$ is rhomboidally convex when $t$ is an integer; then that it is rhomboidally convex when $0 < t < 1$. Combining the two we obtain the general result.

*Step 1.* To prove that $F_m|_{\mathbf{Z}^2}$ is rhomboidally convex for $m = 2, 3, \dots$, we just need to note that this follows from the implication (B) ⇒ (C) in Theorem 11.6.2. Indeed, introducing two matrices

$$A = \begin{pmatrix} 1 & 1 \\ c & 0 \end{pmatrix} \text{ and } B = \begin{pmatrix} m & m \\ cm & 0 \end{pmatrix},$$

where $c = -1, 0, 1$, and writing $g$ for $F_m|_{\mathbf{Z}^2}$ for brevity, we see that $g_{A,x_0,y_0} = f_{B,mx_0,my_0}$, which is supermodular in view of (B). Hence $g$ satisfies the first condition, (11.6.2) in (A); by symmetry of the conditions, it is rhomboidally convex.

*Step 2.* The second step is to prove that $F_t$ is rhomboidally convex for $0 < t < 1$. We look at (11.6.2) with $a = (1, 0)$, $b = (1, 1)$. (The case

$b = (1, -1)$ is similar, and the case $b = a$, meaning separate $(\mathbf{Z} \times \mathbf{R})$-convexity, is no problem.) Thus, writing $F$ for $\mathbf{can}(f)$, we have to consider an inequality

$$F(t(x+2), t(y+1)) + F(tx, ty) \geqslant F(t(x+1), ty) + F(t(x+1), t(y+1)),$$

involving a parallelogram with vertices at

$$(tx, ty),\ (t(x+1), ty),\ (t(x+1), t(y+1)),\ \text{and}\ (t(x+2), t(y+1)).$$

It may happen that $F$ is affine in that parallelogram, in which case we have equality, or it may happen that there is a fold line, keeping the function convex. Such a fold line is of one of the four kinds $x =$ constant, $y =$ constant, $x + y =$ constant, or $x - y =$ constant. In all these cases, the inequality is preserved or improved compared with the affine case when there is no fold line. In fact, if we have a vertical fold line, one of the terms on the left hand side is increased; if we have a horizontal fold line or a fold line with $x - y =$ constant, a term on the right and a term on the left is increased by the same amount; finally, if we have a fold line with $x+y =$ constant, then either one of the terms on the left-hand side increases or one term on each side increases but that on the left by more. □

## 11.12 Conditions for digital straightness

Combining what we have learned about digital straightness so far we obtain the following result.

**Theorem 11.12.1.** *Let $f \in \mathscr{F}(\mathbf{Z}, \mathbf{Z})$, assume that $|D_1 f| \leqslant 1$, and consider the following properties.*

(A) *The graph of $f$ has the chord property;*
(B) *Both $f$ and $-f$ are convex extensible;*
(C) *The graph of $f$ is $\mathbf{Z}^2$-convex;*
(D) *The inequality $|(D_b D_a f)(x)| \leqslant 1$ holds for all $(x, a, b) \in \mathbf{Z} \times \dot{\mathbf{N}} \times \dot{\mathbf{N}}$;*
(E) *The inequality $|(J_{a,b} f)(x)| < 1$ holds for all $(x, a, b) \in \mathbf{Z} \times \dot{\mathbf{N}} \times \dot{\mathbf{N}}$;*
(F) *The binary word $D_1 f \colon \mathbf{Z} \to \mathbf{Z}$ is balanced;*
(G) *The function $f$ defines a refined digital hyperplane in $\mathbf{Z}^2$ in the sense of Kiselman (2004b);*
(H) *The function $f$ defines a digital straight line in the sense of Reveillès (1991; see Section 8.8).*

*All conditions* (A), (B), (C), (D), (E), (F) *and* (G) *are equivalent, and they are implied by* (H).

**Proof.** The equivalences follow on combining Theorems 8.6.1, 8.7.1 and 11.1.1.

That (G) $\not\Rightarrow$ (H) is clear from the Heaviside function $Y_0\colon \mathbf{Z} \to \mathbf{Z}$, with $Y_0(x) = 0$ for $x \leqslant -1$ and $Y_0(x) = 1$ for $x \geqslant 0$. See also Example 11.14.4. □

We now consider functions defined on a finite interval only:

**Theorem 11.12.2.** *Let $f\colon [c,d]_{\mathbf{Z}} \to \mathbf{Z}$ be defined on a finite interval $[c,d]_{\mathbf{Z}}$ and assume that $|D_1 f(x)| \leqslant 1$ for all $x$ such that $c \leqslant x \leqslant d-1$. Then the following properties are all equivalent.*

(A) *The graph of $f$ has the chord property;*
(B) *Both $f$ and $-f$ are convex extensible;*
(C) *The graph of $f$ is $\mathbf{Z}^2$-convex;*
(D) *The inequalty $|(D_b D_a f)(x)| \leqslant 1$ holds for all $(x,a,b) \in \mathbf{Z} \times \dot{\mathbf{N}} \times \dot{\mathbf{N}}$ such that $c \leqslant x < x+a+b \leqslant d$;*
(E) *The inequality $|(J_{a,b} f)(x)| < 1$ holds for all $(x,a,b) \in \mathbf{Z} \times \dot{\mathbf{N}} \times \dot{\mathbf{N}}$ such that $c \leqslant x < x+a+b \leqslant d$;*
(F) *The binary word $D_1 f\colon [c,d-1]_{\mathbf{Z}} \to \mathbf{Z}$ is balanced;*
(G) *The function $f$ defines a subset of a refined digital hyperplane in $\mathbf{Z}^2$ in the sense of Kiselman (2004b);*
(H) *The function $f$ defines a subset of a digital straight line in the sense of Reveillès.*

**Proof.** In view of Theorem 11.12.1 it is enough to prove that (H) is implied by one of the other conditions. From Theorem 8.10.1 it follows that the graph of a function satisfying (D) is contained in an open strip $S(\alpha,\beta,\gamma)$ of height $\alpha - \beta < 1$. So then it is contained in the digital straight line in the sense of Reveillès $S^*(\alpha,\beta,\beta+1)$ (as well as in others). Hence (H) holds. □

We also note the following result on locality of the various properties. Let us say that a property of functions $f \in \mathscr{F}(A, \mathbf{Z}_!)$, where $A$ is an arbitrary subinterval of $\mathbf{Z}$, is ***local*** if it is true that it has the property if and only if all its restrictions $f|_{[c,d]_{\mathbf{Z}}}$ to finite intervals $[c,d]_{\mathbf{Z}} \subset A$ have the property.

**Proposition 11.12.3.** *The properties* (A), (B), (C), (D), (E), (F) *and* (G) *of Theorems 11.12.1 and 11.12.2, understood respectively for functions defined on $\mathbf{Z}$ and on subintervals of $\mathbf{Z}$, are local properties. The property* (H) *is not local.*

**Proof.** For properties (A)–(F) this is obvious. For property (G) we just return to (D) for example. That (H) is not local follows on considering the Heaviside function. □

## 11.13 What is the result when we digitize a Euclidean line?

To complete Theorem 11.12.1 we may ask for a more detailed description of the digital lines that can occur when we digitize a Euclidean line in the plane by the floor function. So let $\beta_1 x + \beta_2 y = d$ be the equation of a line in $\mathbf{R}^2$, with $\beta_1$ and $\beta_2$ not both zero. By a simple change of coordinates we can assume that $\beta_2 \neq 0$ and that $\gamma = -\beta_1/\beta_2$ satisfies $0 \leqslant \gamma \leqslant 1$, implying that it is enough to study the function $F(x) = \gamma x + c$, where $c = d/\beta_2$. We define $f\colon \mathbf{Z} \to \mathbf{Z}$ by $f(x) = \lfloor F(x) \rfloor$, $x \in \mathbf{Z}$. The chain code $D_1 f$ of $f$ is a sum of Kronecker deltas,

$$D_1 f = \sum_{a \in A} \delta_a, \tag{11.13.1}$$

where $A$ is a subset of $\mathbf{Z}$.

If $\gamma = 0$ we have a horizontal line: $f(x)$ is the constant function with value $\lfloor c \rfloor \in \mathbf{Z}$ and $A$ is empty. If $\gamma = 1$, we have the diagonal line $F(x) = x + c$, and $f(x) = \lfloor x + c \rfloor = x + \lfloor c \rfloor$, $x \in \mathbf{Z}$; now $A = \mathbf{Z}$.

If $\gamma = \frac{1}{2}$ we get $f(x) = \lfloor \frac{1}{2}x + c \rfloor$, so that $f(x) = \frac{1}{2}x + \lfloor c \rfloor$ when $x$ is even; $f(x) = \frac{1}{2}(x-1) + \lfloor \frac{1}{2} + c \rfloor$ when $x$ is odd. It follows that $A$ consists either of all even or of all odd integers.

**Proposition 11.13.1.** *Let $F\colon \mathbf{R} \to \mathbf{R}$ be an affine function, $F(x) = \gamma x + c$ with $-1 \leqslant \gamma \leqslant 1$. Define a discretization $f\colon \mathbf{Z} \to \mathbf{Z}$ of $F$ by $f(x) = \lfloor F(x) \rfloor$, $x \in \mathbf{Z}$, and define $f_\gamma\colon \mathbf{Z} \to \mathbf{R}$ by $f_\gamma(x) = f(x) - \gamma x$. Then the following three properties are equivalent.*

(A) *$\gamma$ is rational;*
(B) *$f_\gamma$ is a periodic function;*
(C) *$\mathbf{im}(f)$ is a periodic set;*
(D) *$D_1 f$, defined by $(D_1 f)(x) = f(x+1) - f(x)$, is a periodic function.*

**Proof.** (A) $\Rightarrow$ (B). If $\gamma = p/q$ for integers $p$, $q$ with $q > 0$, then

$$f_\gamma(x+q) = f(x) + p - \gamma x - p = f_\gamma(x), \qquad x \in \mathbf{Z},$$

so $f_\gamma$ has period $q$.

(B) $\Rightarrow$ (C). This is obvious.

(C) $\Rightarrow$ (D). If $f_\gamma$ is periodic, then also $D_1 f_\gamma$ and $D_1 f = D_1 f_\gamma + \gamma$ are periodic.

(D) $\Rightarrow$ (A). Finally, if $D_1 f$ has a period $q$, then so has $f_\gamma$, and

$$\sum_{y=0}^{q-1} (D_1 f_\gamma)(y) = f_\gamma(q) - f_\gamma(0) = 0.$$

so that $q\gamma = f(q) - f(0)$ is an integer. □

So the projection of the graph of $f$ on the first coordinate is a periodic set while the projection on the second coordinate, equal to $\mathbf{im}(f)$ is aperiodic for irrational $\gamma$.

When we continue this study, we can restrict the slope $\gamma$ to satisfy $0 < \gamma < \frac{1}{2}$. If $\frac{1}{2} < \gamma < 1$, we can study instead the function $G(x) = F(x) - \frac{1}{2}x$. Then $g = \lfloor G \rfloor|_{\mathbf{Z}}$ satisfies $g(x) = f(x) - \frac{1}{2}x$ if $x \in \mathbf{Z}$ is even and $g(x) = \lfloor F(x) + \frac{1}{2} \rfloor - \frac{1}{2}(x+1)$ if $x$ is odd.

**Proposition 11.13.2.** *Let $F\colon \mathbf{R} \to \mathbf{R}$ be an affine function $F(x) = \gamma x + c$, where now $0 < \gamma \leqslant \frac{1}{2}$, $c \in \mathbf{R}$. Define $f = \lfloor F \rfloor|_{\mathbf{Z}}$. Then the set $A$ in* (11.13.1) *never has neighbors, i.e., $a$ and $a+1$ can never both be elements of $A$.*

*The distance $m$ between a point $a \in A$ and the next largest element $b \in A$ must belong to the open interval*

$$I_\gamma = \{s \in \mathbf{R};\ 1/\gamma - 1 < s < 1/\gamma + 1\}.$$

*It follows that, if $1/\gamma$ is an integer, then the distance must be equal to $1/\gamma$, the only integer in that interval; if $1/\gamma$ is not an integer, then $I_\gamma$ contains exactly two integers, viz. $\lfloor 1/\gamma \rfloor$ and $\lceil 1/\gamma \rceil$.*

**Proof.** It is enough to study a situation with

$$f(0) = 0, \quad f(1) = \cdots = f(m) = 1, \quad f(m+1) = 2,$$

implying that $(D_1 f)(0) = (D_1 f)(m) = 1$, and $(D_1 f)(x) = 0$ for $1 \leqslant x \leqslant m-1$. Since we always have $f(x) \leqslant F(x) < f(x) + 1$ for $x \in \mathbf{Z}$, it follows that

$$0 \leqslant c < 1, \quad 1 \leqslant \gamma + c < 2, \quad 1 \leqslant \gamma m + c < 2, \quad 2 \leqslant \gamma(m+1) + c < 3.$$

Combining the last two inequalities we obtain

$$\frac{2-c}{\gamma} - 1 \leqslant m < \frac{2-c}{\gamma},$$

a half-open interval of length one containing exactly one integer for each $c$. But $c$ can vary as shown by the first two inequalities above, which, combined with what was just said, gives the inequality $1-\gamma \leqslant c < 1$, so that

$$\frac{1}{\gamma} - 1 < m < \frac{1}{\gamma} + 1,$$

i.e., that $m \in I_\gamma$. If $1/\gamma$ is an integer, $I_\gamma$ contains only this integer; if not, the interval contains two integers. That both of these can actually occur follows from the example in Exercise 11.1 below.

If there are two consecutive points $a$ and $a+1$ in $A$, then we must have $m = 1$ and the slope is larger than $\frac{1}{2}$ contrary to assumption. □

If $1/\gamma$ is not an integer, the distances between two consequtive points in $A$ must vary between the two possibilities $\lfloor 1/\gamma \rfloor$ and $\lceil 1/\gamma \rceil$ in a proportion that makes the average slope over large intervals close to $1/\gamma$.

## 11.14 Discretizations of a function

We define here the lower and upper discretization of a function with real values or extended real values:

**Definition 11.14.1.** Given any function $F\colon \mathbf{R}^n \to \mathbf{R}_!$ we define two functions, $\operatorname{discr}_\triangledown(F)$ and $\operatorname{discr}^\vartriangle(F)$, by

$$\operatorname{discr}_\triangledown(F) = \lfloor F|_{\mathbf{Z}^n} \rfloor : \mathbf{Z}^n \to \mathbf{Z}_!, \quad \operatorname{discr}^\vartriangle(F) = \lceil F|_{\mathbf{Z}^n} \rceil : \mathbf{Z}^n \to \mathbf{Z}_!.$$

They are called the ***lower discretization of*** $f$ and the ***upper discretization of*** $f$, respectively. □

### 11.14.1 *Convexity of a discretized function*

**Proposition 11.14.2.** *For any* $(\mathbf{Z}^n \times \mathbf{Z})$*-convex function* $f\colon \mathbf{Z}^n \to \mathbf{Z}_!$ *and any* $x \in \mathbf{Z}^n$ *we have one of the following cases for* $F = \mathbf{cvxe}(f)$*,* $\lfloor F \rfloor = \operatorname{discr}_\triangledown(F)$ *and* $\lceil F \rceil = \operatorname{discr}^\vartriangle(F)$*.*

(A) $F(x) = f(x)$*. Then* $\lfloor F(x) \rfloor = \lceil F(x) \rceil = f(x)$*;*

(B) $f(x) - 1 < F(x) < f(x)$*. Then* $f(x) - 1 = \lfloor F(x) \rfloor < \lceil F(x) \rceil = f(x)$*;*

(C) $F(x) = f(x) - 1$*. Then* $\lfloor F(x) \rfloor = \lceil F(x) \rceil = f(x) - 1$*.*

*All three cases can occur as we shall see in Example 11.14.4 (cases* (A) *and* (C)*), and Example 11.14.5 (case* (B)*).*

**Proof.** We note that we always have

$$\lfloor F|_{\mathbf{Z}^n} \rfloor \leqslant F|_{\mathbf{Z}^n} \leqslant \lceil F|_{\mathbf{Z}^n} \rceil \leqslant f.$$

If $f$ is convex extensible we also know that $\mathbf{epi}^{\text{finite}}_{\text{strict}}(F) \subset \mathbf{cvxh}(\mathbf{epi}^{\text{finite}}(f))$, which leads to $f-1 \leqslant F|_{\mathbf{Z}^n}$. Hence, for $(\mathbf{Z}^n \times \mathbf{Z})$-convex functions we have

$$f-1 \leqslant \lfloor F|_{\mathbf{Z}^n} \rfloor \leqslant F|_{\mathbf{Z}^n} \leqslant \lceil F|_{\mathbf{Z}^n} \rceil \leqslant f.$$

From this we easily deduce the conclusion in the proposition. □

**Theorem 11.14.3.** *Assume that* $F\colon \mathbf{R} \to \mathbf{R}_!$ *is convex. Then* $\operatorname{discr}^{\triangle}(F)$ *and* $\operatorname{discr}_{\triangledown}(F)$ *are both* $(\mathbf{Z} \times \mathbf{Z})$*-convex.*

**Proof.** That the upper discretization is $(\mathbf{Z} \times \mathbf{Z})$-convex is an immediate consequence of the definitions; we just have to observe that

$$\mathbf{epi}^{\text{finite}}(\operatorname{discr}^{\triangle}(F)) = \mathbf{epi}^{\text{finite}}(F) \cap \mathbf{Z}^2.$$

It is perhaps surprising that also the lower discretization is $(\mathbf{Z} \times \mathbf{Z})$-convex. In general we have, writing $g = \operatorname{discr}^{\triangle}(F)$ and $h = \operatorname{discr}_{\triangledown}(F)$, that $h(x) = g(x)$ when $F(x) \in \mathbf{Z}_!$ and $h(x) = g(x) - 1$ when $F(x) \in \mathbf{R} \smallsetminus \mathbf{Z}$. Since both cases can occur, it is not obvious that (11.1.2) for $g$ implies the same inequality for $h$.

However, we always have $h \leqslant F < h+1$ at points where $F$ is finite, so that

$$h(s) \leqslant F(s) \leqslant (1-\lambda)F(p) + \lambda F(q) < (1-\lambda)h(p) + \lambda h(q) + 1,$$

assuming $h(p)$ and $h(q)$ to be finite. Since for any integer $m$, the inequality $m < t+1$ is equivalent to $m \leqslant \lceil t \rceil$, we see that (11.1.2) holds for $h$.

If one of $h(p)$, $h(q)$ is equal to $+\infty$, the inequality certainly holds; if one is equal to $-\infty$ while the other is less than $+\infty$, then also $h(s) = -\infty$ and the inequality holds as well. □

**Example 11.14.4.** Define $h(x) = 0$ for $x < 0$, $h(x) = 1$ for $x \geqslant 0$, $x \in \mathbf{Z}$ and $H = \mathbf{cvxe}(h)$. Then $H = 0$ everywhere, and $\operatorname{discr}^{\triangle}(H) = \operatorname{discr}_{\triangledown}(H) = 0$. We have $H = h - 1$ on the natural numbers. The function $h = Y_0$ is the Heaviside function, defined in (3.3.5). The word $D_1 h$ was called a skew Sturmian word by Morse & Hedlund (1940:8)—it is not periodic but ultimately periodic. This function satisfies $|D_b D_a h| \leqslant 1$ as well as $|J_{a,b} h| < 1$ for all $a, b \in \dot{\mathbf{N}}$. In fact, $J_{a,b}h(x)$ can be any rational number in $]-1, 1[$, and $D_b D_a h(x)$ can assume any of the values $-1, 0, 1$. But the graph of $h$ is not a discrete straight line in the sense of Reveillès; see Section 8.8 for the definition. It is, however, a refined digital hyperplane in the sense Definition 8.9.1.

For more general examples, see case (A) in Theorem 8.9.2. Also in this case the convex envelope of the corresponding function $f$ is affine: $F(x) = \alpha x + \beta$, $x \in \mathbf{R}$. □

**Example 11.14.5.** Define $F(x) = \alpha x$, $x \in \mathbf{R}$, for an irrational number $\alpha$, and let $f = \operatorname{discr}^{\triangle}(F)$, $g = \operatorname{discr}_{\triangledown}(F)$. Then $f$ and $g$ are convex extensible and we note that $g = f - 1$ except at the origin, where $f(0) = g(0) = 0$. The convex hull of the finite epigraph of $g$ is the open half plane $\{x \in \mathbf{R}^2;\ y > \alpha x - 1\}$; that of $f$ is the open half plane $\{x \in \mathbf{R}^2;\ y > \alpha x\}$ with the point $(0, 0)$ added. Neither is closed, so both $f$ and $g$ have irregular points as defined in Subsection 11.14.2 below. In fact, all points are irregular for $g$, all but the origin are irregular for $f$. The convex envelope of $f$ is $F$, that of $g$ is $F - 1$, so that $\mathbf{cvxe}(f) = \mathbf{cvxe}(g) + 1 = F$.

We note that in this example, $\operatorname{discr}^{\triangle}(\mathbf{cvxe}(f)) = f$ but

$$\operatorname{discr}^{\triangle}(\mathbf{cvxe}(g)) = \operatorname{discr}^{\triangle}(F - 1) = f - 1,$$

which takes the value $-1$ at the origin. Thus we have $\operatorname{discr}^{\triangle}(\mathbf{cvxe}(g)) = g - 1$ at the origin.

Both the graph of $f$ and that of $g$ can be described as discrete straight lines in the sense of Reveillès: the graph of $f$ is defined by $0 \leqslant -\alpha x + y < 1$; that of $g$ by $0 \leqslant \alpha x - y < 1$.

This example is related to the functions appearing in the theorem of Klein mentioned in the introduction of Chapter 8, but there is an essential difference: here we define $f$ and $g$ at all integers, whereas in Klein's theorem we used only their values for $x \geqslant 1$. □

The discretization operators may be used for rescaling of convex extensible functions. Let us define, given two positive numbers $\alpha$, $\beta$, and a function $F \in \mathscr{F}(\mathbf{R}^n, \mathbf{R}_!)$, its rescaling $F_\alpha^\beta(x) = \beta F(x/\alpha)$, $x \in \mathbf{R}^n$. Then for $f \in \mathscr{F}(\mathbf{Z}^n, \mathbf{Z}_!)$ we define its rescalings

$$f_\alpha^{\beta,\triangle *}(x) = \operatorname{discr}^{\triangle}\big((\mathbf{cvxe}(f))_\alpha^\beta\big) \text{ and } f_{\alpha,\triangledown}^{\beta}(x) = \operatorname{discr}_{\triangledown}\big((\mathbf{cvxe}(f))_\alpha^\beta\big)$$

for $x \in \mathbf{Z}^n$. Both functions are convex extensible. If $f$ is $(\mathbf{Z}^n \times \mathbf{Z})$-convex, they are reasonable candidates for rescaled functions of $f$.

### 11.14.2 *Regular and irregular points*

Given $f \colon \mathbf{Z} \to \mathbf{R}_!$, the convex set $\mathbf{cvxh}(\mathbf{epi}^{\text{finite}}(f))$ is either empty or the convex hull of a denumerably infinite set in the plane. It may or may not be closed.

We define $C(s) = \big\{(s, y) \in \mathbf{cvxh}(\mathbf{epi}^{\text{finite}}(f))\big\}$, $s \in \mathbf{R}$. This set may be empty or equal to a straight line; if not, it is a vertical ray with endpoint $(s, F(s))$, where $F = \mathbf{cvxe}(f)$. We shall say that $s \in \mathbf{R}$ is a ***regular point for*** $f$ if $C(s)$ is closed, and that $s$ is an ***irregular point for*** $f$ if $C(s)$ is not closed.

**Proposition 11.14.6.** *Let $f \in \mathscr{F}(Z, \mathbf{Z}_!)$ and write $F$ for* $\mathbf{cvxe}(f)$. *If $F(s) \geqslant f(s)$ for a point $s \in \mathbf{Z}$ (case* (A) *in Proposition 11.14.2), then $s$ is regular. The converse does not hold. For a convex extensible function $f$, $s \in \mathbf{Z}$ regular implies $F(s) > f(s) - 1$ (case* (A) *or* (B) *in Proposition 11.14.2). The converse does not hold. Thus for convex extensible functions we have for all integer points $s$,*

$$F(s) \geqslant f(s) \;\Rightarrow\; s \text{ is regular} \;\Rightarrow\; F(s) > f(s) - 1 \;\Leftrightarrow\; \lceil F(s) \rceil = f(s).$$

**Proof.** For the second assertion, cf. Proposition 11.14.2. That the converse does not hold is clear from Example 11.14.5. □

**Proposition 11.14.7.** *Let $f$ be a function in $\mathscr{F}(\mathbf{Z}, \mathbf{R}_!)$. If the convex hull of $\mathbf{epi}^{\text{finite}}(f)$ is closed, then all points are regular. Conversely, if $F = \mathbf{cvxe}(f) > -\infty$ and all points are regular, then*

$$\mathbf{cvxh}(\mathbf{epi}^{\text{finite}}(f)) = \mathbf{epi}^{\text{finite}}(F) = \overline{\mathbf{epi}^{\text{finite}}_{\text{strict}}(F)},$$

*a closed set.* □

If $f$ is allowed to take the value $-\infty$, regularity does not imply that $\mathbf{cvxh}(\mathbf{epi}^{\text{finite}}(f))$ is closed:

**Example 11.14.8.** Define $f \colon \mathbf{Z} \to \mathbf{Z}_!$ by $f(x) = -\infty$ for $x \leqslant -1$, $f(0) = 0$, and $f(x) = +\infty$ for $x \geqslant 1$. Then $f$ is $(\mathbf{Z} \times \mathbf{Z})$-convex and every point is regular, but $\mathbf{cvxh}(\mathbf{epi}^{\text{finite}}(f))$ is not closed: the point $(0, -1)$ belongs to its closure but not to the set itself. □

**Proposition 11.14.9.** *Assume that $f \in \mathscr{F}(\mathbf{Z}, \mathbf{Z}_!)$ is $(\mathbf{Z} \times \mathbf{Z})$-convex, i.e., that it has the property that all points in $\mathbf{cvxh}(\mathbf{epi}^{\text{finite}}(f))$ with integer coordinates belong to $\mathbf{epi}^{\text{finite}}(f)$ (see Definition 10.3.3), and that $a, b \in \mathbf{R}$, $a \leqslant b$, are regular points for $f$. Then all points in $[a, b]$ are regular for $f$. In particular we have $\lceil \mathbf{cvxe}(f) \rceil = f$ on $[a, b]_{\mathbf{Z}}$.*

**Proof.** Assume that $a$ and $b$ are regular, and that $s$ is irregular for a $(\mathbf{Z} \times \mathbf{Z})$-convex function $f \in \mathscr{F}(\mathbf{Z}, \mathbf{Z}_!)$, $a, b, s \in \mathbf{R}$, $a < s < b$. We shall reach a contradiction. Let $H$ be the affine function which agrees with $F = \mathbf{cvxe}(f)$ at $a$ and $b$. Then $H(s) > F(s)$, but on the other hand there must exist points $r_j, t_j \in \mathbf{R}$ with $r_j < s < t_j$, such that the affine function $H_j$ which agrees with $F$ at $r_j$ and $t_j$ has the property that $H_j(s)$ tends to $F(s)$. However, we can take $r_j$ and $t_j$ as integers. This is because $F$ is not an arbitrary convex function but the convex envelope of a function which is $+\infty$ on $\mathbf{R} \smallsetminus \mathbf{Z}$.

Indices $j$ such that $r_j \leqslant a < b \leqslant t_j$ cannot contribute to this convergence. Indeed, for these indices we must have $H_j(a) \geqslant H(a)$ and $H_j(b) \geqslant H(b)$ so that also $H_j(s) \geqslant H(s) > F(s)$, which prevents convergence to $F(s)$.

Also indices such that $a < r_j \leqslant t_j < b$ cannot contribute to the convergence. Indeed, since the $r_j$ and $t_j$ are integers, there are only finitely many different functions $H_j$ for such indices, and for all of them we have $H_j(s) > F(s)$.

Finally we need to look at indices $j$ such that $a < r_j < b \leqslant t_j$ or $r_j \leqslant a < t_j < b$. For an index such that $a < r_j < b \leqslant t_j$ we must have $H_j(b) \geqslant H(b) = F(b)$. Let $K_j$ be the affine function which agrees with $H_j$ at $r_j$ and with $H$ at $b$. Then there are only finitely many different values $K_j(s)$, and they are all strictly larger than $F(s)$, so, since $H_j(s) \geqslant K_j(s)$, these indices cannot contribute to the convergence. The case $r_j \leqslant a < t_j < b$ is symmetric.

Thus in all cases we have found a contradiction.

For the last assertion, see Proposition 11.14.6. □

**Proposition 11.14.10.** *Suppose that $s \in \mathbf{Z}$ is an irregular point for a function $f\colon \mathbf{Z} \to \mathbf{Z}_!$. Then the boundary of the finite epigraph of $F =$ $\mathbf{cvxe}(f)$ is a polygon with finitely or infinitely many vertices at integer points, and either all points in $[s, +\infty[$ are irregular and $F$ agrees with an affine function on that interval, or all points in $]-\infty, s]$ are irregular and $F$ equals the restriction of an affine function there.*

**Proof.** The previous proposition shows that there cannot be regular points both to the left and to the right of an irregular point. That $F$ agrees with an affine function on an interval consisting of irregular points follows easily. □

For $(\mathbf{Z} \times \mathbf{Z})$-convex functions $f\colon \mathbf{Z} \to \mathbf{Z}_!$ we have a priori six cases when comparing $f$ and $F = \mathbf{cvxe}(f)$ at a point $s \in \mathbf{Z}$:

(1) $C(s) = \emptyset$, a closed set. Then $F(s) = f(s) = +\infty$.
(2) $C(s) = \{s\} \times \mathbf{R}$, a closed set. Then $F(s) = f(s) = -\infty$.
(3) $C(s)$ is closed and $F(s)$ is an integer. Then the endpoint $(s, F(s))$ of $C(s)$ belongs to $\mathbf{cvxh}(\mathbf{epi}^{\text{finite}}(f))$ and $F(s) = \lceil F(s) \rceil = f(s)$. Example: $f(x) = \lceil \frac{1}{2}x \rceil$, $s$ any even integer.
(4) $C(s)$ is closed and $F(s) \in \mathbf{R} \smallsetminus \mathbf{Z}$. Then the endpoint $(s, F(s))$ belongs to $\mathbf{cvxh}(\mathbf{epi}^{\text{finite}}(f))$ and $F(s) < \lceil F(s) \rceil = f(s)$. Example: $f(x) = \lceil \frac{1}{2}x \rceil$,

$s$ any odd integer.

(5) $C(s)$ is not closed and $F(s)$ is an integer. Then the endpoint $(s, F(s))$ of $C(s)$ does not belong to $\mathbf{cvxh}(\mathbf{epi}^{\text{finite}}(f))$ and $F(s) = \lceil F(s) \rceil = f(s) - 1 < f(s)$. Example: $f(x) = 0$ for $x \leqslant -1$ and $f(x) = 1$ for $x \geqslant 0$, $s$ any natural number. See Example 11.14.4.

(6) $C(s)$ is not closed and $F(s) \in \mathbf{R} \smallsetminus \mathbf{Z}$. Then the endpoint $(s, F(s))$ of $C(s)$ does not belong to $\mathbf{cvxh}(\mathbf{epi}^{\text{finite}}(f))$ and $F(s) < \lceil F(s) \rceil = f(s)$. Example: $f(x) = \lceil x/\sqrt{2} \rceil$, $F(x) = x/\sqrt{2}$, $s$ any nonzero integer. See Example 11.14.5.

For such functions, $s$ is regular in cases 1, 2, 3, and 4 and irregular in cases 5 and 6.

Note that $\lceil F(s) \rceil = f(s)$ in all cases except 5, and that $\lceil F(s) \rceil = f(s) - 1 < f(s)$ in case 5.

## 11.15 Extending convex extensible functions

We may extend a function defined and finite on a finite interval $[c, d]_{\mathbf{Z}}$ if it satisfies the conditions $D_b D_a f \geqslant -1$ whenever this expression has a sense. The function obviously has a convex extension with values in $\mathbf{R}_!$; the problem is to find an extension with finite values. Since the conditions are now onesided, we have even more freedom in the choice of extension.

**Theorem 11.15.1.** *Let $f\colon [c, d]_{\mathbf{Z}} \to \mathbf{Z}$ be given such that $D_b D_a f(x) \geqslant -1$ for all $a, b, x$ for which the expression is defined, i.e., for $c \leqslant x \leqslant d - a - b$ with $x \in \mathbf{Z}$ and $a, b \in \dot{\mathbf{N}}$. Then $f$ can be extended to a function $g\colon \mathbf{Z} \to \mathbf{Z}$ such that $D_b D_a g(x) \geqslant -1$ for all $x \in \mathbf{Z}$ and all $a, b \in \dot{\mathbf{N}}$.*

**Proof.** We shall define $g$ as follows.

$$g(x) = \begin{cases} f(c) + \beta(x - c), & x \leqslant c - 1; \\ f(x), & c \leqslant x \leqslant d; \\ f(d) + \gamma(x - d), & x \geqslant d + 1. \end{cases}$$

Here

$$\beta = \inf_{t \in \mathbf{Z}} (D_1 f(t);\ c \leqslant t \leqslant d - 1) \in \mathbf{Z}$$

and

$$\gamma = \sup_{t \in \mathbf{Z}} (D_1 f(t);\ c \leqslant t \leqslant d - 1) \in \mathbf{Z}.$$

This means that we define $g$ by affine functions outside the given interval $[c,d]_{\mathbf{Z}}$.

We shall prove by induction that $g$ satisfies $D_b D_a g \geqslant -1$. This is true by hypothesis if $c \leqslant x < x+a+b \leqslant d$. Assume that it is true for $p \leqslant x < x+a+b \leqslant q$ for a certain $p$ and a certain $q$ with $p \leqslant c$ and $q \geqslant d$, and let us prove that it is true if we augment $q$ by one unit as well as if we decrease $p$ by one unit. It is enough to consider $p \leqslant x < x+a+b = q+1$, i.e., to pass from $[p,q]_{\mathbf{Z}}$ to $[p,q+1]_{\mathbf{Z}}$. We shall then compare $D_b D_a g(x)$ with $D_{b-1} D_a g(x)$, i.e., we move $x+b$ and $x+a+b$ one step to the left to get the rightmost point inside the interval $[p,q]_{\mathbf{Z}}$. Then

$$\begin{aligned} D_b D_a g(x) &= g(x) - g(x+a) - g(x+b) + g(q+1) \\ &= g(x) - g(x+a) - g(x+b) + g(q) + \gamma \\ &\geqslant g(x) - g(x+a) - g(x+b-1) + g(q) = D_{b-1} D_a g(x) \geqslant -1 \end{aligned}$$

in view of our choice of $\gamma$, which guarantees that $g(q+1) = g(q) + \gamma$ while $g(x+b) \leqslant g(x+b-1) + \gamma$. □

## 11.16 Possible future studies

For which $A$ are all $A$-laterally convex functions convex extensible? As we have seen, for $n = 2$, this is so if $\Lambda(A) = \Psi(\Phi(A))$ contains $R$ (the implication (A) ⇒ (C) in Theorem 11.6.1.

Here $\Phi$ and $\Psi$ are defined in (13.4.3) and (13.4.4), and $R$ is the subset of $\mathbf{Z}^2 \times \mathbf{Z}^2$ used to define rhomboidal convexity; see Definition 11.5.1.

Conversely, if every $\Lambda(A)$-laterally convex is convex extensible, then $\Lambda(A)$ contains the set $\Delta = \{(a,a)\};\ a \in \mathbf{Z}^2\}$ (Theorem 11.6.1, (C) ⇒ (D)).

So we have one sufficient condition, $\Lambda(A) \supset R$, and one necessary condition, $\Delta \subset \Lambda(A)$, but I have not found a condition which is both necessary and sufficient.

## 11.17 Notes

### 11.17.1 *Digital lines have many expressions*

Some of the equivalences in Theorem 11.12.1 have a long history. Morse & Hedlund (1940) proved that Sturmian words (aperiodic words of minimal complexity, meaning that the number of distinct factors of length $n$ is equal to $n+1$) are balanced, and conversely. That balance of a binary word is equivalent to the property of being a mechanical word is proved in the case of irrational slope by Lothaire (2002: Theorem 2.1.13).

For a thorough treatment of digital lines with an irrational slope and their relation to continued fractions, see (Uscka-Wehlou 2007, 2008a, 2008b, 2009a, 2009b, 2009c, 2009d). Clifton Callender explores the relation between Sturmian words and music. He mentions on his web site Hania Uscka-Wehlou's work (2009b) on digital lines of irrational slope, which "has a natural interpretation in terms of rhythmic canons. My own work teases out these implications and extends her results to hierarchical periodicities manifest in terms of canonic structure." See also (Callender 2013) and (Voss 1991).

### 11.17.2 *Another definition of convexity*

A kind of convexity in $\mathbf{Z}^2$ which is not of the type $Y$-convexity as defined in Definition 10.3.1 has been introduced and studied by Sara Brunetti, Péter Balázs and collaborators; see (Balázs & Brunetti 2016, 2017) and (Brunetti et al. 2019).

### 11.17.3 *Digital hyperplanes*

The pioneer in the field of digital hyperplanes has been Jean-Pierre Reveillès (1991). The more general refined hyperplanes defined here were introduced in my paper (2004b).

### 11.17.4 *Recognizing digital lines and finding the convex hull*

Hübler et al. (1981) determined the convex hull of a finite set in the plane.

For a study of the problem of recognizing digital lines and calculating their thickness, see (Buzer 2006); for the same problem for planes and hyperplanes, see (Coeurjolly & Brimkov 2006).

### 11.17.5 *Terms used*

The Finnish, Estonian and Breton languages use eight primitive words for the eight compass directions N, NE, E, SE, S, SW, W, NW, thus simple words also for northeast, southeast, southwest, and northwest, where English uses compounds. In Finnish they are in this order: *pohjoinen, koillinen, itä, kaakko, etelä, lounas, länsi, luode*, thus *lounas* 'southwest' is not obtained from the words for south and west. In Estonian the words are similar, except that south is *lõuna* and southwest is *edel*, a permutation of the corresponding Finnish words. (Can anybody explain this permutation?)

## 11.18 Exercise

11.1. For which points $a = (a_1, a_2) \in \mathbf{Z}^2$ with $a_1, a_2 \geqslant 1$ is it true that the set $A = \{0, a\}$ is $(\mathbf{Z} \times \mathbf{Z})$-convex?

# Chapter 12

# Three problems in discrete optimization

## 12.1 Introduction

THREE PROBLEMS IN DISCRETE OPTIMIZATION will be presented here. We first present, in the next section, the three problems; then in Sections 12.3–12.5, solutions to the problems. In Sections 12.6–12.15, a more detailed analysis of the first problem, that on marginal functions, will be given.

We shall define, using difference operators, classes of functions defined on the set of points with integer coordinates which are preserved under the formation of marginal functions. The duality between classes of functions with certain convexity properties and families of second-order difference operators plays an important role and is explained using notions from mathematical morphology. Dualities of this kind are more systematically studied in Chapter 13.

We continue here the study of lateral convexity, which we defined in Chapter 10 using second-order difference operators (a special kind of convolution operators). We prove that this concept of convexity is perfectly adapted for proving that the marginal function of a real-valued function defined on the set of points with integer coordinates remains in the same class.

## 12.2 Three fundamental problems

We have mentioned in Section 9.21 three fundamental properties in convex analysis in vector spaces over the field of real numbers. In this section we show that they do not hold in a naive approach to digital convexity, and then indicate some of the attempts at saving them. Most definitions and some of the results are given also in higher dimension.

In all these cases, the most obvious definition of a convex function of discrete variables leads to severe difficulties.

The difficulties encountered may be used as tests for any class of functions proposed to serve as convex functions on a discrete space like $\mathbf{Z}^n$, the set of all points in $\mathbf{R}^n$ with integer coordinates.

In particular we shall see that the property of having a convex extension is not sufficient. Also, since it is sometimes claimed[1] that the submodular functions[2] constitute a discrete analogue of convexity, we note that submodularity is also insufficient for each of the three propertiesfoot studied.

### 12.2.1 *Problem 1: Wild marginal functions*

| | | | | | ↑ | | | | | | | |
|---|---|---|---|---|---|---|---|---|---|---|---|---|
| | 8 | 7 | 6 | 5 | 4 | 3 | 2 | 1 | 0 | 1 | 2 | |
| | 6 | 5 | 4 | 3 | 2 | 1 | 0 | 1 | 2 | 3 | 4 | |
| — | 4 | 3 | 2 | 1 | 0 | 1 | 2 | 3 | 4 | 5 | 6 | → |
| | 2 | 1 | 0 | 1 | 2 | 3 | 4 | 5 | 6 | 7 | 8 | |
| | 0 | 1 | 2 | 3 | 4 | 5 | 6 | 7 | 8 | 9 | 10 | |
| | 2 | 3 | 4 | 5 | 6 | 7 | 8 | 9 | 10 | 11 | 12 | |
| | | | | | \| | | | | | | | |
| $f_{\mathrm{marg}}$ | 0 | 1 | 0 | 1 | 0 | 1 | 0 | 1 | 0 | 1 | 0 | |

Figure 12.1. The marginal function $f_{\mathrm{marg}}$ of the function $f$ given by $f(x, y) = |x - 2y|$ is not convex in any sense.

Look at

$$(12.2.1) \qquad f(x, y) = |x - 2my|, \qquad (x, y) \in \mathbf{Z}^2,$$

where $m$ is a positive integer. Its marginal function

$$f_{\mathrm{marg}}(x) = \inf_{y \in \mathbf{Z}} f(x, y), \qquad x \in \mathbf{Z},$$

is equal to $|x|$ for $-m \leqslant x \leqslant m$ (then the infimum is attained for $y = 0$), and is given for other values of $x$ by the fact that it is periodic with period $2m$. Thus $f_{\mathrm{marg}}$ is a function with teeth that are as large as we like. Still,

[1] For instance by Iwata (2010:2944).

[2] See the definition in Section 11.3.

we are tempted to accept $f$ as a convex function: it is the restriction to $\mathbf{Z}^2$ of a convex function defined on $\mathbf{R}^2$.

We can take $m = 1$ and define $f(x, y) = |x - 2y|$, $(x, y) \in \mathbf{Z}^2$, a convex extensible function, and let $f_{\text{marg}}(x) = \inf_{y \in \mathbf{Z}} f(x, y)$, $x \in \mathbf{Z}$, be its marginal function (see Figure 12.1 on page 260). We see that $f_{\text{marg}}$ is not convex in any sense of the word.

The function $f(x, y) = -x^2$, $(x, y) \in \mathbf{Z}^2$, is submodular and its marginal function is $f_{\text{marg}}(x) = -x^2$, $x \in \mathbf{Z}$, so for submodular functions there are even simpler counterexamples.

In Example 11.7.1 we have seen examples of functions which are relevant for the study of convexity properties of marginal functions.

### 12.2.2 *Problem 2: Local minima are not global*

Look at the function

$$g(x, y) = x + f(x, y) = x + |x - 2my|, \qquad (x, y) \in \mathbf{Z}^2,$$

a modification of (12.2.1), $m$ being a positive integer as large as we please. Then we have $g(x, y) \geqslant g(0, 0) = 0$ when $x \geqslant -m$, thus when $(x, y)$ belongs to a large set containing the origin, but in spite of this, we have

$$g(-m - 1, -1) = -2 < g(0, 0).$$

Still $g$ is the restriction to $\mathbf{Z}^2$ of a convex function on $\mathbf{R}^2$.

No finite set can serve as a neighborhood to guarantee that a function which is convex extensible has a global minimum. This can be seen as follows.

Given any finite subset $P$ of $\mathbf{R}^2$ there exists a convex function $F\colon \mathbf{R}^2 \to \mathbf{R}$ which is unbounded from below but is such that $F(p_1, p_2) \geqslant F(0, 0)$ for all $(p_1, p_2) \in P$. Indeed, we may define $F(x, y) = |y - \gamma x| - \beta x$, $(x, y) \in \mathbf{R}^2$, where the constants $\gamma$ and $\beta$ are to be chosen. We see that $F$ is negative in a sector $(\gamma - \beta)x < y < (\gamma + \beta)x$, and that $F$ is nonnegative outside this sector. It is now clear that $\gamma$ and $\beta$ can be chosen so that the sector is nonempty and so that $P$ is in its complement.

The function $f(x, y) = \min(0, a - x)$, $(x, y) \in \mathbf{Z}^2$, is submodular for any $a \in \mathbf{N}$ and it satisfies $f(x, y) \geqslant f(0, 0)$ for all $(x, y) \in \mathbf{Z}^2$ with $x \leqslant a$, thus in a large neighborhood of the origin if $a$ is large, but it is unbounded from below. Hence for submodular functions there are very simple counterexamples.

### 12.2.3 *Problem 3: No separating plane*

Now look at the two sets

$$A = \{(x, y, z) \in \mathbf{Z}^3;\ z \leqslant 1 - 2|x + y - 1|\},$$
$$B = \{(x, y, z) \in \mathbf{Z}^3;\ 2|x - y| \leqslant z\};$$

cf. a similar example in (Murota 2003:15). They are disjoint. A separating plane must have the equation $z = H(x, y)$ for some affine function $H\colon \mathbf{R}^2 \to \mathbf{R}$. But this leads to

$$1 - 2|x + y - 1| \leqslant H(x, y) \leqslant 2|x - y|, \qquad (x, y) \in \mathbf{Z}^2.$$

Then these inequalities must hold also for $x = y = \frac{1}{2}$, which is a contradiction.

The sets in $\mathbf{R}^3$ defined by the same inequalities are not disjoint: the point $\left(\frac{1}{2}, \frac{1}{2}, z\right)$ belongs to both sets for all $z \in [0, 1]$.

The sets

$$A = \{(x, y, z) \in \mathbf{Z}^3;\ z \leqslant -f(x, y)\},\ B = \{(x, y, z) \in \mathbf{Z}^3;\ g(x, y) \leqslant z\}$$

are disjoint if $f$ and $g$ are real-valued and $0 < f + g$. If we just want submodular functions, we may take $f(x, y) = -x^2$, $g(x, y) = x^2 + 1$. There is then no separating hyperplane.

## 12.3 Solution to Problem 1: Marginal functions

**Theorem 12.3.1.** *Assume that $f\colon \mathbf{Z}^2 \to \mathbf{R}_!$ is rhomboidally convex (Definition 11.5.1) and that $f > -\infty$. Then its marginal function*

$$f_{\text{marg}}(x) = \inf_{y \in \mathbf{Z}} f(x, y), \qquad x \in \mathbf{Z},$$

*is $(\mathbf{Z} \times \mathbf{R})$-convex in the sense of Definition 10.3.3.*

**Proof.** Let us first assume that the function is such that the infimum defining $h$ is attained for every $x$.

It then suffices to prove:

$$\text{if } h(0) = f(0, 0) = 0, \text{ then } f(-1, y) \dotplus f(1, z) \geqslant 0 \text{ for all } y, z \in \mathbf{Z}.$$

We shall use induction on $|z - y| \in \mathbf{N}$.

We first note that if $z = y$, then, by separate $(\mathbf{Z} \times \mathbf{R})$-convexity,

$$f(-1, y) \dotplus f(1, y) \geqslant 2f(0, y) \geqslant 2f(0, 0) \geqslant 0.$$

If $|z-y|=1$, say $z=y+1$, then by rhomboidal convexity

$$f(-1,y) \dotplus f(1,y+1) \geqslant f(0,y) \dotplus f(0,y+1) \geqslant 2f(0,0) \geqslant 0.$$

If $|z-y|=2$, say $z=y+2$, then we can use $(\mathbf{Z}\times\mathbf{R})$-convexity along diagonals to conclude that $f(-1,y) \dotplus f(1,y+2) \geqslant 2f(0,y+1) \geqslant 0$.

If $|z-y|=3$, say $z=y+3$, then by rhomboidal convexity

$$f(-1,y) \dotplus f(1,y+3) \geqslant f(-1,y+1) \dotplus f(1,y+2),$$

which brings us back to the case where $z=y+1$.

If finally $|z-y|=k\geqslant 4$, say $z\geqslant y+k$, then again by rhomboidal convexity we have

$$f(-1,y) \dotplus f(1,z) \geqslant f(-1,z-2) \dotplus f(1,y+2),$$

which brings us to the case when $|z-y|=k-4$. Hence the induction works. This proves the desired result when the infimum is attained.

For $\theta>0$ define $f_\theta(x,y)=f(x,y) \dotplus \theta y^2$. Then $f_\theta(x,y)\to+\infty$ as $y\to\pm\infty$ and the infimum is attained. Hence the marginal function $f_{\theta\,\mathrm{marg}}$ of $f_\theta$ is $(\mathbf{Z}\times\mathbf{R})$-convex. This implies that also its limit $f_{\mathrm{marg}}=\lim_{\theta\to 0} f_{\theta\,\mathrm{marg}}$ is $(\mathbf{Z}\times\mathbf{R})$-convex. □

One may ask for a characterization of the functions $f$ which have convex marginal functions. However, this seems not to be a reasonable problem, since for any function $\varphi\in\mathscr{F}(\mathbf{Z},\mathbf{Z})$ the function $f(x,y)=|y-\varphi(x)|$ has marginal function zero and still can be very irregular. It is more reasonable to ask for a characterization of the functions $f$ such that the marginal function of $f(x,y)-\beta y$ is $(\mathbf{Z}\times\mathbf{R})$-convex for all choices of real numbers $\beta$. This we can do under the condition that $f$ is $(\mathbf{Z}\times\mathbf{R})$-convex in the second variable. The result is related to the partial Fenchel transform defined in Definition 9.17.10, i.e., we transform in the second variable only:

$$f^\diamond(x,\eta)=\sup_{y\in\mathbf{Z}}\big(\eta y-f(x,y)\big),\qquad (x,\eta)\in\mathbf{Z}\times\mathbf{R}. \tag{12.3.1}$$

**Example 12.3.2.** Let $\varphi\in\mathscr{F}(\mathbf{Z},\mathbf{R})$ be any function and define $f(x,y)=|y-\varphi(x)|$, $(x,y)\in\mathbf{Z}^2$. Then its marginal function is $f_{\mathrm{marg}}(x)=d(\varphi(x),\mathbf{Z})$, the distance between $\varphi(x)$ and the integers, defined by

$$d(t,\mathbf{Z})=\min\big(\mathbf{frac}(t),1-\mathbf{frac}(t)\big)\text{ for any real number } t,$$

where $\mathbf{frac}(t)=t-\lfloor t\rfloor$ is the fractional part of $t$. We see that $f_{\mathrm{marg}}$ is $(\mathbf{Z}\times\mathbf{R})$-convex if and only if $d(t,\mathbf{Z})$ is constant. In particular, if $\varphi\in\mathscr{F}(\mathbf{Z},\mathbf{Z})$, $f_{\mathrm{marg}}$ is $(\mathbf{Z}\times\mathbf{R})$-convex even though $\varphi$ can be very irregular, as already noted.

On the other hand, defining

$$f_\beta(x,y) = f(x,y) + \beta y = |y - \varphi(x)| + \beta y, \qquad (x,y) \in \mathbf{Z}^2,$$

and taking $\varphi$ so that it maps $\mathbf{Z}$ into $\mathbf{Z}$, then

$$(f_{\pm 1})_{\text{marg}}(x) = \inf_{y \in \mathbf{Z}} (f(x,y) \pm y) = \pm\varphi(x)$$

is $(\mathbf{Z} \times \mathbf{R})$-convex for both choices of signs if and only if $\varphi$ is the restriction to $\mathbf{Z}$ of an affine function. This shows that imposing the condition on $f_\beta$ and not just on $f = f_0$ restricts the class of functions considerably, even though we use only two values of $\beta$. □

**Example 12.3.3.** Let $\varphi\colon \mathbf{R} \to \mathbf{R}$ be a convex function, let $\gamma$ be an irrational number, $\theta$ a real number, and define

$$F_{\gamma,\theta}(x,y) = \varphi(y - \gamma x) + \theta x^2, \qquad (x,y) \in \mathbf{R}^2,$$

and let $f_{\gamma,\theta}$ be its restriction to $\mathbf{Z}^2$. Then $F_{\gamma,\theta}$ is convex if and only if $\theta \geqslant 0$. Moreover $f_{\gamma,\theta}$ is $(\mathbf{Z}^2 \times \mathbf{R})$-convex in the sense of Definition 10.3.3 also if and only if $\theta \geqslant 0$. The marginal function of $f_{\gamma,\theta}$ is

$$f_{\gamma,\theta\,\text{marg}}(x) = \varphi(d(\gamma x, \mathbf{Z})) + \theta x^2, \qquad x \in \mathbf{Z},$$

where $d(t, \mathbf{Z})$ is the distance from $t \in \mathbf{R}$ to $\mathbf{Z}$ as in Example 12.3.2. Define $\beta = \mathbf{frac}(\gamma) = \gamma - \lfloor\gamma\rfloor$, the fractional part of $\gamma$, and $\tau = \min(\beta, 1-\beta)$. We note that $d(\gamma x, \mathbf{Z}) = d(\beta x, \mathbf{Z}) = d(\tau x, \mathbf{Z})$, so $f_{\gamma,\theta\,\text{marg}} = f_{\beta,\theta\,\text{marg}} = f_{\tau,\theta\,\text{marg}}$. The marginal function $f_{\gamma,\theta\,\text{marg}}$ is $(\mathbf{Z} \times \mathbf{R})$-convex if and only if $\theta \geqslant \varphi(\frac{1}{2}) - \varphi(\frac{1}{2} - \tau)$. This is what is required of the term $\theta x^2$ to compensate for the lack of convexity in the first term. Only the values of $\varphi$ in the interval $[0, \frac{1}{2}]$ influence the behavior of the marginal function. For $\varphi(t) = |t|$, the condition becomes $\theta \geqslant \tau$; for $\varphi(t) = t^2$, the condition is, as to be expected, weaker: $\theta \geqslant \tau(1-\tau)$. □

**Theorem 12.3.4.** *Let $f\colon \mathbf{Z}^2 \to \mathbf{R}$ be $(\mathbf{Z}^2 \times \mathbf{R})$-convex in the second variable. Then the following three properties are equivalent.*

(A) *The marginal function*

$$h_\beta(x) = \inf_{y \in \mathbf{Z}} (f(x,y) - \beta y) = -f^\diamond(x, \beta), \qquad x \in \mathbf{Z},$$

*is $(\mathbf{Z} \times \mathbf{R})$-convex for all $\beta \in \mathbf{R}$ such that $h_\beta > -\infty$ (here $f^\diamond$ is defined in Definition 9.17.10);*

(B) *The canonical extension* $\mathbf{can}(f)$ *satisfies Jensen's inequality*

$$(12.3.2)\qquad \begin{aligned}&(\mathbf{can}(f))(x'',y'')\\ &\qquad\leqslant (1-\lambda)(\mathbf{can}(f))(x,y)+\lambda(\mathbf{can}(f))(x',y')\end{aligned}$$

*for all points* $(x,y),(x',y')\in\mathbf{Z}^2$, $(x'',y'')=(1-\lambda)(x,y)+\lambda(x',y')$, *and all* $\lambda\in[0,1]$ *such that* $x''=(1-\lambda)x+\lambda x'\in\mathbf{Z}$.

(C) *The canonical extension* $\mathbf{can}(f)$ *satisfies Jensen's inequality* (12.3.2) *if* $x'-x=2$ *and* $\lambda=\frac{1}{2}$.

**Proof.** (A) $\Rightarrow$ (B). If $x=x'$, property (B) means just that $f$ is $(\mathbf{Z}\times\mathbf{R})$-convex in the second variable, which we assume.

If $|x-x'|=1$, the inequality is only required when $(x'',y'')$ is equal to $(x,y)$ or $(x',y')$ and is therefore trivially satisfied.

We define $f_\beta(x,y)=f(x,y)-\beta y$ for all $(x,y)\in\mathbf{Z}^2$ and all $\beta\in\mathbf{R}$. Given any points $(x,y),(x',y')\in\mathbf{Z}^2$ with $|x-x'|\geqslant 2$ and $(x'',y'')=(1-\lambda)(x,y)+\lambda(x',y')$ such that $x''\in\mathbf{Z}$, we choose $\beta=D_{(0,1)}f(x'',\lfloor y''\rfloor)$. Then

$$f_\beta(x'',\lfloor y''\rfloor)=f_\beta(x'',\lfloor y''\rfloor+1)=f_\beta(x'',\lceil y''\rceil)$$

and also $(\mathbf{can}(f_\beta))(x'',y'')$ and $h_\beta(x'')$ are equal to this value.

Now property (A) implies that

$$\begin{aligned}0&\leqslant(1-\lambda)h_\beta(x)-h_\beta(x'')+\lambda h_\beta(x')\\ &\leqslant(1-\lambda)f_\beta(x,y)-(\mathbf{can}(f_\beta))(x'',y'')+\lambda f_\beta(x',y').\end{aligned}$$

This proves (12.3.2).

(B) $\Rightarrow$ (C). This is obvious since (C) is a special case of (B).

(C) $\Rightarrow$ (A). If (C) holds, then for all $(x,y),(x',y')\in\mathbf{Z}^2$, $(x'',y'')\in\mathbf{Z}\times\mathbf{R}$ with $x=x''-1$ and $x'=x''+1$ we have

$$\begin{aligned}0&\leqslant\tfrac{1}{2}f(x,y)-(\mathbf{can}(f))(x'',\tfrac{1}{2}y+\tfrac{1}{2}y')+\tfrac{1}{2}f(x',y')\\ &\leqslant\tfrac{1}{2}f(x''-1,y)-h_0(x'')+\tfrac{1}{2}f(x''+1,y').\end{aligned}$$

Taking the infimumum over all $y$ and $y'$ we obtain (A) in the special case $\beta=0$. However, this implies the general case. □

**Remark 12.3.5.** In condition (C) we can have $(x',y')=(x+2,y\pm1)$, $(x'',y'')=(x+1,y\pm\frac{1}{2})$ but not $(x',y')=(x\pm1,y+2)$, $(x'',y'')=(x\pm\frac{1}{2},y+1)$. This means that (C) implies (11.6.2) but not (11.6.3). That this asymmetry is essential is shown by the fact that, if $f$ satisfies (B), then so does $g(x,y)=f(x,y-\gamma x)$ for any integer $\gamma$; condition (11.6.3) is not invariant under this transformation. □

By permuting the variables we get the following corollary, which yields a characterization of rhomboidal convexity.

**Corollary 12.3.6.** *A function $f \in \mathscr{F}(\mathbf{Z}^2, \mathbf{R})$ is rhomboidally convex if and only if it is separately $(\mathbf{Z} \times \mathbf{R})$-convex and the marginal functions*

$$h_\beta(x) = \inf_{y \in \mathbf{Z}} (f(x,y) - \beta y) \text{ and } k_\gamma(x) = \inf_{y \in \mathbf{Z}} (f(y,x) - \gamma y)$$

*are $(\mathbf{Z} \times \mathbf{R})$-convex for all real numbers $\beta$ and $\gamma$ such that $h_\beta$ and $k_\gamma$ are $> -\infty$.*

**Proof.** It is enough to remark that condition (B) in the theorem and its symmetric counterpart together are equivalent to condition (K) in Theorem 11.6.2. □

## 12.4 Solution to Problem 2: Local minima

For a local minimum, the key question is to find a suitable neighborhood. For optimal results we should let the neighborhood depend on the function we study. (It is not a neighborhood for any topology.)

**Definition 12.4.1.** Given a function $f\colon \mathbf{Z}^2 \to \mathbf{R}$ and a point $p \in \mathbf{Z}^2$, we shall say that a point $q \in \mathbf{Z}^2$, $q \neq p$, is a ***fold-line-neighbor*** of $p$ with respect to $f$ if $[p,q]$ is a fold line of $\mathbf{cvxe}(f)$ (see Definition 9.13.5) and if the only points with integer coordinates in the segment $[p,q]$ are $p$ and $q$. We shall write $N_f(p)$ for the set of all fold-line-neighbors of $p$ with respect to $f$. □

If the function $f$ is integrally convex, we know that $\mathbf{cvxe}(f) = \mathbf{can}(f)$. It follows that, for such a function, a point can have at most eight neighbors; they are all elements of the unit sphere for the $l^\infty$-metric,

$$S^\infty(p,1) = \{x \in \mathbf{Z}^2;\ \|x-p\|_\infty = 1\}.$$

In fact it can have any number of neighbors up to eight except 1. (If a point does not have more than one fold-line-neighbor, $f$ is the restriction of an affine function near the point and there are no neighbors at all.)

However, a general convex function may have infinitely many neighbors. A trivial example is the restriction to $\mathbf{Z}^2$ of the Euclidean norm,

$$f(x) = \|x\|_2 = \sqrt{x_1^2 + x_2^2}, \qquad x \in \mathbf{Z}^2,$$

where every point $q$ in $\mathbf{Z}^2$ such that $[0,q] = \{0,q\}$ is a fold-line-neighbor of the origin. A less trivial example is presented in Exercise 12.13.

**Lemma 12.4.2.** *Let $f\colon \mathbf{Z}^2 \to \mathbf{R}$ be $(\mathbf{Z}^2 \times \mathbf{R})$-convex and fix a point $p \in \mathbf{Z}^2$. If $f$ is $\geqslant f(p)$ in $N_f(p)$, then* $\mathbf{cvxe}(f) \geqslant f(p)$ *in the convex hull of $N_f(p) \cup \{a\}$ (taken in $\mathbf{R}^2$).*

**Proof.** If $N_f(p)$ is finite and $x \in \mathbf{cvxh}(N_f(p) \cup \{a\})$, then there exists three or fewer points in $N_f(p) \cup \{p\}$, the convex hull of which contains $x$ and such that $\mathbf{cvxe}(f)$ is affine in the convex hull of these points. Hence $\mathbf{cvxe}(f) \geqslant f(p)$ also in this convex hull and consequently in all of $\mathbf{cvxh}(N_f(p) \cup \{p\})$.

If $N_f(p)$ is infinite, we let $F_r$ be the convex envelope of the function $f$ restricted to points $x \in B^{\infty}_{\leqslant}(p,r) = \{x \in \mathbf{Z}^2;\quad \|x-p\|_\infty \leqslant r\}$. By the previous result we have $F_r \geqslant f(p)$ in the convex hull of $N_f(p) \cap B^{\infty}_{\leqslant}(p,r)$. The convex envelope $\mathbf{cvxe}(f)$ of $f$ is the limit of $F_r$ as $r \to +\infty$; we conclude that $\mathbf{cvxe}(f) \geqslant f(p)$ in the union of $N_f(p) \cap B^{\infty}_{\leqslant}(p,r)$ when $r$ varies, i.e., in all of $N_f(p)$. □

**Theorem 12.4.3.** *Let $f\colon \mathbf{Z}^2 \to \mathbf{R}$ be a $(\mathbf{Z}^2 \times \mathbf{R})$-convex function. Assume that the convex hull of the set $\{x \in \mathbf{Z}^2;\ f(x) \geqslant f(p)\}$ contains $p$ in its interior, taken with respect to the usual topology of $\mathbf{R}^2$. Then $f(p)$ is the global minimum of $f$, i.e., $f(p) = \inf_{y \in \mathbf{Z}^2} f(y)$.*

*We can distinguish three cases:*

(A) *If $N_f(p)$ has at least three points, then the convex hull of $N_f(p)$ contains $p$ in its interior, and it suffices to assume that $f(q) \geqslant f(p)$ for all $q \in N_f(p)$.*

(B) *If $N_f(p)$ has two elements, then* $\mathbf{cvxh}\big(N_f(p)\big)$ *must nevertheless contain $p$, and it suffices to assume, in addition to $\inf_{q \in N_f(p)} f(q) \geqslant f(p)$, that one has $f(x), f(y) \geqslant f(p)$ for two points $x, y$ such that* $\mathbf{cvxh}\big(N_f(p) \cup \{x\} \cup \{y\}\big)$ *contains $p$ in its interior.*

(C) *Finally, if $p$ does not have any fold-line-neighbor, it is enough to assume that there exist two points $x, y$ such that $x - p$ and $y - p$ are linearly independent and such that $f(x), f(y) \geqslant f(p)$.*

**Proof.** The theorem follows, because in all cases the function $\mathbf{cvxe}(f)$ is convex in $\mathbf{R}^2$ and has a local minimum at $p$. □

Although the result is valid for all $(\mathbf{Z}^2 \times \mathbf{R})$-convex functions, it is most interesting for the rhomboidally convex functions, because a neighbor must then be one of the eight points in $S^\infty(p,1)$, so we need only check the condition there. By way of contrast, in general there may be infinitely many neighbors as already pointed out, and even if there are only finitely many, they may be arbitrarily many and lie far away from $p$; this makes the assumption hard to check. We thus have:

**Corollary 12.4.4.** *Let* $f\colon \mathbf{Z}^2 \to \mathbf{R}$ *be rhomboidally convex. If* $f(x) \geqslant f(p)$ *for all eight points in* $S^\infty(p,1)$*, then* $f \geqslant f(p)$ *everywhere.*

For rhomboidally convex functions, the whole set $S^\infty(p,1)$ can thus serve as neighborhood. Miller (1971:168) proved this for functions which are discretely convex in his sense, which means that the functions shall satisfy

$$(12.4.1) \qquad \inf_{y \in N(x)} f(y) \leqslant (1-t)f(a) + tf(b), \qquad t \in [0,1], \quad a,b \in \mathbf{Z}^n,$$

where we define $x = (1-t)a + tb$ and where $N(x)$ is the integer neigborhood of $x$ defined in Definition 10.4.1. Favati & Tardella (1990:9) proved it for integrally convex functions. However, to get an optimal result, we have defined here a neighborhood depending on the function. As we shall see, no smaller neighborhood will do.

Case (B) in Theorem 12.4.3 is illustrated by the function $f(x) = |x_1|$, $x \in \mathbf{Z}^2$. Here the origin has two fold-line-neigbors, $(0,1)$ and $(0,-1)$. The origin is in their convex hull, but not in its interior, so two more points are needed.

The set of neighborhoods defined in the theorem is minimal in the following precise sense.

**Proposition 12.4.5.** *Given any rhomboidally convex function* $f\colon \mathbf{Z}^2 \to \mathbf{R}$*, a point* $p \in \mathbf{Z}^2$ *and a proper subset* $B$ *of* $N_f(p)$*, there is an affine function* $H\colon \mathbf{R}^2 \to \mathbf{R}$ *such that* $g = f - H\big|_{\mathbf{Z}^2}$ *satisfies*

$$\inf_{q \in B} g(q) \geqslant g(p) \text{ and } \inf_{y \in N_f(p)} g(y) < g(p).$$

**Proof.** That the sets indicated in cases (B) and (C) of Theorem 12.4.3 are optimal is evident. Let us consider case (A). Then there are at least three and at most eight points in $N_f(p)$; let us denote them by $q^{(1)}, \dots, q^{(m)}$ in cyclic order. For any point $q^{(k)} \in N_f(p)$, we define an affine function $H$ such that $H(p) = f(p)$ and $H(q^{(k-1)}) = f(q^{(k-1)})$ and $H(q^{(k+1)}) = f(q^{(k+1)})$, counting indices modulo $m$. This defines $H$ uniquely, and we note that $H(q^{(j)}) \leqslant f(q^{(j)})$ for all $j \neq k$, while the fact that $[a, q^{(k)}]$ is actually a fold line implies that $H(q^{(k)}) > f(q^{(k)})$. Now $g = f - H\big|_{\mathbf{Z}^2}$ has all the properties mentioned in the statement of the proposition. □

The result on local and global minima yields a new characterization of rhomboidal convexity:

**Theorem 12.4.6.** *A function $f\colon \mathbf{Z}^2 \to \mathbf{R}$ is rhomboidally convex if and only if $f$ is separately $(\mathbf{Z} \times \mathbf{R})$-convex and, for every $p \in \mathbf{Z}^2$ and every affine function $H$ which is a maximal minorant of $f$ in $p+\{0,1\}^2$, we have $H \leqslant f$ in all of $\mathbf{Z}^2$.*

**Proof.** If $f$ is rhomboidally convex it is certainly separately $(\mathbf{Z}\times\mathbf{R})$-convex. Suppose that $f \geqslant 0$ in $\{0,1\}^2$ and that zero is a maximal minorant, which implies that we must have $f = 0$ at three of the four points of $\{0,1\}^2$, say at $(0,0)$, $(1,0)$ and $(0,1)$. Then we also know that $f(1,1) \geqslant 0$. We have to prove that $f \geqslant 0$ in all of $\mathbf{Z}^2$. Separate $(\mathbf{Z} \times \mathbf{R})$-convexity implies that $f$ is nonnegative at $(-1,0)$ and $(0,-1)$ (and also at $(0,2)$ and $(2,0)$, but that is not needed now). In addition to this, rhomboidal convexity implies that $f$ is nonnegative at $(-1,-1)$, $(-1,1)$, and $(1,-1)$. Summing up, we see that $f$ is nonnegative at all eight points of $S^\infty(p,1)$, where $p = (0,0)$. Thus the theorem on local minima of rhomboidally convex functions implies that $f \geqslant 0$ everywhere.

Conversely, suppose that $f$ is separately $(\mathbf{Z} \times \mathbf{R})$-convex and satisfies the indicated property for affine minorants. We shall then prove that $f$ is rhomboidally convex. It suffices to prove that, if $f(0,0) = f(1,0) = f(1,1) = 0$, then $f(2,1) \geqslant 0$. Now, if $f(0,1)$ happens to be nonnegative, the condition on affine minorants yields the result, for $f$ then has zero as a maximal affine minorant in $\{0,1\}^2$, hence everywhere. On the other hand, if $f(0,1)$ is negative, then the separate $(\mathbf{Z}\times\mathbf{R})$-convexity implies that $f(2,1)$ is positive. We are done. □

## 12.5 Solution to Problem 3: Separating hyperplanes

The third problem we consider, the one on separating planes, receives a satisfying solution in the class of integrally convex functions, defined in Definition 10.6.1.

**Theorem 12.5.1.** *Let $f, g\colon \mathbf{Z}^n \to \mathbf{R}_!$ be two arbitrary functions. Let us define two sets $A$ and $B$ in $\mathbf{Z}^n \times \mathbf{R}$ as*

$$A = \{(x,t) \in \mathbf{Z}^n \times \mathbf{R};\ t \leqslant -f(x)\},\ B = \{(x,t) \in \mathbf{Z}^n \times \mathbf{R};\ g(x) \leqslant t\}.$$

*Assume that one of $f$ and $g$ is less than $+\infty$ everywhere. Then the following three properties are equivalent.*

(1) *There exists a separating plane;*
(2) *There exists a non-vertical separating plane;*

(3) $0 \leqslant \mathbf{cvxe}(f) \dotplus \mathbf{cvxe}(g)$ *in* $\mathbf{R}^2$.

*In particular, if $f$ and $g$ are integrally convex, a separating plane exists if and only if* $0 \leqslant \mathbf{can}(f) + \mathbf{can}(g)$.

**Proof.** We know that (A) implies (E) in Theorem 9.15.3. But if one of the functions is less than $+\infty$ the converse also holds: any separating plane must be non-vertical.

When a function is integrally convex, we know that its convex envelope is equal to its canonical extension. □

The canonical extension is affine in each of the four triangles into which a square $a + [0, 1]^2$ is divided by its diagonals. Therefore, to prove an inequality like $0 \leqslant \mathbf{can}(f) + \mathbf{can}(g)$ in all of $\mathbf{R}^2$, it is enough to do so at the points of $\mathbf{Z}^2 \cup (\mathbf{Z} + \frac{1}{2})^2$, i.e., at only one additional point in each square $p + [0, 1[^2$.

We define $f_\odot : \mathbf{Z}^2 \cup (\mathbf{Z} + \frac{1}{2})^2 \to \mathbf{R}$ for $p \in \mathbf{Z}^2$ by

$$\begin{aligned} &f_\odot(p) = f(p) \text{ and} \\ &f_\odot(p + (\tfrac{1}{2}, \tfrac{1}{2})) \\ &\qquad = \tfrac{1}{2} \min\big(f(p) + f(p + (1, 1)), f(p + (1, 0)) + f(p + (0, 1))\big). \end{aligned}$$

Thus $f_\odot + g_\odot \geqslant 0$ if and only if $\mathbf{can}(f) + \mathbf{can}(g) \geqslant 0$. With this definition we have a corollary:

**Corollary 12.5.2.** *Let $f, g \colon \mathbf{Z}^2 \to \mathbf{R}$, define $A$ and $B$ be as in the theorem, and assume in addition that $f$ and $g$ are integrally convex. Then there exists a separating plane in $\mathbf{R}^3$ if and only if $0 \leqslant f_\odot + g_\odot$ in $\mathbf{Z}^2 \cup (\mathbf{Z} + \frac{1}{2})^2$.* □

**Corollary 12.5.3.** *Let $f, g \colon \mathbf{Z}^2 \to \mathbf{R}$ and assume in addition that their modules (defined in Section 11.3) are of the same sign everywhere in the non-strict sense that*

$$[D_{(0,1)}D_{(1,0)}f(p)] \cdot [D_{(0,1)}D_{(1,0)}g(p)] \geqslant 0$$

*at every point $p \in \mathbf{Z}^2$. Then there exists a plane separating the sets $A$ and $B$ defined as in the theorem if and only if $f + g \geqslant 0$.*

**Proof.** The hypothesis implies that $0 \leqslant f_\odot + g_\odot$ if and only if $0 \leqslant f + g$, because the fold lines, if any, go in the same direction. □

**Theorem 12.5.4.** *For every dimension $n$ there exists a subset $P_n$ of $\mathbf{Q}^n$ such that $P_n \cap [0, 1[^n$ is finite, $P_n + a = P_n$ for every $a \in \mathbf{Z}^n$, and* *which has the property that for all functions* $f, g\colon \mathbf{Z}^n \to \mathbf{R}_!$, *the infimum of* $\mathbf{can}(f) \dotplus \mathbf{can}(g)$ *over the cube* $C(a) = a + [0,1]^n$ *is equal to its infimum over the finite set* $P_n \cap C(a)$.

**Proof.** Since $\mathbf{can}(f)$ is, in each cube $[0,1]^n$, the convex envelope of a function defined at $2^n$ points, it is the maximum of a finite family of affine functions. Where it is finite, the function $\mathbf{can}(f) \dotplus \mathbf{can}(g)$ is affine outside the union of the fold hyperplanes of $\mathbf{can}(f)$ and $\mathbf{can}(g)$. The infimum of $\mathbf{can}(f) \dotplus \mathbf{can}(g)$ must be attained at some point which is in the intersection of a fold hyperplane of $\mathbf{can}(f)$ and a fold hyperplane of $\mathbf{can}(g)$. These two fold planes typically intersect in a plane of codimension two. However, in this plane, the sum is piecewise affine, and the sum attains its infimum in some plane of codimension three. We go on like this and finally arrive at planes of codimension $n$, i.e., at points. Therefore we only have to determine which are the possible points of intersection of a finite number of hyperplanes which are fold hyperplanes of either $\mathbf{can}(f)$ or $\mathbf{can}(g)$. These fold hyperplanes have rational coefficients since they must pass through enough points with integer coordinates. We define $P_n$ to be the set of all points of intersection obtained in this way. □

We shall determine $P_n$ for $n = 2$ and $n = 3$ in the next two subsections. It turns out that $P_2$ has two points in $[0,1[^2$ and that $P_3$ has thirteen points in $[0,1[^3$.

### 12.5.1 *Separating planes in dimension three*

When studying functions of two integer variables, i.e., with graphs in $\mathbf{Z}^3$, we note the following.

The canonical extension is affine in each of the four triangles into which a square $p + [0,1]^2$ is divided by its diagonals. Therefore, to prove an inequality like $0 \leqslant \mathbf{can}(f) + \mathbf{can}(g)$ in all of $\mathbf{R}^2$, it is enough to do so at the points of $\mathbf{Z}^2 \cup (\mathbf{Z} + \frac{1}{2})^2$, i.e., at only one additional point in each square $p + [0,1[^2$.

We find that $P_2$ is the set of all points in $\mathbf{R}^2$ with integer coordinates or both coordinates equal to half-integers, thus $P_2 = \mathbf{Z}^2 \cup (\mathbf{Z} + \frac{1}{2})^2$. We note that, for all $p \in \mathbf{Z}^2$,

$$(\mathbf{can}(f))(p) = f(p) \text{ and}$$

$$(\mathbf{can}(f))(p + (\tfrac{1}{2}, \tfrac{1}{2})) = \tfrac{1}{2}\big[\big(f(p) + f(p + (1,1))\big) \wedge \big(f(p + (1,0)) + f(p + (0,1))\big)\big].$$

Thus $\mathbf{can}(f) + \mathbf{can}(g) \geqslant 0$ everywhere if the inequality holds in $P_2$:

**Proposition 12.5.5.** *For any functions* $f, g\colon \{0,1\}^2 \to \mathbf{R}_!$, *the infimum of* $\mathbf{cvxe}(f) \dotplus \mathbf{cvxe}(g)$ *over* $[0,1]^2$ *is the same as the infimum over* $P_2 \cap [0,1]^2$.

**Proof.** The only fold lines that can appear are the lines are the diagonals in a square and the lines with equation $x_1 = c$ or $x_2 = c$ for some integer $c$. □

**Theorem 12.5.6.** *Let* $f, g\colon \mathbf{Z}^2 \to \mathbf{R}$, *let us define* $A$ *and* $B$ *be as in Theorem 12.5.1, and assume in addition that* $f$ *and* $g$ *are integrally convex. Then there exists a separating plane in* $\mathbf{R}^3$ *if and only if* $0 \leqslant \mathbf{can}(f) \dotplus \mathbf{can}(g)$ *in* $P_2$. □

**Theorem 12.5.7.** *Let* $f, g\colon \mathbf{Z}^2 \to \mathbf{R}$ *and assume in addition that their modules are of the same sign everywhere in the non-strict sense that*

$$[D_{(0,1)}D_{(1,0)}f(p)] \cdot [D_{(0,1)}D_{(1,0)}g(p)] \geqslant 0$$

*at every point* $p \in \mathbf{Z}^2$. *Then there exists a plane separating the sets* $A$ *and* $B$ *defined in Theorem 12.5.1 if and only if* $f + g \geqslant 0$.

**Proof.** The hypothesis implies that $0 \leqslant \mathbf{can}(f) + \mathbf{can}(g)$ if and only if $0 \leqslant f + g$, because the fold lines, if any, go in the same direction in each square $p + [0,1]^2$. Thus the points $p + (\frac{1}{2}, \frac{1}{2})$ are not needed. □

### 12.5.2 *Separating planes in dimension four*

We find that $P_3$ is the set in $\mathbf{R}^3$ consisting of the following thirteen points in each cube $p + [0,1[^3$, $p \in \mathbf{Z}^3$.

(1) the point $p$;
(2) the three points $p + (0, \frac{1}{2}, \frac{1}{2})$, $p + (\frac{1}{2}, 0, \frac{1}{2})$, $p + (\frac{1}{2}, \frac{1}{2}, 0)$;
(3) the point $p + (\frac{1}{2}, \frac{1}{2}, \frac{1}{2})$;
(4) the eight points $p + \frac{1}{3}(i, j, k)$, where $i, j, k = 1, 2$.

Thus in the unit cube $[0,1]^3$ we have 8 points of the first kind, 6 points of the second kind, 1 point of the third kind, and 8 points of the fourth kind, a total of 23 points, i.e., the 8 integer points plus 15 more. In the half-open cube $[0,1[^3$ there are 13 points, $3+1+8 = 12$ in addition to the only integer point.

**Proposition 12.5.8.** *For any functions* $f, g\colon \{0,1\}^3 \to \mathbf{R}$, *the infimum of* $\mathbf{cvxe}(f) + \mathbf{cvxe}(g)$ *over* $[0,1]^3$ *is the same as the infimum over* $P_3 \cap [0,1]^3$.

**Theorem 12.5.9.** *Let $f, g\colon \mathbf{Z}^3 \to \mathbf{R}_!$, and let again $A$ and $B$ be as in Theorem 12.5.1, and assume in addition that $f$ and $g$ are integrally convex. Then there exists a separating plane in $\mathbf{R}^4$ if and only if $0 \leqslant \mathbf{can}(f) \dotplus \mathbf{can}(g)$ in $P_3$.* □

**Lemma 12.5.10.** *In the set $\{0,1\}^3$ of eight points, there are*

$$\binom{8}{3} = \frac{8!}{3!5!} = 56$$

*triples. Taking their convex hulls, we get 56 triangles. Of these, 24 lie in a face of the cube $[0,1]^3$ and are therefore contained in a plane with equation $x_1 = c$ or $x_2 = c$ or $x_3 = c$ for some integer $c$. Then there are 24 more which contain one edge of the cube, one face diagonal, and one space diagonal, and therefore lie in a plane with equation $x_1 \pm x_2 = c$ or an equation which arises out of these two by a permutation of the variables. Finally, there are eight triangles which contain no edges of the cube, three face diagonals, and no space diagonals. These lie in a plane with an equation $x_1 \pm x_2 \pm x_3 = c$, where $c$ is an integer.* □

The triangles can have $\sigma = 0$ or 1 space diagonals and $\varphi = 1$ or 3 face diagonals—see Table 12.1 on page 274. The number of segments that are edges in the cube is $3 - \sigma - \varphi$.

**Lemma 12.5.11.** *Let $f\colon \{0,1\}^3 \to \mathbf{R}$ be an arbitrary function. Then every fold plane of $F = \mathbf{cvxe}(f)$ has an equation of form $\gamma_1 x_1 + \gamma_2 x_2 + \gamma_3 x_3 = a$, where $a$ in an integer and $\gamma_j \in \{-1,0,1\}$.*

**Proof.** It is clear that $F$ can have a fold plane only if it passes through three points in $\{0,1\}^3$. In view of Lemma 12.5.10 the fold plane must have an equation of the indicated type. (The 27 elements in $\{-1,0,1\}$ define 13 different types of equations to be considered.) □

**Proof of Proposition 12.5.8.** The piecewise affine function $F+G$, where we write $F$ and $G$ for $\mathbf{cvxe}(f)$ and $\mathbf{cvxe}(g)$, is affine outside the union of the fold planes of $F$ and $G$. The infimum of $F+G$ must be attained at some point which is in the intersection of two fold planes of $F$ or $G$. These two fold planes typically intersect in a line. However, in this line, the sum is piecewise affine, and the sum attains its infimum at some point in the line which lies in another fold plane of either $F$ and $G$. Therefore we only have to determine which are the possible points of intersection of three planes of the form mentioned in Lemma 12.5.11. It is not difficult to check that

Table 12.1. Triangles in the cube $[0,1]^3$, in total 56 of them. The number of triangles having 0 or 1 space diagonals is listed, as well as the number of triangles having 1 or 3 face diagonals in the cube:

| *Number of face diagonals* | 1 | 3 | $\Sigma$ |
|---|---|---|---|
| *No space diagonal* | 24 | 8 | 32 |
| *One space diagonal* | 24 | 0 | 24 |
| $\Sigma$ | 48 | 8 | 56 |

these intersection points must be of the form mentioned in the statement of the proposition. □

That the points in items 1 and 2 in the begining of the present subsection (which all have at least one integer coordinate) are necessary for the conclusion in Theorem 12.5.9 to hold is clear from examples in one and two dimensions. That the points mentioned in items 3 and 4 are also necessary follows from two examples given below.

The following example shows that the point $(\frac{1}{2}, \frac{1}{2}, \frac{1}{2})$ cannot be omitted in Theorem 12.5.9. Table 12.1 here. Triangles in the cube ...

**Example 12.5.12.** Define $f\colon \{0,1\}^3 \to \mathbf{R}$ as $-1$ at $(1,1,1)$ and equal to zero at the other points. Then

$$F(x) = (\mathbf{cvxe}(f))(x) = \max(-x_1, -x_2, -x_3), \qquad x \in [0,1]^3.$$

Define $g\colon \{0,1\}^3 \to \mathbf{R}$, $g(x) = (x_1 + x_2 + x_3 - \frac{3}{2})^+$. Its convex envelope $G = \mathbf{cvxe}(g)$ is given by the same expression. We note that $F + G$ takes the value $-\frac{1}{2}$ at the point $(\frac{1}{2}, \frac{1}{2}, \frac{1}{2})$. Actually $F + G$ is negative in the set

$$\{x \in [0,1]^3;\ \max(-x_1, -x_2, -x_3) + \max(0, x_1 + x_2 + x_3 - \tfrac{3}{2}) < 0\},$$

which can also be described as

$$\{x \in [0,1]^3;\ x_1, x_2, x_3 > 0,\ x_1 + x_2 < \tfrac{3}{2},\ x_2 + x_3 < \tfrac{3}{2},\ x_3 + x_1 < 1\}$$

and is the interior of the convex hull of the five-set

$$\{(0,0,0), (0,0,1), (0,1,0), (1,0,0), (\tfrac{3}{4}, \tfrac{3}{4}, \tfrac{3}{4})\}.$$

It contains the point $(\frac{1}{2}, \frac{1}{2}, \frac{1}{2})$. In fact, $F + G \geqslant -\frac{1}{2}$ everywhere, and the infimum is attained at $(\frac{1}{2}, \frac{1}{2}, \frac{1}{2})$ and only there, proving that this point cannot be removed in the hypothesis of Theorem 12.5.9. □

The next example shows that the point $(\frac{1}{3}, \frac{1}{3}, \frac{1}{3})$ cannot be omitted in Proposition 12.5.8. By symmetry the same is true of all the other seven points mentioned in item 4 in the beginning of the present subsection.

**Example 12.5.13.** Define $f\colon \{0,1\}^3 \to \mathbf{R}$ as $-1$ at $(1,1,1)$ and equal to zero at the other points. Then

$$F(x) = (\mathbf{cvxh}(f))(x) = \max(-x_1, -x_2, -x_3), \qquad x \in [0,1]^3.$$

Then define $g\colon \{0,1\}^3 \to \mathbf{R}$, $g(x) = (x_1+x_2+x_3-1)^+$. Its convex envelope $G = \mathbf{cvxe}(g)$ is given by the same expression. We note that $F+G$ takes the value $-\frac{1}{3}$ at the point $(\frac{1}{3}, \frac{1}{3}, \frac{1}{3})$. Actually $F+G$ is negative in the set

$$\left\{x \in [0,1]^3;\ \max(-x_1, -x_2, -x_3) + \max(0, x_1+x_2+x_3-1) < 0\right\},$$

which can also be described as

$$\left\{x \in [0,1]^3;\ x_1, x_2, x_3 > 0,\ x_1+x_2 < 1,\ x_2+x_3 < 1,\ x_3+x_1 < 1\right\}$$

and is the interior of the convex hull of the five-set

$$\left\{(0,0,0), (0,0,1), (0,1,0), (1,0,0), (\tfrac{1}{2}, \tfrac{1}{2}, \tfrac{1}{2})\right\}.$$

It contains the point $(\frac{1}{3}, \frac{1}{3}, \frac{1}{3})$ but no point in $(\frac{1}{2}\mathbf{Z})^3$. In fact, $F+G \geqslant -\frac{1}{3}$ everywhere, and the infimum is attained at $(\frac{1}{3}, \frac{1}{3}, \frac{1}{3})$ and only there, proving that this point cannot be removed. □

## 12.6 The marginal function of a function of integer variables

The question remains which kind of convexity we shall use. Let us formulate the problem explicitly.

**Problem.** *Define, for $n = 1, 2, \dots$, classes $\mathscr{R}_n$ of functions defined in $\mathbf{Z}^n$ such that $\mathscr{R}_1 = CVX(\mathbf{R},\mathbf{R})|_{\mathbf{Z}}$ and such that the successive marginal functions $h_{n-1}, h_{n-2}, \dots, h_1$ of any function $f \in \mathscr{R}_n$ defined by $h_n = f$,*

$$h_k(x) = \inf_{t \in \mathbf{Z}} h_{k+1}(x,t), \qquad x = (x_1, \dots, x_k) \in \mathbf{Z}^k, \quad k = n-1, \dots, 1,$$

*belong to $\mathscr{R}_k$ whenever they do not take the value $-\infty$.* □

We should also require that the classes are large enough so as to avoid trivial results, e.g., by taking $\mathscr{R}_n$ as the set of all functions $f\colon \mathbf{Z}^n \to \mathbf{R}$ such that $f(x_1, \dots, x_n) = g(x_1)$ for some $g \in CVX(\mathbf{R},\mathbf{R})|_{\mathbf{Z}}$.

We shall define in the sequel classes of functions defined on the integer points which solve the problem completely; see Theorem 12.9.1. Since this theorem is very general, we have formulated a corollary (Corollary 12.9.3), which is perhaps easier to apply. The functions of interest are the $A$-laterally convex functions, where $A$ is a subset of $\mathbf{Z}^n \times \mathbf{Z}^n$, and which were defined in Definition 10.7.2. This subset $A$ determines a family of second-order difference operators; there is a duality between such families and classes of functions with certain convexity properties, which we explained in Section 10.8 using basic notions of mathematical morphology.

Moreover, we shall prove that the classes obtained are optimal in a natural sense (see Example 11.7.1 and Section 12.10).

The most obvious attempt at defining a convex function of integer variables, i.e., taking $\mathscr{R}_n = CVX(\mathbf{R}^n, \mathbf{R})|_{\mathbf{Z}^n}$, fails in a very conspicuous way, even in low dimensions, as we have seen in Subsection 12.2.1. Our conclusion is that the property of being convex extensible is too weak to be of use in this context. In view of this observation, I have studied a class of functions defined on $\mathbf{Z} \times \mathbf{Z}$ which is suitable for this and other important properties in convexity theory; see my papers (2008a, 2010a).

We shall extend this study to higher dimensions, i.e., to functions on $\mathbf{Z}^n \times \mathbf{Z}^m$.

In the case of two integer variables, there are several equivalent ways to define integral convexity. In (Kiselman 2008a) integral convexity was introduced using difference operators. From this characterization it is obvious that the class is closed under addition.

## 12.7 Convolution and convex extensibility

We recall that the discrete convolution product $f * g$ of two functions $f, g\colon \mathbf{Z}^n \to \mathbf{R}$ was defined in Definition 3.3.2 as

$$(f * g)(x) = \sum_{y \in \mathbf{Z}^n} f(x-y)g(y), \qquad x \in \mathbf{Z}^n, \tag{12.7.1}$$

assuming some kind of convergence.

We define for $p = (p^{(1)}, \dots, p^{(k)}) \in (\mathbf{R}^n)^k$ and $\lambda = (\lambda_1, \dots, \lambda_k) \in \mathbf{R}^k$ satisfying $\lambda_j \geqslant 0$, $\sum_{j=1}^k \lambda_j = 1$, and $\sum_{j=1}^k \lambda_j p^{(j)} = 0$,

$$\mu_{p,\lambda} = \sum_{j=1}^k \lambda_j \delta_{p^{(j)}}.$$

Here $\delta_a$ denotes the Kronecker delta placed at $a$, defined in (3.3.4).

The convex envelope of a function defined on a subset $A$ of $\mathbf{R}^n$ is given by

$$\mathbf{cvxe}(f)(x) = \inf_{p,\lambda} \big(\mu_{p,\lambda} * f\big)(x) = \inf_{p,\lambda} \sum_{y \in A} \mu_{p,\lambda}(x-y) f(y), \qquad x \in \mathbf{R}^n.$$

(In view of Carathéodory's theorem it suffices to take $k = n + 1$.)

This implies that convex extensibility of a function $f$ defined on a subset $A$ of $\mathbf{R}^n$ can be characterized by means of an infinite family of convolution operators, viz.

$$((\mu_{p,\lambda} - \delta_0) * f)(x) = \sum_{y \in A} \mu_{p,\lambda}(x - y) f(y) - f(x) \geqslant 0, \qquad x \in A,$$

for all $p$ and $\lambda$ of the kind mentioned.

When $n = 1$, the convex-extensible functions are those which satisfy the inequality $(\mu_{p,\lambda} - \delta_0) * f \geqslant 0$ for $p = (-1, 1)$ and $\lambda = (\frac{1}{2}, \frac{1}{2})$, thus defining the class using a single convolution operator.

## 12.8 The set where the infimum is attained

We shall first study the relation between $A$-lateral convexity and the interval (possibly empty) where the infimum defining the marginal function is attained.

**Theorem 12.8.1.** *Let us define, for any function* $f\colon \mathbf{Z}^n \to \mathbf{R}$,

$$\begin{aligned} &M_f(x_1, \dots, x_{n-1}) \\ &\qquad = M_f(x') = \{b \in \mathbf{Z};\ f(x_1, \dots, x_{n-1}, b) = \inf\nolimits_{t \in \mathbf{Z}} f(x_1, \dots, x_{n-1}, t)\}, \end{aligned}$$

*where* $x' = (x_1, \dots, x_{n-1}) \in \mathbf{Z}^{n-1}$. *We also define*

$$f_\beta(x) = f(x) - \beta x_n, \qquad x = (x_1, \dots, x_n) \in \mathbf{Z}^n, \quad \beta \in \mathbf{R}.$$

*Now fix an element* $a = (a', a_n)$ *of* $\mathbf{Z}^n$, *where* $a' = (a_1, \dots, a_{n-1})$ *and* $a_n \geqslant 0$, *and define*

$$A = \{(e^{(n)}, e^{(n)}), ((a', a_n), e^{(n)}), ((-a', a_n), e^{(n)})\},$$

*a subset of* $(\mathbf{R}^n)^2$ *with three elements. Then* $f$ *is* $A$*-laterally convex if and only if*
*(1)* $t \mapsto f(x', t)$ *is convex extensible for every* $x'$ *and*
*(2) a certain Lipschitz property holds: for all* $x' \in \mathbf{Z}^{n-1}$, *all* $\beta \in \mathbf{R}$, *and all* $a = (a', a_n)$:

$$M_{f_\beta}(x' + a') \subset M_{f_\beta}(x') + [-a_n, a_n]_{\mathbf{Z}}. \tag{12.8.1}$$

**Proof.** Assume first that $f$ is $A$-laterally convex. Since $A$ contains $(e^{(n)}, e^{(n)})$, $\mathbf{Z} \ni t \mapsto f(x', t)$ is convex extensible for every $x'$.

We note that for a function which is convex extensible in the last variable,

(12.8.2) $b \in M_f(x')$ if and only if $D_{e^{(n)}} f(x', b-1) \leqslant 0 \leqslant D_{e^{(n)}} f(x', b)$.

Moreover

(12.8.3) $b, b+1 \in M_f(x')$ if and only if $D_{e^{(n)}} f(x', b) = 0$.

Let now $f$ satisfy $D_a D_{e^{(n)}} f \geqslant 0$ and consider two points $x'$ and $x' + a'$ in $\mathbf{Z}^{n-1}$. Then for any $b \in M_f(x')$ we have, since also $((-a', a_n), e^{(n)})$ is in $A$,

(12.8.4)
$$\begin{aligned} D_{e^{(n)}} f(x' + a', b - a_n - 1) &\leqslant D_{e^{(n)}} f(x', b-1) \leqslant 0 \\ &\leqslant D_{e^{(n)}} f(x', b) \leqslant D_{e^{(n)}} f(x' + a', b + a_n), \end{aligned}$$

which implies that there is a point $c \in [b - a_n, b + a_n]_{\mathbf{Z}}$ with

$$D_{e^{(n)}} f(x' + a', c - 1) \leqslant 0 \leqslant D_{e^{(n)}} f(x' + a', c).$$

In view of (12.8.2), this means that $c \in M_f(x' + a')$. We have proved that $b \in c + [-a_n, a_n]_{\mathbf{Z}} \subset M_f(x' + a') + [-a_n, a_n]_{\mathbf{Z}}$, and, since $b$ was any point in $M_f(x')$, that $M_f(x') \subset M_f(x' + a') + [-a_n, a_n]_{\mathbf{Z}}$. We are done, since the whole argument holds also for $f_\beta$.

Conversely, suppose that the function $f$ satisfies $D_{e^{(n)}} D_{e^{(n)}} f \geqslant 0$ but is not $A$-laterally convex. Then it does not satisfy one of the two inequalities $D_a D_{e^{(n)}} f \geqslant 0$ and $D_{(-a', a_n)} D_{e^{(n)}} f \geqslant 0$. It suffices to consider one of these cases. We thus assume that there exist $(x', b) \in \mathbf{Z}^{n-1} \times \mathbf{Z}$ such that $D_{e^{(n)}} f(x' + a', b + a_n) < D_{e^{(n)}} f(x', b)$. We shall reach a contradiction to the Lipschitz property (12.8.1).

We take a real number $\beta$ such that

(12.8.5) $D_{e^{(n)}} f(x' + a', b + a_n) < \beta < D_{e^{(n)}} f(x', b)$.

If we rewrite this for the function $f_\beta$, for which $D_{e^{(n)}} f_\beta = D_{e^{(n)}} f - \beta$, we obtain

(12.8.6) $D_{e^{(n)}} f_\beta(x' + a', b + a_n) < 0 < D_{e^{(n)}} f_\beta(x', b)$,

which implies that

$$M_{f_\beta}(x' + a') \subset [b + a_n + 1, +\infty[_{\mathbf{Z}} \text{ and that } M_{f_\beta}(x') \subset ]-\infty, b]_{\mathbf{Z}}.$$

Hence

$$M_{f_\beta}(x' + a') + [-a_n, a_n]_{\mathbf{Z}} \subset [b + 1, +\infty[_{\mathbf{Z}}$$

and

$$M_{f_\beta}(x') + [-a_n, a_n]_{\mathbf{Z}} \subset ]-\infty, b + a_n]_{\mathbf{Z}}.$$

Thus $M_{f_\beta}(x' + a')$ is not contained in $M_{f_\beta}(x') + [-a_n, a_n]_{\mathbf{Z}}$ unless it is empty, and $M_{f_\beta}(x')$ is not contained in $M_{f_\beta}(x' + a') + [-a_n, a_n]_{\mathbf{Z}}$ unless it is empty. As soon as one of them is nonempty, we get a contradiction to the Lipschitz property (12.8.1).

So the case when both sets are empty remains to be considered—so far, there is no contradiction in this case. Since $M_{f_\beta}(x' + a')$ is now empty by hypothesis, the function $t \mapsto D_{e^{(n)}} f_\beta(x' + a', t)$ can never change sign, and since $D_{e^{(n)}} f_\beta(x'+a', b+a_n)$ is negative, we must have $D_{e^{(n)}} f_\beta(x'+a', t) < 0$ for all $t \in \mathbf{Z}$. Now define $\gamma = D_{e^{(n)}} f(x', b) > \beta$. Then, by (12.8.3), $M_{f_\gamma}(x')$ is certainly nonempty; it contains $b$ and $b + 1$. And since $\gamma > \beta$ we have

$$D_{e^{(n)}} f_\gamma(x' + a', t) = D_{e^{(n)}} f_\beta(x' + a', t) + \beta - \gamma < D_{e^{(n)}} f_\beta(x' + a', t) < 0$$

for all $t \in \mathbf{Z}$, so that (12.8.2) shows that $M_{f_\gamma}(x' + a')$ is empty. This contradicts the inclusion $M_{f_\gamma}(x') \subset M_{f_\gamma}(x' + a') + [-a_n, a_n]_{\mathbf{Z}}$. We are done. □

By permuting the variables we easily obtain the following corollary.

**Corollary 12.8.2.** *Given a function $f\colon \mathbf{Z}^n \to \mathbf{R}$, we define, for $1 \leqslant j \leqslant n$ and $x' = (x_1, \dots, x_{j-1}, x_{j+1}, \dots, x_n) \in \mathbf{Z}^{n-1}$,*

$$\begin{aligned} M_{j,f}(x_1, \dots, x_{j-1}, x_{j+1}, \dots, x_{n-1}) &= M_{j,f}(x') \\ &= \big\{b \in \mathbf{Z};\ f(x_1, \dots, x_{j-1}, b, x_{j+1}, \dots, x_{n-1}, x_n) = \inf\nolimits_{x_j \in \mathbf{Z}} f(x)\big\}. \end{aligned}$$

*We also define*

$$f_{j,\beta}(x) = f(x) - \beta x_j,\ x = (x_1, \dots, x_n) \in \mathbf{Z}^n,\ j = 1, \dots, n,\ \beta \in \mathbf{R}.$$

*Fix a set $A$ which contains $(a, e^{(j)})$ and $(\bar{a}, e^{(j)})$, where*

$$\bar{a} = 2a_j e^{(j)} - a = (-a_1, \dots, a_j, \dots, -a_n),$$

*and satisfies (10.8.4) and (10.9.1). If $f$ is $A$-laterally convex, then $f$ is convex extensible in each variable separately and we have*

$$M_{j,f_{j,\beta}}(x' + a') \subset M_{j,f_{j,\beta}}(x') + [-a_j, a_j]_{\mathbf{Z}}, \qquad x' \in \mathbf{Z}^{n-1},$$

*where now $a' = (a_1, \dots, a_{j-1}, a_{j+1}, \dots, a_n)$ and similarly for $x'$.* □

## 12.9 Lateral convexity of marginal functions

### 12.9.1 *Arbitrary dimensions*

In my paper (2008a: Theorem 3.1), it was shown that for integrally convex functions of two integer variables, the marginal function is convex extensible. We shall now study the marginal function of $A$-laterally convex functions in arbitrary dimension and for more general choices of $A$. We defined $\Lambda$ as the composition $\Psi \circ \Phi$ in Definition 10.7.3.

**Theorem 12.9.1.** *Let $A \subset \mathbf{Z}^{n-1} \times \mathbf{Z}^{n-1}$ and $B \subset \mathbf{Z}^n \times \mathbf{Z}^n$ be given. We assume that* (10.8.4) *and* (10.9.1) *hold for both $A$ and $B$. Assume also that*

(12.9.1) $$\text{If } (a, b) \in A \text{ and } s \in [-1, 1]_{\mathbf{Z}}, \text{ then } ((a, s), (b, 0)) \text{ belongs to } \Lambda(B);$$

*that*

(12.9.2) $$\text{If there exists } c \in \mathbf{Z}^{n-1} \text{ such that } (a, c) \in A, \text{ then } ((a, 1), e^{(n)}) \in \Lambda(B);$$

*and finally that*

(12.9.3) $$\text{If } ((a, 1), e^{(n)}) \in B, \text{ then } ((-a, 1), e^{(n)}) \in \Lambda(B).$$

*If $f\colon \mathbf{Z}^n \to \mathbf{R}$ is $B$-laterally convex, then its marginal function*

$$h(x) = \inf_{t \in \mathbf{Z}} f(x, t), \qquad x \in \mathbf{Z}^{n-1},$$

*is $A$-laterally convex, provided that it does not take the value $-\infty$.*

**Lemma 12.9.2.** *Let $A$ and $B$ satisfy the hypotheses in Theorem 12.9.1. Then*

(12.9.4) $$\text{If } (a, b) \in A, \text{ then } ((a, -1), (b, -1)), ((a, 1), (b, 1)) \in \Lambda(B).$$

**Proof.** From the conditions (12.9.1) and (12.9.2) we know that both $((a, 1), (b, 0))$ and $((a, 1), e^{(n)})$ belong to $\Lambda(B)$. By partial addition $+_{\{1\}}$ we conclude that so does $((a, 1), (b, 1))$.

From the condition (12.9.2) we know that $((a, 1), e^{(n)})$ and, consequently, in view of (12.9.3), also $((-a, 1), e^{(n)})$ belongs to $\Lambda(B)$. So does the opposite pair $-((-a, 1), e^{(n)}) = ((a, -1), -e^{(n)})$.

By condition (12.9.1) we find that $((a, -1), (b, 0))$ is in $\Lambda(B)$, and we now only have to form the partial sum

$$((a, -1), -e^{(n)}) +_{\{1\}} ((a, -1), (b, 0)) = ((a, -1), (b, -1))$$

to conclude. □

By this lemma and (12.9.1) we know that if $A$ and $B$ satisfy the hypotheses of the theorem and if $(a, b) \in A$, then there are pairs of the form $((a, s), (b, t))$ in $\Lambda(B)$ with $-1 \leqslant s, t \leqslant 1$ and the sum $s + t$ taking any of the five values $-2, -1, 0, 1, 2$.

**Proof of Theorem 12.9.1.** It is enough to prove the theorem for functions such that the infimum defining $h$ is attained at a unique point. Indeed, if $t \mapsto f(x, t)$ is convex extensible, then for any positive $\theta > 0$, the infimum defining the marginal function $h_\theta$ of $f_\theta(x, t) = f(x, t) + \theta t^2$ is attained at a unique integer $t = \varphi_\theta(x)$, and $h_\theta$ tends to $h$ as $\theta \to 0$, preserving the $A$-lateral convexity of $h_\theta$. We observe that $f_\theta$ is $B$-laterally convex with $f$ provided that $(e^{(n)}, e^{(n)}) \in B$, which we assume. We may therefore suppose that $h(x) = f(x, \varphi(x))$ for some function $\varphi\colon \mathbf{Z}^{n-1} \to \mathbf{Z}$. Moreover, we know that $\varphi$ is Lipschitz in the sense that

$$|\varphi(x + a) - \varphi(x)| \leqslant 1, \qquad x \in \mathbf{Z}^{n-1}, \tag{12.9.5}$$

for certain values of $a \in \mathbf{Z}^{n-1}$, viz. when $((a, 1), e^{(n)})$ and $((-a, 1), e^{(n)})$ both belong to $\Lambda(B)$. For this to happen, it is enough that there exists a $c$ such that $(a, c) \in A$.

Similarly, we know that

$$|\varphi(x + b) - \varphi(x)| \leqslant 1, \qquad x \in \mathbf{Z}^{n-1}, \tag{12.9.6}$$

for certain values of $b \in \mathbf{Z}^{n-1}$, viz. when $((b, 1), e^{(n)})$ and $((-b, 1), e^{(n)})$ both belong to $\Lambda(B)$. For this it is enough that there exists a $d$ such that $(b, d) \in A$.

In particular, if $(a, b)$ is in $A$, we can take $c = b$ and $d = a$ above to conclude that the two Lipschitz conditions (12.9.5) and (12.9.6) hold.

We have

$$\begin{aligned} D_b D_a h(x) &= f(x + a + b, \varphi(x + a + b)) \\ &\quad - f(x + a, \varphi(x + a)) - f(x + b, \varphi(x + b)) + f(x, \varphi(x)). \end{aligned} \tag{12.9.7}$$

The formula holds for all $x, a, b \in \mathbf{Z}^{n-1}$, but we shall need it only when $(a, b) \in A$. We shall compare (12.9.7) with

$$\begin{aligned} D_{(b,t)} D_{(a,s)} f(x, \varphi(x)) &= f(x + a + b, \varphi(x) + s + t) \\ &\quad - f(x + a, \varphi(x) + s) - f(x + b, \varphi(x) + t) + f(x, \varphi(x)) \end{aligned} \tag{12.9.8}$$

for suitable $s$ and $t$. This expression is nonnegative if $((a, s), (b, t)) \in \Lambda(B)$.

By the definition of $\varphi$ we have

$$-f(x + a, \varphi(x + a)) \geqslant -f(x + a, s) \text{ and } -f(x + b, \varphi(x + b)) \geqslant -f(x + b, t)$$

for any $s$ and $t$, so we get $D_b D_a h(x) \geqslant D_{(b,t)} D_{(a,s)} f(x, \varphi(x))$ as soon as $s + t = \varphi(x + a + b) - \varphi(x)$.

In view of (12.9.5) and (12.9.6), which, as we have remarked, are applicable,

$$|\varphi(x + a + b) - \varphi(x)| \leqslant |\varphi(x + a + b) - \varphi(x + a)| + |\varphi(x + a) - \varphi(x)| \leqslant 2,$$

and we know from Lemma 12.9.2 that there are numbers $s, t$ such that

$$s + t = \varphi(x + a + b) - \varphi(x) \text{ and } ((a, s), (b, t)) \in \Lambda(B).$$

We are done. □

By iteration we easily obtain the following result.

**Corollary 12.9.3.** *Let us define* $B^{(0)} = \{(0, 0)\}$, $B^{(1)} = \{(1, 1)\}$, *and generally* $B^{(n)} \subset \mathbf{Z}^n \times \mathbf{Z}^n$ *such that* $B^{(n-1)}$ *and* $B^{(n)}$ *satisfy the conditions for* $A$ *and* $B$ *in Theorem 12.9.1 for* $n \geqslant 2$. *If* $f: \mathbf{Z}^n \to \mathbf{R}$ *is a given* $B^{(n)}$*-laterally convex function, then the successive marginal functions* $h_n = f$,

$$h_k(x) = \inf_{t \in \mathbf{Z}} h_{k+1}(x, t), \qquad x = (x_1, \ldots, x_k) \in \mathbf{Z}^k, \quad k = n - 1, \ldots, 1,$$

*are* $B^{(k)}$*-laterally convex, provided that* $h_1 > -\infty$. *In particular, the marginal function* $h_1$ *of one variable is* $\{(1, 1)\}$*-laterally convex, equivalently convex extensible.* □

In condition (12.9.1) it is often preferable to replace the pair

$$((a, -1), (b, 0)) \text{ by its opposite } ((-a, 1), (-b, 0)),$$

which determines the same condition. This is in order to be able to continue as in Corollary 12.9.3, where the last component should be nonnegative—this is needed in Theorem 12.8.1. We denote the set $B$ so constructed by $\Xi^n(A)$. We can now define $B^{(n)} = \Xi^n(B^{(n-1)})$ and get Corollary 12.9.3 to work.

Thus taking $\mathscr{R}_n$ as the set of all $B^{(n)}$-laterally convex functions such that the marginal functions $h_1$ do not take the value $-\infty$ gives a satisfactory solution to the problem in Section 12.6.

### 12.9.2 *The case of two variables*

Let us look in more detail at the construction of $\Xi^2(A)$. Then the corollary is about three functions: $h_2 = f$ defined on $\mathbf{Z}^2$, $h_1(x) = \inf_{y \in \mathbf{Z}} f(x, y)$

defined on $\mathbf{Z}^1$, and the constant $h_0 = \inf_{(x,y)\in\mathbf{Z}^2} f(x,y)$ defined as a function on $\mathbf{Z}^0 = \{0\}$. But here we do not say anything about the marginal function $k_1(y) = \inf_{x\in\mathbf{Z}} f(x,y)$. To do so, we should permute the variables. However, it turns out, perhaps surprisingly, that this is not necessary, for the conditions are symmetric in the two variables.

If we start with $A = \{(1,1)\} \subset \mathbf{Z}^1 \times \mathbf{Z}^1$ in one variable, the construction in Theorem 12.9.1 yields, in order,

$(e^{(1)}, e^{(1)})$, $(e^{(2)}, e^{(2)})$, applying (10.8.4);

$((1,-1),(1,0))$, $((1,1),(1,0))$, applying (12.9.1);

$((1,1), e^{(2)})$, applying (12.9.2); and

$((-1,1), e^{(2)})$, applying (12.9.3).

However, as already remarked, we should replace $((1,-1),(1,0))$ by $((-1,1),(-1,0))$. We thus obtain

$$B = \{(e^{(1)}, e^{(1)}),\ (e^{(2)}, e^{(2)}),\ ((-1,1),(-1,0)),\\ ((1,1),(1,0)),\ ((-1,1),(0,1)),\ ((1,1),(0,1))\},$$

This means that the two one-dimensional conditions and the four lozenge conditions are all satisfied, while the two diagonal conditions need not be listed since they follow from the others (Remark 11.5.2). We see now that if we permute the variables, the conditions remain the same.

We see that the set $B = \Xi^2(A) \subset \mathbf{Z}^2 \times \mathbf{Z}^2$, which defines rhomboidal convexity and corresponds to the six conditions (11.6.7), (11.6.8), (11.6.9), (11.6.10) and (11.6.11), consists of 6 pairs, and that $\Xi^3(B)$ consists of $6^2 = 36$ pairs.

### 12.9.3 *Symmetric and asymmetric conditions*

The condition on a function to have a convex-extensible marginal function is asymmetric. Indeed, the function $f(x,y) = (2x-y)^2$, $(x,y) \in \mathbf{Z}^2$, has a convex-extensible marginal function, whereas the marginal function of its reflection $g(x,y) = (2y-x)^2$ does not. Therefore a symmetric condition can never be necessary and sufficient.

For functions $f$ such that $D_{(0,1)}D_{(0,1)}f \geqslant 0$, a necessary and sufficient condition for all functions $f_\beta(x,y) = f(x,y) - \beta y$, $\beta \in \mathbf{R}$, to have a convex-extensible marginal function is that all conditions

$$D_{(1,q)}D_{(1,p)}f \geqslant 0, \qquad p,q \in \mathbf{Z},$$

shall be satisfied.[3] These are infinitely many conditions as opposed to the six conditions obtained in our construction: $\Xi^2(A)$ has six elements.

We conclude that there is a choice between a sufficient condition which is finite and symmetric but not necessary, and a sufficient and necessary condition which is infinite—and by necessity asymmetric.

## 12.10 Necessity of lateral convexity

As can be guessed from Example 11.7.1 for $\alpha = \frac{1}{2}$ and $\alpha = \frac{3}{2}$, the convexity property we have defined is essentially best possible. Before showing this, two remarks are in order.

Let $\varphi\colon \mathbf{Z}^2 \to \mathbf{Z}$ be any function such that $\mathbf{Z} \ni y \mapsto \varphi(x,y) \in \mathbf{Z}$ is a surjection for every $x \in \mathbf{Z}$. Then the function $g(x,y) = f(x,\varphi(x,y))$, $(x,y) \in \mathbf{Z}^2$, has the same marginal function as $f$. In particular, the values on a vertical line can be arbitrarily scrambled. It follows that no reasonable conclusion concerning regularity of $f$ can be drawn from knowledge of its marginal function. But if we consider the marginal functions $h_\beta$ of the tilted functions $f_\beta(x,y) = f(x,y) - \beta y$, $\beta \in \mathbf{R}$, things are different.

For simplicity we now restrict attention to functions of two variables $(x,y) \in \mathbf{Z}^2$. We defined in Definition 9.17.10 the partial Fenchel transform of a function $f\colon \mathbf{Z} \times \mathbf{Z} \to \mathbf{R}_!$ by

$$f^\circ(x,\eta) = \sup_{y\in\mathbf{Z}} \big(\eta y - f(x,y)\big), \qquad (x,\eta) \in \mathbf{Z} \times \mathbf{R},$$

to be compared with the complete Fenchel transform,

$$\widetilde{f}(\xi,\eta) = \sup_{(x,y)\in\mathbf{Z}^2} \big(\xi x + \eta y - f(x,y)\big), \qquad (\xi,\eta) \in \mathbf{R} \times \mathbf{R}.$$

Thus the marginal function of $f$ is $h(x) = -f^\circ(x,0)$. Since the third transform $f^{\circ\circ\circ}$ is equal to the first, the second transform $f^{\circ\circ}$ has the same marginal function as $f$. Therefore, again, it is not reasonable to expect that, from knowledge of a marginal function, one can conclude anything about $f$, only about its minorant $f^{\circ\circ}$.

**Proposition 12.10.1.** *Let $f\colon \mathbf{Z} \times \mathbf{Z} \to \mathbf{R}$ be such that the marginal function $h_\beta$ of $f_\beta(x,y) = f(x,y) - \beta y$ is convex extensible for all real numbers $\beta$. Then $f^{\circ\circ}$ satisfies the one-variable conditions* (11.6.7), *the diagonal conditions* (11.6.12), *and the horizontal lozenge conditions* (11.6.8) *and* (11.6.9).

[3] It is sufficient here to require this condition only for two values of $q$, viz. $q = p$ and $q = p + 1$. Still, these are infinitely many in number.

**Proof.** For brevity, let us write $g$ instead of $f^{\circ\circ}$.

By replacing $g(x,y)$ by $g(x,y)+\theta y^2$, $\theta > 0$, we may assume that the infimum of $y \mapsto g_\beta(x,y) = g(x,y) - \beta y$ is always attained at some point. Afterwards we let $\theta$ tend to zero; the properties are stable under this operation.

The vertical condition $g(x,y-1)+g(x,y+1) \geqslant 2g(x,y)$ is always satisfied by assumption.

Consider next the horizontal condition $g(x-1,y)+g(x+1,y) \geqslant 2g(x,y)$ for fixed $x$ and $y$ and define $\beta = g(x,y+1)-g(x,y)$. Then $h_\beta(x) = g_\beta(x,y) = g_\beta(x,y+1)$, and if $h_\beta$ is convex extensible, we get

$$\begin{aligned} g_\beta(x-1,y)+g_\beta(x+1,z) &\geqslant h_\beta(x-1)+h_\beta(x+1) \\ &\geqslant 2h_\beta(x) = 2g_\beta(x,y) = 2g_\beta(x,y+1). \end{aligned}$$

Taking $z=y$ we see that the horizontal one-variable condition is satisfied; taking $z = y+1$ we see that the right horizontal lozenge condition is satisfied; and taking $z=y+2$ we see that one of the diagonal conditions is satisfied. For the left horizontal lozenge condition and the other diagonal condition we can argue similarly. □

**Theorem 12.10.2.** *Let $f\colon \mathbf{Z}\times\mathbf{Z}\to\mathbf{R}$ satisfy the one-variable conditions* (11.6.7). *Define two marginal functions by*

$$h_\beta(x) = \inf_{y\in\mathbf{Z}} (f(x,y)-\beta y), \qquad x\in\mathbf{Z},\ \beta\in\mathbf{R},$$

*and*

$$k_\gamma(y) = \inf_{x\in\mathbf{Z}} (f(x,y)-\gamma x), \qquad y\in\mathbf{Z},\ \gamma\in\mathbf{R}.$$

*Assume that $h_\beta$ and $k_\gamma$ are convex extensible for all real numbers $\gamma$ and $\beta$. Then $f$ is rhomboidally convex.*

**Proof.** We apply Proposition 12.10.1 to $f$ and to $(x,y)\mapsto f(y,x)$. □

## 12.11 Possible future studies

Since, as observed in Remark 9.21.2, the operation of taking the marginal function is a special case of infimal convolution, it may be of interest to extend this study of discrete convexity to more general infimal convolutions.

It may also be interesting to consider functions $f\colon \mathbf{Z}^n\to\mathbf{Z}$ with integer values and their marginal functions. Then convex extensibility of the marginal function $f_{\text{marg}}\colon \mathbf{Z}^m \to \mathbf{Z}_!$ is too strong a condition. Instead it is relevant to require that the functions are $(\mathbf{Z}^m \times \mathbf{Z})$-convex, meaning that there exists a convex subset $C$ of $\mathbf{R}^m \times \mathbf{R}$ such that

$$C \cap (\mathbf{Z}^m \times \mathbf{Z}) = \mathbf{epi}^{\text{finite}}(f) = \big\{(x,t) \in \mathbf{Z}^m \times \mathbf{Z};\ t \geqslant f(x)\big\}.$$

A step in this direction is taken in my manuscript (2021).

## 12.12 Notes

Many of the results presented in this chapter were briefly summarized in my paper (2007b) and then published in my paper (2008a). A survey of the results in Subsections 12.5.1, 12.5.2 was published in (2010c).

For other convexity properties see Subsection 10.14.1.

## 12.13 Exercises

12.1. Prove that the marginal function $h$ of

$$f(x,y) = \sqrt{x^2+y^2} - (\sqrt{2}-1)y, \qquad (x,y) \in \mathbf{Z}^2,$$

obtained by tilting the Euclidean norm, is not midpoint convex.

12.2. Let

$$G(x,y) = \sqrt{x^2+y^2} + \big((|y|-1)^+\big)^2,$$

for $(x,y) \in \mathbf{R}^2$, and let $g = G|_{\mathbf{Z}^2}$ be its restriction to the integer points in the plane. Prove that all points $(k,1) \in \mathbf{Z}^2$, $k \in \mathbf{N}$, are $g$-neighbors of the origin.

12.3. Take $\gamma = 2/5$ and $c = 0$ in Proposition 11.13.2. Prove that in this case $A = (5\mathbf{Z}+2) \cup (5\mathbf{Z}+4)$, which implies that $\lfloor 1/\gamma \rfloor = \lfloor 5/2 \rfloor = 2$ as well as $\lceil 5/2 \rceil = 3$ actually occur as values of the integer $m$ in $I_\gamma$.

12.4 Let $F$ and $f$ be as in Proposition 11.13.2.

(a) Prove that the slope $\gamma$ of the line defined by $F$ is determined by knowledge of $f(x)$ with $x = p, q$ for $p, q \in \mathbf{Z}$ with arbitrarily large differences $q-p$, for instance if the values are known for all $p, q \in \mathbf{N}$.

(b) Determine how well the slope $\gamma$ is determined if the values of the discretized function $f$ are known for a bounded interval $[a,b]_{\mathbf{Z}}$ only.

## Chapter 13

# Duality of convolution operators

### 13.1 Introduction

DUALITY is a term which represents a collection of ideas where two sets of mathematical objects confront each other. A fundamental example is the dual of a normed space, where the linear forms on the space operate on its points. A most successful duality is that between the space $\mathscr{D}(\Omega)$ of test functions (smooth functions of compact support) and its dual $\mathscr{D}'(\Omega)$ of distributions. The distributions that are not defined by a locally integrable function live like ghosts in the dark, perceptible only through their actions on test functions.

Typically some but not all properties of an object are preserved under a duality. A function which is nonzero only in a set of Lebesgue measure zero defines the zero distribution. The support function of a set forgets about holes in the set but the information retained enables us to reconstruct the closed convex hull of the set.

Mathematical morphology can be quite helpful in providing guiding concepts and ideas in the study of discrete convolution operators and other topics, like discrete optimization. We shall now apply these ideas to the duality of convolution operators.

### 13.2 Properties of convolution

If $G$ is an abelian semigroup we defined the discrete convolution product $h = f * g$ of two functions defined on $G$ and with real or complex values in Definition 3.3.2 on page 46.

### 13.2.1 *Associativity is problematic*

It is impossible to define a general associative convolution product, as shown by the following simple example.

**Example 13.2.1.** Define on $G = \mathbf{Z}$ three functions: $f = 1$ identically; $g$ the difference operator $\delta_0 - \delta_1$; and $h$ equal to the Heaviside function $Y_0$, defined by $h(x) = 0$ for $x \leqslant -1$, $h(x) = 1$ for $x \geqslant 0$. Then $f * g$ and $g * h$ are well defined: $f * g = 0$; $g * h = \delta_0$. It follows that $(f * g) * h = 0$ while $f * (g * h) = f \neq 0$. (We get a warning here: $f * h$ cannot be defined to have finite values—neither can $(f * |g|) * h$, nor $f * (|g| * h)$.) □

Via the Fourier transformation this example is the same as Laurent Schwartz's famous result (1954) that it is impossible to define an associative multiplication for distributions. (We leave aside the question whether there might exist an interesting non-associative convolution algebra.)

### 13.2.2 *Convolution in some special cases*

The convolution product $f * 1$ of $f \in l^1(\mathbf{N})$ with the constant 1 is the constant function with value $\sum f(x)$. The study of summation methods is tantamount to studying definitions of $f * 1$. They could be generalized to convolutions $f * g$ for other functions $g$.

For summation methods, Hardy's book (1949) is a classic. Among these methods, the best known is Cesàro summability $(\mathrm{C}, 1)$, which means that one takes the limit

$$\lim_{x \to +\infty} \frac{(f * g)(x)}{g(x)}$$

with $g(x) = x + 1$, $x \in \mathbf{N}$. More generally, this formula conserves the sum of an absolutely convergent series if $g\colon \mathbf{N} \to \mathbf{R}$ is increasing, $g(0) > 0$, and such that $\lim_{x\to+\infty} g(x-a)/g(x) = 1$ for all $a \in \mathbf{N}$. The formula assigns a sum to a wider class of series.

All efforts to sum a divergent series—like those of Mejlbo (1968) and Cegrell & Kiselman (1970)—could be generalized to studying more general convolution products.

If $G$ is an abelian group, we have the well-known inequality

$$\|f * g\|_1 \leqslant \|f\|_1 \cdot \|g\|_1, \qquad f, g \in \mathscr{F}(G, \mathbf{R}), \tag{13.2.1}$$

making $l^1(G)$, the space of summable functions, into a convolution algebra, and more generally

$$\|f * g\|_p \leqslant \|f\|_1 \cdot \|g\|_p, \qquad f, g \in \mathscr{F}(G, \mathbf{R}), \quad 1 \leqslant p \leqslant \infty. \tag{13.2.2}$$

### 13.2.3 *Associativity does hold under stringent conditions*

We note that $\|f\|_1$ is finite for all $f \in \mathscr{F}_{\text{finite}}(G, \mathbf{R})$. In this context, commutativity and associativity follow easily. But we will need more general results, with functions that are not summable or even unbounded, just with finite values but possibly growing very fast.

While we cannot hope for a general convolution algebra, we can confront a family of functions with another family, which is an instance of duality. The simplest example of this is that $\mathscr{F}_{\text{finite}}(G, \mathbf{R})$ can work against $\mathscr{F}(G, \mathbf{R})$ and conversely.

Provided that we accept $+\infty$ as a value, we can define the convolution product $|f| * |g|$ of the absolute values of any two functions $f, g \in \mathscr{F}(G, \mathbf{R})$. The sum is simply defined to be the supremum of all finite partial sums:

$$(|f| * |g|)(z) = \sup_{A \in \mathscr{P}_{\text{finite}}(G)} \sum_{\substack{x, y \in A \\ x + y = z}} |f(x)| \cdot |g(y)| \leqslant +\infty.$$

Therefore it makes sense to require that $|f| * |g|$ be finite everywhere. If this is so, then also $f * g$ is well defined as a function with finite values; the sum defining it being absolutely convergent at every point. More precisely, we have the following result concerning commutativity and associativity.

**Proposition 13.2.2.** *Let $G$ be an abelian group and $f_j \colon G \to \mathbf{R}$, $j = 1, 2, 3$, three functions.*

(1) *If $|f_1| * |f_2|$ is finite everywhere, then $f_1 * f_2$ is defined everywhere as an absolutely converging series, $f_1 * f_2 = f_2 * f_1$, and $|f_1 * f_2| \leqslant |f_1| * |f_2|$.*
(2) *If $(|f_1| * |f_2|) * |f_3|$ is finite everywhere, then $(f_1 * f_2) * f_3$ and $f_1 * (f_2 * f_3)$ are defined everywhere as absolutely converging series, and they are equal.*

First a word about convergence. We shall say that, given a function $F \colon G \to \mathbf{R}$, a series $\sum_{x \in G} F(x)$ ***converges to a sum*** $s$ if for every $\theta > 0$ there exists a finite subset $A$ of $G$ such that, for every finite subset $B \supset A$, we have

$$\left| \sum_{x \in B} F(x) - s \right| \leqslant \theta.$$

We will write this as

$$\lim_{A \to G} \sum_{x \in A} F(x) = s, \text{ or just } \sum_{x \in G} F(x) = s.$$

A well-known necessary and sufficient condition for convergence is ***Cauchy's criterion***:

*For every $\theta > 0$ there exists a finite subset $A$ of $G$ such that, for every finite set $C \subset \complement A$, we have $\left|\sum_{x \in C} F(x)\right| \leqslant \theta$.*

If $\left|\sum_{x \in C} F(x)\right| \leqslant \theta$ for all finite sets $C$ which are disjoint from $A$, then also $\sum_{x \notin A} F(x)$ is well defined and satisfies $\left|\sum_{x \notin A} F(x)\right| \leqslant \theta$.

**Proof.** (1) We denote by $g_j$ the functions defined as $f_j$ in a finite subset $A_j$ of $G$ and as 0 in $G \smallsetminus A_j$, $j = 1, 2, 3$. Clearly $|g_j| \leqslant |f_j|$ and $|g_j| * |g_k| \leqslant |f_j| * |f_k|$.

We know that $g_1 * g_2 = g_2 * g_1$ and that $|f_1| * |f_2| = |f_2| * |f_1|$.

For the first part of the theorem it is easy to see that Cauchy's criterion is satisfied. For any finite subset $B$ of $G$ we have

$$\sum_{y \in B} (f_1(y) - g_1(y)) g_2(x - y) = \sum_{y \in B \smallsetminus A_1} f_1(y) g_2(x - y),$$

which implies that the absolute value of the last sum is majorized by

$$\sum_{y \notin A_1} |f_1(y)| \cdot |f_2(x - y)|,$$

which is smaller than $\theta$ if $A_1$ is large. Hence $\lim_{A_1 \to G} (g_1 * g_2) = f_1 * g_2$.

We go on and find that

$$\sum_{y \in B} f_1(x - y)(f_2(y) - g_2(y)) = \sum_{y \in B \smallsetminus A_2} f_1(x - y) f_2(y),$$

with absolute values

$$\left| \sum_{y \in B \smallsetminus A_2} f_1(x - y)(f_2(y) - g_2(y)) \right| \leqslant \sum_{y \notin A_2} |f_1(x - y)| \cdot |f_2(y)| \leqslant \theta,$$

which shows that $\lim_{A_2 \to G} (f_1 * g_2) = f_1 * f_2$. We conclude that the itererated limit is as we wished for:

$$\lim_{A_2 \to G} \lim_{A_1 \to G} (g_1 * g_2) = f_1 * f_2.$$

Since $g_1 * g_2 = g_2 * g_1$, commutativity follows.

(2) + We know that $(g_1 * g_2) * g_3 = g_1 * (g_2 * g_3)$, and also that $(|f_1| * |f_2|) * |f_3| = |f_1| * (|f_2| * |f_3|)$.

Now, since $(|f_1| * |f_2|) * |f_3|$ is finite everywere, we see that also $|f_1| * |f_2|$ is finite everywhere; similarly for all the $|f_j| * |f_k|$. We have now

$$\sum_{y \in B} (f_1 * f_2)(x - y)(f_3(y) - g_3(y)) = \sum_{y \in B \smallsetminus A_3} (f_1 * f_2)(x - y) f_3(y),$$

with absolute values majorized by

$$\sum_{y \notin A_3} |(f_1 * f_2)(x-y)| \cdot |f_3(y)|,$$

which by hypothesis can be arbitrality small when $A_3$ is large. So the limit $\lim_{A_3 \to G}((f_1 * f_2) * g_3)$ is equal to $(f_1 * f_2) * f_3$, and we conclude for the iterated limit that

$$\lim_{A_3 \to G} \lim_{A_2 \to G} \lim_{A_1 \to G}((g_1 * g_2) * g_3) = (f_1 * f_2) * f_3.$$

Since the $g_j$ satisfy the associative law, so do the $f_j$ We are done. □

## 13.3 Families of functions under duality

**Definition 13.3.1.** Given an abelian semigroup $G$ we define $\Theta(\mathscr{G}, \mathscr{H}, \mathscr{K})$ as the set

(13.3.1) $$\{f \colon G \to \mathbf{R};\ \text{for all } g \in \mathscr{G}, |f| * |g| \in \mathscr{H} \text{ and } f * g \in \mathscr{K}\},$$

where $\mathscr{G}, \mathscr{H}$ and $\mathscr{K}$ are subsets of $\mathscr{F}(G, \mathbf{R})$, thus defining a transformation $\Theta \colon \mathscr{P}(\mathscr{F}(G,\mathbf{R})) \times \mathscr{P}(\mathscr{F}(G,\mathbf{R})) \times \mathscr{P}(\mathscr{F}(G,\mathbf{R})) \to \mathscr{P}(\mathscr{F}(G,\mathbf{R}))$. □

Clearly $\Theta(\mathscr{G}, \mathscr{H}, \mathscr{K})$ is decreasing in $\mathscr{G}$ and increasing in $\mathscr{H}$ and $\mathscr{K}$.

Three particular cases will be of interest:

**Definition 13.3.2.** If we take both $\mathscr{H}$ and $\mathscr{K}$ equal to $\mathscr{F}(G, \mathbf{R})$, then $\Theta(\mathscr{G}, \mathscr{H}, \mathscr{K})$ is equal tp

(13.3.2) $$\Gamma(\mathscr{G}) = \{f \in \mathscr{F}(G,\mathbf{R});\ \text{for all } g \in \mathscr{G}, |f| * |g| < +\infty\},$$

where the formula defines a mapping $\Gamma \colon \mathscr{P}(\mathscr{F}(G,\mathbf{R})) \to \mathscr{P}(\mathscr{F}(G,\mathbf{R}))$.

□

**Definition 13.3.3.** If $\mathscr{H} = \mathscr{F}(G, \mathbf{R})$ and

$$\mathscr{K} = \{f \in \mathscr{F}(G,\mathbf{R});\ f \geqslant 0\},$$

then $\Theta(\mathscr{G}, \mathscr{H}, \mathscr{K})$ is equal to $\Gamma^{\geqslant 0}(\mathscr{G})$, which in turn is equal to

(13.3.3) $$\{f \in \mathscr{F}(G,\mathbf{R});\ \text{for all } g \in \mathscr{G}, |f| * |g| < +\infty \text{ and } f * g \geqslant 0\},$$

where the formula defines $\Gamma^{\geqslant 0} \colon \mathscr{P}(\mathscr{F}(G,\mathbf{R})) \to \mathscr{P}(\mathscr{F}(G,\mathbf{R}))$. □

**Definition 13.3.4.** If $\mathscr{H} = \mathscr{F}(\mathbf{R}, \mathbf{R})$ and $\mathscr{K} = \{0\}$, then $\Theta(\mathscr{G}, \mathscr{H}, \mathscr{K})$ is equal to $\Gamma^0(\mathscr{G})$, which is equal to

(13.3.4) $$\{f \in \mathscr{F}(\mathbf{R},\mathbf{R});\ \text{for all } g \in \mathscr{G}, |f| * |g| < +\infty \text{ and } f * g = 0\},$$

defining $\Gamma^0 \colon \mathscr{P}(\mathscr{F}(\mathbf{R},\mathbf{R})) \to \mathscr{P}(\mathscr{F}(\mathbf{R},\mathbf{R}))$. □

Clearly $\Gamma$ and $\Gamma^0$ are decreasing, so that the compositions $\Gamma \circ \Gamma$, $\Gamma^0 \circ \Gamma$, $\Gamma \circ \Gamma^0$, $\Gamma^0 \circ \Gamma^0$ are all increasing. Together with the fact that $\Gamma^0 \leqslant \Gamma$ these facts give rise to order relations

$$\begin{array}{ccc} & \Gamma \circ \Gamma^0 & \\ / & & \backslash \\ \Gamma \circ \Gamma & & \Gamma^0 \circ \Gamma^0 \\ \backslash & & / \\ & \Gamma^0 \circ \Gamma. & \end{array}$$

The composition $\Gamma^0 \circ \Gamma$ maps all families to the singleton set of the zero function and is therefore idempotent. The composition $\Gamma \circ \Gamma^0$ is in general not idempotent.

**Proposition 13.3.5.** *The mapping* $\Gamma$ *is decreasing and satisfies*

$$\bigcap_{j\in J} \Gamma(\mathscr{G}_j) = \Gamma(\textstyle\bigcup_{j\in J} \mathscr{G}_j) \tag{13.3.5}$$

*for all families* $(\mathscr{G}_j)_{j\in J}$*, from which is follows that* $\mathbf{id} \leqslant \Gamma \circ \Gamma$.

*As a consequence, the pair* $(\Gamma, \Gamma)$ *is a Galois correspondence.*

*Similarly for* $\Gamma^{\geqslant 0}$ *and* $\Gamma^0$.

**Proof.** Clearly $\Gamma$ is decreasing. Like all increasing mappings it satisfies $\bigcap_j \Gamma(\mathscr{G}_j) \supset \Gamma(\bigcup_j \mathscr{G}_j)$. The inclusion in the other direction follows from the definition.

If $|g| * |h| < +\infty$, then $g \in \Gamma(\{h\})$. If $|g| * |h| < +\infty$ for all $h \in \mathscr{H}$, then $g \in \bigcap_{h\in\mathscr{H}} \Gamma(\{h\}) = \Gamma(\bigcup_{h\in\mathscr{H}} \{h\}) = \Gamma(\mathscr{H})$. If $|g| * |h| < +\infty$ for all $g \in \mathscr{G}$ and all $h \in \mathscr{H}$, then $\mathscr{G} \subset \Gamma(\mathscr{H})$. Finally, taking $\mathscr{H} = \Gamma(\mathscr{G})$ we get $\mathscr{G} \subset \Gamma(\Gamma(\mathscr{G}))$, proving $\mathbf{id} \leqslant \Gamma \circ \Gamma$. □

We shall determine some classes of functions which are fixed points for the corresponding cleistomorphisms.

When studying convolution equations $\mu * u = f$ with $\mu$ of finite support, it is convenient to introduce two variants of $\Gamma^{\geqslant 0}$ in (13.3.3):

$$\Gamma_0^{\geqslant 0}(\mathscr{G}) = \{u \in \mathscr{F}(G, \mathbf{R});\ \text{for all } \mu \in \mathscr{G},\ \mu * u \geqslant 0\} \tag{13.3.6}$$

for $\mathscr{G} \subset \mathscr{F}_{\text{finite}}(G, \mathbf{R})$.

$$\Gamma_1^{\geqslant 0}(\mathscr{H}) = \{\mu \in \mathscr{F}_{\text{finite}}(G, \mathbf{R});\ \text{for all } u \in \mathscr{H},\ \mu * u \geqslant 0\} \tag{13.3.7}$$

for $\mathscr{H} \subset \mathscr{F}(G, \mathbf{R})$. The first is just the restriction of $\Gamma^{\geqslant 0}$ to the set of functions with finite support. Now $(\Gamma_0^{\geqslant 0}, \Gamma_1^{\geqslant 0})$ is a Galois correspondence.

The mapping $\Phi\colon \mathscr{P}(G\times G) \to \mathscr{P}(\mathscr{F}(G, \mathbf{R}))$ defined in Definition 10.7.3 is the same as $\Gamma_0^{\geqslant 0}$ when all functions $\mu$ in $\mathscr{G}$ are of the form $\mu = \delta_0 - \delta_a - \delta_b + \delta_{a+b}$, and $\Gamma_1^{\geqslant 0}(\mu)$ is equal to $\Psi(\mu)$ for these functions, where $\Psi\colon \mathscr{P}(G \times G) \to \mathscr{P}(\mathscr{F}(G \times G))$.

**Example 13.3.6.** Let us define $\mathscr{G}_0 = \mathscr{F}_{\text{finite}}(G, \mathbf{R})$ and $\mathscr{G}_1 = \mathscr{F}(G, \mathbf{R})$. Then $\Gamma(\mathscr{G}_0) = \mathscr{G}_1$ and $\Gamma(\mathscr{G}_1) = \mathscr{G}_0$; it follows that $(\Gamma^{\circ 2})(\mathscr{G}_j) = \mathscr{G}_j$, $j = 0, 1$. So both $\mathscr{G}_0$ and $\mathscr{G}_1$ are fixed points for $\Gamma^{\circ 2}$. □

Several other types of duality may be studied.

We note that $\Gamma(\mathscr{G})$ is a vector subspace of $\mathscr{F}(G, \mathbf{R})$ which contains $\mathscr{F}_{\text{finite}}(G, \mathbf{R})$. Also

$$\mathscr{G} \subset \mathscr{G} + \mathscr{F}_{\text{finite}}(G, \mathbf{R}) \subset \Gamma^{\circ 2}(\mathscr{G}),$$

which implies that

$$\Gamma(\mathscr{G}) = \Gamma(\mathscr{G} + \mathscr{F}_{\text{finite}}(G, \mathbf{R})) = \Gamma(\Gamma^{\circ 2}(\mathscr{G})) = \Gamma^{\circ 3}(\mathscr{G}).$$

This implies that $\Gamma$ can be defined in the quotient space

$$\mathscr{F}(G, \mathbf{R})/\mathscr{F}_{\text{finite}}(G, \mathbf{R}),$$

a space analogous to the space of singularities $\mathscr{D}'(\Omega)/\mathscr{E}(\Omega)$, the space of distributions modulo the subspace of smooth functions. Also the more elementary singularity space $C^0(\Omega)/C^\infty(\Omega)$ is interesting.

From Proposition 5.7.2 we see that the lower opposite inverse to $\Gamma$ is equal to $\Gamma$ (this inverse is defined in (5.2.5) and denoted by $\overline{\cdot\Gamma}$ there).

**Example 13.3.7.** Let $\mathscr{G}$ be the subfamily of $\mathscr{F}(G, \mathbf{R})$ whose only member is the constant function with the value 1. Then $\Gamma(\mathscr{G}) = l^1(G)$, the set of functions $h\colon G \to \mathbf{R}$ such that $\sum_{x\in G} |h(x)|$ is finite. Moreover $(\Gamma \circ \Gamma)(\mathscr{G}) = l^\infty(G)$, the family of all bounded functions $h\colon G \to \mathbf{R}$. The higher powers $\Gamma^{\circ k}(\mathscr{G})$ are alternatively equal to $l^1(G)$ (odd $k \geqslant 1$) and $l^\infty(G)$ (even $k \geqslant 2$). So both $l^1(G)$ and $l^\infty(G)$ are fixed points for $\Gamma \circ \Gamma$. □

### 13.3.1 *Duality for functions of one variable*

We may ask for descriptions of $\Gamma^0(\mathscr{G})$, defined by (13.3.4), for any given class of functions $\mathscr{G}$. Also of interest is the second power $(\Gamma^0)^{\circ 2} = \Gamma^0 \circ \Gamma^0$ of $\Gamma^0$, the cleistomorphism associated to $\Gamma^0$.

Similarly we can study $\Gamma^{\geqslant 0}(\mathscr{G})$ and $(\Gamma^{\geqslant 0})^{\circ 2}(\mathscr{G})$ defined by (13.3.3).

In particular we shall take $\mathscr{G}$ as a singleton, $\mathscr{G} = \{\beta\}$ for some $\beta \in \mathscr{F}_{\text{finite}}(\mathbf{Z}, \mathbf{R})$. We may always assume that $\inf(x;\ \beta(x) \neq 0)$ is equal to 0, and that $\beta = \delta_0 - \gamma$ for some $\gamma \in \mathscr{F}(\mathbf{Z}, \mathbf{R})$ which vanishes for all points $x \leqslant 0$. Then $\gamma^{*j}$ has its support in $[jq, +\infty[ \subset [j, +\infty[$ if the support of $\gamma$ is contained in $[q, +\infty[$, $q \geqslant 1$. A convolution inverse to $\beta$ is given by

$$\beta^{*(-1)} = (\delta_0 - \gamma)^{*(-1)} = \delta_0 + \gamma + (\gamma * \gamma) + (\gamma * \gamma * \gamma) + \cdots = \sum_{j=0}^{\infty} \gamma^{*j}.$$

It is the only inverse of $\beta$ which is zero for large negative arguments. The series converges nicely, but $\beta^{*(-1)}$ usually does not have finite support. Given any function $\mu$, we may ask whether $\beta^{*(-1)} * \mu$ has finite support.

We obtain

$$\Gamma^0(\{\beta\}) = \{f \in \mathscr{F}(\mathbf{Z}, \mathbf{R});\ \beta * f = 0\} = \{f;\ \gamma * f = f\},$$

which is a kind of periodicity for $f$.

We find that $(\Gamma^0)^{\circ 2}(\{\beta\})$ is equal to the set of all $\mu \in \mathscr{F}(\mathbf{Z}, \mathbf{R})$ such that $\mu * f = 0$ for all $f \in \mathscr{F}(\mathbf{Z}, \mathbf{R})$ which satisfy the conditions $|\mu| * |f| < +\infty$, $|\beta| * |f| < +\infty$ and $\beta * f = 0$.

For brevity, let us write $\rho$ for $-\sum_{j=1}^{\infty} \gamma^{*j}$, so that the inverse $\beta^{*(-1)}$ of $\beta = \delta_0 - \gamma$ is $\delta_0 - \rho$. Let now $\mu \in \mathscr{F}_{\text{finite}}(\mathbf{Z}, \mathbf{R})$ be given. In order for $\beta^{*(-1)} * \mu = \mu - \rho * \mu$ to have finite support, it is necessary and sufficient that $(\rho * \mu)(x)$ vanish for large positive values of $x$. The functions $\rho$ that can occur are those which have a certain periodicity property: $\rho * \gamma = \rho + \gamma$.

If $(\rho * \mu)(x) = 0$ for large values of $x$, then $\mu = \beta * \nu$ for some $\nu$ with finite support, and this implies that $\Gamma^0(\{\mu\})$ contains $\Gamma^0(\{\beta\})$. In general I find it difficult to say more now in this general framework, but for special choices of $\beta$ we can analyze the situation completely.

Let $Y_a \colon \mathbf{Z} \to \mathbf{Z}$ be the Heaviside function which is zero for $x \leqslant a - 1$ and one for $x \geqslant a$. Then $Y_a * (\delta_0 - \delta_1) = \delta_a$. So we can solve the equation $(\delta_0 - \delta_1) * u = f$ for an $f$ which vanishes for large negative arguments by taking $u = Y_0 * f$. We also define

$$V_a(x) = x - a + 1, \qquad x \in \mathbf{Z}, \tag{13.3.8}$$

the restriction of an affine function, and note that $V_a^+ * (\delta_0 - \delta_1) = Y_a$, thus $V_a^+ * (\delta_0 - \delta_1)^{*2} = \delta_a$.

The convolution product $Y_0 * Y_0$ is an inverse of $(\delta_0 - \delta_1)^{*2}$; we have $(Y_0 * Y_0)(x) = V_0^+(x)$, where $V_0(x) = x + 1$, $x \in \mathbf{Z}$.

**Example 13.3.8.** A first example is $\beta = \delta_0 - \delta_1$, $\gamma = \delta_1$. Then $(\beta * f)(x) = f(x) - f(x-1)$, and $\Gamma^0(\{\beta\})$ is equal to the set of all constant functions.

For $((\Gamma^0)^{\circ 2})(\{\delta_0 - \delta_1\})$ the following three properties are equivalent.

(A1) $\mu \in (\Gamma^0)^{\circ 2}(\{\beta\})$;

(A2) $\mu * 1 = 0$;

(A3) $\mu = \beta * \nu$ for some $\nu \in \mathscr{F}_{\text{finite}}(\mathbf{Z}, \mathbf{R})$.

To see that the second property implies the third we note that $\beta^{*(-1)} = Y_0$ and that, for any given $\mu \in \mathscr{F}_{\text{finite}}(\mathbf{Z}, \mathbf{R})$,

$$(\beta^{*(-1)} * \mu)(x) = (Y_0 * \mu)(x) = \sum_{y=0}^{\infty} \mu(x-y),$$

a sum which is equal to $(\mu * 1)(x)$ for large positive values of $x$, and which vanishes for large $x$ if and only if $\mu * 1 = 0$. □

**Example 13.3.9.** Take now $\beta = (\delta_0 - \delta_1)^{*2}$, implying that $\gamma = \delta_0 - \beta = 2\delta_1 - \delta_2$. Then $(\beta * f)(x) = f(x) - 2f(x-1) + f(x-2)$, implying that the functions $f$ which satisfy $\beta * f = 0$ are the restrictions to $\mathbf{Z}$ of the affine functions, and that $\Gamma^0(\{\beta\})$ is equal to the set of all restrictions of affine functions.

For $(\Gamma^0)^{\circ 2}(\{\beta\})$ the following three properties are equivalent.

(B1) $\mu \in (\Gamma^0)^{\circ 2}(\{\beta\})$;

(B2) $\mu * f = 0$ for all functions $f$ which are restrictions to $\mathbf{Z}$ of an affine function;

(B3) $\mu = \beta * \nu$ for some $\nu \in \mathscr{F}_{\text{finite}}(\mathbf{Z}, \mathbf{R})$.

To see that the second property implies the third, we first note that $V_0^+ = \beta^{*(-1)}$ and that (B2) implies that $\mu * V_0 = 0$ for the function $V_0$ defined above; $V_0(x) = x + 1$. For any function $\mu \in \mathscr{F}_{\text{finite}}(\mathbf{Z}, \mathbf{R})$,

$$(\beta^{*(-1)} * \mu)(x) = (V_0^+ * \mu)(x) = \sum_{y=0}^{\infty} V_0^+(x-y)\mu(y), \qquad x \in \mathbf{Z},$$

a sum which is equal to $(V_0 * \mu)(x)$ for large positive values of $x$ and vanishes for large negative values of $x$. Thus $(\beta^{*(-1)} * \mu) - (V_0 * \mu) \cdot Y_0$ has finite support. We denote it by $\nu$, so that

$$\beta^{*(-1)} * \mu = (V_0 * \mu) \cdot Y_0 + \nu$$

and, if $V_0 * \mu = 0$,

$$\mu = \beta * \beta^{*(-1)} * \mu = \beta * \nu.$$

So property (B2) implies property (B3). □

**Example 13.3.10.** Third, we take again $\beta = (\delta_0 - \delta_1)^{*2}$ and now ask for $\Gamma_0^{\geqslant 0}(\{\beta\})$ and $\Gamma_1^{\geqslant 0}(\Gamma_0^{\geqslant 0}(\{\beta\}))$. The first set is the set $CVX(\mathbf{R},\mathbf{R})|_{\mathbf{Z}}$ of all real-valued convex extensible functions, and the second is the set of all functions $\mu$ of the form $\beta * \sigma$ where $\sigma$ has finite support and satisfies $\sigma \geqslant 0$. Since $\Gamma_1^{\geqslant 0} \circ \Gamma_0^{\geqslant 0}$ is idempotent, we conclude that

$$(\Gamma_1^{\geqslant 0} \circ \Gamma_0^{\geqslant 0})^{\circ k}(\beta) = (\Gamma_1^{\geqslant 0} \circ \Gamma_0^{\geqslant 0})(\beta), \qquad k = 1, 2, \ldots,$$

and

$$\Gamma_0^{\geqslant 0} \circ (\Gamma_1^{\geqslant 0} \circ \Gamma_0^{\geqslant 0})^{\circ k}(\beta) = CVX(\mathbf{R},\mathbf{R})|_{\mathbf{Z}}, \qquad k = 0, 1, \ldots.$$

More precisely, let us consider the following three conditions.

(C1) $\mu \in \Gamma_1^{\geqslant 0}(\Gamma_0^{\geqslant 0}(\{\beta\}))$.

(C2) $\mu = \beta * \sigma$ for some $\sigma \in \mathscr{F}_{\text{finite}}(\mathbf{Z},\mathbf{R})$ with only nonnegative values.

(C3) $\mu$ is a finite positive sum of translations of $\beta$:

$$\mu = \sum_{j \in \mathbf{Z}} \lambda_j (\beta * \delta_j),$$

with $\lambda_j \geqslant 0$ and only finitely many of them nonzero.

That (C2) and (C3) are equivalent is easy to see. That (C2) implies (C1) is also easy: If (C2) holds we obtain

$$\mu * f = (\beta * \sigma) * f = \sigma * (\beta * f) \geqslant 0$$

for all convex extensible functions $f$, so $\mu$ belongs to $\Gamma_1^{\geqslant 0}(\Gamma_0^{\geqslant 0}(\{\beta\}))$, i.e., (C1) holds.

Conversely, suppose now that (C1) holds. We define $\sigma$ as the function which vanishes for large negative values of the argument and is such that $\beta * \sigma = \mu$. Then, since $\mu * V_a^+ \geqslant 0$ for the elementary convex extensible functions $V_a^+(x) = (x - a + 1)^+$, where $V_a(x) = x - a + 1$ as in (13.3.8), we obtain

$$0 \leqslant (\mu * V_a^+)(x) = ((\beta * \sigma) * V_a^+)(x) = (\sigma * (\beta * V_a^+))(x)$$
$$= (\sigma * \delta_a)(x) = \sigma(x - a).$$

(Note that the associativity formula $(\beta * \sigma) * V_a^+ = \sigma * (\beta * V_a^+)$ holds, since both factors $\sigma$ and $V_a^+$ vanish for large negative arguments.)

So clearly $\sigma \geqslant 0$ everywhere; secondly, if $x - a$ is a large positive number, then

$$(\mu * V_a^+)(x) = \sum_{y \leqslant x-a} \mu(y)(x - y - a + 1) = \sum_{y \in \mathbf{Z}} \mu(y)(x - y - a + 1)$$

vanishes; consequently, so does $\sigma$ for large positive arguments. This proves that $\sigma$ has finite support. □

### 13.3.2 *Duality for functions of several variables*

We recall the definition of Jensen functions and generalized Jensen functions in Definitions 9.9.2 and 9.9.3, respectively. The sets of these functions are denoted by $\mathscr{J}(\mathbf{R}^n)$ and $\mathscr{J}_{\text{gen}}(\mathbf{R}^n)$. We shall now describe the two sets.

**Proposition 13.3.11.** $\Gamma_0^{\geqslant 0}(\mathscr{J}(\mathbf{Z}^n)) = CVX(\mathbf{R}^n, \mathbf{R})|_{\mathbf{Z}^n}$, *where* $\mathscr{J}(\mathbf{R}^n)$ *is as in Definition 9.9.2.*

**Proof.** This is just the well-known statement that a function $f\colon \mathbf{Z}^n \to \mathbf{R}$ satisfies $\mu * f \geqslant 0$ for all Jensen functions $\mu \in \mathscr{J}(\mathbf{Z}^n)$ if and only if it has a convex extension. □

**Proposition 13.3.12.** $(\Gamma_1^{\geqslant 0} \circ \Gamma_0^{\geqslant 0})(\mathscr{J}(\mathbf{R}^n)) = \mathscr{J}_{\text{gen}}(\mathbf{R}^n)$. *Continuing, we have*
$(\Gamma_1^{\geqslant 0} \circ \Gamma_0^{\geqslant 0})^{\circ k} = \Gamma_1^{\geqslant 0} \circ \Gamma_0^{\geqslant 0}$ *and* $\Gamma_0^{\geqslant 0} \circ (\Gamma_1^{\geqslant 0} \circ \Gamma_0^{\geqslant 0})^{\circ k} = \Gamma_0^{\geqslant}, \quad k = 1, 2, \dots$ .

**Proof.** By definition, $(\Gamma_1^{\geqslant 0} \circ \Gamma_0^{\geqslant 0})(\mathscr{J}(\mathbf{R}^n))$ is equal to

$$\{\mu \in \mathscr{F}_{\text{finite}}(\mathbf{Z}^n, \mathbf{R});\ \mu * f \geqslant 0 \text{ for all } f \in CVX(\mathbf{R}^n, \mathbf{R})|_{\mathbf{Z}^n}\}.$$

If $\mu$ belongs to $\Gamma_1^{\geqslant 0}(CVX(\mathbf{R}^n, \mathbf{R})|_{\mathbf{Z}^n})$, then $\mu * f \geqslant 0$ for all functions $f$ in $CVX(\mathbf{R}^n, \mathbf{R})$, in particular if $f$ is an elementary convex function as defined in Definition 10.10.1: the conditions $(\alpha')$, $(\beta')$, $(\gamma')$ and $(\delta')$ in Definition 9.9.2 are satisfied. Thus $\Gamma_1^{\geqslant 0}(CVX(\mathbf{R}^n, \mathbf{R})|_{\mathbf{Z}^n}) \subset \mathscr{J}_{\text{gen}}(\mathbf{R}^n)$.

In the other direction, assume now that $\mu$ belongs to $\mathscr{J}_{\text{gen}}(\mathbf{R}^n)$. Then $\mu * g \geqslant 0$ for every elementary convex function, and therefore also $\mu * f \geqslant 0$ if $f$ is a finite sum of elementary convex functions. We can now apply Theorem 10.10.5 to the function $f_\theta$ defined as $f_\theta(x) = f(x) + \theta\|x\|_2^2$, with $\theta > 0$, a function of fast growth, to prove that $\mu * f_\theta \geqslant 0$ even if $f$ is a denumerably infinite sum of elementary convex functions: at every given point $x$, the convolution $(\mu * f_\theta)(x)$ agrees with the convolution of $\mu$ with a finite partial sum of the infinite sum of elementary convex functions. Letting $\theta$ tend to zero we see that also $\mu * f \geqslant 0$. □

## 13.4 Duality between classes of functions and second-order difference operators

Given an abelian semigroup $G$ we defined in Definition 10.7.3 two mappings $\Phi$ and $\Psi$ which are special cases of the mappings $\Gamma_1^{\geqslant 0}$ and $\Gamma_0^{\geqslant 0}$ defined in Section 13.2. We recall the definitions:

$$\Phi\colon \mathscr{P}(G \times G) \to \mathscr{P}(\mathscr{F}(G, \mathbf{R})) \tag{13.4.1}$$

Table 13.1. Convolution operators of several kinds. For the meaning of the symbol $\otimes$, see page 299.

| $\odot$ \ T | $\bigwedge$ | $\bigvee$ | $\sum$ | $\prod$ | $\int$ |
|---|---|---|---|---|---|
| $\wedge$ | $\otimes$ | | | | |
| $\vee$ | | $\otimes$ | | | |
| $+$ | infimal c. | supremal c. | $\otimes$ | | $\otimes$ |
| $\cdot$ | | | discrete c. | $\otimes$ | non-discrete c. |

and

$$\Psi\colon \mathscr{P}(\mathscr{F}(G,\mathbf{R})) \to \mathscr{P}(G\times G), \tag{13.4.2}$$

defined by

$$\Phi(A) = \big\{f \in \mathscr{F}(G,\mathbf{R});\ D_bD_af \geqslant 0 \text{ for all } (a,b)\in A\big\}, \tag{13.4.3}$$

defined for $A \in \mathscr{P}(G\times G)$, and

$$\Psi(\mathscr{G}) = \big\{(a,b)\in G\times G;\ D_bD_af \geqslant 0 \text{ for all } f\in\mathscr{G}\big\}, \tag{13.4.4}$$

defined for $\mathscr{G} \in \mathscr{P}(\mathscr{F}(G,\mathbf{R}))$. This is just the operations $\Gamma_1^{\geqslant 0}$ and $\Gamma_0^{\geqslant 0}$ with elements of the special form

$$g = \delta_0 - \delta_{-a} - \delta_{-b} + \delta_{-a-b}$$

and indexed by the pair $(a,b)$: $g*f = D_bD_af$. The definitions are motivated by the fact that these special convolution operators play an important role for discrete convexity.

## 13.5 More general convolution operators

We have already seen infimal and supremal convolution, defined in (5.8.2) and (5.8.1) and discrete convolution in Section 13.2. All convolution operators are of the kind

$$h(x) = \underset{y\in G}{\mathrm{T}}\,\big(f(x-y)\odot g(y)\big), \qquad x\in G, \tag{13.5.1}$$

where T is an operator like summation or integration, and $\odot$ is a binary operator. We can take for T any of the operators $\bigwedge$, $\bigvee$, $\sum$, $\prod$ and $\int$; and for $\odot$ any of the operators $\wedge$, $\vee$, $+$, $\cdot$, $+^p$. Here $+^p$ is defined by $x +^p y = (x^p + y^p)^{1/p}$, $x, y \geqslant 0$, $1 \leqslant p < +\infty$. We have $+^p = \vee$ when $p=\infty$. We call them ***of type*** $(\mathrm{T},\odot)$; thus $(\sum,\cdot)$ is discrete convolution.

A common trait is that the arguments for $f$ and $g$ sum up to the argument of $h$.

A table of the various choices can look as in Table 13.1 on page 298.

To be precise, we should use $\dot{+}$ for infimal convolution and $\underset{\cdot}{+}$ for supremal convolution.

The first two, $\bigwedge$ and $\bigvee$, are defined for all functions, while the others are defined only under some condition guaranteeing convergence.

The diagonal choices, marked by $\otimes$ in the table, are not interesting. Indeed, they produce only constant functions. For instance we have:

$$\bigwedge_y (f(x-y) \wedge g(y)) \text{ is the constant } \bigwedge_{x\in G} f(x) \wedge \bigwedge_{y\in G} g(y);$$

$$\bigvee_{y\in G} (f(y) \vee g(y)) \text{ is equal to the constant } \bigvee_{x\in G} f(x) \vee \bigvee_{y\in G} g(y);$$

and

$$\sum_{y\in G} (f(x-y) + g(y)) = \sum_{x\in G} f(x) + \sum_{y\in G} g(y),$$

assuming that $f$ and $g$ vanish outside a finite set.

Some operations can be easily transformed to others, like

$$\log\left(\textstyle\bigwedge_y (f(x-y)\cdot g(y))\right) = (\log f \sqcap \log g)(x), \quad x \in G, \quad f, g \geqslant 0,$$

infimal convolution of the logarithms of $f$ and $g$. Also $(\bigvee, +)$ and $(\bigwedge, +)$ can be obtained from one another by changing the order relation.

Some can be obtained from others by a passage to a limit, for instance the usual discrete convolution product of functions of the form $\mathrm{e}^{-pf}$ converges as $p$ tends to infinity:

$$\left(\textstyle\sum_{y\in G} \mathrm{e}^{-pf(x-y)} \cdot \mathrm{e}^{-pg(y)}\right)^{1/p} \to \exp\left(-(f \sqcap g)(x)\right) \text{ as } p \to +\infty,\ x \in G.$$

Cf. formula (3.4.5) on page 48.

We can also study pointwise products $fg$ instead of the convolutions $f * g$. Then a well-known type of duality is when $\mathscr{H}$ is the set of functions $h \in \mathscr{F}(G, \mathbf{R})$ of absolute value bounded by 1 and $\mathscr{G}$ is the closed unit ball for the $l^p$ norm, $1 < p < +\infty$. The set corresponding to $\Gamma(\mathscr{G})$ under this operation is the closed unit ball for the $l^q$ norm, $1/p + 1/q = 1$.

**Remark 13.5.1.** In this context it can be mentioned that there is a family of operations which generalize both convolution and pointwise multiplication. Define, for functions $f, g\colon \mathbf{R}^n \times \mathbf{R}^n \to \mathbf{R}$,

$$h(x,z) = (f \odot g)(x,z) = \sum_{y\in\mathbf{R}^n} f(x,y)g(y,z), \quad (x,z) \in \mathbf{R}^n \times \mathbf{R}^n. \tag{13.5.2}$$

This is just matrix multiplication if the supports are finite.

A special case is when $f(x, y) = f_0(x)\delta_x(y)$ and $g(y, z) = g_0(y)\delta_y(z)$, where $\delta_y$ denotes the Kronecker delta placed at $y$. Then we see that $h(x, x) = f_0(x)g_0(x)$, while $h(x, z) = 0$ for $z \neq x$. Thus we obtain pointwise multiplication of $f_0$ and $g_0$.

Another special case, assuming now that $G$ is an abelian group, is when $f(x, y) = f_1(x - y)$ and $g(y, z) = g_1(y - z)$. Then $h(x, z) = (f_1 * g_1)(x - z)$ and in particular $h(x, 0) = (f_1 * g_1)(x)$, the convolution product of $f_1$ and $g_1$.

In connection with this operation we could study functions

$$f\colon \mathbf{R}^n \times \mathbf{R}^m \to \mathbf{R}, \quad g\colon \mathbf{R}^m \times \mathbf{R}^p \to \mathbf{R}, \quad h = f \odot g\colon \mathbf{R}^n \times \mathbf{R}^p \to \mathbf{R},$$

where

$$h(x, z) = (f \odot g)(x, z) = \sum_{y \in \mathbf{R}^m} f(x, y)g(y, z), \qquad (x, z) \in \mathbf{R}^n \times \mathbf{R}^p,$$

a somewhat more general operation. □

## 13.6 Discrete convexity defined by convolution

Let us take up again the definition of $\Gamma_0^{\geqslant 0}$, defined by equation (13.3.6); see also the definitions of $\Phi$ and $\Psi$ in Definition 10.7.3 and formulas (13.4.1) and (13.4.2). We shall say that a function $f \in \mathscr{F}(G, \mathbf{R})$ is ***$\mathscr{M}$-positive*** if it satisfies the inequality $f * \mu \geqslant 0$ for all $\mu \in \mathscr{M}$. Here $\mathscr{M}$ can be any subset of $\mathscr{F}_{\text{finite}}(G, \mathbf{R})$, including the case of a singleton subset $\{\mu\}$.

**Example 13.6.1.** The simplest example is when $\mu = \delta_0$. Then $f$ is $\{\mu\}$-positive if and only if $f \geqslant 0$. A more interesting—and guiding—example is when is $G = \mathbf{Z}$ and $\mu = \frac{1}{2}\delta_{-1} - \delta_0 + \frac{1}{2}\delta_1$. Then $f * \mu \geqslant 0$ if and only if $f$ is convex extendible. □

Since constant functions on $\mathbf{R}^n$ are convex, it is natural to require that all constants be $\mathscr{M}$-positive. Writing $\mu^+ = \mu \vee 0$ and $\mu^- = (-\mu \vee 0)$, we see that this is true of and only if $\mu^+ * 1 = \mu^- * 1$, i.e., if and only if the two have the same total mass for all $\mu \in \mathscr{M}$. The total mass of $\mu^+$ is its $l^1$ norm $\|\mu^+\|_1$, which is equal to the value of the constant function $\mu^+ * 1$.

Also all linear functions on $\mathbf{R}^n$ are convex. We see that this is true for $\mathscr{M}$-positive functions defined on $\mathbf{R}^n$ if and only if $\mu^+$ and $\mu^-$ have the same barycenter as defined in equation (9.9.5), except in the uninteresting case of $\mu^+ = \mu^- = 0$.

We conclude that $\mu * g = 0$ for all affine functions $g$ on $\mathbf{R}^n$ iff $\mu^+$ and $\mu^-$ have the same mass and the same barycenter (cf. conditions $(\beta')$ and $(\gamma')$ in Definition 9.9.3).

Convex functions defined on $\mathbf{R}^n$ have the property that $f \vee g$ is convex if both $f$ and $g$ are. This would be a desirable property. We first need the following lemma.

**Lemma 13.6.2.** *Let $\mu \in \mathscr{F}_{\text{finite}}(\mathbf{R}^n, \mathbf{R})$ be a nonzero function. Then there is a natural number $m$ such that $\mu * p = 0$ for all polynomials $p$ of order at most $m-1$ and a polynomial $q$ of order $m$ such that $\mu * q$ is the constant* 1.

**Proof.** If $\mu * p = 0$ for all polynomials we shall prove that $\mu$ is zero. The Fourier transform of $\mu$ is

$$\hat{\mu}(\zeta) = \sum \mu(x) \mathrm{e}^{-\mathrm{i}\zeta\cdot x}, \qquad \zeta \in \mathbf{C}^n,$$

an entire function. We differentiate it and get

$$\frac{\partial^{k_1}}{\partial \zeta^{k_1}} \cdots \frac{\partial^{k_n}}{\partial \zeta^{k_n}} \hat{\mu}(\zeta) = \sum \mu(x) \mathrm{e}^{-\mathrm{i}\zeta\cdot x} (-\mathrm{i})^{\|k\|_1} x_1^{k_1} \cdots x_n^{k_n}, \quad \zeta \in \mathbf{C}^n, \ k \in \mathbf{N}^n.$$

So if $\mu * p = 0$ for all monomials $p(x) = x^k = x_1^{k_1} \cdots x_n^{k_n}$, then all derivates of $\hat{\mu}$ vanish at the origin, which implies that $\mu$ is zero. This proves that for every nonzero $\mu$ there is an integer $m$ such that $\mu * p$ vanishes for polynomials of order $< m$ but not for all polynomials of order $m$. This also implies that $\mu * q$ is a constant for every polynomial of order $m$ since lower-order terms in $q$ do not influence the value of the convolution product. The constant is nonzero for some choice of $q$ and can be taken to be 1. □

**Theorem 13.6.3.** *Let $\mathscr{M}$ be any subset of $\mathscr{F}_{\text{finite}}(\mathbf{R}^n, \mathbf{R})$, the elements of which are nonzero and annihilate all affine functions. Suppose that* supp $\mu^-$ *is a singleton set for all elements $\mu$ of $\mathscr{M}$. If $(f_j)_{j \in J}$ is any family of $\mathscr{M}$-positive functions, then also $f = \sup_{j\in J} f_j$ is $\mathscr{M}$-positive.*

*Conversely, if $\mu$ is a nonzero function which annihilates all affine functions and is such that $f \vee g$ is $\{\mu\}$-positive for all $\{\mu\}$-positive functions $f$ and $g$, then the support of $\mu^-$ is a singleton.*

**Proof.** For every $j \in J$ we have

$$\mu^- * f_j \leqslant \mu^+ * f_j \leqslant \mu^+ * f.$$

We now take the supremum over all $j$ and obtain

$$\sup_j (\mu^- * f_j) \leqslant \mu^+ * f.$$

The general inequality $\mu^- * f \geqslant \sup_j(\mu^- * f_j)$ is an equality when the support of $\mu^-$ is a singleton set. So we have

$$\mu^- * f = \sup_j(\mu^- * f_j) \leqslant \mu^+ * f,$$

proving that $f$ is $\mathscr{M}$-positive.

For the converse we take $m$ and $q$ according to Lemma 13.6.2. (We can note that $m$ must be at least 2 under the hypotheses, but can be arbitrarily large.)

Let $a$ and $b \neq a$ be such that $\mu(a) < 0$ and $\mu(b) < 0$. Without loss of generality we may suppose that $\mu(a) = \min \mu$. Take

$$f = \delta_{-a} + Aq \text{ and } g = \delta_{-b} + Aq,$$

where $q$ is the polynomial we get from Lemma 13.6.2 and $A$ is a constant to be determined. Then

$$f \vee g = (\delta_{-a} \vee \delta_{-b}) + Aq = \delta_{-a} + \delta_{-b} + Aq.$$

We have

$$(\mu * f)(x) = \mu(a + x) + A, \quad (\mu * g)(x) = \mu(b + x) + A,$$

as well as

$$(\mu * (f \vee g))(x) = \mu(a + x) + \mu(b + x) + A, \qquad x \in \mathbf{R}^n.$$

To prove the proposition it suffices to have

$\mu(a + x) \geqslant -A$, $\mu(b + x) \geqslant -A$ for every $x \in \mathbf{R}^n$ and $\mu(a) + \mu(b) < -A$.

We now choose $A = -\mu(a) = -\min \mu$. The inequalities then become

$\mu(a+x) \geqslant \mu(a)$, $\mu(b+x) \geqslant \mu(a)$ for every $x \in \mathbf{R}^n$ and $\mu(a)+\mu(b) < \mu(a)$, which are all true. □

## 13.7 Possible future studies

Kiselman and Samieinia studied in (2010, 2017) a duality between classes of functions and classes of difference operators of second order. This is a fundamental kind of duality and is subject to treatment using concepts from mathematical morphology. This study could be continued and extended.

As already mentioned in Section 12.11 on page 285, the marginal functions of functions with integer values remain to be studied, as well as generalizations where the functions are defined on general discrete sets, not just on points with integer coordinates. This idea is developed in my manuscript (2021).

Definition 13.5.2 appears in my paper (2005) but is not developed systematically there. I believe a more detailed study could be worthwhile.

## 13.8 Notes

### 13.8.1 *Ubiquity of duality*

Duality of the kind we have studied here appears in many branches of mathematics, too many to be mentioned here.

Duality as discussed here and in Chapter 9 was presented in my paper (2019b)

The class of M- and L-convex functions form two distinct classes of discrete convexity which are conjugate to each other under the Fenchel transformation. Similarly we have a conjugacy relation between the classes of M$^\natural$- and L$^\natural$-convex functions. An M$^\natural$-convex function is supermodular, whereas an L$^\natural$-convex function is submodular (Murota 2003:145, 189). On the other hand, in dimensions $n \geqslant 3$, it is not in general true that the Fenchel transform of a supermodular function is submodular; however, the converse is always true. Thus, these two concepts, submodularity and supermodularity, are not symmetric under this transformation. On the other hand, adding an exchange axiom to supermodularity and linearity to submodularity makes the behavior under the Fenchel transformation different. Moreover, an L-convex function is submodular and integrally convex (Fujishige & Murota 2000: Theorem 2). Also an M-convex function is integrally convex (Murota & Shioura 1999: Lemma 5.4). Various classes of discrete convexity are compared in (Murota & Shioura 2001). For conjugacy relations between M-convex and L-convex functions of real variables, see (Murota & Shioura 2004).

An elementary duality is that between lines in $\mathbf{R}^2$ of equation $y = \alpha x + \beta$ and the parameter space of the coefficients $(\alpha, \beta) \in \mathbf{R}^2$. The line passes through a given point $(p, q)$ if and only if $q = \alpha p + \beta$, thus defining a straight line in the dual plane of all pairs $(\alpha, \beta)$. This classical duality is often called the *Hough transformation* in the context of digital geometry, and is described in, e.g., (Chassery et al. 2006).

In connection with Example 13.6.1 we could mention that also subharmonic functions on discrete structures have been studied, for instance in my paper (2005).

### 13.8.2 *Convexity of marginal functions*

In my paper (2008a) I proved convex extensibility of the marginal function of a function $f \colon \mathbf{Z} \times \mathbf{Z} \to \mathbf{R}$. In our joint paper (2017), Shiva Samieinia and I proved results to the effect that a function defined in $\mathbf{Z}^n$ and with real values possesses a convex extensible marginal function.

### 13.8.3 *The supremum of a family of functions*

Adama Arouna Koné's doctoral dissertation (2016), presented on 2016 January 14, contains a result like Proposition 13.6.3 but where the converse is stated and proved under two special hypotheses, viz. that the minimum of the function $\mu$ is attained at one point only and that $m = 2$ (Koné 2014:79, Proposition 4.3.11). Here these hypotheses have been eliminated.

## 13.9 Exercises

13.1. Determine the lower and upper inverses of
(a) $\Gamma$ (Definition 13.3.2),
(b) $\Gamma^0$ (Definition 13.3.4),
(c) $\Gamma_0^{\geqslant 0}$ (Equation (13.3.6)), and
(d) $\Gamma_1^{\geqslant 0}$ (Equation (13.3.7)).

13.2. Define $f(x, y) = |x - 2y|$, $(x, y) \in \mathbf{Z}^2$. Its marginal function $h(x) = \inf_{y \in \mathbf{Z}} f(x, y)$ takes the value 0 when $x \in 2\mathbf{Z}$, at odd points the value 1. We have $\mu * f \geqslant 0$ for all $\mu \in \mathscr{I}_{\mathrm{gen}}(\mathbf{Z}^2)$. Determine $\Psi(\{h\}) = \{(a, b) \in \mathbf{Z} \times \mathbf{Z};\ D_b D_a h \geqslant 0\}$.

13.3. Formulate the statement dual to
*To be, or not to be, that is the question* (Shakespeare 1868:547)
Prove (at least in some special cases) the dual.

13.4. Explain the instance of Ursula's Duality expressed in the book title *The Left Hand of Darkness.*

# Chapter 14

# Topology

## 14.1 Introduction

TOPOLOGY is a highly developed branch of mathematics. It was not created with discrete phenomena as a primary goal. However, in spite of the fact that the topology of $\mathbf{Z}^n$ as a subset of $\mathbf{R}^n$ is discrete and therefore a completely uninteresting topology, there are many situations where discrete sets can carry interesting topologies. Such topologies will be the topic of the next chapter. As a preparation, we present in the present chapter fundamental definitions and results from topology. The chapter can be skipped by a reader who is already familiar with the main concepts.

## 14.2 Mappings

We recall Definition 2.6.1, where we associated with $f\colon X \to Y$ two mappings $f^*\colon \mathscr{P}(Y) \to \mathscr{P}(X)$ and $f_*\colon \mathscr{P}(X) \to \mathscr{P}(Y)$, defined by the formulas

$$\begin{aligned} f^*(B) &= \{x \in X;\ f(x) \in B\}, \qquad && B \subset Y; \\ f_*(A) &= \{f(x) \in Y;\ x \in A\}, \qquad && A \subset X. \end{aligned} \tag{14.2.1}$$

We also recall Definition 5.10.1, where, given three sets $X$, $Y$ and $Z$ and a mapping $f\colon X \to Y$, we defined a mapping $\overleftarrow{f}\colon \mathscr{F}(Y,Z) \to \mathscr{F}(X,Z)$, the pullback, by the formula $\overleftarrow{f}(g) = g \circ f$, $g \in \mathscr{F}(Y,Z)$. The two definitions are compatible in a natural way if $Z = \mathbf{R}$: if $g \in \mathscr{F}(Y,\{0,1\})$ is the characteristic function of a set $B \subset Y$, $g = \chi_B$, then $\overleftarrow{f}(\chi_B) = \chi_{f^*(B)}$, where we use the notation from (14.2.1) in the right-hand side. Thus we see that the pullback of $\chi_B$ is the characteristic function of the inverse image $f^*(B)$ of $B$.

We note that $f^*$ is a homomorphism of Boolean algebras: it satisfies

$$f^*(B_1 \cup B_2) = f^*(B_1) \cup f^*(B_2),$$
$$f^*(B_1 \cap B_2) = f^*(B_1) \cap f^*(B_2). \tag{14.2.2}$$

and

$$f^*(B_1 \smallsetminus B_2) = f^*(B_1) \smallsetminus f^*(B_2), \text{ in particular } f^*(\complement B) = \complement f^*(B).$$

The formulas (14.2.2) can be generalized to infinite unions and intersections, implying that $f^*$ is both a dilation and an erosion:

$$f^*\big(\bigcup B_j\big) = \bigcup f^*(B_j), \qquad f^*\big(\bigcap B_j\big) = \bigcap f^*(B_j).$$

The homomorphism $f^*$ is injective if and only if $f$ is surjective, and surjective if and only if $f$ is injective.

The mapping $f_*$ is not so well-behaved: it always satisfies

$$f_*(A_1 \cup A_2) = f_*(A_1) \cup f_*(A_2) \text{ and } f_*\big(\bigcup A_j\big) = \bigcup f_*(A_j);$$

it is thus a dilation. But for intersections we have only $f_*(A_1 \cap A_2) \subset f_*(A_1) \cap f_*(A_2)$; $f_*$ is in general not an erosion. Also there is in general no inclusion relation between $f_*(\complement A)$ and $\complement f_*(A)$.

We note that $A \subset f^*(B)$ if and only if $f_*(A) \subset B$, i.e., expressed with the help of hypographs and epigraphs, that $(\mathbf{hypo}\, f^*)^{\smile} = \mathbf{epi}(f_*)$. By Corollary 5.3.4 this implies that $(f^*)^{[-1]} = f_*$ and that $(f_*)_{[-1]} = f^*$; $f_*$ is the dilation corresponding to the erosion $f^*$. Moreover $f^* \circ f_* \circ f^* = f^*$ and $f_* \circ f^* \circ f_* = f_*$. It follows that $\big((f^*)^{[-1]}\big)_{[-1]} = f^* = \big((f^*)_{[-1]}\big)^{[-1]}$ and $\big((f_*)_{[-1]}\big)^{[-1]} = f_*$; cf. Proposition 5.6.1.

Proposition 5.4.1 shows that

$$f^* \circ f_* \geqslant \mathbf{id}_{\mathscr{P}(X)} \text{ and } f_* \circ f^* \leqslant \mathbf{id}_{\mathscr{P}(Y)}.$$

(These formulas can of course also be proved directly.) The first formula means that $f^*(f_*(A)) \supset A$; equality holds for all $A$ if and only if $f$ is injective. The second means that $f_*(f^*(B)) = B \cap \mathbf{im}(f) \subset B$; equality holds for all $B$ if and only if $f$ is surjective. Since $f^*$ is also a dilation, Theorem 5.5.1 implies that $f^* \circ (f^*)_{[-1]} \leqslant \mathbf{id}_{\mathscr{P}(X)}$. Proposition 5.4.1 implies that

$$(f_*)^{[-1]} \circ f_* \leqslant \mathbf{id}_{\mathscr{P}(X)} \text{ and } (f^*)_{[-1]} \circ f^* \geqslant \mathbf{id}_{\mathscr{P}(Y)}.$$

Finally we note that

$$(f^*)_{[-1]} = \complement \circ f_* \circ \complement = (f_*)^{\text{dual}} = \big((f^*)^{[-1]}\big)^{\text{dual}}.$$

(There seems to be no analogous formula for $(f_*)^{[-1]}$.)

We can now pass to mappings on a higher level by taking the upper and lower star again. There are four mappings on the next level:

$$(f^*)^*, (f_*)_* \colon \mathscr{P}(\mathscr{P}(X)) \to \mathscr{P}(\mathscr{P}(Y)) \text{ and}$$
$$(f_*)^*, (f^*)_* \colon \mathscr{P}(\mathscr{P}(Y)) \to \mathscr{P}(\mathscr{P}(X));$$

two go in the same direction as $f$ and two in the opposite direction. Of these four, two will be used in Section 14.4 to transport topologies, viz. $(f^*)^*$ and $(f^*)_*$.

## 14.3 Defining topologies

A ***topology*** on a set $X$ is a collection $\tau = \mathscr{U}(X)$ of subsets of $X$—thus an element of $\mathscr{P}(\mathscr{P}(X))$—which is closed under the formation of arbitrary unions and finite intersections. The elements of $\mathscr{U}(X)$ are called ***open sets***; thus any union of open sets is open and any finite intersection of open sets is open. In particular, the union and the intersection of the empty family is open, so $\emptyset$ and $X$ are always open subsets of $X$. A set equipped with a topology will be called a ***topological space*** or, in this context, just a ***space***.

The union of all open sets contained in a given set $A$ is open and called the ***interior*** of $A$, denoted by $A^\circ$. The complement of an open set is called ***closed***. It follows that the instersection of all closed set containing a set $A$ is closed; it is said to be the ***closure*** of $A$ and is denoted by $\overline{A}$.

The interior and the closure are related by the formula $\complement A^\circ = \overline{\complement A}$.

We shall say that a subset $B$ of a set $A$ is ***dense*** if the closure of $B$ contains $A$.

The set-theoretical difference $\overline{A} \smallsetminus A^\circ$ is called the ***boundary*** of $A$, denoted by $\partial A$.

In a metric space there is a topology defined by the metric: it consists of all unions of balls $B_<(c, r)$ with $c \in X$, $r > 0$.

A topology can be given in several different ways. As mentioned, the family $\mathscr{C}(X)$ of all closed sets is closed under the formation of arbitrary intersections and finite unions. We may also impose these conditions as axioms, and then define a set to be open if its complement is closed. A topology can be equivalently defined using open or closed sets.

Another concept is that of neighborhood. If a topology $\mathscr{U}(X)$ is given, we say that a set $V$ is a ***neighborhood*** of a point $x$ if there exists an open set $U$ such that $x \in U \subset V$. The families $\mathscr{V}(x)$, $x \in X$, of all neighborhoods of points in $X$ satisfy the following conditions:

$$\begin{aligned}
&\text{If } V \in \mathscr{V}(x), \text{ then } x \in V;\\
&\text{If } V \in \mathscr{V}(x) \text{ and } W \supset V, \text{ then } W \in \mathscr{V}(x);\\
&\text{If } V, W \in \mathscr{V}(x), \text{ then } V \cap W \in \mathscr{V}(x);\\
&\text{If } V \in \mathscr{V}(x), \text{ then there exists } W \in \mathscr{V}(x) \text{ such that}\\
&\qquad V \in \mathscr{V}(y) \text{ for every } y \in W.
\end{aligned} \tag{14.3.1}$$

These properties are easy to verify if the topology is given and the neighborhoods are defined as above. On the other hand, if we have a collection $\mathscr{V}(x)$ for every $x \in X$ satisfying (14.3.1) and define a set $U$ to be open if it belongs to $\mathscr{V}(x)$ for every $x \in U$, then we get a topology for which the neighborhoods are the given ones.

We can also define a topology using cleistomorphisms. If a topology is given, then we have a cleistomorphism in $\mathscr{P}(X)$ by mapping $A$ to its closure $\overline{A}$. Then this operator satisfies

$$\overline{\varnothing} = \varnothing \text{ and } \overline{A \cup B} = \overline{A} \cup \overline{B}. \tag{14.3.2}$$

Conversely, if a cleistomorphism is defined and satisfies these two conditions, we can define a set to be closed if $\overline{A} = A$; we then get a topology, a topology for which the topological operator of taking the closure is the originally given one.

The operation $A \mapsto A^\circ$ is increasing, idempotent, and antiextensive. It is thus an anoiktomorphism, which of course means that it is a cleistomorphism if we reverse the order, i.e., if we define $A \leqslant B$ to mean $A \supset B$. In addition to being an anoiktomorphism, it satisfies

$$X^\circ = X \text{ and } (A \cap B)^\circ = A^\circ \cap B^\circ. \tag{14.3.3}$$

Conversely, we may take these properties as axioms and define a set to be open if it is in the image of the operator. Then we get a topology. The operation of taking the interior of a set for this topology is equal to the original operator.

Summing up, we have five equivalent ways to define a topology: using open sets, closed sets, neighborhoods, taking the topological closure, taking the interior.

If we have two topologies $\mathscr{U}_1(X)$ and $\mathscr{U}_2(X)$ on the same set $X$ we say that the first is ***weaker*** or ***coarser*** than the second, and that the second is ***stronger*** or ***finer*** than the first, if $\mathscr{U}_1(X) \subset \mathscr{U}_2(X)$. Expressed in terms of cleistomorphisms, this means that $c_2 \leqslant c_1$ if $c_j$ denotes the cleistomorphism associated with $\mathscr{U}_j(X)$, $j = 1, 2$. The weakest topology is the ***chaotic topology*** $\{\emptyset, X\}$ and the strongest is the ***discrete topology*** $\mathscr{P}(X)$. The closure of a nonempty set in the chaotic topology is always the whole space, wheras the closure of a set in the discrete topology is the set itself.

If $\mathbf{card}(X) = m$, then $\mathbf{card}(\mathscr{P}(X)) = 2^m$ and $\mathbf{card}(\mathscr{P}(\mathscr{P}(X))) = 2^{2^m}$. However, since the empty set and the whole space are always open, there are at most $2^m - 2$ subsets of $X$ for which it is possible to define them as not open. So a set with $m$ elements can have at most $2^{2^m - 2}$ topologies. For $m = 2$ this number is 4, for $m = 3$, it is 64. A two-point space can have four topologies: in addition to the chaotic and the discrete we have $\{\emptyset, \{a\}, \{a, b\}\}$, where $\{a\}$ is open and $\{b\}$ is closed. It is called ***Sierpiński space***. Then also $\{\emptyset, \{b\}, \{a, b\}\}$, which appears after a change of notation—thus there are only three different topologies.

## 14.4 Transporting topologies

If $f\colon X \to Y$ is a mapping from a set $X$ into a topological space $Y$ we can transport the topology on $Y$ to $X$ by defining a subset of $X$ to be open if and only if it is of the form $f^*(U)$ for some open subset $U$ of $Y$. Because $f^*$ is both a dilation and an erosion, it is clear that the family of all sets

$$(f^*)_*(\mathscr{U}(Y)) = \{f^*(U);\ U \in \mathscr{U}(Y)\}$$

is a topology. Here we have used the notation introduced in Definition 2.6.1 at the next higher level: $f^*\colon \mathscr{P}(Y) \to \mathscr{P}(X)$, $(f^*)_*\colon \mathscr{P}(\mathscr{P}(Y)) \to \mathscr{P}(\mathscr{P}(X))$. For brevity we shall denote $(f^*)_*(\tau) = (f^*)_*(\mathscr{U}(Y))$ by $\overleftarrow{f}(\tau) = \overleftarrow{f}(\mathscr{U}(Y))$, the ***pullback*** of $\tau = \mathscr{U}(Y)$.

If $\lambda\colon \mathscr{P}(Y) \to \mathscr{P}(Y)$ is a cleistomorphism in $Y$, then $\overleftarrow{\lambda} = f^* \circ \lambda \circ f_*$ is a cleistomorphism in $X$, and if $\lambda$ satisfies the topological axioms (14.3.2), $\lambda(\emptyset) = \emptyset$ and $\lambda(B_1 \cup B_2) = \lambda(B_1) \cup \lambda(B_2)$, then $\overleftarrow{\lambda}$ does the same. Thus we can transport topological closure operators from $Y$ to $X$. One can verify that the transported open sets correspond to the transported closure operator.

A common situation is when $X$ is a subset of $Y$ and $f$ is the inclusion mapping. Then we say that the topology $\overleftarrow{f}(\mathscr{U}(Y))$ defined on $X$ is the ***induced topology***. We see that $U$ is open in $X$ if and only if $U = V \cap Y$ for some open set $V$ in $Y$; we also see that the closure operator $\overleftarrow{\lambda}$ in $X$ is defined as $\overleftarrow{\lambda}(A) = \lambda(A) \cap X$.

If $X$ is a topological space and $f\colon X \to Y$ a mapping of $X$ into a set $Y$, we can of course consider the family $\{f_*(A);\ A \in \mathscr{U}(X)\} \subset \mathscr{P}(Y)$. However, this family is usually not a topology on $Y$, since $f_*$ is not an erosion. Instead we use again $f^*$ and declare a subset $B$ of $Y$ to be open if $f^*(B)$ is open in $X$. And we can verify that this is indeed a topology on $Y$; we shall denote it by

$$f_{\rightarrow}(\sigma) = f_{\rightarrow}(\mathscr{U}(X)) = (f^*)^*(\mathscr{U}(X)) = \{B \in \mathscr{P}(Y);\ f^*(B) \in \mathscr{U}(X)\},$$

the ***pushforward*** of the topology $\sigma = \mathscr{U}(X)$ on $X$.

A common instance of this definition is when $Y$ is a quotient set of $X$, i.e., when we have an equivalence relation $\sim$ in $X$ and let $Y = X/{\sim}$ be the set of all equivalence classes in $X$ with respect to the relation. The mapping $f$ associates to each element in $X$ its equivalence class in $Y$. Then a subset $B$ of $Y$ is open in $Y$ with respect to the topology we have pushed forward from $X$ if and only if the union of all equivalence classes in $B$ is open in $X$. The topology obtained in this way on $X/{\sim}$ is called the ***quotient topology***.

If $f\colon X \to Y$ is injective, and if we have a topology on $X$, push it forward to $Y$ and then pull it back to $X$, the new topology agrees with the original one: $f^{\leftarrow}(f_{\rightarrow}(\mathscr{U}(X))) = \mathscr{U}(X)$. Similarly, if $f$ is surjective and we start with a topology on $Y$, pull it back to $X$ and then push it forward to $Y$, we obtain the original topology; $f_{\rightarrow}(f^{\leftarrow}(\mathscr{U}(Y))) = \mathscr{U}(Y)$. (This works so well because we did not use $f_*$ but $f^*$ in the definition of $f_{\rightarrow}$.)

However, if we have a cleistomorphism $\kappa$ in $X$, we cannot define a cleistomorphism in $Y$ by $\kappa_{\rightarrow} = f_* \circ \kappa \circ f^*$ (Exercise 14.12).

## 14.5 Continuous mappings

Let $f\colon X \to Y$ be a mapping of a topological space $X$ into a topological space $Y$ and let $x$ be a point in $X$. We say that $f$ is ***continuous at*** $x$ if $f^*(V)$ is a neighborhood of $x$ for every neighborhood $V$ of $f(x)$. It is called ***continuous*** if it is continuous at every point in $X$. We can translate this well-known concept into the language of open sets, closed sets, and cleistomorphisms. One can prove that $f$ is continuous if and only if $f^*(U) \in \mathscr{U}(X)$ for every $U \in \mathscr{U}(Y)$; in other words if and only if the topology $f^{\leftarrow}(\mathscr{U}(Y))$ is weaker than the topology $\mathscr{U}(X)$.

## 14.6 Connectedness

The family of all open and closed sets of a topological space $X$ is a Boolean algebra. This algebra must contain the two sets $\varnothing$ and $X$, for they are always both open and closed. (If $X$ is empty, there is of course only one such set.) A topological space is said to be ***connected*** if the only sets which are both open and closed are the empty set and the whole space.[1]

A subset of a topological space is called ***connected*** if it is connected as a topological space with the induced topology.

A ***component*** of a topological space is a nonempty connected subset which is maximal with respect to inclusion.

A connected subset which is both open and closed is a component. It is easy to prove that the closure of a connected subset is connected. Therefore all components are closed. They need not be open (Exercise 14.12).

**Example 14.6.1.** We associate to every point $x$ in a topological space $X$ the component $\gamma(x)$ of $X$ to which $x$ belongs. Thus $\gamma \colon X \to \mathscr{P}(X)$, more precisely $\gamma \colon X \to \mathscr{C}(X)$, the family of closed subsets. The reader is invited to explore the properties of this mapping. □

**Proposition 14.6.2.** *Let $f \colon X \to Y$ be a continuous mapping of a topological space $X$ into a topological space $Y$. If $X$ is connected, then so is $f_*(X) = \mathbf{im}(f)$.*

**Proof.** Let $B$ be a closed and open subset of $\mathbf{im}(f)$. Then $f^*(B)$ is closed and open in $X$. Hence $f^*(B)$ is either empty or equal to $X$. Therefore $f_*(f^*(B)) = B \cap \mathbf{im}(f) = B$ is either empty or equal to $\mathbf{im}(f)$. This means that $\mathbf{im}(f)$ is connected. □

**Corollary 14.6.3.** *Let $f \colon X \to Y$ be a mapping of a topological space $X$ into a set $Y$. Equip $Y$ with the strongest topology such that $f$ is continuous. Suppose that $X$ is connected. Then $\mathbf{im}(f)$ is connected, and the points in $Y \smallsetminus \mathbf{im}(f)$ are isolated. In particular, any quotient space of a connected topological space is connected.*

**Proof.** For any point $y \in Y$ not in the image of $f$, the inverse image $f^*(\{y\})$ is empty, thus both open and closed. This means that $Y \smallsetminus \mathbf{im}(f)$ has the discrete topology and its components are just the singleton sets. □

[1] Hence the empty set is connected, which agrees with Bourbaki's definition (1961: I:§11:1). Adrien Douady (personal communication, 2000-06-26) claimed that it would be more bourbakistic to require that there are exactly two sets which are open and closed, and that therefore the empty set should not be connected. However, I have decided to follow Bourbaki here.

Of the four topologies that can live on a space consisting of two points, only three are connected, and out of these, only two are different in the sense that they cannot be obtained from another one by renaming.

## 14.7 Quotient spaces

In particular we shall use Corollary 14.6.3 with $X = \mathbf{R}$ and $Y = \mathbf{Z}$ to define connected topologies on the digital line $\mathbf{Z}$. Let $f\colon \mathbf{R} \to \mathbf{Z}$ be a surjective mapping. Then $\mathbf{Z}$ equipped with the strongest topology such that $f$ is continuous is a connected topological space. Thus we consider $\mathbf{Z}$ as a quotient space of $\mathbf{R}$, not as a subspace. Now there exist very many surjective mappings $\mathbf{R} \to \mathbf{Z}$. It is not unnatural to restrict attention to increasing surjections $f\colon \mathbf{R} \to \mathbf{Z}$. Then $f^*(\{n\})$ is an interval for every integer $n$; denote its endpoints by $a_n$ and $b_n \geqslant a_n$, so that

$$]a_n, b_n[ \subset f^*(\{n\}) \subset [a_n, b_n].$$

We can normalize the situation to $a_n = n - \frac{1}{2}$, $b_n = n + \frac{1}{2}$; this does not change the topology on $\mathbf{Z}$. Then $f(x) = \lfloor x + \frac{1}{2} \rfloor$ for all $x \in \mathbf{R} \smallsetminus (\mathbf{Z} + \frac{1}{2})$ while $f(n + \frac{1}{2}) = n$ or $f(n + \frac{1}{2}) = n + 1$ for $n \in \mathbf{Z}$. The topology is therefore determined if we know for which $n$ we have $f(n + \frac{1}{2}) = n$. For every subset $A$ of $\mathbf{Z}$ we get a topology on $\mathbf{Z}$ by declaring that $f(n + \frac{1}{2})$ shall be equal to $n$ for $n \in A$ and that, for all other real numbers $x$, we shall have $f(x) = \lfloor x + \frac{1}{2} \rfloor$. Thus $A$ describes faithfully all topologies obtained from increasing surjections—the others are just too many.

This can be explained as follows. It is natural to think of $\mathbf{Z}$ as an approximation of the real line $\mathbf{R}$ and to consider mappings $f\colon \mathbf{R} \to \mathbf{Z}$ expressing this idea. We may define $f(x)$ to be the integer closest to $x$; this is well-defined unless $x$ is a half-integer: $f(x) = \lfloor x + \frac{1}{2} \rfloor$ when $x \in \mathbf{R} \smallsetminus \left(\mathbf{Z} + \frac{1}{2}\right)$. So when $x = n + \frac{1}{2}$ we have a choice for each $n$: shall we define $f\left(n + \frac{1}{2}\right) = n$ or $f\left(n + \frac{1}{2}\right) = n+1$? If we choose the first alternative for every $n$, thus putting $f^*(\{n\}) = \left]n - \frac{1}{2}, n + \frac{1}{2}\right]$, the topology defined in Corollary 14.6.3 is called the ***right topology*** on $\mathbf{Z}$; if we choose the second, we obtain the ***left topology*** on $\mathbf{Z}$; cf. (Bourbaki 1961:I:§1: Exerc. 2).

Another choice is to always choose an even integer as the best approximant of a half-integer. Then the closed interval $\left[-\frac{1}{2}, \frac{1}{2}\right]$ is mapped to 0, so $\{0\}$ is closed, whereas the inverse image of 1 is the open interval $\left]\frac{1}{2}, \frac{3}{2}\right[$, so that $\{1\}$ is open.

**Definition 14.7.1.** The ***Khalimsky topology*** on $\mathbf{Z}$ is the topology for which the open sets are all three-sets $\{2k - 1, 2k, 2k + 1\}$, $k \in \mathbf{Z}$, their intersections and all unions of these. □

This topology was introduced by E. D. Halimskiĭ (Efim Khalimsky). The integer line **Z** with this topology is called the ***Khalimsky line.***

The Khalimsky line is connected, but the complement of any point is disconnected. Among all the topologies defined by increasing surjections $f\colon \mathbf{R} \to \mathbf{Z}$ only two have this property: the one just defined and the one obtained by translating everything by one step. For the left topology, for instance, all subsets are connected.

## 14.8 Separation axioms

**Definition 14.8.1.** The relation $x \in \overline{\{y\}}$ between two points in a topological space will be called the ***specialization preorder*** and will be denoted by $x \preccurlyeq y$.

The intersection of all neighborhoods of a point $y$ in a topological space will be denoted by $N(y)$. □

That $\preccurlyeq$ is a preorder is easy to see. We always have $x \in \overline{\{x\}}$. That $x \preccurlyeq y$ and $y \preccurlyeq z$ means that $x \in \overline{\{y\}}$ and $y \in \overline{\{z\}}$, implying that $x \in \overline{\overline{\{z\}}} = \overline{\{z\}}$.

We note that $x \in \overline{\{y\}}$ if and only if $y \in N(x)$.

A ***Kolmogorov space*** (Bourbaki 1961:I:§1: Exerc. 2, b), p. 142), also called a $T_0$***-space***, is a topological space such that $x \in N(y)$ and $y \in N(x)$ only if $x = y$, thus precisely when the specialization preorder is an order (satisfies Definition 2.7.3). It is quite reasonable to impose this axiom; if $x$ belongs to $N(y)$ and vice versa, then $x$ and $y$ are indistinguishable from the point of view of topology: we cannot distinguish points from knowledge of the open sets to which they belong. We should therefore identify them and consider a quotient space.

So every topology $\tau$ on a set $X$ defines a preorder in $X$, to be denoted by $P(\tau)$, and this preorder is an order if and only if $\tau$ satisfies Kolmogorov's axiom.

Conversely, every preorder $\preccurlyeq$ in a set $X$ defines a topology $T(\preccurlyeq)$ on $X$ whose smallest neighborhoods are $N(x) = \{y \in X;\ y \preccurlyeq x\}$. This means that a subset $A$ of $X$ is open if and only if, for all $a \in A$ and all $y \in X$, $a \preccurlyeq y$ implies $y \in A$.

If we start with a preorder $\preccurlyeq$, define the topology $T(\preccurlyeq)$, and then define the preorder $P(T(\preccurlyeq))$, then we get back to the original preorder: $\preccurlyeq$ is equal to $P(T(\preccurlyeq))$; in other words $P \circ T$ is the identity on the set of all preorders.

Table 14.1. Continuous mappings vs. increasing mappings.

| *Mappings $X \to Y$ between topological spaces* | *Mappings $X \to Y$ between preordered sets* |
|---|---|
| $X$ has the discrete topology<br>$\Downarrow$<br>all mappings are continuous | $X$ has the discrete order<br>$\Downarrow$<br>all mappings are increasing |
| $Y$ has the chaotic topology<br>$\Downarrow$<br>all mappings are continuous | $Y$ has the chaotic preorder<br>$\Downarrow$<br>all mappings are increasing |
| $X$ has the chaotic topology and $Y$ has a Kolmogorov topology<br>$\Downarrow$<br>only the constants are continuous | $X$ has the chaotic preorder and $Y$ is ordered<br>$\Downarrow$<br>only the constants are increasing |
| $Y$ has the discrete topology and $X$ is connected<br>$\Downarrow$<br>only the constants are continuous | $Y$ has the discrete order and $X$ is connected<br>$\Downarrow$<br>only the constants are increasing |

However, if we start with a topology $\tau$, construct the preorder $P(\tau)$, then the topology $T(P(\tau))$ defined by that preorder is not equal to $\tau$ in general; $T \circ P$ is not the identity. However, $T(P(\tau)) = \tau$ if and only if $\tau$ defines a smallest-neighborhood space as defined below in Definition 15.2.1.

The separation axiom $T_1$ states that $N(x) = \{x\}$. It is too strong to be of interest for the spaces considered here. The specialization preorder in this case is the discrete order: we have $x \preccurlyeq y$ if and only if $x = y$.

Two points $x$ and $y$ in a topological space $Y$ are said to be ***adjacent*** if $x \neq y$ and $\{x, y\}$ is connected. We note that $\{x, y\}$ is connected if and only if either $x \in N(y)$ or $y \in N(x)$. Hence two points are adjacent if and only if they are different and comparable for the specialization preorder.

An even stronger separation axiom is $T_2$, which states that, given two points $x$ and $y \neq x$, there exists neighborhoods $V$ of $x$ and $W$ of $y$ which are disjoint. A topological space satisfying the $T_2$ axiom is also called a ***Hausdorff space***.

## 14.9 Continuous mappings vs. increasing mappings

To exhibit some of the analogies between topological spaces and preordered sets, we list in Table 14.1 on page 314 some properties of continuous and increasing mappings.

## 14.10 Matheron's hit-or-miss topology

The family $\mathscr{C}(X)$ of all closed subsets of a topological space $X$ can be equipped with a topology as follows. We declare a family $\mathscr{A}$ of closed subsets of $X$ to be open in $\mathscr{C}(X)$ if there are finitely many subsets $U_1, \dots, U_m, K$ of $X$, the $U_j$ being open and $K$ being compact, such that every $A \in \mathscr{A}$ meets each of the $U_j$ but is disjoint from $K$. Then a set is declared to be open if it is a union of such families $\mathscr{A}$. The whole space is open (take $U_1 = X$ and $K$ empty), and the empty set is open (take the union of no open sets). This topology was constructed by Matheron (1975:3) and is now known as ***Matheron's hit-or-miss topology***. See also (Serra 1982:75–76).

It turns out that $\mathscr{C}(\mathbf{R}^n)$ provided with this topology is a compact Hausdorff space (Matheron 1975: Theorem 1-2-1; Serra 1982: Theorem 3-18).

A sequence $(C_j)_{j \in \mathbf{N}}$ of closed sets converges to $C$ if and only if, for all open sets $U_1, \dots, U_m$, and all compact sets $K$, if $C$ meets all the $U_j$ and is disjoint from $K$, then the same is true for all the $C_j$ except for a finite number of them (Matheron 1975:6).

## 14.11 Notes

The preimage $f^*(B)$ is sometimes denoted by $f^{-1}(B)$, and the direct image $f_*(A)$ sometimes by just $f(A)$. It is however convenient to have a special notation for $f_* \colon \mathscr{P}(X) \to \mathscr{P}(Y)$, so that it is not confused with $f \colon X \to Y$; similarly, $f^*$ is not the pointwise inverse of $f$.

The specialization preorder was introduced by Aleksandrov (1937:503), and the name *specialization order* was used by Kong et al. (1991:905).

## 14.12 Exercises

14.1. Give an example of a topological space with a component which is not open.

14.2. Prove that if $f: X \to Y$ is a mapping of a set $X$ into a set $Y$, then the cleistomorphism $f^*/^*f^*$ is equal to $f^* \circ f_*$. Similarly $f_*/_*f_* = f_* \circ f^*$ (an anoiktomorphism; cf. Theorem 6.2.1).

14.3. Prove that, if $\lambda: \mathscr{P}(Y) \to \mathscr{P}(Y)$ is a cleistomorphism in $Y$, then $\kappa = f^* \circ \lambda \circ f_*$ is a cleistomorphism in $X$ (see Definitiion 5.10.1), and that, if $\lambda$ satisfies the topological axioms (14.3.2), then $\kappa$ does the same.

14.4. Show that if we have a cleistomorphism $\kappa$ in $X$ and a mapping $f: X \to Y$, then we cannot define a cleistomorphism in $Y$ by $\lambda = f_* \circ \kappa \circ f^*$. In general $\lambda$ will be neither extensive nor idempotent. Construct examples which show this.

14.5. Express the continuity of a mapping $f: X \to Y$ from one topological space $X$ into another, $Y$, in terms of the families of all closed sets $\mathscr{C}(X)$ and $\mathscr{C}(Y)$ as well as in terms of the closure operators in $X$ and $Y$.

14.6. We have seen that a two-point space can have three different topologies. How many different topologies are there on a three-point space? How many are connected?

14.7. Describe, for the usual topology in $\mathbf{R}^n$, the upper and lower inverses of the four mappings $\kappa$, $\alpha$, $\eta$ and $\theta$ which are defined for $A \in \mathscr{P}(\mathbf{R}^n$ by

$$\kappa(A) = \overline{A}, \quad \alpha(A) = A^\circ, \quad \eta(A) = \left(\overline{A}\right)^\circ, \quad \theta(A) = \overline{A^\circ}.$$

Find also the second inverses.

# Chapter 15

# The Khalimsky topology

## 15.1 Introduction

THE KHALIMSKY TOPOLOGY is a remarkable topology that introduces continuity in situations where one might imagine only the uninteresting discrete topology. The leading idea is to consider **Z** not as a subspace of **R** but as a quotient space.

In this chapter we first introduce spaces which possess a complete symmetry between open and closed sets, called smallest-neighborhood spaces. We explain what continuity means in such spaces and then take a look at the most important example among them, viz. the Khalimsky plane. In particular we shall consider the Jordan curve theorem and Brouwer's fixed-point theorem in this new setting.

## 15.2 Smallest-neighborhood spaces

In a topological space the union of any family of open sets is open. It may happen that also the intersection of any family of open sets is open.

**Definition 15.2.1.** A topological space such that the intersection of any family of open sets is open will be called a ***smallest-neighborhood space***. □

These spaces were introduced by Pavel Sergeevič Aleksandrov[1] in his seminal paper (1935). Another suitable name could therefore be a ***P. S. Aleksandrov space***. The study of cell spaces by A. W. Tucker (1936) is analogous, with a duality between open and closed sets.

[1]Not to be confused with Aleksandr Danilovič Aleksandrov, for whom other spaces are named.

It is equivalent to require that the union of an arbitrary family of closed sets be closed. The intersection $N(x)$ of all neighborhoods of a point $x$ is a neighborhood for all $x$ if and only if the space is a smallest-neighborhood space.

Let now $N\colon X \to \mathscr{P}(X)$ be a mapping. When is the family of all sets $N(x)$, $x \in X$, the family of smallest neighborhoods in a smallest-neighborhood space? In other words, if we define $\mathscr{V}(x)$ as the family of all sets $V$ such that $V \supset N(x)$, when is $(\mathscr{V}(x))_{x\in X}$ the family of neighborhoods in some topological space? Looking at the properties the families of all neighborhoods shall satisfy in order to define a topology, equation (14.3.1), we see that this is true if and only if $x$ belongs to $N(x)$ for all $x$ (the first property in (14.3.1)) and $N$ is increasing for the specialization preorder, i.e., that $y \preccurlyeq x$ implies $N(y) \subset N(x)$. Indeed, the second and third properties follow easily from the definition, and the fourth means just that $N$ is increasing.

Aleksandrov (1935, 1937) introduced the term *espace discret, diskreter Raum* 'discrete space' for a topological space such that the intersection of any family of open sets is open. The intersection of all closed sets containing a set $M$ he called its *Hülle* 'hull', and denoted it by $\overline{M}$ or $AM$ (this is the usual closure of $M$). The intersection of all open sets containing a set $M$ he called its *Stern* 'star' and denoted it by $OM$ (1937:504). He noted that the operation of taking the star of a set is a cleistomorphism satisfying the two extra conditions $\overline{\varnothing} = \varnothing$ and $\overline{A \cup B} = \overline{A} \cup \overline{B}$ (see the discussion in Section 2.11, in particular formula (2.11.3), and therefore defines a topology, which he called *réciproque* (1935) or *dual* (1937). The closed sets of a smallest-neighborhood space satisfies the axioms of the open sets of a topology, so there is a complete symmetry between the two topologies in such a space. So, to any topology $\tau$ of this kind, there is a dual topology $\tau^*$ the open sets of which are the closed sets for the original topology.

A set $A$ in a smallest-neighborhood space is closed if and only if, for every $a$ in $A$, the subset $\overline{\{a\}}$ is contained in $A$. This is a convenient property, which is not true in $\mathbf{R}^n$.

It is easy to see that a mapping $f\colon X \to Y$ between two smallest-neighborhood spaces is continuous if and only if it is increasing for the specialization preorder, defined in Section 14.8. Thus continuity in these spaces is actually order theoretic, and the smallest-neighborhood spaces are actually special cases of preordered sets. This means that the rich theory of preordered sets can be put to work here.

Alexandrov's choice of terms is understandable, but nowadays it is not possible to use the term *discrete space* in Aleksandrov's sense, since the discrete topology in modern usage refers only to the topology where every set is open, the strongest of all topologies. This is why I propose to call a discrete space in Aleksandrov's sense a *smallest-neighborhood space* or a *P. S. Aleksandrov space.*

The closed points, i.e., the points $x$ such that $\overline{\{x\}} = \{x\}$, Aleksandrov (1937:504) called *Eckpunkte* 'vertices', and the open points, i.e., the points $x$ such that the singleton $\{x\}$ is open, he called *Grundpunkte* 'base points'.

We can define a topology on the digital line $\mathbf{Z}$ by declaring all odd points to be open, thus $N(2k+1) = \{2k+1\}$, and all even points to have a smallest neighborhood $N(2k) = \{2k-1, 2k, 2k+1\}$. It follows that the even points are closed, for the complement of an even point $2k$ is the union of all $N(x)$ with $x \neq 2k$, thus an open set. This is the Khalimsky topology already defined in Definition 14.7.1. Thus in the Khalimsky topology the even points are vertices and the odd points are base points in Aleksandrov's terminology. In the specialization order, the vertices lie higher than the base points ($1 \prec 0$, i.e., $1 \in N(0)$, and $1 \prec 2$, i.e., $1 \in N(2)$).

A ***Khalimsky interval*** is an interval $[a,b]_{\mathbf{Z}} = [a,b]_{\mathbf{R}} \cap \mathbf{Z}$ equipped with the topology induced by the Khalimsky topology on $\mathbf{Z}$. A ***Khalimsky circle*** is a quotient space $\mathbf{Z}_m = \mathbf{Z}/m\mathbf{Z}$ of the Khalimsky line for some even integer $m \geqslant 4$. (If $m$ is odd, the quotient space receives the chaotic topology, which is not interesting.)

The ***Khalimsky plane*** is the Cartesian product of two Khalimsky lines, and, more generally, ***Khalimsky space*** is the Cartesian product of $n$ copies of $\mathbf{Z}$. Equivalently, we can define Khalimsky $n$-space by declaring the set $\{x \in \mathbf{Z}^n;\ \|x-p\|_\infty \leqslant 1\}$ to be open for any point $p \in (2\mathbf{Z})^n$ and then taking all intersections of such sets as open sets, then all unions of such intersections. The points $(2k, 2m)$ with $(k,m) \in \mathbf{Z}^2$ are closed; the points $(2k+1, 2m+1)$ are open. These points are called ***pure***; the other points, with one even and one odd coordinate, are neither open nor closed and are called ***mixed***. See Figure 15.1 on page 320.

There are, however, other topologies in $\mathbf{Z}^2$ which are of interest: we may declare $\{x \in \mathbf{Z}^2;\ \|x-p\|_1 \leqslant 1\}$ to be open for any $p$ such that $\sum p_j \in 2\mathbf{Z}$ as well as all unions of such sets. This topology is called the Wyse–Marcus topology after Wyse, Marcus et al. (1970). The Khalimsky topology and the Wyse–Marcus topology just defined are not comparable: none is stronger than the other. However, they are related, for if we rotate the Khalimsky plane by 45° and delete all points which are not open or closed, we obtain the Wyse–Marcus topology.

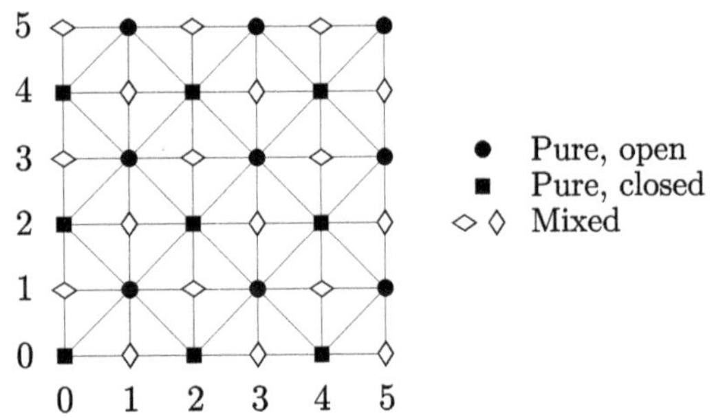


Figure 15.1. A part of the Khalimsky plane. (Courtesy Erik Melin)

## 15.3 Continuity for the Khalimsky topology

A function $f\colon X \to Y$ from one smallest-neighborhood space into another is continuous at a point $x$ if and only if the inverse image of $N_Y(f(x))$ contains $N_X(x)$, or, equivalently, the direct image of $N_X(x)$ is contained in $N_Y(f(x))$:

$$N_X(x) \subset f^*(N_Y(y)), \text{ equivalently } f_*(N_X(x)) \subset N_Y(f(x)). \tag{15.3.1}$$

(See Examples 4.4.3 and 5.2.3.) Here $N_X(x)$ and $N_Y(y)$ denote the smallest neighborhood of $x \in X$ and $y \in Y$, respectively. If we apply this to the case when $X = Y = \mathbf{Z}$, it means the following. If $x \in \mathbf{Z}$ is odd, the property always holds; if $x$ is even and $f(x)$ is odd, it means that $f(x \pm 1) = f(x)$, and if $x$ is even and also $f(x)$ is even, it means that $|f(x \pm 1) - f(x)| \leqslant 1$. In particular, a continuous function is Lip-1, but it must sometimes have intervals of constancy, viz. every time it takes an odd value at an even point (and hence also when it takes an even value at an odd point).

We observe that the following functions are continuous.

(1) $\mathbf{Z} \ni x \mapsto a \in \mathbf{Z}$, where $a$ is a constant;

(2) $\mathbf{Z} \ni x \mapsto \pm x + c \in \mathbf{Z}$, where $c$ is an even constant;

(3) $\max(f, g)$ and $\min(f, g)$ if $f$ and $g$ are continuous.

Actually every continuous function on a bounded Khalimsky interval can be obtained by applying the rules (1), (2), (3) a finite number of times. Note that the function $x \mapsto x + 1$ is discontinuous.

The number of Lip-1 functions $f\colon [0, n-1] \to \mathbf{Z}$ with $f(0) = 0$ is $3^{n-1}$, $n \in \dot{\mathbf{N}}$. Of these, $\frac{1}{2}\big((1+\sqrt{2})^n + (1-\sqrt{2})^n\big)$ are continuous for the Khalimsky

Table 15.1. Parity of the three numbers $x_1$, $x_2$, and $f(x_1, x_2)$.

| *Parity of* $x_1$ | *Parity of* $x_2$ | *Parity of* $f(x)$ | *Condition* |
|---|---|---|---|
| Odd | Odd | Odd | None |
| Odd | Odd | Even | None |
| Odd | Even | Odd | $y_1 = x_1, \lvert y_2 - x_2\rvert \leqslant 1 \Rightarrow f(y) = f(x)$ |
| Odd | Even | Even | $y_1 = x_1, \lvert y_2 - x_2\rvert \leqslant 1 \Rightarrow \lvert f(y) - f(x)\rvert \leqslant 1$ |
| Even | Odd | Odd | $\lvert y_1 - x_1\rvert \leqslant 1, y_2 = x_2 \Rightarrow f(y) = f(x)$ |
| Even | Odd | Even | $\lvert y_1 - x_1\rvert \leqslant 1, y_2 = x_2 \Rightarrow \lvert f(y) - f(x)\rvert \leqslant 1$ |
| Even | Even | Odd | $\lVert y - x\rVert_\infty \leqslant 1 \Rightarrow f(y) = f(x)$ |
| Even | Even | Even | $\lVert y - x\rVert_\infty \leqslant 1 \Rightarrow \lvert f(y) - f(x)\rvert \leqslant 1$ |

topology. Let us denote these numbers by $Kh_n$. They satisfy the recursion relation $Kh_n = 2Kh_{n-1} + Kh_{n-2}$, $n \geqslant 3$, and the ratio $Kh_n/Kh_{n-1}$ tends to $1 + \sqrt{2} < 3$ as $n$ tends to infinity. (Samieinia 2010b: Theorem 2.3.)

Functions with a smaller codomain than $\mathbf{Z}$, like $f\colon [0, n-1] \to [0, +\infty[$ and $f\colon [0, n-1] \to [0, m-1]$, are of course fewer in number than those with $\mathbf{Z}$ as their codomain; see (Sameinia 2010c; 2011).

For functions of two variables, $f\colon \mathbf{Z}^2 \to \mathbf{Z}$, formula (15.3.1) means the following. We list the eight possible parities of the triple $(x_1, x_2, f(x_1, x_2))$ in $\mathbf{Z}^3$:

Table 15.1 above lists what it means for $f$ to be continuous at a particular point $x$. However, the description becomes much simpler if we use the specialization order $\preccurlyeq$, for then a continuous function is just an increasing function. We know that, in $\mathbf{Z}$,

$$\cdots \prec \text{-2} \succ \text{-1} \prec 0 \succ 1 \prec 2 \succ 3 \prec 4 \succ \cdots, \tag{15.3.2}$$

where we write $s \prec t$ when $s \preccurlyeq t$ and $s \neq t$, and $s \succ t$ when $s \succcurlyeq t$ and $s \neq t$ for $s, t \in \mathbf{Z}$.

In $\mathbf{Z}^2$, $(0,0) \succ (1,0) \succ (1,1)$ and $(0,0) \succ (0,1) \succ (1,1)$ and, in general,

$$(2m, 2n) \succ (2m+1, 2n),$$
$$(2m, 2n+1) \succ (2m+1, 2n+1) \text{ for all } m, n \in \mathbf{Z}.$$

So continuity at $x$ boils down to $x \preccurlyeq y \Rightarrow f(x) \preccurlyeq f(y)$ for all $y \in \mathbf{Z}^2$; continuity everywhere to the same implication but now for all $x, y \in \mathbf{Z}^2$. For example, if both components of $x$ are odd, the only $y$ which satisfies $x \preccurlyeq y$ is $y = x$, so $f(x) \preccurlyeq f(y)$ holds automatically. If, on the other hand both components of $x$ are even, then $\{y; x \preccurlyeq y\} = B^\infty_\leqslant(x, 1)$ for the $l^\infty$ norm, and if $f(x)$ in addition is odd, then $f(x) \preccurlyeq f(y)$ holds only for $f(y) = f(x)$, so $f$ must be constant on $B^\infty_\leqslant(x, 1)$.

We note that if $x, y \in \mathbf{Z}$ and $x \preccurlyeq y$, then $|x - y| \leqslant 1$. Conversely, if $|x - y| \leqslant 1$, then either $x \preccurlyeq y$ or $y \preccurlyeq x$. Hence $|x - y| \leqslant 1$ implies $|f(x) - f(y)| \leqslant 1$ for any continuous function $f\colon \mathbf{Z} \to \mathbf{Z}$, and we see that $f$ is Lip-1. In two variables we have the same conclusion. In the proof of this fact we shall need the following notation. For any two points $x, y \in \mathbf{Z}^2$ we define $q(x, y) = (x_1, y_2)$. The four points $x, y, q(x, y), q(y, x)$ thus form a rectangle (perhaps degenerate); if $y_j = x_j \pm 1$, $j = 1, 2$, they form a square with side length 1.

**Theorem 15.3.1.** *A continuous function $f\colon \mathbf{Z}^2 \to \mathbf{Z}$ is Lip-1 for the $l^\infty$ norm.*

*This conclusion also holds for any continuous function $f\colon X \to \mathbf{Z}$, where $X$ is a connected subset of $\mathbf{Z}^2$ such that $q(x, y)$, $q(y, x) \in X$ for all $x, y \in X$ such that $y_j = x_j \pm 1$, $j = 1, 2$.*

**Proof.** It is enough to prove that if $\|x - y\|_\infty \leqslant 1$, then $|f(x) - f(y)| \leqslant 1$. If $x, y \in X$ with $x \preccurlyeq y$, then $f(x) \preccurlyeq f(y)$, which in turn implies that $|f(x) - f(y)| \leqslant 1$.

Assume now that $\|x - y\|_\infty \leqslant 1$ but that we do not have $x \preccurlyeq y$, nor $y \preccurlyeq x$. Then we have essentially the case $x = (1, 0)$, $y = (0, 1)$. We have $q(y, x) = (0, 0) \succcurlyeq x, y \succcurlyeq (1, 1) = q(x, y)$, which implies that $f(0, 0) \succcurlyeq f(x), f(y) \succcurlyeq f(1, 1)$. (By hypothesis both $(0, 0)$ and $(1, 1)$ belong to $X$.) Hence $|f(0, 0) - f(1, 1)| \leqslant 1$. Also, $|f(0, 0) - f(x)| \leqslant 1$ and $|f(x) - f(1, 1)| \leqslant 1$, which implies that $f(x)$ is equal to $f(0, 0)$ or to $f(1, 1)$. Similarly, $f(y) = f(0, 0)$ or $f(y) = f(1, 1)$. Therefore $|f(x) - f(y)| \leqslant 1$. □

This result holds for many subsets $X$ of $\mathbf{Z}^2$, but not for all:

**Example 15.3.2.** Let $X = \{x \in \mathbf{Z}^2; \|x\|_1 \leqslant 1\}$ , the $l^1$-ball in $\mathbf{Z}^2$ and define $f(x) = x_1 - x_2$. Then $X$ is connected for the Khalimsky topology and $f$ is continuous, but $f$ is not Lip-1 for the $l^\infty$ norm. We note that, if we take $x = (1, 0) \in X$ and $y = (0, 1) \in X$, then $q(y, x) = (0, 0) \in X$ while $q(x, y) = (1, 1) \notin X$. So both $q(x, y)$ and $q(y, x)$ need to be elements of $X$. □

**Theorem 15.3.3.** *A function $f\colon \mathbf{Z}^2 \to \mathbf{Z}$ is continuous if and only if it is separately continuous. More generally, the equivalence holds for any function $f\colon X \to \mathbf{Z}$ where $X$ is a subset of $\mathbf{Z}^2$ such that one of $q(x,y)$, $q(y,x)$ belongs to $X$ if $y_j = x_j \pm 1$ and $x \preccurlyeq y$.*

**Proof.** Assume that $f$ is separately continuous and that $x \preccurlyeq y$. Then we shall prove that $f(x) \preccurlyeq f(y)$. If $x_1 = y_1$, then $x_2 \preccurlyeq y_2$, and the inequality $f(x) \preccurlyeq f(y)$ follows from the separate continuity of the function $x_2 \mapsto f(x)$ for a fixed $x_1$. The conclusion is similar if $x_2 = y_2$; then the continuity of $x_1 \mapsto f(x)$ for a fixed $x_2$ does the job. The case when $x_1 \neq y_1$ and $x_2 \neq y_2$ remains to be considered. Then $y_j = x_j \pm 1$. One of the points $q(x,y)$ and $q(y,x)$ belongs to $X$; let $z$ be one of them that does. Then clearly $x \preccurlyeq z \preccurlyeq y$, which in view of the separate continuity implies $f(x) \preccurlyeq f(z)$ and $f(z) \preccurlyeq f(y)$, and we are done. □

**Example 15.3.4.** Let $X = \{0\} \cup \{x \in \mathbf{Z}^2; |x_1| = |x_2| = 1\}$. This five-set is connected for the Khalimsky topology. Every function $f\colon X \to \mathbf{Z}$ is separately continuous, but not all of them are continuous. With $x = (0,0)$ and $y = (1,1)$ we see that none of $q(x,y) = (0,1)$ and $q(y,x) = (1,0)$ belongs to $X$. □

Like in real analysis there is an intermediate-value theorem for the Khalimsky line:

**Theorem 15.3.5.** *Let two continuous functions $f, g\colon I \to \mathbf{Z}$ be given on a Khalimsky interval $I = [a,b]_{\mathbf{Z}}$. Assume that there are points $p, r \in I$ with $f(p) \geqslant g(p)$ and $f(r) \leqslant g(r)$. Then there exists a point $q$, intermediate between $p$ and $r$, such that $f(q) = g(q)$.*

**Proof.** Without loss of generality we may assume that $p \leqslant r$. Define

$$M = \{x \in \mathbf{Z};\ p \leqslant x \leqslant r \text{ and } f(x) \geqslant g(x)\}.$$

Clearly $p \in M$, so $M$ is not empty. Let $q = \max_{x\in M} x$. If $q = r$, then $f(r) = g(r)$ and we are done; if not, $q + 1 \leqslant r$. Then we must have $f(q) \geqslant g(q)$ and $f(q+1) < g(q+1)$. We claim that $f(q) = g(q)$. If this were not true, we would have $f(q) \geqslant g(q)+1$ and $f(q+1) \leqslant g(q+1)-1$. Because of the Lipschitz property, the only possibility then would be $f(q) = g(q)+1$ and $f(q+1) = g(q+1) - 1$. But also this situation is impossible. If $q$ and $f(q)$ are of different parity, then $f(q+1) = f(q)$, which would require a jump of two units in $g$. If on the other hand $q$ and $f(q)$ are of the same parity, then $q$ and $g(q)$ are of different parity, so that $g(q+1) = g(q)$, requiring a jump of two units in $f$. This contradiction leaves us with the only possibility $f(q) = g(q)$. □

## 15.4 Digitization of straight lines in the Khalimsky plane

In this section we shall prove a modification of Azriel Rosenfeld's Theorems 8.2.2 and 8.2.4 when $\mathbf{Z}^2$ is provided with the Khalimsky topology. It is a satisfying result in that it proves that the digitization of any real line or line segment is homeomorphic to the Khalimsky line or a Khalimsky interval, respectively.

We start with a modification of the Rosenfeld cross $R(p)$:

**Definition 15.4.1.** Let us define a cross $R_\times(p)$ in $\mathbf{R}^2$ as the set of all points $p + (t, \pm t)$, where $p \in \mathbf{Z}^2$ and $-\frac{1}{2} < t \leqslant \frac{1}{2}$. We then define the pure digitization of a set $A \subset \mathbf{R}^2$ as the set

$$\mathbf{dig}_\mathrm{p}(A) = \{x \in \mathbf{Z}^2;\ x \text{ is pure and } R_\times(x) \cap A \neq \varnothing\}.$$

Finally we define $\mathbf{dig}_\mathrm{kh}(A)$ as the closure of $\mathbf{dig}_\mathrm{p}(A)$ under the cleistomorphism $\kappa_M = \varepsilon_M \circ \delta_M$, where $M = \{(0,0), (1,0), (0,1), (1,1)\}$. □

The role of the set $\kappa_M$ is to add some mixed points close to the pure digitization.

**Theorem 15.4.2.** *If $L$ is a straight line in the Euclidean plane $\mathbf{R}^2$, then $\mathbf{dig}_\mathrm{kh}(L)$ is homeomorphic to the Khalimsky line. If $[a, b]$ is a rectilinear segment with $a, b \in \mathbf{R}^2$ and $a \neq b$, then $\mathbf{dig}_\mathrm{kh}([a, b])$ is homeomorphic to a Khalimsky segment.*

**Proof.** For reasons of symmetry it is enough to consider lines of equation $x_2 = \gamma x_1 + c$ with $0 \leqslant \gamma \leqslant 1$, and, since moreover lines with $\gamma$ equal to 0 or 1 are easy to treat, we assume now that $0 < \gamma < 1$.

Let us first prove that there are no two points $x$ and $y$ in the digitization such that $x_1 = y_1$ and $x_2 \neq y_2$.

If $x$ is mixed, then the two pure points $(x_1 \pm 1, x_2)$ belong to the digitization, so $L$ cannot meet any $R_\times(p)$ with $p_1 = x_1 - 1$ and $p_2 \geqslant x_2 + 2$; nor with $p_1 = x_1 + 1$ and $p_2 \leqslant x_2 - 1$. Hence there cannot be any other point in the digitization with the same first coordinate.

If, on the other hand, $x$ is pure, then $L$ cannot meet any $R_\times(y)$ of a pure point $y$ with $y_1 = x_1$ and $y_2 \geqslant x_2 + 2$ or with $y_2 \leqslant x_2 - 2$. Nor can it meet any $R_\times(p)$ with $p_1 = x_1 - 1$ and $p_2 \geqslant x_2 + 1$ or with $p_1 = x_1 + 1$

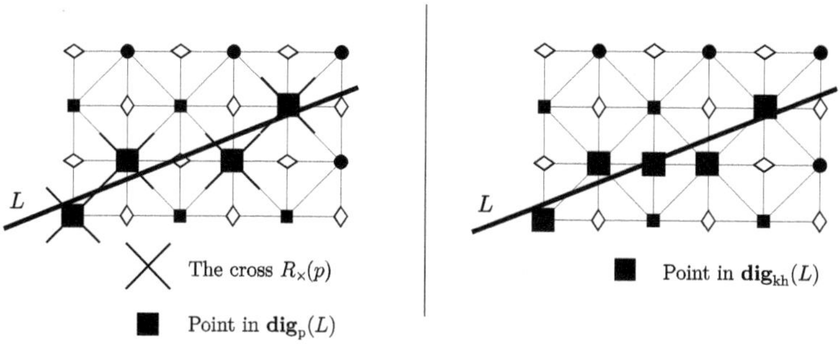


Figure 15.2. Digitization of a straight line in the Khalimsky plane. To the left, the pure digitization $\mathbf{dig}_\mathrm{p}(L)$; to the right, the Khalimsky digitization $\mathbf{dig}_\mathrm{kh}(L)$. One mixed point has been added. (Courtesy Erik Melin)

and $p_2 \leqslant x_2 - 1$. Therefore, also now, there cannot exist any point with the same first coordinate.

We now know that $\mathbf{dig}_\mathrm{kh}(L)$ is the graph of a function $f\colon \mathbf{Z} \to \mathbf{Z}$. It remains to be seen that the mapping $\mathbf{Z} \ni j \mapsto (j, f(j)) \in \mathbf{Z}^2$ is a homeomorphism. The mapping $\mathbf{Z}^2 \ni (j, f(j)) \mapsto j \in \mathbf{Z}$ is the projection which forgets the second component and is automatically continuous, so we only have to prove that $f$ is continuous.

If $x = (j, f(j))$ is a mixed point, then $j$ and $f(j)$ have different parity and $(x_1 + 1, x_2)$ and $(x_1 - 1, x_2)$ belong to $\mathbf{dig}_\mathrm{kh}(L)$ by definition, which means that $f(j \pm 1) = f(j)$ as required.

If $x = (j, f(j))$ is pure, then $j$ and $f(j)$ have the same parity, and $L$ may meet either $R_\times(x+(1,1))$ or $R_\times(x+(2,0))$. In the first case $f(j+1) = f(j)+1$; in the second case $f(j+1) = f(j)$. Similary we can prove that $f(j-1)$ is equal to $f(j)-1$ or to $f(j)$. Thus $f$ is continuous.

The digitization of a segment of positive length is a connected and finite subset of the straight line it determines, and so has the properties claimed. □

## 15.5 Fixed-point theorems

Let us say that a topological space $X$ has the ***fixed-point property*** if every continuous mapping $f\colon X \to X$ possesses a ***fixed point***, i.e., a point $p$ such that $f(p) = p$, thus that the invariance set $\mathbf{invar}(f)$ is nonempty.

The famous Brouwer fixed-point theorem states that a closed ball in $\mathbf{R}^n$

has the fixed-point property. This theorem is a deep result when $n \geqslant 2$ but easy to prove if $n = 1$: we may take

$$p = \sup_{x \in A} \big(x;\ f(x) \geqslant x\big) \tag{15.5.1}$$

with $A = B_{\leqslant}(0,1) = [-1,1] \subset \mathbf{R}$.

If $f\colon [-1,1] \to [-1,1]$ is a mapping from a compact interval into itself which is increasing for the usual order inherited from the real axis, then it also has a fixed point, and formula (15.5.1) again yields a fixed point. This generalizes to complete lattices; if $f\colon L \to L$ is an increasing mapping of a complete lattice $L$ into itself, then the point $p$ defined by (15.5.1) with $A = L$ is a fixed point of $f$. This is Tarski's fixed-point theorem (1955), exhibiting one of the many analogies between continuous mappings and increasing mappings. However, a Khalimsky interval of more than two points is not a lattice.

In this section we shall prove that certain subsets of the Khalimsky plane possess the fixed-point property. Since the Khalimsky line and the Khalimsky plane are ordered sets and the continuous mappings are precisely the increasing mappings, there are theorems from the theory of ordered sets that can be applied here; cf. (Baclawski & Björner 1979). However, the proofs presented here are more direct in the context of digital geometry.

A topological space cannot have the fixed-point property unless it is connected. Indeed, if $U$ and $V$ are two disjoint nonempty open sets whose union is the whole space $X$, then we may define $f(x) = b$ for all $x \in U$ and $f(x) = a$ for all $x \in V$, where $a$ is an arbitrary point in $U$ and $b \in V$. Then the inverse image of any set is open, for it is either empty, equal to $U$, equal to $V$, or equal to $X$. Thus $f$ is certainly continuous, but it has no fixed point.

On the other hand, connectedness is by no means sufficient. The mapping $f(x) = -x$ is continuous in $\mathbf{R}^n$, and in fact in any abelian group with a reasonable topology; it has no fixed point in $X$ if $0 \notin X$. Any nonempty subset $X$ of such a group with $0 \notin X$ and such that $x \in X$ implies $-x \in X$ will be a space such that some continuous selfmapping is without fixed point.

In a finite set $X$ with $m$ points there are $m^m$ selfmappings $X \to X$. Out of these, $(m-1)^m$ do not have fixed points; there are thus $m^m - (m-1)^m$ mappings which have a fixed point. The proportion of mappings with a fixed point is $1 - (1 - 1/m)^m$, which tends to $1 - 1/\mathrm{e}$ (about 63 percent) when $m$ tends to infinity. If we introduce a topology on $X$ we may ask how many of the $m^m$ mappings are continuous. The number of mappings of

Table 15.2. Continuous and discontinuous mappings and mappings with and without the fixed-point property in a space with the fixed-point property and with $m$ elements.

| | *Continuous* | *Discontinuous* | *Sum* |
|---|---|---|---|
| *Fixed point* | $c$ | $m^m - (m-1)^m - c$ | $m^m - (m-1)^m$ |
| *No fixed point* | 0 | $(m-1)^m$ | $(m-1)^m$ |
| *Sum* | $c$ | $m^m - c$ | $m^m$ |

different kinds in a space with the fixed-point property can be listed as in Table 15.2 above. It contains only one unknown, the number of continuous mappings, denoted by $c$.

We shall now prove that a continuous mapping of a Khalimsky interval into itself has a fixed point. Later we shall consider subsets of the Khalimsky plane, which we shall attack using an implicit-function theorem.

**Theorem 15.5.1.** *Let us define a subset $C_{\#}(\mathbf{Z}, \mathbf{Z})$ of the set of all continuous mappings of $\mathbf{Z}$ into itself,*

$$C_{\#}(\mathbf{Z}, \mathbf{Z}) = \{f \in C(\mathbf{Z}, \mathbf{Z});\ \exists p, r \in \mathbf{Z} \text{ such that } f(p) \geqslant p \text{ and } f(r) \leqslant r\}.$$

*Then $f \in C(\mathbf{Z}, \mathbf{Z})$ has a fixed point if and only if $f \in C_{\#}(\mathbf{Z}, \mathbf{Z})$.*

**Proof.** If $f$ has a fixed point, it is obvious that $f$ belongs to $C_{\#}(\mathbf{Z}, \mathbf{Z})$.

The converse is a special case of the intermediate-value theorem (Theorem 15.3.5), taking $g$ as the identity. □

**Corollary 15.5.2.** *Every bounded Khalimsky interval has the fixed-point property.*

**Proof.** Let $f\colon I \to I$ be a continuous mapping, where $I = [a, b]_{\mathbf{Z}}$ is a bounded interval. Extend $f$ to a mapping $g\colon \mathbf{Z} \to \mathbf{Z}$ by defining $g(x) = f(a)$ for $x < a$ and $g(x) = f(b)$ for $x > b$. Then it is easy to check that $g$ is continuous, and it is obvious that $g$ belongs to $C_{\#}(\mathbf{Z}, \mathbf{Z})$. Thus it has a fixed point $q \in \mathbf{Z}$, but as $q \in \mathbf{im}(g) \subset I$, $q$ is a fixed point also of $f$. □

**Example 15.5.3.** For a Khalimsky interval $\{a, a+1\}$ consisting of two points only, there are four mappings: the two constant mappings, the identity, and the one interchanging $a$ and $a+1$. The first three obviously have a fixed point; the fourth does not. But it is discontinuous. Thus the statistics in Table 15.2 looks like in Table 15.3 on page 328 ($m = 2$, $c = 3$). □

Table 15.3. Continuous and discontinuous mappings and mappings with and with out the fixed-point property in a Khalimskky interval $\{a, a+1\}$.

| | *Continuous* | *Discontinuous* | *Sum* |
|---|---|---|---|
| *Fixed point* | 3 | 0 | 3 |
| *No fixed point* | 0 | 1 | 1 |
| *Sum* | 3 | 1 | 4 |

**Theorem 15.5.4.** *Every bounded Khalimsky rectangle $I \times J \subset \mathbf{Z}^2$ has the fixed-point property.*

We shall prove this result using an implicit-function theorem.

In real analysis there is an implicit-function theorem which says the following. If $f$ is a real-valued function of class $C^1$ defined in an open subset $\Omega$ of $\mathbf{R}^2$ with a zero $a \in \Omega$, and its derivative with respect to the first variable, $\partial f/\partial x_1$, is non-zero at $a$, then there is a function $g$ of one variable defined near $a_2$ such that $f(g(x_2), x_2) = 0$ for all $x_2$ near $a_2$. This function $g$ is also of class $C^1$. One says that the equation $f(x_1, x_2) = 0$ defines $x_1$ implicitly as a function of $x_2$; the equation $f(x_1, x_2) = 0$ is equivalent to the explicit formula $x_1 = g(x_2)$ near $a$. The result is local in the sense that we only assert something in a (perhaps very small) neighborhood of $a$.

In the digital plane we have an implicit-function theorem of a similar kind. We shall solve the implicit equation $f(x_1, x_2) = x_1$. (In real analysis, this is equivalent to $f(x_1, x_2) - x_1 = 0$, but the left-hand side $f(x_1, x_2) - x_1$ is not continuous in general in the digital case even if $f$ is.)

**Theorem 15.5.5.** *Let us define an order, and hence a topology, in the space $C(\mathbf{Z}, \mathbf{Z})$ of continuous mappings $\mathbf{Z} \to \mathbf{Z}$ by declaring that $f \preccurlyeq g$ if and only if $f(x) \preccurlyeq g(x)$ for all $x \in \mathbf{Z}$. Let us also define a mapping $W\colon C(\mathbf{Z}, \mathbf{Z}) \to \mathbf{Z}_!$ by*

$$W(g) = \sup_{x \in \mathbf{Z}}(x;\ g(x) \geqslant x), \qquad g \in C(\mathbf{Z}, \mathbf{Z}). \tag{15.5.2}$$

*The mapping $W$ is continuous in $C_\#(\mathbf{Z}, \mathbf{Z})$ where it is finite.*

If the set of fixed points of a mapping $g \in C_\#(\mathbf{Z}, \mathbf{Z})$ is bounded to the right, then $W(g)$ is its largest fixed point; otherwise $W(g) = +\infty$.

**Lemma 15.5.6.** *The mapping $W$ defined by* (15.5.2) *satisfies*

$$g(a) \geqslant a \text{ if and only if } W(g) \geqslant a, \qquad a \in \mathbf{Z}, \quad g \in C_\#(\mathbf{Z}, \mathbf{Z}).$$

**Proof.** The implication to the right follows easily from the definition of $W$. For the other implication we use the Lip-1 property of $g$: if $g(a) < a$, then $g(x) < x$ for all $x \geqslant a$, which implies that $W(g) < a$. □

**Proof of Theorem 15.5.5.** We shall prove that $f \preccurlyeq g$ implies $W(f) \preccurlyeq W(g)$. Fix two functions $f$ and $g$ in $C_{\#}(\mathbf{Z}, \mathbf{Z})$ satisfying $f \preccurlyeq g$. We consider the two cases $W(f)$ odd, $W(f)$ even.

*Case 1.* Assume that $W(f)$ is odd. Without loss of generality we may assume that $W(f) = 1$. Then $f(1) \geqslant 1$ and $f(2) \leqslant 1$, since $f(2) \geqslant 2$ would imply $W(f) \geqslant 2$. We conclude that $f(1) = 1$ and that $f(2) = 1$ or $f(2) = 0$ (cf. (15.3.2)). We shall prove that $W(g)$ is 0, 1, or 2. We write the inequality $0 \leqslant W(g) \leqslant 2$ as $0 \leqslant W(g) < 3$ and then translate it using Lemma 15.5.6 as $0 \leqslant g(0)$ and $g(3) < 3$. We know that $g(1) \succcurlyeq f(1) = 1$, which implies that $g(1) \geqslant 0$ and that $g(0) \geqslant 0$. Hence it only remains to be proved that $g(3) < 3$.

*Case 1.1.* Assume that $f(2) = 1$. Then $f(1) = f(2) = f(3) = 1$, so that $g(3) \succcurlyeq f(3) = 1$ and $g(3) \leqslant 2 < 3$.

*Case 1.2.* Assume that $f(2) = 0$. Then $g(2) \succcurlyeq f(2) = 0$, which implies that $g(2) = 0$; by the Lipschitz property $g(3) \leqslant 1 < 3$.

*Case 2.* If $W(f)$ is even, we may assume that $W(f) = 0$ without loss of generality. We shall then prove that $W(g) = 0$. We write this as $0 \leqslant W(g) < 1$ and translate it using Lemma 15.5.6 as $0 \leqslant g(0)$ and $g(1) < 1$. From $W(f) = 0$ it follows that $f(0) = 0$ and that $f(1) = 0$ or $f(1) = -1$. Since $0 = f(0) \preccurlyeq g(0)$, it follows that $g(0) = 0$. It remains to be proved that $g(1) < 1$.

*Case 2.1.* Assume that $f(1) = -1$. Then $g(1) \succcurlyeq f(1) = -1$, which implies that $g(1) \leqslant 0 < 1$.

*Case 2.2.* Assume finally that $f(1) = 0$. Then $g(1) \succcurlyeq f(1) = 0$, so $g(1) = 0 < 1$. □

**Corollary 15.5.7.** *Let $f\colon I \times J \to I$ be a continuous function defined in a rectangle $I \times J$, where $I$ and $J$ are Khalimsky intervals, $I$ being bounded. Then the function $w$ defined by*

$$w(y) = \sup_{x \in I}(x;\ f(x, y) \geqslant x), \qquad y \in J, \tag{15.5.3}$$

*is continuous in $J$. The function $w$ satisfies $f(w(y), y) = w(y)$.*

**Proof.** The function $w$ is given by $w(y) = W(x \mapsto f(x, y))$. It depends continuously on the function $x \mapsto f(x, y)$, which in turn is a continuous function of $y$. □

**Example 15.5.8.** It is natural to ask if the function

$$w(y) = \sup_{x \in I} \big(x;\ f(x,y) \geqslant g(x,y)\big), \qquad y \in J,$$

is continuous for more general choices of functions $f$ and $g$. That this is not always so is shown by the example $I = J = [-a, a]$, $f(x,y) = y$, $g(x,y) = x \wedge 0$ for $(x,y) \in I \times J$, thus replacing $x$ by $x \wedge 0$ in (15.5.3) above. Then $w(-1) = -1$ and $w(0) = a$. Thus $w$ is discontinuous as soon as $a \geqslant 1$. □

**Proof of Theorem 15.5.4.** A mapping $f$ has two components, $f_1$ and $f_2$. As in (15.5.3) we let $w(y)$ denote the largest fixed point of the partial mapping $x \mapsto f_1(x,y)$ for a fixed $y$, $w(y) = W(x \mapsto f_1(x,y))$. So $f_1(w(y),y) = w(y)$ for all $y \in J$. We then form the composition $k(y) = f_2(w(y),y)$. As a composition of continuous mappings it is continuous, and by the one-dimensional theorem it possesses a fixed point $q$, $k(q) = q$. Collecting what we have, we see that $f_1(w(q),q) = w(q)$ and that $k(q) = f_2(w(q),q) = q$, which means that we have proved that $f(w(q),q) = (f_1(w(q),q), f_2(w(q),q)) = (w(q),q)$, thus that $(w(q),q)$ is a fixed point. □

**Example 15.5.9.** The Khalimsky square $\{0,1\}^2 \subset \mathbf{Z}^2$. There are $m^m = 4^4 = 256$ selfmappings of $\{0,1\}^2$, of which $(m-1)^m = 3^4 = 81$ do not have fixed points. The proportion of mappings without a fixed point is $81/256 \approx 0.3164$, as expected slightly lower than the limit $1/e \approx 0.3679$. The remaining $256 - 81 = 175$ have a fixed point. Of the 16 mappings $\{0,1\}^2 \to \{0,1\}$, 6 are continuous. There are therefore $6^2 = 36$ continuous mappings $\{0,1\}^2 \to \{0,1\}^2$, and we know already that they all have fixed points.

The table of different kinds of mappings in a Khalimsky square therefore looks as in Table 15.4 on page 331 ($m = 4$, $c = 6^2$). In this simple case we can of course see directly that all continuous mappings have a fixed point. Indeed, of the 6 continuous mappings $\{0,1\}^2 \to \{0,1\}$, five map $(0,0)$ to 0; the remaining one is the constant 1. Therefore, of the 36 continuous mappings $\{0,1\}^2 \to \{0,1\}^2$, $(0,0)$ is a fixed point except when one of the components is the constant 1. Thus they all have a fixed point. □

It is easy to generalize the implicit-function theorem and the fixed-point theorem to somewhat more general sets. We formulate an example of the latter:

Table 15.4. Continuous and discontinuous mappings and mappings with and without the fixed-point property in a Khalimsky square $\{a, a+1\}^2$.

| | *Continuous* | *Discontinuous* | *Sum* |
|---|---|---|---|
| *Fixed point* | 36 | 139 | 175 |
| *No fixed point* | 0 | 81 | 81 |
| *Sum* | 36 | 220 | 256 |

**Theorem 15.5.10.** *Let $X$ be a subset of $\mathbf{Z}^2$ defined as*

$$X = \{(x, y) \in \mathbf{Z} \times J;\ \beta(y) \leqslant x \leqslant \gamma(y)\},$$

*where $J$ is a bounded Khalimsky interval, and $\beta$ and $\gamma$ two continuous functions defined on $J$. Assume that $\beta(y) < \gamma(y)$ for all $y \in J$. Then $X$ has the fixed-point property.*

**Proof.** Take an interval $I = [a, b]_{\mathbf{Z}}$ which is so large that $\beta(y), \gamma(y) \in I$ for all $y \in J$. Then extend $f\colon X \to X$ to a mapping $g\colon I \times J \to I \times J$ by defining $g(x, y) = f(\beta(y), y)$ when $a \leqslant x \leqslant \beta(y) - 1$ and $g(x, y) = f(\gamma(y), y)$ when $\gamma(y) + 1 \leqslant x \leqslant b$. Then $g$ is continuous. It must have a fixed point in $I \times J$ according to Theorem 15.5.5. However, since $g$ maps $I \times J$ into $X$, this fixed point must actually lie in $X$ and be a fixed point of $f$. □

The result on separate continuity (Theorem 15.3.3) makes it easy to go up in dimension.

## 15.6 Jordan curve theorems

Efim Khalimsky's digital Jordan curve theorem states that the complement of a Jordan curve in the digital plane equipped with the Khalimsky topology has exactly two components. We present a short proof of this theorem using induction on the Euclidean length of the curve. We also prove that the theorem holds with another topology on the digital plane but then only for a restricted class of Jordan curves.

The classical Jordan curve theorem says that the complement of a Jordan curve in the Euclidean plane $\mathbf{R}^2$ consists of exactly two components. Efim Khalimsky's digital Jordan curve theorem states the same thing for the digital plane $\mathbf{Z}^2$. Of course we must use a suitable definition of the concept of digital Jordan curve, as well as a suitable topology on $\mathbf{Z}^2$. In this case $\mathbf{Z}^2$ is given the Cartesian product topology of two copies of the digital line $\mathbf{Z}$ equipped with the Khalimsky topology.

A proof of Khalimsky's theorem was published in 1990 by Khalimsky et al. (1990). They refer to earlier proofs by Khalimsky, E. D. Halimskiĭ (1970, 1977). Our first purpose here is to present a short proof.

The idea of the proof is simple. For the smallest Jordan curves (having four or eight points) the conclusion of the theorem can be proved by inspection. For any other Jordan curve $J$ we construct another Jordan curve $J'$ which has shorter Euclidean length and such that the complement of $J'$ has as many components as the complement of $J$. Since the possible Euclidean lengths form a discrete set, this procedure will lead to one of the smallest Jordan curves, for which the conclusion is already established. The construction of $J'$ can intuitively be described as follows: attack $J$ where its curvature is maximal and shorten it there; it cannot offer resistance from within.

We then consider a topology on $\mathbf{Z}^2$ which is not a product topology. In contrast to the Khalimsky topology it has the property that every point is either open or closed—there are no mixed points as in the Khalimsky plane. We prove that the Jordan curve theorem holds for this topology for a restricted class of Jordan curves.

Finally, we introduce a topology on the hexagonal grid and observe that the Jordan curve theorem does not hold for this topology.

It should be mentioned that there are similar results for curves in $\mathbf{Z}^2$ without any topology, see (Capraro 2013).

### 15.6.1 *Adjacency*

If $X$ and $Y$ are topological spaces and $f\colon X \to Y$ a continuous mapping, then the image $f(A)$ of a connected subset $A$ of $X$ is connected. We may apply this result to a situation where we define a topology on a set $Y$ using a mapping $f\colon X \to Y$ from a topological space $X$; see Corollary 14.6.3.

Two points $x$ and $y$ in a topological space $Y$ are said to be ***adjacent*** if $x \neq y$ and $\{x, y\}$ is connected. We note that $\{x, y\}$ is connected if and only if either $x \in N(y)$ or $y \in N(x)$. We shall say that two points $x, z$ are ***second adjacent*** if $x \neq z$; $x$ and $z$ are not adjacent; and there exists a third point $y \in Y$ such that $x$ and $y$ are adjacent, and such that $y$ and $z$ are adjacent.

### 15.6.2 *Paths, arcs and Jordan curves in the Khalimsky plane*

Khalimsky et al. (1990: 3.1) used the following definitions of paths and arcs in the Khalimsky plane. We just extend them here to any topological space. We modify slightly their definition of a Jordan curve (1990:5.1).

**Definition 15.6.1.** Let $Y$ be any topological space. A ***path*** in $Y$ is a continuous image of a Khalimsky interval. An ***arc*** is a homeomorphic image of a Khalimsky interval. A ***Jordan curve*** in $Y$ is a homeomorphic image of a Khalimsky circle. □

Some Jordan curves are too narrow. We impose a condition on them to make their interior fatter:

**Definition 15.6.2.** Let $J$ be a Jordan curve in a topological space $Y$. We shall say that $J$ is a ***strict Jordan curve*** if every point in $J$ is second adjacent to exactly two points in $J$. □

We note that if $x, z \in J$ are second adjacent, then the intermediary $y$ required by the definition need not belong to $J$. Thus the concept of strict Jordan curve is not intrinsic.

A three-set $\{x, y, z\}$ such that all three points are adjacent to each other can be a path but never an arc. This follows from the fact that in a Khalimsky interval $[a, b]_{\mathbf{Z}}$, the endpoints $a$ and $b$ are not adjacent unless $b = a + 1$. Let us say that a three-set $\{x, y, z\}$ in a topological space is a ***forbidden triangle*** if all points are adjacent to each other. The absence of forbidden triangles is therefore a necessary condition for arcs and consequently for Jordan curves, and it is often easy to check.

Different topologies may induce the same adjacency structure. However, when the adjacency structure of a finite set is that of a Khalimsky circle, the topology of the set must also be that of a Khalimsky circle. More precisely we have the following result.

**Theorem 15.6.3.** *Given a subset $J$ of a topological space $Y$, the following conditions are equivalent.*

(A) *$J$ is a Jordan curve.*
(B) *$J$ has at least four points, and for every $a \in J$, $J \smallsetminus \{a\}$ is homeomorphic to a Khalimsky interval.*
(C) *$J$ is finite, connected, with cardinality at least 4, and each of its elements has exactly two adjacent points.*

(D) *$J$ has the adjacency structure of a Khalimsky circle, i.e., $J = \{x_1, x_2, \ldots, x_m\}$ for some $m \geqslant 4$ and for each $j = 1, \ldots, m$, the two points $x_{j-1}$ and $x_{j+1}$ but no other points are adjacent to $x_j$. (Here we count indices modulo $m$.)*

**Proof.** If (A) holds, then for every $a \in J$, $J \smallsetminus \{a\}$ is homeomorphic to a Khalimsky circle minus one point, thus to a Khalimsky interval. Conversely, suppose that (B) holds and consider $J \smallsetminus \{a\}$ and $J \smallsetminus \{b\}$ for two points $a, b \in J$ which are not adjacent. Then we have homeomorphisms of $J \smallsetminus \{a\}$ and $J \smallsetminus \{b\}$ into a Khalimsky circle $\mathbf{Z}_m$. We can modify them by rotating the circle so that the two mappings agree on $J \smallsetminus \{a, b\}$. Then they define a local homeomorphism of $J$ onto $\mathbf{Z}_m$, thus a homeomorphism; we have proved (A).

It is clear that (A) implies (C) and (D).

Suppose now that (C) holds. Then call an arbitrary point $x_1$ and one of its adjacent points $x_2$ and then go on, always choosing $x_{j+1}$ after $x_j$ so that $x_{j+1}$ is adjacent to $x_j$ but not equal to any of the already chosen $x_1, \ldots, x_{j-1}$. After a while we must arrive at a situation where there are no new points left, i.e., we arrive at $x_m$ and the two points adjacent to $x_m$ are $x_{m-1}$ and a point which has already been chosen, say $x_k$. A priori $k$ may be any of $1, 2, \ldots, m-2$, but in fact the only possibility is $k = 1$: any other choice would mean that $x_k$ has three adjacent points contrary to the assumption. Thus we have proved (D). Conversely, (D) obviously implies (C) since (D) is just a more detailed version of (C).

It remains to be seen that (D) implies (A). First of all it is clear that, assuming (D), $N(x)$ can never have more than three elements—a fourth element would mean that $x$ has at least three adjacent points. So $N(x_j) \subset \{x_{j-1}, x_j, x_{j+1}\}$. Consider three points $x_{j-1}, x_j, x_{j+1}$. Then either $x_{j-1} \in N(x_j)$ or $x_j \in N(x_{j-1})$, and $x_j \in N(x_{j+1})$ or $x_{j+1} \in N(x_j)$. However, these alternatives cannot be chosen at will, for if $x_{j-1} \in N(x_j)$ and $x_j \in N(x_{j+1})$, then $x_{j-1} \in N(x_{j+1})$ so that $x_{j-1}$ has three adjacent points, viz. $x_{j-2}$, $x_j$, and $x_{j+1}$, contrary to the hypothesis. (Incidentally, this proves already that $Y$ is a Kolmogorov space.) Consider now the case $x_{j-1} \in N(x_j)$. Then $x_{j+1} \in N(x_j)$, so that $N(x_j) \supset \{x_{j-1}, x_j, x_{j+1}\}$. On the other hand we know already that $N(x_j)$ has at most three elements; we conclude that $N(x_j) = \{x_{j-1}, x_j, x_{j+1}\}$. By the same argument, $N(x_{j+2}) = \{x_{j+1}, x_{j+2}, x_{j+3}\}$. Therefore $N(x_{j+1}) = \{x_{j+1}\}$, and we have proved that $Y$ is a Khalimsky circle where points with indices of the same parity as $j$ have neighborhoods consisting of three points, and points with indices of the other parity are open. The other possibility, viz. that $x_j \in N(x_{j-1})$, can be reduced to the former by just shifting the indices one step. □

It follows from property (C) that Jordan curves can never be contained in each other, that is, if $J$ and $K$ are Jordan curves and $J \subset K$, then $J = K$.

A point on a Jordan curve $J$ consisting of at least six points has at least two second adjacent points; with the order introduced in property (D), $x_{j-2}$ and $x_{j+2}$ are second adjacent to $x_j$ and $x_{j-2} \neq x_{j+2}$ when $m > 4$. Then $x_{j\pm 1}$ serve as intermediaries, but there may also exist other intermediaries. When a Jordan curve is not strict, then some point, say $x_j$, has at least one second adjacent point in addition to $x_{j-2}$ and $x_{j+2}$, say $x_k$. Then an intermediary $b$ such that $x_j$ and $b$ as well as $b$ and $x_k$ are adjacent cannot belong to $J$.

Suppose now that $Y$ is a metric space with metric $d$. Since every arc $\Gamma$ is homeomorphic either to $[0, m-1]_{\mathbf{Z}}$ or to $[1, m]_{\mathbf{Z}}$ for some $m$, it can be indexed as $\{x_1, \ldots, x_m\}$, where the indices are uniquely determined except for inversion. We may define its length as

$$\operatorname{length}(\Gamma) = \sum_{1}^{m-1} d(x_{j+1}, x_j).$$

Similarly, a Jordan curve can be indexed as $\{x_1, \ldots, x_m\}$, where the indices are uniquely determined up to inversion and circular permutations, and its length can defined as

$$\operatorname{length}(J) = \sum_{1}^{m} d(x_{j+1}, x_j),$$

where we count the indices modulo $m$.

### 15.6.3 *Khalimsky's digital Jordan curve theorem*

The Khalimsky topology of the digital plane is the Cartesian product topology of two copies of the Khalimsky line $\mathbf{Z}$. A singleton set $\{x\} = \{(x_1, x_2)\}$ in the product $\mathbf{Z}^2 = \mathbf{Z} \times \mathbf{Z}$ is closed if and only if both $\{x_1\}$ and $\{x_2\}$ are closed, thus if and only if both $x_1$ and $x_2$ are even; similarly $\{x\}$ is open if and only if both coordinates are odd.

Perhaps the quickest way to describe the Khalimsky topology $\tau_\infty$ on $\mathbf{Z}^2$ is this: We first declare the nine-set

$$U_\infty = \{x \in \mathbf{Z}^2;\ \|x\|_\infty \leqslant 1\} = \{(0,0), \pm(1,0), \pm(1,1), \pm(0,1), \pm(-1,1)\}$$

to be open, as well as all translates $U_\infty + p$ with $p \in 2\mathbf{Z} \times 2\mathbf{Z}$. Then all intersections of such translates are open, as well as all unions of the sets so obtained. As a consequence, $\{(1,-1),(1,0),(1,1)\}$, the intersection of $U_\infty$ and $U_\infty + (2,0)$, and $\{(1,1)\}$, the intersection of $U_\infty$ and $U_\infty + (2,2)$, are open sets, and $\{(0,0)\}$ is a closed set. The sets $\{(1,0)\}$ and $\{(0,1)\}$ are mixed, i.e., neither open nor closed.

**Theorem 15.6.4.** *Given a subset $J$ of $\mathbf{Z}^2$ equipped with the Khalimsky topology, the conditions* (A), (B), (C), *and* (D) *of Theorem 4.3 are all equivalent to the following.*

(E) $J = \{x^{(1)}, x^{(2)}, \dots, x^{(m)}\}$ *for some* $m \geqslant 4$ *and for all* $j$, $x^{(j-1)}$ *and* $x^{(j+1)}$ *and no other points are adjacent to* $x^{(j)}$*; moreover each path* $\{x^{(j-1)}, x^{(j)}, x^{(j+1)}\}$ *turns at* $x^{(j)}$ *by* $45°$ *or* $90°$ *or not at all if* $x^{(j)}$ *is a pure point, and goes straight ahead if* $x^{(j)}$ *is mixed.*

**Proof.** If (D) holds, we see that $J$ cannot turn at a mixed point and cannot turn $135°$ at a pure point—otherwise we would have a forbidden triangle. So (D) implies (E). Conversely, (E) is just a more precise version of (D), so (E) implies (D). □

In this subsection we shall measure the lengths of Jordan curves using the Euclidean metric, $d(x,y) = \|x-y\|_2$. As we will see, it is not possible to use $\|\cdot\|_1$ or $\|\cdot\|_\infty$ in the proof of the Jordan curve theorem given here.

The smallest possible Jordan curve is the four-set

$$J_4 = \big\{x \in \mathbf{Z}^2;\ \|x-(1,0)\|_1 = 1\big\} = \{(0,0),(1,-1),(2,0),(1,1)\}.$$

We add all translates of $J_4$ by a vector $p \in \mathbf{Z}^2$ with $p_1 + p_2$ even and call these the *Jordan curves of type* $J_4$.

There is also a Jordan curve having eight points,

$$J_8 = \big\{x \in \mathbf{Z}^2;\ \|x\|_\infty = 1\big\} = U_\infty \smallsetminus \{(0,0)\}. \tag{15.6.1}$$

This curve and all its translates by a vector $p \in \mathbf{Z}^2$ with $p_1 + p_2$ even we call the ***Jordan curves of type*** $J_8$.

Let us agree to call the three-set

$$T = \{(1,1),(0,0),(1,-1)\} \tag{15.6.2}$$

and rotations of $T$ by $90°$, $180°$ and $270°$, as well as all translates of these sets by vectors $p \in \mathbf{Z}^2$ with $p_1 + p_2$ even, a ***removable triangle***. It turns out that elimination of removable triangles is a convenient way to reduce Jordan curves, as shown by the following lemma.

**Lemma 15.6.5.** *Let $J$ be a Jordan curve in the Khalimsky plane and assume that $J$ contains the three-set $T$ defined by* (15.6.2). *Define*

$$J' = (J \setminus \{(0,0)\}) \cup \{(1,0)\}.$$

*Then either $J = J_4$ or else $J'$ is a Jordan curve such that $\complement J'$ and $\complement J$ have the same number of components, and with*

$$\operatorname{length}(J') = \operatorname{length}(J) - 2\sqrt{2} + 2.$$

**Proof.** Assume first that $(2,0) \in J$, thus that $J \supset J_4$. Then necessarily $J = J_4$.

Next we suppose that $(2,0) \notin J$. Then $J'$ is a Jordan curve: $J'$ is a set where the new point $(1,0)$ plays exactly the same role topologically as the old point $(0,0)$ in $J$. Thus also $J'$ is homeomorphic to a Khalimsky circle.

Finally we must check that the number of components in $\complement J'$ is the same as that of $\complement J'$. Indeed, $(1,0)$ and $(2,0)$ belong to the same component of $\complement J$, and $(0,0)$ and $(-1,0)$ belong to the same component of $\complement J'$. □

**Theorem 15.6.6** (Khalimsky's Jordan curve theorem). *Let us equip the digital plane $\mathbf{Z}^2$ with the Khalimsky topology $\tau_\infty$. Then for any Jordan curve $J$ in $\mathbf{Z}^2$, the complement $\complement J = \mathbf{Z}^2 \setminus J$ consists of exactly two components.*

**Proof.** The complement of $J_4$ consists of $A = \{(1,0)\}$ and the set $B$ of all points $x$ with $|x_1 - 1| + |x_2| > 1$. It is obvious that these two sets are connected. Moreover, $A$ is closed and open in $\complement J_4$, so it is a component. Therefore, also $B = \complement J_4 \setminus A$ is closed and open in $\complement J_4$ and also a component. The proof for $J_8$ is similar.

Thus we know that the conclusion of the theorem holds for Jordan curves of types $J_4$ and $J_8$.

Next we shall prove that if $J$ is not of the kind already treated, then there exists a Jordan curve $J'$ of strictly smaller Euclidean length such that $\complement J$ and $\complement J'$ have the same number of components. After a finite number of steps, we must arrive at a situation where the hypothesis is no longer satisfied, which means that we have a Jordan curve of type $J_4$ or $J_8$, for which the complement has two components as we already proved.

The construction of $J'$ is as follows. First we may assume, after having applied the procedure described in Lemma 15.6.5 a finite number of times, that $J$ contains no removable triangles. Define

$$a_2 = \inf(x_2;\ x \in J).$$

Thus $x_2 \geqslant a_2$ for all points $x \in J$ with equality for at least one $x$. Consider a horizontal interval

$$H = \{(b_1, a_2)\} + \{(0,0), (1,0), \ldots, (p,0)\},$$

where $b_1$ is chosen so that $H$ consists of points in $J$ with ordinate equal to $a_2$ and is maximal with respect to inclusion. The maximality implies that the two points $(b_1 - 1, a_2)$ and $(b_1 + p + 1, a_2)$ do not belong to $J$. Then we see that $p$ must be an even number, but we cannot have $p = 0$, since that would imply that $J$ contained a removable triangle, contrary to the assumption. Thus $H$ contains at least three points. Moreover, at the endpoints of $H$, $J$ must turn upwards. Indeed, since $(b_1 - 1, a_2)$ does not belong to $J$, exactly one of the points $(b_1 - 1, a_2 + 1)$, $(b_1, a_2 + 1)$ belongs to $J$; when we go left from $(b_1, a_2)$, the curve must turn upwards by either 45° or 90°; it cannot turn downwards. Similarly, the curve turns upwards by 45° or 90° when we go right from the last point in $H$, viz. from $(b_1 + p, a_2)$.

We now consider the set $\mathscr{I}$ of all maximal horizontal intervals $I$ in $J$ such that $J$ turns upwards at the endpoints of $I$. The previous argument served just to prove that there exists such an interval. Now there exists an interval $K \in \mathscr{I}$ of smallest length,

$$K = \{y\} + \{(0,0), (1,0), \ldots, (q,0)\},$$

containing $q + 1$ points for some even number $q \geqslant 2$. We shall assume that $K$ is of smallest length also among all intervals that can be obtained from the intervals in $\mathscr{I}$ by rotating them 90°, 180° or 270°.

To simplify the notation we may assume (after a translation if necessary) that $y = (0,0)$, so that

$$K = \{(0,0), (1,0), \ldots, (q,0)\} = [(0,0), (q,0)] \cap \mathbf{Z}^2.$$

*Case 1.* $J$ turns upwards by 45° at both ends of $K$. This means that $(-1, 1)$ and $(q + 1, 1)$ both belong to $J$. In this case, we define

$$J' = (J \smallsetminus K) \cup (K + (0,1)).$$

This operation shortens the Euclidean length by $2\sqrt{2} - 2$ (but it does not shorten the $l^\infty$ length). We note that the interval $K + (0,1)$ is disjoint from $J$; otherwise some point in $K$ would have three adjacent points. Moreover $K + (0,2)$ must be disjoint from $J$. Indeed, if $(K + (0,2)) \cap J$ were nonempty, then either $J$ would contain a removable triangle (contrary to our assumption) or there would exist a subinterval $K'$ of $K + (0,2)$ contained in $J$ and such that $J$ turns upwards at its endpoints; thus $K' \in \mathscr{I}$. This subinterval must have fewer than $q + 1$ points, since

$(0,2)$ and $(q,2)$ cannot belong to $J$—otherwise there would be a removable triangle in $J$. A shorter interval is impossible, since $K$ is by assumption an interval in $\mathscr{I}$ of shortest length. One checks that $J'$ is a Jordan curve. Indeed, the points of $K+(0,1)$ play the same role topologically in $J'$ as do the points of $K$ in $J$. The number of components in the complement of $J'$ is the same as for $J$.

*Case 2.* $J$ turns upwards by $90°$ at one end of $K$. Assume that $(0,1) \in J$, the case $(q,1) \in J$ being symmetric. Then also $(0,2) \in J$. We consider the subcases 2.1 and 2.2.

*Case 2.1.* $(2,2) \notin J$. We cut off a corner, i.e., we remove $(0,1)$, $(0,0)$, $(1,0)$, and add $(1,1)$. This operation shortens the Euclidean length by $4-2\sqrt{2}$ (but $J'$ has the same $l^1$-length as $J$). Since $(1,1)$ and $(2,2)$ belong to the same component of $\complement J$, and $(0,1)$, $(0,0)$, $(1,0)$, and $(-1,0)$ belong to the same component of $\complement J'$, the number of components in the respective complements are the same.

*Case 2.2.* $(2,2) \in J$. We consider four subcases, 2.2.1.1, 2.2.1.2, 2.2.2.1 and 2.2.2.2.

*Case 2.2.1.1.* $(2,1) \in J$, $(1,2) \in J$. Then $J$ contains a Jordan curve of type $J_8$, more precisely $J \supset (1,1) + J_8$. So $J$ must be equal to that curve.

*Case 2.2.1.2.* $(2,1) \in J$, $(1,2) \notin J$. Remove the five points $(0,1)$, $(0,0)$, $(1,0)$, $(2,0)$, $(2,1)$, and add $(1,2)$. Thus $J'$ is shorter by 4. We can check that $J'$ has all desired properties.

*Case 2.2.2.1.* $(2,1) \notin J$, $(1,2) \in J$. Turn $90°$ to reduce to case 2.2.1.2.

*Case 2.2.2.2.* $(2,1) \notin J$, $(1,2) \notin J$. This case cannot occur since $q$ is smallest possible. To see this, define $I'$ as the set of all points $(2,2), (3,2), \dots, (q',2) \in J$ with $q'$ as large as possible. If $J$ turns upwards at $(q',2)$, then $I'$ belongs to $\mathscr{I}$ with $q' < q$, which contradicts the definition of $K$ and $q$. If on the other hand $J$ turns downwards at $(q',2)$, then there exists a vertical interval consisting of three points, which becomes an interval in $\mathscr{I}$ if we turn it $90°$, thus again contradicting the definition of $\mathscr{I}$, and so bringing the proof to an end. □

### 15.6.4 *The Jordan curve theorem for the Wyse–Marcus topology*

We define a topology $\tau_1$ on $\mathbf{Z}^2$ by first declaring the five-set

$$U_1 = \{x \in \mathbf{Z}^2;\ \|x\|_1 \leqslant 1\} = \{(0,0), \pm(1,0), \pm(0,1)\}$$

to be open, then all translates $U+p$ with $p \in \mathbf{Z}^2$, $p_1+p_2 \in 2\mathbf{Z}$ to be open, as well as all unions of such translates. This implies that $\{(1,0)\}$ is open, and that the origin is closed. In fact, all points $x \in \mathbf{Z}^2$ with $x_1+x_2 \in 2\mathbf{Z}$ are closed, and all points with $x_1+x_2 \notin 2\mathbf{Z}$ are open; there are no mixed points. This topology was described by Frank Oliver Wyse (Wyse, Marcus et al. 1970) in response to a question posed by Dan Marcus. See also (Rosenfeld 1979:624).

The topology $\tau_1$ is not a product topology of topologies on $\mathbf{Z}$. It cannot be described as a quotient of the topology in $\mathbf{R}^2$; see Exercise 15.8.

The four-set

$$J_4 = \{(0,0),(1,0),(1,1),(0,1)\}$$

is a Jordan curve for $\tau_1$. However, it is not strict, for a point in $J_4$ has only one second adjacent point. Its complement is connected, so the Jordan curve theorem does not hold. The set $J_8$ defined by (15.6.1) is a Jordan curve and its complement has exactly two components. Also $J_8$ is not strict, because the point $(1,0)$ has three second adjacent points, viz. $(0,1)$, $(0,-1)$ and $(-1,0)$ ($(0,0)$ is adjacent to $(1,0)$ and $(-1,0)$).

Another example is the twelve-set

$$J_{12} = \{(0,0),(1,0),(2,0),(2,1),(3,1),(3,2),$$
$$(3,3),(2,3),(1,3),(1,2),(0,2),(0,1)\}.$$

It is a Jordan curve, not strict, and its complement has three components, viz. an infinite component and the two singleton sets $\{(1,1)\}$ and $\{(2,2)\}$.

**Theorem 15.6.7.** *Let $\mathbf{Z}^2$ be equipped with the topology $\tau_1$ just defined. Then the complement of every strict Jordan curve has exactly two components.*

**Proof.** For the proof we shall use the fact that $\mathbf{Z}^2$ equipped with the topology $\tau_1$ is homeomorphic to the subspace of all pure points in the Khalimsky plane. This fact was used also by Kong et al. (1991:915).

Let $X$ be the digital plane $\mathbf{Z}^2$ with the topology $\tau_1$, and $Y$ the Khalimsky plane ($\mathbf{Z}^2$ with the topology $\tau_\infty$). Consider the mapping $\varphi\colon X \to Y$ defined by $\varphi(x) = (x_1 - x_2, x_1 + x_2)$. Its image $\varphi(X)$ is the set of all pure points in $Y$, and if we equip it with the topology induced by $Y$ it is homeomorphic to $X$. Moreover, the image of any Jordan curve $J$ in $X$ is a Jordan curve in $Y$. Therefore $Y \smallsetminus \varphi(J)$ has exactly two components by Theorem 15.6.6. We claim that $\varphi(X) \smallsetminus \varphi(J)$ has exactly two components. It is clear that this set has at least two components, so the problem is to prove that a component $A$ of $Y \smallsetminus \varphi(J)$ gives rise to a connected set $A \cap \varphi(X)$, i.e., that the pure points in $A$ form a connected set.

To this end, assume that $a, a' \in A \cap \varphi(X)$, and consider an arc

$$\{a = a^{(0)}, a^{(1)}, \ldots, a^{(s)} = a'\}$$

contained in $Y \smallsetminus \varphi(J)$. (Connectedness in $Y$ is the same as arcwise connectedness; cf. (Khalimsky et al. 1990: Theorem 3.2).) We shall prove that this arc can be replaced by another consisting only of pure points. So assume that $a^{(j)}$ is a mixed point. Then its predecessor $a^{(j-1)}$ and its successor $a^{(j+1)}$ are both pure points. Without loss of generality, we may assume that $a^{(j-1)} = (0,0)$, $a^{(j)} = (0,1)$, and $a^{(j+1)} = (0,2)$. We may then replace $a^{(j)}$ by one of the pure points $(-1,1)$, $(1,1)$, because both of them cannot belong to $\varphi(J)$. To see this, suppose that $(1,1), (-1,1) \in \varphi(J)$. Then $(-1,1)$ would be a second adjacent point to $(1,1)$, and this point has, by hypothesis, exactly two second adjacent points in $\varphi(J)$ (considering everything in the space $\varphi(X)$). However, none of them can be equal to $(-1,1)$, for the only possible intermediaries would then be $(0,0)$ and $(0,2)$, none of which belongs to $\varphi(J)$. (In a strict Jordan curve, one of the possible intermediaries to a second adjacent point must belong to the curve.) This contradiction shows that not both of $(1,1)$ and $(-1,1)$ can belong to $\varphi(J)$. Thus we may define $b = (-1,1)$ or $b = (1,1)$ so that $b \notin \varphi(J)$ and observe that

$$\{a^{(0)}, \ldots, a^{(j-1)}, b, a^{(j+1)}, \ldots, a^{(s)}\}$$

is an arc with a mixed point replaced by a pure point. After finitely many such replacements we obtain an arc connecting $a$ and $a'$ and consisting only of pure points. This shows that $\varphi(X) \smallsetminus \varphi(J)$ has at most as many components as $Y \smallsetminus \varphi(J)$; therefore exactly two components, and then the same is true of $X \smallsetminus J$. □

### 15.6.5 *Tessellation by hexagons*

Let us now take a look at the hexagonal grid; see Figure 1.5 on page 11. To define it, it is convenient to identify the Euclidean plane $\mathbf{R}^2$ with the complex plane $\mathbf{C}$ and use complex multiplication. We define $\omega = -\frac{1}{2} + \frac{1}{2}\sqrt{3}\,\mathrm{i}$. This is a third root of unity ($\omega^3 = 1$), and $\omega^2 = -1-\omega = -\frac{1}{2} - \frac{1}{2}\sqrt{3}\,\mathrm{i}$. The hexagonal grid $H$ is now the set of all points $x + \omega y \in \mathbf{C}$ with $x, y \in \mathbf{Z}$. So topologically it can be identified with $\mathbf{Z}^2$ ($x + \omega y$ can be identified with $x + \mathrm{i}y$), but for metrical purposes it is convenient not to do so.

We now introduce a topology $\tau_2$ on $H$ by first declaring

$$\begin{aligned} U_2 &= \{x + \omega y;\ x, y \in \mathbf{Z},\ \|x + \omega y\|_2 \leqslant 1\} \\ &= \{0,\ 1,\ 1 + \omega,\ \omega,\ -1,\ -1 - \omega,\ -\omega\} \end{aligned}$$

to be open, as well as all translates $U_2 + x + \omega y$ of this set with $x, y \in \mathbf{Z}$, $x + y \in 3\mathbf{Z}$. The set $U_2$ consists of seven points, the origin and the six points in $H$ with Euclidean distance 1 to the origin. Then all intersections of such open sets are also open. This implies that 1 is an open point and that the origin is a closed point. Every point is either open or closed; there are no mixed points. Actually one third of the points are closed (those of the form $x + \omega y$ with $x, y \in \mathbf{Z}$ and $x + y \in 3\mathbf{Z}$), and two thirds of them are open (those with $x + y \notin 3\mathbf{Z}$). This topology cannot be obtained as a product topology of some copies of $\mathbf{Z}$; nor can it be obtained as a quotient topology of the usual topology in $\mathbf{R}^2$; see Exercise 15.8.

The Jordan curve theorem does not hold for $H$ with this topology. Indeed there are many curves which are so narrow that their interior is a totally disconnected set.

Let us look at a few examples. The set

$$J_4 = \{0,\ 1,\ 2 + \omega,\ 1 + \omega\}$$

is a Jordan curve whose complement is connected. The set

$$J_8 = \{0,\ -\omega,\ 1 - \omega,\ 2 - \omega,\ 3,\ 3 + \omega,\ 2 + \omega,\ 1 + \omega\}$$

is a Jordan curve whose complement has three components, viz. $\{1\}$, $\{2\}$, and an infinite component. We can now easily go on:

$$J_{12} = \{0,\ -\omega,\ 1-\omega,\ 2-\omega,\ 3,\ 4,\ 5+\omega,\ 5+2\omega,\ 4+2\omega,\ 3+2\omega,\ 2+\omega,\ 1+\omega\}$$

is a Jordan curve and its complement has five components: an infinite component plus the four points $1$, $2$, $3 + \omega$, $4 + \omega$, which are open and closed in $\complement J_{12}$. Certainly this $J_{12}$ is not strict. I do not know whether the Jordan curve theorem holds for strict Jordan curves (see Definition 15.6.2).

## 15.7 Notes

For the topological constructions mentioned here, see publications by Kong & Rosenfeld (1989), Kong et al. (1991), Kronheimer (1992), and Kopperman et al. (1998). The geometry of digital spaces, including the Khalimsky topology, is discussed in the books by Gabor T. Herman (1998) and Vladimir Kovalevsky (2006, 2008). For the uniqueness of the Khalimsky topology in $\mathbf{Z}^n$, see (Kong 2003).

For the digital Jordan curve theorem, see (E. D. Halimskiĭ 1970, 1977) and (Khalimsky et al. 1990).

The proof presented here for the Khalimsky topology as well as for the Wyse–Marcus topology was published in my paper (2000); the negative result mentioned here for the hexagonal tessellation did not appear there.

There is also a Jordan theorem for surfaces in three dimensions; see (Malgouyres 1997) and (Šlapal 2020b).

For a popular survey of the Khalimsky Jordan curve theorem, see also my papers (2003, 2008b).

For more recent developments of topologies similar to the Khalimsky topology and related constructions, see, among others, publications by Erik Melin (2004, 2005a, 2005b, 2007, 2008a, 2008c, 2009), Atsushi Imiya (2011) and Josef Šlapal (2009, 2013, 2014). In his article (2009), Šlapal proves a digital Jordan curve theorem with respect to pretopologies on $\mathbf{Z}^2$, continued in (2014), where he studies nine interrelated pretopologies and Jordan curve theorems obtained for these. In (2013), he studies an Aleksandrov topology the quotient topologies of which include the Khalimsky topology and the Wyse–Marcus topology.

Of particular interest is a duality between cleistomorphisms and sets of paths studied by Josef Šlapal (2018, 2020a, 2021a). In his paper (2021b) he establishes a bijection between cleistomorphisms and certain preorders in a category.

Dimension is a fundamental concept in topology, but not easy to define in a discrete setting. For a definition in digital spaces, see, e.g., (Brimkov et al. 2006) and the discussion there.

Theorem 15.4.2 is due to Erik Melin (2005a: Theorem 4.5).

## 15.8 Exercises

15.1. How many selfmappings of an interval $I \subset \mathbf{Z}$ are of class Lip-1 and how many of these are continuous for the Khalimsky topology when
(a) $I$ has two elements?
(b) $I$ has three elements?
(c) $I$ has four elements?

15.2. Prove that there are six continuous mappings $\{0,1\}^2 \to \{0,1\}$ and hence that there are 36 continuous mappings $\{0,1\}^2 \to \{0,1\}^2$. Compare with mappings $\{0,1,2,3\} \to \{0,1\}$, which also have a domain with four elements.

15.3. In a topological space $X$, define $\kappa\colon \mathscr{P}(X) \to \mathscr{P}(X)$ by $\kappa(A) = \overline{A}$, the closure of $A$, and $\lambda\colon \mathscr{P}(X) \to \mathscr{C}(X)$, where $\mathscr{C}(X)$ is the family of all closed subsets of $X$, also by $\lambda(A) = \overline{A}$. (Thus $\kappa$ and $\lambda$ have the same images but different codomains.)

(a) Prove that $\lambda$ is a dilation while $\kappa$ is a dilation if and only if $X$ is a smallest neighborhood space.

(b) Determine the lower and upper inverses of $\kappa$ and $\lambda$ when $X = \mathbf{R}^n$ with the usual topology.

(c) Determine the lower and upper inverses of $\kappa$ and $\lambda$ when $X = \mathbf{Z}^n$ with the Khalimsky topology.

15.4. Describe explicitly the two mappings $A \mapsto \overline{A}$, $A \mapsto A^\circ$ for the Khalimsky topology in $\mathbf{Z}$.

15.5. Describe the upper and lower inverses of the mappings in the previous exercise. Also the second inverses. Compare with Exercise 14.12.

15.6. We have seen that the Khalimsky topology in $\mathbf{Z}^2$ is the product topology of those in two copies of $\mathbf{Z}$. Prove that this topology in $\mathbf{Z}^2$ can also be obtained as a quotient topology of the usual topology in $\mathbf{R}^2$.

15.7. Prove that the Wyse–Marcus topology $\tau_1$ in $\mathbf{Z}^2$ defined in Subsection 15.6.4 cannot be obtained as a quotient topology of the usual topology in $\mathbf{R}^2$.

The same task for the topology described in Subsection 15.6.5 for the hexagonal tessellation of the plane.

15.8. If you are a experienced topologist you can put on your shoes first and afterwards your socks. What are the consequences for the Khalimsky topology when you use this skill?

# Chapter 16

# Distance transformations

## 16.1 Defining distances

DISTANCE TRANSFORMS of digital images are a useful tool in image analysis and image processing for many applications regarding shape, in particular in industrial and medical applications. They have been around since the 1960s.

The distance transform of a set (or shape, or image) is a function on the image carrier. Outside the set, the value of the distance transform at a certain pixel is defined to be the distance from that pixel to the set. Inside the set, it is often defined as the distance to the complement, but we shall find it convenient to define it instead as minus the distance to the complement, for a very simple reason: the distance transform of a convex set in $\mathbf{R}^n$ is a convex function with this definition.

Distances can be measured in many ways as explained in Section 7.3. We shall define more distances on $\mathbf{Z}^n$ in Section 16.5.

## 16.2 Translation-invariant distances

Whenever $X$ is an abelian group it is of particular interest to use ***translation-invariant distances***, i.e., those which satisfy

$$d(x-a, y-a) = d(x,y) \text{ for all } a, x, y \in X.$$

We note that in any abelian group with a translation-invariant metric we have the relations

$$B_<(c_1,r_1) + B_<(c_2,r_2) \subset B_\leqslant(c_1,r_1) + B_<(c_2,r_2) \subset B_<(c_3,r_3);$$
$$B_\leqslant(c_1,r_1) + B_\leqslant(c_2,r_2) \subset B_\leqslant(c_3,r_3),$$

where $c_1 + c_2 = c_3$, $r_1 + r_2 = r_3$. In a vector space over $\mathbf{R}$, with $d$ defined by $d(x, y) = \|x - y\|$ using some norm $\|\cdot\|$, the inclusions here are actually equalities if $r_1, r_2 > 0$.

**Lemma 16.2.1.** *Any translation-invariant distance $d$ on an abelian group $G$ defines a function $d_0(x) = d(x, 0)$ on $X$ which is positive definite:*

(16.2.1) $$d_0(x) \geqslant 0 \text{ with equality precisely when } x = 0;$$

*and symmetric:*

(16.2.2) $$d_0(-x) = d_0(x) \qquad x \in X.$$

*Conversely, a function $d_0$ which satisfies (16.2.1) and (16.2.2) defines a translation-invariant function $d(x, y) = d_0(x - y)$, which is a metric if and only if $d_0$ is subadditive:*

(16.2.3) $$d_0(x + y) \leqslant d_0(x) + d_0(y) \text{ for all } x, y \in X.$$

**Proof.** The properties of being positive definite and symmetric are easy to treat. If $d$ satisfies the triangle inequality, we can write, using the triangle inequality and the translation invariance,

$$\begin{aligned} d_0(x + y) &= d(x + y, 0) \leqslant d(x + y, y) + d(y, 0) \\ &= d(x, 0) + d(y, 0) = d_0(x) + d_0(y). \end{aligned}$$

Conversely, if $d_0$ is subadditive,

$$d(x, z) = d_0(x - z) \leqslant d_0(x - y) + d_0(y - z) = d(x, y) + d(y, z),$$

proving the triangle inequality. □

Infimal convolution, denoted by $\sqcap$ and defined in Definition 3.4.1, can serve to give a characterization here:

**Lemma 16.2.2.** *Let $d$ be a translation-invariant distance on an abelian group $G$ and $d_0$ a function on $G$ related to $d$ as in Lemma 16.2.1. Then $d_0 \sqcap d_0 = d_0$. Conversely, if a function $g\colon G \to \mathbf{R}$ is nonnegative and zero only at the origin, is symmetric, and satisfies $g \sqcap g = g$, then $d(x, y) = g(x - y)$ is a metric.*

The proof is easy.

## 16.3 Distance transforms

**Definition 16.3.1.** In a metric space $X$ we define the ***distance transform*** $\mathbf{dt}_A$ of a subset $A$ of $X$ by

$$\mathbf{dt}_A(x) = -\inf_{y \notin A} d(x, y)$$

for $x \in A$, and

$$\mathbf{dt}_X(x) = \inf_{y \in A} d(x, y)$$

for $x \in X \smallsetminus A$. □

Note the symmetry: $\mathbf{dt}_{X \smallsetminus A} = -\mathbf{dt}_A$.

**Lemma 16.3.2.** *The distance transform satisfies*

$$\mathbf{dt}_A(x) = \begin{cases} -\sup\big(r;\ B_<(x,r) \subset A\big), & x \in A; \\ \sup\big(r;\ B_<(x,r) \subset \complement A\big), & x \in X \smallsetminus A. \end{cases} \tag{16.3.1}$$

**Proof.** If $x \notin A$, $y \in A$, and $B_<(x,r) \subset \complement A$, then $r \leqslant d(x,y)$. This shows that the supremum of all such $r$ cannot exceed the infimum of all $d(x,y)$ with $y \in A$. On the other hand, if $r_1$ is larger than the supremum of all $r$ such that $B_<(x,r)$ does not meet $A$, then there exists a point $z \in A$ such that $d(x,z) < r_1$, which shows that the infimum of $d(x,z)$ over all such $z$ is less than $r_1$, hence less than or equal to $\sup r$. This proves the result when $x \notin A$. The case when $x \in A$ is similar. □

Thus when $x$ does not belong to $A$, the value of the distance transform $\mathbf{dt}_A(x)$ is what we naturally understand by the distance from $x$ to $A$, but we complement this idea by defining also the distance transform inside $A$. It is then natural to take it negative there, so that $A$ is approximately the set where the transform is negative. As already remarked it is convenient that, in a normed space, the distance transform of a convex set is a convex function; see Corollary 16.8.3.

**Remark 16.3.3.** Distance transforms are actually lower inverses: the value $\mathbf{dt}_A(x)$ for a fixed $x \in X \smallsetminus A$ evaluated at $\complement A$ is the lower inverse $g_{[-1]}(\complement A)$ of the mapping $g\colon \mathbf{R} \to \mathscr{P}(X)$, defined by $g(r) = B_<(x,r)$. □

The ***distance transformation*** $A \mapsto \mathbf{dt}_A$ is decreasing in the sense that $\mathbf{dt}_A(x) \geqslant \mathbf{dt}_B(x)$ for all $x \in X$ if $A \subset B$. In the two extreme cases $A = \varnothing$ and $A = X$ we have $\mathbf{dt}_\varnothing = +\infty$ and $\mathbf{dt}_X = -\infty$. In all other cases $\mathbf{dt}_A$ is real-valued, $\mathbf{dt}_A\colon X \to \mathbf{R}$.

Every real-valued function can be written as the difference between two nonnegative functions: $f = f^+ - f^-$, where $f^+ = \max(f, 0)$ and $f^- = \max(-f, 0)$. In particular, $\mathbf{dt}_A = (\mathbf{dt}_A)^+ - (\mathbf{dt}_A)^-$. The function $(\mathbf{dt}_A)^-$ is sometimes called the ***quench function*** of $A$.

**Proposition 16.3.4.** *If $A$ is a subset of a metric space $X$ other than $\varnothing$ and $X$, then $(\mathbf{dt}_A)^+$ and $(\mathbf{dt}_A)^-$ have Lipschitz constant as defined by (7.3.5) equal to 1 with respect to $d$:*

$$(16.3.2) \qquad |(\mathbf{dt}_A)^+(x) - (\mathbf{dt}_A)^+(y)| \leqslant d(x, y), \qquad x, y \in X,$$

*and similarly for $(\mathbf{dt}_A)^-$. (In particular the restrictions $\mathbf{dt}_A\big|_A$ and $\mathbf{dt}_A\big|_{\complement A}$ have Lipschitz constant 1.)*

*As a consequence, $\mathbf{dt}_A$ has Lipschitz constant 2.*

*If $X$ is a vector space with distance $d(x, y) = \|x - y\|$ defined by a norm, then the Lipschitz constant is 1.*

Let us say for brevity that a function is ***Lip*-1** if it has Lipschitz constant 1, i.e., when it satisfies (16.3.2).

**Proof.** The restriction to $A$ of $\mathbf{dt}_A$ is the supremum of a family of Lip-1 functions $x \mapsto d(x, y)$; it is easy to prove that this operation preserves the Lipschitz constant. It follows that $|(\mathbf{dt}_A)^-(a) - (\mathbf{dt}_A)^-(b)| \leqslant d(a, b)$ if $a, b \in A$. If $a, b \notin A$ the function takes the value zero at both points.

Now take $a \in A$ and $b \in X \smallsetminus A$ and define $r = -\mathbf{dt}_A(a) \geqslant 0$ and $s = \mathbf{dt}_A(b) \geqslant 0$. Then the strict ball $B_<(a, r)$ is contained in $A$, and the strict ball $B_<(b, s)$ is contained in $\complement A$, so that $r, s \leqslant d(a, b)$. The two balls are disjoint. In general this only implies that $\max(r, s) \leqslant d(a, b)$, but in a normed vector space case the stronger inequality $s + r \leqslant d(a, b)$ follows, thus that $0 \leqslant \mathbf{dt}_A(b) - \mathbf{dt}_A(a) = s + r \leqslant d(a, b)$; proving that the Lipschitz constant is 1 in this case.

Returning to the general case, we note that, when $a \in A$ and $b \notin A$, we have $0 = -\mathbf{dt}_A(b)^- \leqslant \mathbf{dt}_A(a)^- - \mathbf{dt}_A(b)^- = r \leqslant d(a, b)$ and the Lipschitz property of $(\mathbf{dt}_A)^-$ is established. Passing to the complement, we obtain the result for $(\mathbf{dt}_A)^+ = (\mathbf{dt}_{X \smallsetminus A})^-$. The difference $\mathbf{dt}_A = (\mathbf{dt}_A)^+ - (\mathbf{dt}_A)^-$ thus has Lipschitz constant at most 2. □

The Lipschitz constant 2 in the proposition cannot be improved as can be seen from the simplest of examples:

**Example 16.3.5.** Take $X = \mathbf{Z}$ with the usual metric and $A = \{0\}$. Then $\mathbf{dt}_A(0) = -1$ and $\mathbf{dt}_A(1) = 1$, so that the Lipschitz constant is actually 2. □

However, in the distance transform there is a jump 2 only when we go from a point in $A$ to a point in $X \smallsetminus A$. This indicates that it might be possible to adjust the distance transform in $A$ by an additive constant so that the modified function is Lip-1. Define for $\varnothing \neq A \neq X$ the quantity

$$s(A) = \inf_{b \notin A} \inf_{a \in A} d(a, b) \in [0, +\infty[.$$

We have $\mathbf{dt}_A \leqslant -s(A)$ in $A$ and $\mathbf{dt}_A \geqslant s(A)$ in the complement of $A$, which means that we can lift the values of the distance transform in $A$ by the amount $s(A)$ and lower the distance transform in $X \smallsetminus A$ by $s(A)$. If $s(A)$ is positive, the modified transform, which satisfies $\mathbf{dt}_A + s(A) \leqslant 0$ in $A$ and $\mathbf{dt}_A - s(A) \geqslant 0$ in the complement, thus has a Lipschitz constant $\leqslant (2 - 2s(A)) \vee 1$, which might be 1.[1] In Example 16.3.5 we have $s(A) = 1$, so this simple procedure does work. But we can very well have $s(A) = 0$ so that no such modification is possible:

**Example 16.3.6.** Let

$$X = \{(x, 0);\ x \in \dot{\mathbf{N}}\} \cup \{(x, y) \in \dot{\mathbf{N}} \times \mathbf{R};\ y = 1/x\}$$

with the Euclidean distance and $A = \{(x, 0);\ x \in \dot{\mathbf{N}}\}$. Then the distance between $(a, 0)$ and $(a, 1/a)$ is $1/a$ for $a \in \dot{\mathbf{N}}$, so that $s(A) = 0$. And the distance between $(1, 0)$ and $(1, 1)$ is 1, so that the Lipschitz constant is 2 and cannot be improved by modifying with $s(A)$. □

A more sophisticated approach is therefore necessary, and this is what we pursue in the following theorem.

**Theorem 16.3.7.** *For a subset $A$ in a metric space $X$ we define two quantities*

$$\theta(A) = \sup \big(r + s - d(a, b)\big) \ \text{and}\ \Theta(A) = \inf \big(r + s + d(a, b)\big),$$

*where the supremum and infimum are taken over all points $a \in A$, $b \notin A$, and all balls $B_<(a, r)$, $B_<(b, s)$, $r, s > 0$, that are disjoint and maximal with this property. We define a modified distance transform, lifted in $A$ by an amount $\omega$, as*

$$\mathbf{dt}_A^\omega(x) = \begin{cases} \mathbf{dt}_A(x) + \omega, & x \in A; \\ \mathbf{dt}_A(x), & x \in X \smallsetminus A. \end{cases}$$

[1] We could even modify by larger amounts, but not without losing the property that the transform shall be $\leqslant 0$ in $A$ and $\geqslant 0$ in $\complement A$.

*Then if $A$ and $X \smallsetminus A$ are nonempty and $\theta(A) \leqslant \Theta(A)$, there exists a real number $\omega$ such that $\mathbf{dt}_A^\omega$ is Lip-1; we can take $\omega$ as any number satisfying $\theta(A) \leqslant \omega \leqslant \Theta(A)$.*

*Conversely, if the function $\mathbf{dt}_A^\omega$ is Lip-1 for some $\omega$, then $\theta(A) \leqslant \Theta(A)$.*

It is also possible to change the transform both in $A$ and $\complement A$: We can take $\mathbf{dt}'_A = \mathbf{dt}_A + \omega_0$ in $A$ and $\mathbf{dt}'_A = \mathbf{dt}_A - \omega_1$ in the complement provided that the total change $\omega = \omega_0 + \omega_1$ satisfies the conditions.

**Proof.** We have to prove that

$$|\mathbf{dt}_A^\omega(b) - \mathbf{dt}_A^\omega(a)| \leqslant d(a,b), \qquad a, b \in X. \tag{16.3.3}$$

The proof of Proposition 16.3.4 shows that the inequality holds if $a, b \in A$ as well as if $a, b \notin A$. The only question left is when $a \in A$ and $b \notin A$. Then we have

$$|\mathbf{dt}_A^\omega(b) - \mathbf{dt}_A^\omega(a)| = |s - (-r + \omega)| = |r + s - \omega|, \tag{16.3.4}$$

where $r$ and $s$ are defined as in the proof of Proposition 16.3.4. If $\theta(A) \leqslant \omega \leqslant \Theta(a)$ we know that for all $a, b, r, s$ under consideration we have $r + s - d(a,b) \leqslant \theta(A)$ and $r + s + d(a,b) \geqslant \Theta(A)$. With any $\omega$ betwen $\theta(A)$ and $\Theta(A)$ we thus get

$$r + s - d(a,b) \leqslant \omega \leqslant r + s + d(a,b),$$

which may be written as $|r+s-\omega| \leqslant d(a,b)$. With this inequality, (16.3.4) implies (16.3.3), which means that the modified distance transform is Lip-1.

Conversely, if (16.3.3) holds for all $a, b, r, s$ that we consider, then the inequality $|r + s - \omega| \leqslant d(a,b)$ must be true, which leads to the inequality $\theta(A) \leqslant \Theta(A)$. □

**Example 16.3.8.** In a vector space we have $\theta(A) = \Theta(A) = 0$ for every nonempty $A \neq X$. Hence $\omega = 0$ is the only choice; this is already clear from Proposition 16.3.4. □

**Example 16.3.9.** In $\mathbf{Z}^n$ with the $l^\infty$-metric, we have $\theta(A) = 1$, $\Theta(A) = 3$ for all $A \neq \emptyset, \mathbf{Z}^n$, so that any $\omega \in [1,3]$ will do. However the choice $\omega = 1$ seems to be the most natural: then the lifted distance function is nonpositive in $A$ and positive in its complement. □

**Example 16.3.10.** We defined in Section 7.4, for any subset $C$ of a metric space $X$, the separation function $\mathbf{sep}_C \colon C \to \mathbf{R}_!$ with the value at $c \in C$ equal to $\inf_x \big(d(x,c);\ x \neq c\big)$. Given two numbers $r_0$ and $r_1$, assume that there are two points $a, b \in X$ such that $d(a,b) \leqslant r_0$, and also two points $p, q \in X$ such that $d(p,q) \geqslant r_1$, $\mathbf{sep}_X(p) \geqslant r_1$, and $\mathbf{sep}_X(q) \geqslant r_1$. Let now $A$ be a set such that $a, p \in A$ while $b, q \notin A$. We then have $\theta(A) \geqslant r_1$ and $\Theta(A) \leqslant 3r_0$, so that $\Theta(A) - \theta(A) \leqslant 3r_0 - r_1$, which implies that no choice of $\omega$ can produce a Lip-1 distance transform if $r_0 < r_1/3$.

As a concrete example we can take, in $\mathbf{R}$, $a = 1$, $b = r_0 = 1$, $p = 1 + r_1$, $q = 2 + 2r_1$. For $r_1 > 3$ we cannot choose $\omega$ satisfying the conditions. □

**Example 16.3.11.** A periodic set like $\mathbf{Z}$ or $b\mathbf{Z}$, $b > 0$, is easy to understand. Let us perturb it a little, taking

$$X = \{a_j = jb + c_j;\ j \in \mathbf{N}\},$$

where $b$ is a positive number and $0 < c_j < b$. For ease in calculation we assume that $c_{j+1} < c_j$. We look at a subset $A \neq \emptyset, X$. Then there are two indices $j$ and $k$ such that $|j - k| = 1$ and $a_j \in A$, $a_k \notin A$. We take $j$ and $k$ minimal with this property.

If $k = j + 1$ we obtain

$$\theta(A) \geqslant a_{j+2} - a_{j+1} = b + c_{j+2} - c_{j+1};$$

$$\Theta(A) \leqslant 3a_{j+1} - 3a_j = 3b + 3c_{j+1} - 3c_j;$$

$$\Theta(A) - \theta(A) \leqslant 2b - c_{j+2} + 4c_{j+1} - 3c_j.$$

If, on the other, hand $k = j - 1$, we obtain

$$\theta(A) \geqslant a_{k+2} - a_{k+1} = b + c_{k+2} - c_{k+1};$$

$$\Theta(A) \leqslant 3a_{k+1} - 3a_k = 3b + 3c_{k+1} - 3c_k;$$

$$\Theta(A) - \theta(A) \leqslant 2b - c_{k+2} + 4c_{k+1} - 3c_k,$$

the same condition as before.

When

$$0 < b < \tfrac{1}{2}c_{j+2} - 2c_{j+1} + \tfrac{3}{2}c_j,$$

no choice of $\omega$ will yield a Lip-1 function. It is easy to find a sequence $(c_j)$ such that the right-hand side is positive, for instance $c_j = 1/(j+1)$. □

Again, infimal convolution appears as a useful tool:

**Proposition 16.3.12.** *Let $G$ be an abelian group with a translation-invariant metric $d(x,y) = d_0(x - y)$, and let $A$ be an arbitrary subset of $G$. Then*

$$(\mathbf{dt}_A)^+ = \max(\mathbf{dt}_A, 0) = (\mathbf{indf}_A) \sqcap d_0;$$

$$(\mathbf{dt}_A)^- = \max(-\mathbf{dt}_A, 0) = (\mathbf{indf}_{\complement A}) \sqcap d_0,$$

*and, taking the difference between the two,*

$$\mathbf{dt}_A = (\mathbf{dt}_A)^+ - (\mathbf{dt}_A)^- = (\mathbf{indf}_A \sqcap d_0) - (\mathbf{indf}_{\complement A} \sqcap d_0).$$

**Proof.** We see that in $X \smallsetminus A$ we have $\mathbf{dt}_A = \mathbf{indf}_A \sqcap d_0$. Passing to the complement we get $\mathbf{dt}_A = -\mathbf{dt}_{\complement A} = -(\mathbf{indf}_{\complement A} \sqcap d_0)$ in $A$. From this the result follows. □

**Proposition 16.3.13.** *Let $G$ be an abelian group with a translation-invariant metric $d$. Then for any subsets $A, B$ of $G$ we have*

$$\begin{aligned}(\mathbf{dt}_{A+B})^+ &= ((\mathbf{dt}_A)^+) \sqcap (\mathbf{indf}_B) = (\mathbf{indf}_A) \sqcap ((\mathbf{dt}_B)^+) \\ &= ((\mathbf{dt}_A)^+) \sqcap ((\mathbf{dt}_B)^+).\end{aligned}$$

**Proof.** We know from the preceding proposition that

$$(\mathbf{dt}_A)^+ = (\mathbf{indf}_A) \sqcap d_0,$$

where $d_0(x) = d(x, 0)$ is the distance from $x$ to the origin. Hence, using freely the associativity and commutativity of infimal convolution as well as the functional equation $d_0 \sqcap d_0 = d_0$ (cf. Lemma 16.2.2),

$$\begin{aligned}((\mathbf{dt}_A)^+) \sqcap ((\mathbf{dt}_B)^+) &= (\mathbf{indf}_A \sqcap d_0) \sqcap (\mathbf{indf}_B \sqcap d_0) \\ &= (\mathbf{indf}_A \sqcap \mathbf{indf}_B) \sqcap d_0 = (\mathbf{indf}_{A+B}) \sqcap d_0 = (\mathbf{dt}_{A+B})^+.\end{aligned}$$

Also

$$\begin{aligned}((\mathbf{dt}_A)^+) \sqcap \mathbf{indf}_B &= (\mathbf{indf}_A \sqcap d_0) \sqcap \mathbf{indf}_B \\ &= (\mathbf{indf}_A \sqcap \mathbf{indf}_B) \sqcap d_0 = (\mathbf{indf}_{A+B}) \sqcap d_0 = (\mathbf{dt}_{A+B})^+,\end{aligned}$$

equalites which complete the proof. □

## 16.4 Distance transforms and sublevel sets

The ***sublevel sets*** of a function $f\colon X \to \mathbf{R}_!$ are the sets of the form $\{x \in X;\ f(x) < s\}$, a ***strict sublevel set***, and $\{x \in X;\ f(x) \leqslant s\}$, a ***non-strict sublevel set***, for some element $s$ of $\mathbf{R}_!$. For brevity we shall denote them by $\{f < s\}$ rather than $\{x \in X;\ f(x) < s\}$ when no misunderstanding seems possible.

**Lemma 16.4.1.** *If $X$ is a metric space with metric $d$, and $\mathbf{dt}_A$ is the distance transform of a subset $A$ of $X$ calculated with the use of $d$, then the closure, interior and boundary of $A$, taken in the topology given by $d$, can all be recovered from knowledge of the sublevel sets of $\mathbf{dt}_A$:*

$$\overline{A} = \{\mathbf{dt}_A \leqslant 0\}, \qquad A^\circ = \{\mathbf{dt}_A < 0\}, \qquad \partial A = \{\mathbf{dt}_A = 0\}.$$

*Moreover $\mathbf{dt}_{\overline{A}} = \mathbf{dt}_A$ in $\complement A$ and $\mathbf{dt}_{A^\circ} = \mathbf{dt}_A$ in $A$.*

**Proof.** To prove the equality $\overline{A} = \{x;\ \mathbf{dt}_A(x) \leqslant 0\}$, first note that if $x \in A$, then $\mathbf{dt}_A(x) \leqslant 0$ by definition. If on the other hand $x \notin A$, then $x \in \overline{A}$ if and only if there are points $y \in A$ such that $d(x, y)$ is arbitrarily small, which happens if and only if $\mathbf{dt}_A(x) = 0$. This proves the first equality. The second equality in the statement follows by passing to the complement, and the third by taking the set-theoretical difference $\overline{A} \smallsetminus A^\circ = \partial A$. □

If $A$ is any subset of $\mathbf{R}^n$ satisfying $B_<(c, r) \subset A \subset B_\leqslant(c, r)$, where $r > 0$ and we use the distance $d(x, y) = \|x - y\|$ defined by some norm $\|\cdot\|$ on $\mathbf{R}^n$, then $\mathbf{dt}_A(x) = \|x - c\| - r$ provided $r > 0$. This simple example shows that we cannot expect to recover $A$ exactly from $\mathbf{dt}_A$; we have to be content with its interior and closure. However, if $X = \mathbf{Z}^n$, then the topology induced by a norm in $\mathbf{R}^n$ is the discrete topology, so that, for any set $A$,

$$\overline{A} = A^\circ = A = \{\mathbf{dt}_A < 0\} = \{\mathbf{dt}_A \leqslant 0\}.$$

The boundary is empty and $\mathbf{dt}_A$ never takes the value zero.

**Proposition 16.4.2.** *Let $G$ be an abelian group with a translation-invariant metric $d$, and let $A$ be an arbitrary subset of $G$. Then for all positive numbers $r$ and $t$ we have*

$$\begin{aligned}\{\mathbf{dt}_A < r\} &= \delta_{B_<(0,r)}(A) = A + B_<(0, r) \subset A + B_\leqslant(0, r) \\ &\subset \{\mathbf{dt}_A \leqslant r\} \subset \{\mathbf{dt}_A < r + t\};\end{aligned}$$

*and*

$$\begin{aligned}\{\mathbf{dt}_A \leqslant -r\} &= \varepsilon_{B_<(0,r)}(A) \supset \varepsilon_{B_\leqslant(0,r)}(A) \\ &\supset \{\mathbf{dt}_A < -r\} \supset \{\mathbf{dt}_A \leqslant -r - t\}.\end{aligned} \tag{16.4.1}$$

This is easy; we omit the proof. (It is enough to prove the first chain of inclusions; the second follows by duality.)

The dilations by the balls $B = B_<(0, r)$, $A + B_<(0, r) = \delta_B(A)$, thus determine the strict sublevel sets of $\mathbf{dt}_A$ for positive values; similarly for the erosions $\varepsilon_B(A)$ and the non-strict sublevel sets of $\mathbf{dt}_A$ for negative values.

**Proposition 16.4.3.** *Let $G$ be an abelian group and $f_j \colon G \mapsto \mathbf{R}_!$, $j = 1, 2$, two arbitrary functions defined on $G$. Define $f_3 = f_1 \sqcap f_2$. Then for all real numbers $r_1, r_2$ and $r_3 = r_1 + r_2$ we have*

$$\{f_1 < r_1\} + \{f_2 < r_2\} \subset \{f_1 < r_1\} + \{f_2 \leqslant r_2\} \subset \{f_3 < r_3\}; \tag{16.4.2}$$

$$\{f_1 \leqslant r_1\} + \{f_2 \leqslant r_2\} \subset \{f_3 \leqslant r_3\}. \tag{16.4.3}$$

*Moreover, for any real number $r_3$ we have*

$$\bigcup_{r_1 \in \mathbf{R}} \left(\{f_1 < r_1\} + \{f_2 < r_3 - r_1\}\right) = \{f_3 < r_3\}. \tag{16.4.4}$$

**Proof.** We always have $(f_1 \sqcap f_2)(x_1+x_2) \leqslant f_1(x_1) \dotplus f_2(x_2)$, so if $f_1(x_1) < r_1$ and $f_2(x_2) \leqslant r_2$, then $f_3(x_1+x_2) < r_1+r_2 = r_3$. This proves (16.4.2). The inclusion in (16.4.3) follows with the same argument.

Formula (16.4.4) follows from (3.4.7). We can also prove it directly as follows. Assume that $f_3(x_3) < r_3$. Then there exist two points $x_1$ and $x_2$ with $x_1+x_2 = x_3$ such that $f_1(x_1)+f_2(x_2) < r_3$. It is now possible to find a number $r_1$ such that $f_1(x_1) < r_1$ and $f_2(x_2) < r_3 - r_1$. This proves one inclusion in (16.4.4); the other follows from (16.4.2). □

When the functions $f_j$ are distance transforms and $r_3$ is positive we can improve (16.4.4):

**Proposition 16.4.4.** *Let $G$ be an abelian group equipped with a translation-invariant metric, and let $A_j$, $j = 1, 2$, be two subsets. Then, for all positive $r$, their distance transforms $f_j = \mathbf{dt}_{A_j}$ satisfy*

$$\begin{aligned}\{(f_1 \sqcap f_2) < r\} &\supset \big(\{f_1 \leqslant 0\} + \{f_2 < r\}\big) \cup \big(\{f_1 < r\} + \{f_2 \leqslant 0\}\big)\\ &= A_1 + A_2 + B_<(0,r).\end{aligned}$$

*When the Lipschitz constant is 1, we have equality throughout.*

**Proof.** That $\{f_1 \sqcap f_2 < r\}$ contains $\{f_1 \leqslant 0\} \cup \{f_2 < r\}$ follows from Proposition 16.4.3. To prove the inclusion in the other direction, let us assume that $(f_1 \sqcap f_2)(x_3) < r$. Then there exists $x_1, x_2$ such that $x_1+x_2 = x_3$ and $f_1(x_1) + f_2(x_2) < r$. If both $f_1(x_1)$ and $f_2(x_2)$ happen to be nonpositive, we are done. The case when $f_1(x_1)$ or $f_2(x_2)$ is positive remains to be considered. Assume first that $f_1(x_1)$ is positive. Then $x_1 \notin A_1$ and we know that, to any given positive $t$, there exists a point $y_1 \in A_1$ such that $d(x_1, y_1) < f_1(x_1)+t$. We may choose $t = r - f_1(x_1) - f_2(x_2)$. Define $y_2 = x_1+x_2-y_1$. Then $f_1(y_1) \leqslant 0$ and the Lipschitz property of $f_2$ implies that

$$f_2(y_2) \leqslant f_2(x_2) + d(y_2, x_2) < f_1(x_1) + f_2(x_2) + t = r.$$

(Note that $y_2 - x_2 = x_1 - y_1$, so that $d(y_2, x_2) = d(y_1, x_1) < f_1(x_1) + t$.) Hence $x_3 = x_1 + x_2 = y_1 + y_2 \in \{f_1 \leqslant 0\} + \{f_2 < r\}$. The case $f_2(x_2) > 0$ is similar.

To prove the last equality we note that $\{f_1 \leqslant 0\}$ is equal to the closure $\overline{A_1}$ of $A_1$ and that $\{f_2 < r\} = A_2 + B_<(0,r)$. However $\overline{A} + B = A + B$ if $B$ is open, so both terms in the union simplify to $A_1 + A_2 + B_<(0,r)$. □

## 16.5 Finitely generated distances

While the Euclidean distance is easy to visualize geometrically, it has certain drawbacks when it comes to calculations: we need to keep in memory a vector rather than a scalar at each pixel; we need more operations per pixel; and, perhaps most importantly, the Euclidean distance is more difficult to use for various morphological operations, such as skeletonizing, than for instance the city-block distance; see (Borgefors 1994) on this topic. For a study of the computation of the Euclidean distance transform in any dimension, see (Ragnemalm 1993) and (Bailey 2004). In the case of the $l^1$- and $l^\infty$-distances, one first defines the distances between neighboring pixels; we shall call them, following Starovoitov (1995:501), ***prime distances***. Then the distance between any two pixels is defined by following a path and taking as the distance the minimum over all admissible paths of the sum of the prime distances. As an example, for the city-block distance the admissible paths consists of horizontal and vertical moves only, and the prime distance between two pixels which share a side is declared to be one. Thus the distance is calculated successively from neighboring pixels, which is convenient both for sequential and parallel computation. This is impossible for the Euclidean distance in spaces of dimension two or more.

It turns out that many metrics used in image analysis are conveniently defined by starting from finitely many prime distances and then extending these using infimal convolution over all grid points. We shall now explain this, following my paper (1996).

In the definition of an infimal convolution the infimum operator acts over an infinite set of points, and therefore sometimes cannot be computed in finitely many steps. However, there are many situations when the infimum is in fact a minimum over a finite set. One such case is when $f$ is bounded from below and $g$ is coercive in the strong sense that all sublevel sets $\{y;\ g(y) \leqslant a\}$, $a \in \mathbf{R}$, are finite. Then in particular the sublevel set $\{y;\ g(y) \leqslant (f \sqcap g)(x) + 1 - \inf f\}$ is finite for every $x$, and it is enough to search for a minimizing $y$ in that set. Even simpler is the case when $g$ is less than $+\infty$ in a finite set $P$ only. Then the infimal convolution with any function $f$ is equal to the minimum

$$(f \sqcap g)(x) = \min_{y \in P} \big(f(x-y) \dotplus g(y)\big), \qquad x \in G.$$

This is indeed the case for the distances we shall consider: here $P$ is a set with just a few elements where the prime distances are defined.

We have seen that subadditive functions are important when it comes to defining metrics (Lemma 16.2.1). Therefore it is of interest to know that subadditivity can be characterized using infimal convolution:

**Lemma 16.5.1.** *A function $f$ on an abelian group is subadditive in the sense of* (16.2.3) *if and only if it satisfies the inequality $f \sqcap f \geqslant f$. If $f(0) = 0$, this is equivalent to the equation $f \sqcap f = f$.*

**Proof.** If $f$ is subadditive we have $f(x-y) \dotplus f(y) \geqslant f(x)$, so taking the infimum over $y$ gives $(f \sqcap f)(x) \geqslant f(x)$. Conversely, $f(x) \dotplus f(y) \geqslant (f \sqcap f)(x+y)$ for all $x, y$, so $f \sqcap f \geqslant f$ implies subadditivity. Finally, we always have $(f \sqcap f)(x) \leqslant f(x) \dotplus f(0)$, so if $f(0) = 0$ it follows that $f \sqcap f \leqslant f$. □

Infimal convolution is a commutative and associative operation on functions (see Proposition 3.4.2), so we can write iterated convolutions as $f \sqcap g \sqcap h$ without using parentheses. A $k$-fold convolution can be defined by

$$(f_1 \sqcap \cdots \sqcap f_k)(x) = \inf \sum_{j=1}^{k}{}^{\bullet} f_j(x^{(j)}), \qquad x \in G, \tag{16.5.1}$$

where the infimum is over all choices of elements $x^{(j)} \in G$ such that $x^{(1)} + \cdots + x^{(k)} = x$, and with the understanding that the sum receives the value $+\infty$ as soon as one of the terms has that value, even in the presence of a value $-\infty$ (see (2.4.2)). In (16.5.1) it is natural to think of a path leading from 0 to $x$ consisting of segments

$$[0, x^{(1)}], [x^{(1)}, x^{(1)} + x^{(2)}], \ldots, [x^{(1)} + \cdots + x^{(k-1)}, x];$$

if $G = \mathbf{Z}^2$ this path can be realized as a polygon in $\mathbf{R}^2$.

If $A$ is a subset of an abelian group $G$, we shall write $\mathbf{N} \cdot A$ for the semigroup generated by $A$:

$$\mathbf{N} \cdot A = \big\{ \textstyle\sum m_j a_j;\ m_j \in \mathbf{N},\ a_j \in A \big\},$$

where all but finitely many of the $m_j$ are zero. Similarly, we shall write $\mathbf{Z} \cdot A$ for the group generated by $A$:

$$\mathbf{Z} \cdot A = \big\{ \textstyle\sum m_j a_j;\ m_j \in \mathbf{Z},\ a_j \in A \big\}.$$

If $A$ is symmetric, $A = -A$, then of course $\mathbf{Z} \cdot A = \mathbf{N} \cdot A$.

It seems plausible that if a repeated convolution $F \sqcap F \sqcap \cdots \sqcap F$ has a limit $f$ as the number of terms tends to infinity, then this limit will satisfy the equation $f \sqcap f = f$. This is actually so under very general hypotheses. We shall even allow infimal convolutions of the more general form $F_1 \sqcap F_2 \sqcap \cdots \sqcap F_k$ with different functions $F_j$:

**Theorem 16.5.2.** *Let $F_j\colon G \to [0, +\infty]$, $j = \dot{\mathbf{N}}$, be functions on an abelian group $G$ satisfying $F_j(0) = 0$. Define a sequence of functions $(H_j)_{j=1}^{\infty}$ by $H_j = F_1 \sqcap F_2 \sqcap \cdots \sqcap F_j$, the infimal convolution of $j$ terms of the different $F_j$ taken in the order indicated. Then the sequence $(H_j)_j$ is decreasing and its limit $\lim H_j = h \geqslant 0$ is subadditive. Moreover $\mathbf{dom}(h) = \mathbf{N}\cdot(\bigcup_j \mathbf{dom}(F_j))$, i.e., $h$ is finite precisely in the semigroup generated by the sum of the various effective domains $\mathbf{dom}(F_j)$.*

It is easy to prove that $h$ is the largest subadditive function which is a minorant to each of the $F_j$.

**Proof.** That the sequence is decreasing is obvious if we take $y = 0$ in the definition of $H_{j+1}$:

$$H_{j+1}(x) = \inf_y \left(H_j(x - y) + F_j(y)\right) \leqslant H_j(x) + F_j(0) = H_j(x).$$

Next we shall prove that $h(x+y) \leqslant h(x)+h(y)$. If one of $h(x), h(y)$ is equal to $+\infty$ there is nothing to prove, so let $x, y$ be given with $h(x), h(y) < +\infty$ and fix a positive number $t$. Then there exist indices $j, k$ such that $H_j(x) \leqslant h(x)+t$ and $H_k(y) \leqslant h(y)+t$. By associativity $H_{j+k} = H_j \sqcap H_k$, so we get

$$h(x + y) \leqslant H_{j+k}(x + y) \leqslant H_j(x) + H_k(y) \leqslant h(x) + h(y) + 2t.$$

Since $t$ is arbitrary, the inequality $h(x + y) \leqslant h(x) + h(y)$ follows. Finally, the statement about $\mathbf{dom}(h)$ is an easy consequence of (3.4.6). □

**Theorem 16.5.3.** *With $F_j$ as in Theorem 16.5.2, assume in addition that there is a translation-invariant metric $d_1$ on $G$ such that $F_j(x) \geqslant d_1(x, 0)$ for all $j$ and all $x \in G$. Then the limit $h$ of the sequence $H_j$ also satisfies this inequality: $h(x) \geqslant d_1(x, 0)$, so that it is positive definite. If the $F_j$ are symmetric, then $h$ is also symmetric and defines a metric $d(x, y) = h(x - y) \geqslant d_1(x, y)$ on the semigroup $\mathbf{dom}(h) = \sum \mathbf{dom}(F_j)$.*

If all the $F_j$ are equal to $F$, then $\mathbf{dom}(h) = \mathbf{N} \cdot \mathbf{dom}(F)$, the semigroup generated by $\mathbf{dom}(F)$.

**Proof.** Define $K(x) = d_1(x, 0)$ and let $K_j$ be the infimal convolution of $j$ terms equal to $K$. From Lemmas 16.2.1 and 16.2.2 it follows that $K \sqcap K = K$ and so all the $K_j$ are equal to $K$. Therefore $F_j \geqslant K$ implies $H_j \geqslant K$, and so also the limit $h$ must satisfy $h(x) \geqslant K(x) = d_1(x, 0)$ for all $x$. This proves the theorem. □

When applying this theorem we could for instance let $d_1$ be equal to $td_{\text{discr}}$, where $t$ is a small positive number and $d_{\text{discr}}$ is the discrete metric defined by (7.3.4). In $\mathbf{Z}^n$ we can also use $d_1(x,y) = t\|x-y\|$ for any norm on $\mathbf{R}^n$.

**Corollary 16.5.4.** *Let $P$ be a finite set in an abelian group $G$ containing the origin, and let $F$ be a function on $G$ with $F(0) = 0$, taking the value $+\infty$ outside $P$ and with finite positive values at all points in $P \smallsetminus \{0\}$. Then $h = \lim H_j$ is a positive definite subadditive function. If $P$ is symmetric and $F(-x) = F(x)$, then $h$ defines a metric on the subgroup $\mathbf{Z} \cdot P = \mathbf{N} \cdot P$ of $G$ generated by $P$.*

**Proof.** Since $P$ is finite, there is a positive number $t$ such that $F(x) \geqslant t$ for all $x \in P$ except $x = 0$. Thus $F(x) \geqslant td_{\text{discr}}(x, 0)$, where $d_{\text{discr}}$ is the discrete metric defined in (7.3.4). We can now apply the theorem. □

**Definition 16.5.5.** Let us say that a metric $d(x,y) = d_0(x-y)$ is ***finitely generated*** if it is constructed as in Corollary 16.5.4. □

It is easy to prove that the Euclidean metric $d(x,y) = \sqrt{\sum(x_j - y_j)^2}$ on $\mathbf{Z}^n$ is a finitely generated distance if and only if $n \leqslant 1$.

It is by no means necessary to assume that $h$ is positively homogeneous in Corollary 16.5.4. In fact, we can let $P = \{0, \pm 1, \pm 2\} \subset \mathbf{Z}$ and define $F(\pm 1) = a$, $F(\pm 2) = b$, where $a$ and $b$ are arbitrary positive numbers. If $b \geqslant 2a$, then $h(x) = a|x|$ for all $x \in \mathbf{Z}$, but if $b < 2a$, then $h(x) = \frac{1}{2}b|x|$ when $x$ is even, $x = 2k$, $k \in \mathbf{Z}$, whereas $h(x) = bk + a > \frac{1}{2}b|x|$ when $x$ is odd, $x = 2k+1$, $k \in \mathbf{N}$. Nevertheless $h$ is subadditive.

A more interesting example is perhaps the following. Let

$$P = \{x \in \mathbf{Z}^2;\ |x_j| \leqslant 1\},$$

and define the prime distances as

$$F(\pm 1, 0) = F(0, \pm 1) = a > 0, \qquad F(\pm 1, \pm 1) = b > 0.$$

Then, if $b \geqslant a$, we get $h(x_1, 0) = a|x_1|$. But if $b < a$, then $h(2,0) = 2b < 2a$, so that $h(2,0) < 2h(1,0) = 2a$. In fact, by the definition of infimal convolution, $h(2,0) \leqslant H_2(2,0) \leqslant F(1,1) + F(1,-1) = b + b$. On the other hand, it is not difficult to see that for any $k$, $H_k(2,0) \geqslant 2b$, so that actually $h(2,0) = 2b$. This is because if we take $k \geqslant 2$ nonzero steps to go from the origin to (2,0), the distance assigned to the path is at least $F(x^{(1)}) + \cdots + F(x^{(k)}) \geqslant kb$.

Some other metrics that have been studied are modifications of this; to define them, put $F(1,0) = a$ and $F(1,1) = b$. Then the choices $(a,b) =$

$(1, \sqrt{2})$ (Montanari 1968); $(a, b) = (2, 3)$ (Hilditch & Rutovitz 1969); and $(a, b) = (3, 4)$ (Borgefors 1984) have all been studied. Next we can increase the size of the neighborhood where prime distances are defined to include the knight's move $(2, 1)$ as an element of $P$. The distance defined by this move only has been studied by Das & Chatterji (1988). It seems more natural, however, to allow also $(1, 0)$ and $(1, 1)$ in $P$. Then a very good choice under certain criteria is $F(1, 0) = 5$, $F(1, 1) = 7$, and $F(2, 1) = 11$ ("the 5-7-11 weighted distance"). This distance was proposed and studied by Gunilla Borgefors (1986). The balls for this metric have 16 sides. We can note that $11/7$ is only slightly smaller than $\|(2, 1)\|_2/\|(1, 1)\|_2$ and that $7/5$ is only slightly smaller than $\|(1, 1)\|_2/\|(1, 0)\|_2$.

We always have $h \leqslant H$, and it may happen that $h(x) < F(x)$ for some pixel $x \in P$. Let for instance $F(1, 0) = a$, $F(2, 1) = c$, and extend $F$ by reflection and permutation of the coordinates. Then

$$h(1, 0) \leqslant H_3(1, 0) \leqslant F(2, 1) + F(1, -2) + F(-2, 1) = 3c,$$

so if $3c < a$ we get $h(1, 0) \leqslant 3c < a = F(1, 0)$. This is undesirable, because we expect the prime distance originally defined between the origin and $(1, 0) \in P$ to survive and to be equal to the distance defined by the minimum over all paths. It is therefore natural to require that $h = F$ everywhere in $P$.

Since $h(x)$ is the limit of an infinite sequence $H_j(x)$, it is reassuring to know that this sequence is in fact stationary in the cases of interest here. It is easy to explicitly give an index $j$ such that $H_j(x)$ is equal to the limit $h(x)$:

**Proposition 16.5.6.** *Let $F$ be as in Corollary 16.5.4. Then the sequence $(H_j)$ is pointwise stationary, i.e., for every $x \in G$ there is an index $i(x)$ such that $H_j(x) = h(x)$ for all $j \geqslant i(x)$.*

**Proof.** We first note that by hypothesis there are two constants $c$ and $C$ such that

$$0 < c \leqslant F(p) \leqslant C, \qquad p \in P \smallsetminus \{0\}.$$

If $x \notin \mathbf{N} \cdot P$, then $f(x) = H_j(x) = +\infty$ for all $j$. If $x \in \mathbf{N} \cdot P$, then there is an index $m(x)$ such that $x \in P + \cdots + P$ with $m(x)$ terms, and $H_{m(x)}(x) \leqslant m(x)C$. For every $j \geqslant m(x)$ there are elements $y^{(1)}, y^{(2)}, \ldots, y^{(j)}$ in $P$ such that $x = y^{(1)} + \cdots + y^{(j)}$ and $H_j(x) = F(y^{(1)}) + \cdots + F(y^{(j)})$ (cf. (16.5.1)). Since $F \geqslant c$ in $P \smallsetminus \{0\}$ we get

$$Cm(x) \geqslant H_{m(x)}(x) \geqslant H_j(x) \geqslant qc,$$

where $q$ is the number of indices $i$ such that $y^{(i)} \neq 0$. Hence the number $q$ of nonzero terms in any representation of $H_j(x)$ with $j \geqslant m(x)$ can never be larger than $Cm(x)/c$. Now define $i(x) = \lfloor Cm(x)/c \rfloor$, using the floor function defined in (2.5.1). If $j \geqslant i(x)$, then $j \geqslant q$ and we have $H_j(x) = H_q(x)$, since, in the formation of $H_j(x)$, at most $q$ of the terms in a sum $x = y^{(1)} + \cdots + y^{(j)}$ can be nonzero if the value of the sum $F(y^{(1)}) + \cdots + F(y^{(j)})$ shall come down to $H_q(x)$. This means that the sequence $H_j(x)$ is stationary starting with $i(x)$. □

**Remark 16.5.7.** It is not true that $H_j(x) = H_{j-1}(x)$ implies that $H_k(x) = H_j(x)$ for all $k \geqslant j$, so $H_j(x) = H_{j-1}(x)$ at a particular point $x$ is not a sufficient criterion. For example, we may define $F(\pm 1) = 1$ and $F(\pm 100) = 101$. Then $H_j(100) = 101$ for $j = 1, \ldots, 99$, but $F(100) = 100$. □

**Remark 16.5.8.** The numbers $C$ and $c$ are structural constants of the prime distances $F(p)$, $p \in P$, and can be taken as

$$C = \max_{p \in P} F(p) \text{ and } c = \min_{p \in P \smallsetminus \{0\}} F(p).$$

The number $m(x)$ can be easily estimated in the most common applications. For instance, for the $l^\infty$-metric we can take $C = c = 1$ and $m(x) = \|x\|_\infty$. Similarly for the $l^1$-metric we can take $C = c = 1$ and $m(x) = \|x\|_1$. It is therefore easy to estimate the index $i(x)$. □

**Corollary 16.5.9.** *The positive part of a distance transform is a limit*

$$(\mathbf{dt}_A)^+ = \lim \left(\mathbf{indf}_A \sqcap F \sqcap F \sqcap \cdots \sqcap F\right),$$

*where the number of terms tends to infinity. This formula can be used in actual calculations: starting from $g_0 = \mathbf{indf}_A$ one calculates $g_j(x) = (g_{j-1} \sqcap F)(x)$ and stops when the criterion of Proposition 16.5.6 is satisfied.* □

## 16.6 Comparing distances

The distances obtained from the $l^1$-metric and the $l^\infty$-metric in $\mathbf{R}^2$ are translation invariant but not rotation invariant. (In the plane a rotation can distort distances by a factor of up to $\sqrt{2}$; in higher dimensions more.) The Euclidean metric is rotation invariant, and it is desirable to construct a finitely generated distance in $\mathbf{Z}^n$ which is reasonably close to being rotation invariant. There are many studies on the problem of defining an optimal distance in a given family of finitely generated distances. Of course the property of being optimal depends on the criteria employed; beauty is in the eye of the beholder. A basic problem is therefore how to measure deviation: we may ask how far the quotient of two quantities is from 1, alternatively how far their difference is from 0. In this section we shall look briefly into this problem and describe four methods of comparing two nonnegative functions, called here $\Lambda$, $\lambda$, $V$ and $B$; the last-mentioned two have been studied earlier.

It is natural to measure the deviation of a function $f\colon X \to [0, +\infty[$ from a given nonnegative function $g$ defined on the same set by the smallest constant $C \in [0, +\infty]$ such the inequalities $f(x) \leqslant Cg(x)$ and $g(x) \leqslant Cf(x)$ hold for all $x \in X$. We introduce a notation for this constant,

$$\Lambda(f, g) = \max\left[\sup_{x\in X} \frac{f(x)}{g(x)}, \sup_{x\in X} \frac{g(x)}{f(x)}\right],$$

where the supremum is taken over all points in the common domain of definition, and where we count $0/0$ as 0 and $t/0$ as $+\infty$ if $t > 0$ (this is to allow for zeros; $\Lambda(f, g)$ is finite only if the two functions have the same zero set). It is noteworthy that $\log \Lambda(f, g) = \|\log f - \log g\|_\infty$ is a distance on a suitable space of functions; in particular it is symmetric.

If $f$ is a multiple of $g$, $f = cg$ for a positive constant $c$, then $\Lambda(f, g) = c \vee c^{-1}$, which can be very large. But we may think of the functions as being essentially the same, and that this should be reflected in the measure. It is therefore natural to introduce the infimum of all distances $\Lambda(cf, g)$, defined as

$$\lambda(f, g) = \inf_{c>0} \Lambda(cf, g), \quad f, g > 0.$$

It is easy to see that

$$\lambda(f, g) = \sqrt{\sup(f/g)\sup(g/f)} = \sqrt{\frac{\sup(f/g)}{\inf(f/g)}}, \qquad f, g > 0.$$

Another, related measure of the deviation was studied by Ben J. H. Verwer (1991). He used the functional

$$V(f, g) = \sup_{x\in X}\left|\frac{f(x)}{g(x)} - 1\right|.$$

However, one might just as well consider $V(g, f)$. It is readily seen that $V(f, g) = \Lambda(f, g) - 1$ when $f \geqslant g$, and that $V(f, g) = 1 - 1/\Lambda(f, g)$ when $f \leqslant g$. In general, $V(f, g)$, $V(g, f)$ as well as $\log \Lambda(f, g)$ lie between two limits,

$$1 - \frac{1}{\Lambda(f, g)} \leqslant V(f, g), V(g, f), \log \Lambda(f, g) \leqslant \Lambda(f, g) - 1,$$

where we have put in also the well-known inequality $1-1/t \leqslant \log t \leqslant t-1$, $t>0$. In particular, $|V(f,g)-\log\Lambda(f,g)| \leqslant (\Lambda(f,g)-1)^2/\Lambda(f,g)$; the same estimate holds of course for $V(g,f)$. We may also note that, although $V$ is not symmetric, it is approximately symmetric when $f$ and $g$ are close, and there is an estimate

$$\frac{V(f,g)}{\Lambda(f,g)} \leqslant V(g,f) \leqslant \Lambda(f,g)V(f,g),$$

which, by the way, may be written as

$$\Lambda(V(f,g),V(g,f)) \leqslant \Lambda(f,g).$$

When $f$ and $g$ are reasonably close, $V(f,g) \approx V(g,f) \approx \log\Lambda(f,g)$. For many purposes either one may be used. Note, however, that $\Lambda(f,g)$ has better functional properties than $V(f,g)$. In particular, as already noted, $\log\Lambda(f,g)$ is a metric, whereas $V(f,g)$ does not satisfy the triangle inequality and is not even symmetric.

In particular we shall compare the finitely generated distances with the Euclidean norm $g=\|\cdot\|_2$. In Exercises 16.9–16.11 we let the reader determine or estimate the deviation from the Euclidean norm of some well-known finitely generated metric. If the prime vectors are $(\pm1,0)$, $(0,\pm1)$ and $(\pm1,\pm1)$ with prime distances $a$ and $b$ respectively, we note that the optimal prime distances for both $\Lambda$ and $V$ are related by $b=a\sqrt{2}$, but that the actual values are slightly different. For $\Lambda(f,\|\cdot\|_2)$, the optimal choice is

$$a=a_0=\sqrt[4]{\frac{2+\sqrt{2}}{4}} \approx 0.9612, \text{ and} \tag{16.6.1}$$
$$b=b_0=a_0\sqrt{2}=\sqrt[4]{2+\sqrt{2}} \approx 1.3593,$$

whereas Verwer (1991:676) found the optimal choice for $V(f,\|\cdot\|_2)$ to be

$$a=a_1 \approx 0.9604 \text{ and } b=b_1=a_1\sqrt{2} \approx 1.3583.$$

The exact values are

$$a_1=\frac{1}{\frac{1}{2}+\sqrt{1-1/\sqrt{2}}} \text{ and } b_1=a_1\sqrt{2}. \tag{16.6.2}$$

One can calculate also the optimal choice for $V(\|\cdot\|_2,f)$, which is

$$a=a_2=\tfrac{1}{2}+\tfrac{1}{4}\sqrt{2+\sqrt{2}} \approx 0.9619 \text{ and } b=b_2=a_2\sqrt{2} \approx 1.3604. \tag{16.6.3}$$

In this case the optimality has a clear geometrical meaning: the vertices of the octagon defined by $f$ protrude as much outside the disk of radius $r=a_2$ as the midpoints of the edges go into the disk.

In her pioneering work on these distances (1984, 1986, 1991), Gunilla Borgefors used instead the functional

$$B(f) = \sup_{x \in \mathbf{Z}^2} \big( |f(x) - \|x\|_2|;\ \|x\|_\infty = 1 \big),$$

and she determined the optimal distances for this measure as follows:

$$B(f) = \tfrac{1}{2}\sqrt{2\sqrt{2} - 2} - \tfrac{1}{2} \approx 0.04491,$$

attained for

$$a = a_3 = \tfrac{1}{2}\sqrt{2\sqrt{2} - 2} + \tfrac{1}{2} \approx 0.9551;$$

$$b = b_3 = \sqrt{2} - \tfrac{1}{2} + \tfrac{1}{2}\sqrt{2\sqrt{2} - 2} \approx 1.3693;$$

note that $b_3 > a_3\sqrt{2}$ here (1986:351). She also determined the optimal values for the functional $B$ when $a$ is restricted to be 1 and $b$ is free to vary; in this case $B(f) = 1/\sqrt{2} - \sqrt{\sqrt{2} - 1} \approx 0.06$ and is attained for

$$a = a_4 = 1 \text{ and } b = b_4 = 1/\sqrt{2} - \sqrt{\sqrt{2} - 1} \approx 1.351;$$

here $b_4 < a_4\sqrt{2}$ (1984:327).

## 16.7 The calculus of balls

In any metric space the inequality $d(a, b) + r \leqslant s$ implies that the strict ball $B_<(a, r)$ is contained in $B_<(b, s)$ and also that $B_\leqslant(a, r) \subset B_\leqslant(b, s)$. In a normed vector space of dimension at least one and equipped with the distance $d(x, y) = \|x - y\|$ defined by the norm, the converse implications hold, provided $r > 0$ ($r \geqslant 0$ for closed balls). In particular, two open balls $B_<(a, r)$ and $B_<(b, s)$ with $r > 0$ are equal if and only if $a = b$ and $r = s$ ($r \geqslant 0$ suffices for closed balls). But in a general metric space very little can be said: from $B_\leqslant(a, r) \subset B_\leqslant(b, s)$ we can only deduce that $d(a, b) \leqslant s$, assuming that $r \geqslant 0$.

We note, however, that if, in an abelian group with a translation-invariant metric, $B_\leqslant(a, r)$ is contained in $B_\leqslant(b, s)$, $s \leqslant r$, and $0 \leqslant r < +\infty$, then $a - b$ belongs to a bounded subgroup. In the most common applications, like normed spaces and groups with a finitely generated distance, the only bounded subgroup is $\{0\}$, thus $a = b$.

The mappings

$$X \times [0, +\infty[ \ni (a, r) \mapsto B_\leqslant(a, r) \in \mathscr{P}(X),$$

$$X \times ]0, +\infty[ \ni (a, r) \mapsto B_<(a, r) \in \mathscr{P}(X)$$

are injective in a normed vector space, meaning that both the center and the radius of a ball are determined by the set itself.

In general they are far from injective and induce complicated equivalence relations in $X \times [0, +\infty[$. Similarly, the inclusion relations $B_{\leqslant}(a,r) \subset B_{\leqslant}(b,s)$ and $B_{<}(a,r) \subset B_{<}(b,s)$ induce preorders in $X \times \mathbf{R}$. We shall now study these relations, first for balls with a common center.

### 16.7.1 *Identifying balls with a common center*

In any metric space $X$ with metric $d$ we fix a point $c$ in $X$ and define two functions which are of interest when defining balls:

$$\text{(16.7.1)} \qquad \begin{aligned} \rho(r) &= \sup_{x\in X} \big(d(x,c);\ d(x,c) \leqslant r\big) \text{ and} \\ \sigma(r) &= \inf_{x\in X} \big(d(x,c);\ d(x,c) \geqslant r\big), \quad r \in \mathbf{R}_!. \end{aligned}$$

The functions depend of course in general on the choice of $c$, but in most applications we shall let $X$ be an abelian group and assume that $d$ is translation invariant. Then the choice of $c$ is immaterial.

We note that the functions $\rho, \sigma\colon \mathbf{R}_! \to \mathbf{R}_!$ are quotients: $\rho = f/_{\star}f$ and $\sigma = f/^{\star}f$, with $f\colon X \to \mathbf{R}$, $f(x) = d(x,c)$.

The functions

$$\text{(16.7.2)} \qquad \begin{aligned} \rho_-(r) &= \sup_{s<r} \rho(s) = \sup_{x\in X} \big(d(x,c);\ d(x,c) < r\big) \text{ and} \\ \sigma_+(r) &= \inf_{s>r} \sigma(s) = \inf_{x\in X} \big(d(x,c);\ d(x,c) > r\big), \qquad r \in \mathbf{R}_!, \end{aligned}$$

are also of interest.[2] It is clear that

$$\rho_- \leqslant \rho \leqslant \mathbf{id}_{\mathbf{R}_!} \leqslant \sigma \leqslant \sigma_+.$$

On the negative half-axis we have $\rho(r) = -\infty$ for $-\infty \leqslant r < 0$ and $\sigma(r) = 0$ for $-\infty \leqslant r \leqslant 0$.

We have

$$\text{(16.7.3)} \qquad B_{\leqslant}(c,r) = B_{\leqslant}(c,\rho(r)) \text{ and } B_{<}(c,r) = B_{<}(c,\sigma(r)).$$

The open interval $]\rho(r), \sigma(r)[$ contains no distances $d(c,x)$: a sphere with center at $c$ and radius in that interval is empty. The functions $\rho$ and $\sigma$ solve the uniqueness problem for balls with a common center: $B_{\leqslant}(c,r) = B_{\leqslant}(c,s)$ if and only if $\rho(r) = \rho(s)$, and $B_{<}(c,r) = B_{<}(c,s)$ if and only if $\sigma(r) = \sigma(s)$. So these are the equivalence relations we were looking for.

[2]We can define also $\rho_+$ and $\sigma_-$, but they will not appear in the sequel.

The mappings $\rho, \sigma\colon \mathbf{R}_! \to \mathbf{R}_!$ are idempotent; in fact $\rho$ is an anoiktomorphism and $\sigma$ is a cleistomorphism. By way of contrast, $\rho_-$ and $\sigma_+$ are in general not idempotent. However, $\rho \circ \rho_- = \rho_-$.

In an abelian group with a translation-invariant distance, the absolute value of the distance transform of a set $A$ at a point $x \in A$, i.e., $|\mathbf{dt}_A(x)| = -\mathbf{dt}_A(x) = r$, satisfies $r = \sigma(r)$. Similarly for $x \notin A$, we have $r = \mathbf{dt}_A(x)$ with $r = \sigma(r)$. Thus the values of $|\mathbf{dt}_A|$ are always contained in the invariance set $\mathbf{invar}(\sigma)$ of $\sigma$.

In a normed vector space of dimension at least one, we always have $\rho(r) = \sigma(r) = r$ for positive $r$, so there is no need to introduce them in the study of such spaces. In $\mathbf{Z}^n$ equipped with the $l^\infty$- or $l^1$-metrics, the distances are always integers, which implies that $\rho(r) = \lfloor r \rfloor$ and $\sigma(r) = \lceil r \rceil$ for positive $r$, and $\rho_- = \rho - \chi_{\mathbf{N}}$ for $r > 0$; $\sigma_+ = \sigma + \chi_{\mathbf{N}}$ for $r \geqslant 0$.

### 16.7.2 *Three concepts of regularity for metrics*

While the functions $\rho$ and $\sigma$ solve the problem of comparing two balls with a common center, more hypotheses will be needed if we are to successfully compare balls with different centers. This is why we are led to an analysis of the triangle inequality and when it can be an equality.

When we use a Euclidean distance in $\mathbf{R}^n$ we have equality in the triangle inequality $d(x,z) \leqslant d(x,y)+d(y,z)$ if and only if the points lie on a straight line with $y$ between $x$ and $z$. For other distances and in other spaces the situation can be different.

Suppose that we have a strict triangle inequality

$$d(x,z) < d(x,y) + d(y,z), \qquad x, y, z \in X, \tag{16.7.4}$$

and want to change it into an equality. This can sometimes be achieved in three different ways:

(1) By increasing the left-hand side of (16.7.4) while keeping both terms on the right-hand side constant;
(2) By replacing one of the two points appearing on the left-hand side without changing the distance between them, but decreasing one of the terms on the right-hand side;
(3) By replacing the point which appears only on the right-hand side, thereby decreasing one of the terms on the right-hand side while keeping the other term there constant as well as the left-hand side.

To require that we can obtain equality is tantamount to requiring the existence of certain points in the space. We now define these three kinds of regularity as well as their weakened forms, where we require only an approximation.

**Definition 16.7.1.** First, we may be able to replace[3] $z$ in (16.7.4) by $\tilde{z}$ in order to increase $d(x,z)$ to $d(x,\tilde{z})$ while keeping $d(y,\tilde{z}) = d(y,z)$ so that

$$d(x,z) \leqslant d(x,\tilde{z}) = d(x,y) + d(y,\tilde{z}) = d(x,y) + d(y,z). \tag{16.7.5}$$

If this is always possible, we shall say that $d$ is ***upper regular for the triangle inequality***.

We shall say that $d$ is ***almost upper regular*** if for every positive $t$ there is a point $\tilde{z}$ such that

$$d(y,z) - t \leqslant d(y,\tilde{z}) \leqslant d(y,z) \text{ and } d(x,\tilde{z}) \geqslant d(x,y) + d(y,\tilde{z}) - t. \tag{16.7.6}$$

□

**Definition 16.7.2.** Second, assuming that $d(x,z) \geqslant d(y,z)$,[4] we replace[5] $x$ by $\tilde{x}$ in order to decrease $d(x,y)$ to $d(\tilde{x},y)$ while keeping $d(\tilde{x},z) = d(x,z)$, so that

$$d(\tilde{x},z) = d(x,z) = d(\tilde{x},y) + d(y,z) \leqslant d(x,y) + d(y,z). \tag{16.7.7}$$

If this is always possible, we shall say that $d$ is ***lower regular for the triangle inequality***.

We shall say that $d$ is ***almost lower regular*** if for every positive $t$, the inequalities

$$d(x,z) \leqslant d(\tilde{x},z) \leqslant d(x,z) + t \text{ and } d(\tilde{x},z) + t \geqslant d(\tilde{x},y) + d(y,z) \tag{16.7.8}$$

hold for some point $\tilde{x}$. □

**Definition 16.7.3.** Third, assuming that $d(x,z) \geqslant d(x,y)$,[6] we replace $y$ (the point on the right-hand side which appears in both terms there) by $\tilde{y}$ in order to decrease $d(y,z)$ to $d(\tilde{y},z)$ while keeping $d(x,\tilde{y}) = d(x,y)$, so that

$$d(x,z) = d(x,\tilde{y}) + d(\tilde{y},z) = d(x,y) + d(\tilde{y},z) \leqslant d(x,y) + d(y,z). \tag{16.7.9}$$

If this is always possible, we shall say that $d$ is ***inner regular for the triangle inequality***.

---

[3] In view of the symmetry it is equivalent to replace $x$.

[4] If $d(x,z) < d(y,z)$, there can be no such point $\tilde{x}$, so assuming that $d(x,z) \geqslant d(y,z)$ makes sense.

[5] Equivalenty, we can replace $z$.

[6] If $d(x,z) < d(x,y)$ there can be no point $\tilde{y}$ with the required properties, so assuming that $d(x,z) \geqslant d(x,y)$ makes sense.

We shall say that $d$ is ***almost inner regular*** if for every positive $t$, we have inequalities

(16.7.10) $d(x,y)-t \leqslant d(x,\tilde{y}) \leqslant d(x,y)$ and $d(x,z) \geqslant d(x,\tilde{y})+d(\tilde{y},z)-t$

for some point $\tilde{y}$. □

It is clear that the norm in a normed vector space is regular for the triangle inequality in all three senses. We may choose $\tilde{z}$ on the straight line defined by $x$ and $y$ for upper regularity if $x \neq y$ and take $\tilde{z}=x=y$ otherwise. For lower regularity we take $\tilde{x}$ on the line defined by $y$ and $z$ if these points are different and as $x$ if $y=z$. For inner regularity we take $\tilde{y}$ on the segment defined by $x$ and $z$. In $\mathbf{Z}^n$, $n \geqslant 2$, none of the six regularities holds for the $l^2$-metric. But it is different for the $l^\infty$-metric and the $l^1$-metric:

**Example 16.7.4.** We take $X=\mathbf{Z}^n$ and study the regularity for the triangle inequality.

**ur**. For upper regularity it is convenient to take $y=0$. The triangle inequality and the equality obtained with $\tilde{z}$ then take the forms

$$\begin{aligned} &\|x-z\| \leqslant \|x\|+\|z\|; \\ &\|x-\tilde{z}\| = \|x\|+\|\tilde{z}\|, \text{ keeping } \|\tilde{z}\|=\|z\|. \end{aligned} \tag{16.7.11}$$

**ur**, $l^\infty$. For upper regularity and the $l^\infty$-norm, we have $\|x\|_\infty+\|z\|_\infty = |x_k|+|z_l|$ for some indices $k$ and $l$. We want to find $\tilde{z}$ such that

$$\|x-\tilde{z}\|_\infty = \|x\|_\infty+\|\tilde{z}\|_\infty.$$

We can choose $\tilde{z}_k = \pm z_l$, so that $\|x-\tilde{z}\|_\infty \geqslant |x_k-\tilde{z}_k| = |x_k|+|z_l| = \|x\|_\infty+\|z\|_\infty$, and $\tilde{z}_j=x_j$ for other indices.

**ur**, $l^1$. For upper regularity and the $l^1$-norm, we can take $\tilde{z}_j=\pm z_j$ with signs chosen so that so that $|x_j-\tilde{z}_j| = |x_j|+|\tilde{z}_j| = |x_j|+|z_j|$ for every index $j$. Thus $\tilde{z}$ and $z$ have the same norm, but the norm of $x-\tilde{z}$ is larger than the norm of $x-z$ and is equal to $\|x\|_1+\|z\|_1$.

**lr**. For lower regularity it is convenient to take $z=0$. The triangle inequality and the equality obtained with $\tilde{x}$ then take the forms

$$\begin{aligned} &\|x\| \leqslant \|x-y\|+\|y\|, \text{ assuming } \|x\| \geqslant \|y\|; \\ &\|\tilde{x}\| = \|\tilde{x}-y\|+\|y\|, \text{ keeping } \|\tilde{x}\|=\|x\|. \end{aligned} \tag{16.7.12}$$

**lr**, $l^\infty$. For lower regularity and the $l^\infty$-norm we have to find $\tilde{x}$ with the same norm as that of $x$ and with equality $\|\tilde{x}\|_\infty = \|\tilde{x}-y\|_\infty+\|y\|_\infty$. There are indices $k$ and $l$ such that $\|x\|_\infty = |x_k|$ and $\|y\|_\infty = |y_l|$. We take $\tilde{x}_l = |x_k| - |y_l|$ and $\tilde{x}_j = y_j$ for other indices. Then $\|\tilde{x} - y\|_\infty = |\tilde{x}_l| = |x_k| - |y_l| = \|x\|_\infty - \|y\|_\infty$ as we wanted.

**lr**, $l^1$. For lower regularity and the $l^1$-norm we have to take $\tilde{x}$ of the same norm as that of $x$, and with $\|\tilde{x} - y\|_1 = \|x\|_1 - \|y\|_1 \geqslant 0$. Take $\tilde{x} = c + y$, with $c_1 = \pm\|c\|_1$ and $c_j = 0$ for $j \geqslant 2$. The sign of $c_1$ shall be chosen to be the sign of $y_1$. Then $|\tilde{x}_1| = |y_1 + c_1| = |y_1| + |c_1| = |y_1| + \|c\|_1$, and $|\tilde{x}_j| = |y_j|$ for the other indices $j$ (if there are any). So we have $\|\tilde{x}\|_1 = \|c\|_1 + \|y\|_1 = \|\tilde{x} - y\|_1 + \|y\|_1$.

**ir**. For inner regularity it is convenient to take $x = 0$. The triangle inequality and the equality obtained with $\tilde{y}$ then take the forms

$$\begin{aligned} &\|z\| \leqslant \|y\| + \|y - z\|, \text{ assuming } \|z\| \geqslant \|y\|; \\ &\|z\| = \|\tilde{y}\| + \|\tilde{y} - z\|, \text{ keeping } \|\tilde{y}\| = \|y\|. \end{aligned} \tag{16.7.13}$$

**ir**, $l^\infty$. For inner regularity and the $l^\infty$-norm we shall find $\tilde{y}$ such that the equality above holds. We choose indices $k$ and $l$ such that $\|y\|_\infty = |y_k|$ and $\|z\|_\infty = |z_l|$. Then $\tilde{y}_l = \pm y_k$ and $\tilde{y}_j = z_j$ for $j \neq l$ gives a solution: $\|\tilde{y} - z\|_\infty = |z_l| - |y_k| = \|z\|_\infty - \|y\|_\infty$ for some choice of sign.

**ir**, $l^1$. For inner regularity and the $l^1$-norm we shall find $\tilde{y}$ such that $\|z\|_1 = \|\tilde{y}\|_1 + \|\tilde{y} - z\|_1$.

We take $\tilde{y} = z + c$, where $c_1 = \pm\|c\|_1 = \pm(\|z\|_1 - \|y\|_1)$; $c_j = 0$ for $j \neq 1$. For some choice of sign we have $|z_1 + c_1| = |z_1| - |c_1|$, so that $\|z\|_1 = \|\tilde{y}\|_1 + \|\tilde{y} - z\|_1$. □

**Example 16.7.5.** That almost regularity is strictly weaker than regularity can be seen from simple examples. Let $X = \mathbf{Q}^2$ with the $l^2$-metric.

We can take first

$$x = (-2, 1), \quad y = (0, 0), \quad z = (1, 1).$$

For upper regularity we should have $d(x, \tilde{z}) = d(x, y) + d(y, z) = \sqrt{5} + \sqrt{2}$, thus $d(x, \tilde{z})^2 = 7 + 2\sqrt{2}$, which is not possible for $\tilde{z}$ with rational coordinates.

We then take

$$x = (0, 0), \quad y = (1, 0), \quad z = (1, 1).$$

For lower regularity we should have $d(\tilde{x}, y) = d(x, z) - d(y, z) = \sqrt{2} - 1$, thus $d(\tilde{x}, y)^2 = 3 - 2\sqrt{2}$, which is not possible for $\tilde{x} \in \mathbf{Q}^2$.

Finally we take

$$x = (0, 0), \quad y = (1, 0), \quad z = (2, 2).$$

For inner regularity we should have $d(\tilde{y}, z) = d(x, z) - d(x, y) = 2\sqrt{2} - 1$, thus $d(\tilde{y}, z)^2 = 9 - 4\sqrt{2}$, which is not possible for any $\tilde{y}$ with rational coordinates.

However, in these three cases we do have the corresponding almost regularity since the rationals are dense in $\mathbf{R}$. □

The distance in $\mathbf{Z}^2$ with prime distances defined on

$$P = \{0, (\pm 1, 0), (0, \pm 1), (\pm 1, \pm 1)\}$$

is upper regular provided $a = F(1, 0) \leqslant b = F(1, 1)$. If $b \geqslant 2a$, then $d(x, y) = a\|x - y\|_1$ for all $x$ and $y$; if $b = a$, then $d(x, y) = a\|x - y\|_\infty$ for all $x$ and $y$; we are therefore in either of two well-known situations. In the remaining cases, when $a < b < 2a$, the balls are octagonal, and we may argue as follows. By symmetry and reflection of the coordinates it is enough to consider points $x, y, z \in \mathbf{Z}^2$ with $0 \leqslant y_2 - x_2 \leqslant y_1 - x_1$. Then $y - x$ can be written as $k(1, 0) + m(1, 1)$ for some uniquely determined integers $k, m \geqslant 0$, and $d(x, y) = ka + mb$. If $z - y$ is in the same sector, thus satisfying $0 \leqslant z_2 - y_2 \leqslant z_1 - y_1$, then $d(x, z) = d(x, y) + d(x, y)$, so we may take $\tilde{z} = z$. If $z - y$ is in one of the other seven sectors we take $\tilde{z} - y = (|z_1 - y_1|, |z_2 - y_2|)$ or $(|z_2 - y_2|, |z_1 - y_1|)$ so that $\tilde{z} - y$ is in the same sector as $x - y$ and we know that we have equality in the triangle inequality (16.7.5).

However, this distance is not lower regular if $a < b < 2a$ as can be shown by simple examples. It seems therefore that lower regularity is too stringent a criterion.

### 16.7.3 *Identifying balls with different centers*

We study here when a ball is contained in another one, without assuming that they have a common center. The conditions turn out to be different for strict and non-strict ball.

**Theorem 16.7.6.** *Let $G$ be an abelian group with metric $d$. Assume that $d$ is translation invariant and almost upper regular for the triangle inequality. Then for all $r, s \geqslant 0$, $B_\leqslant(a, r)$ is contained in $B_\leqslant(b, s)$ if and only if $d(a, b) + \rho(r) \leqslant \rho(s)$.*

**Proof.** By the definition of $\rho$, $B_\leqslant(a, r) = B_\leqslant(a, \rho(r))$. The inequality $d(a, b) + \rho(r) \leqslant \rho(s)$ implies that $B_\leqslant(a, \rho(r)) \subset B_\leqslant(b, \rho(s))$. Hence $B_\leqslant(a, r) \subset B_\leqslant(b, s)$.

For the other implication we argue as follows, assuming that $B_{\leqslant}(a,r)$ is contained in $B_{\leqslant}(b,s)$. Take a point $z$ in the first ball. Then $d(z,a) \leqslant \rho(r)$. Assuming almost upper regularity we have, given any positive number $t$, a point $\tilde{z}$ such that $d(a,z)-t \leqslant d(a,\tilde{z}) \leqslant d(a,z)$, implying that $\tilde{z} \in B_{\leqslant}(b,s)$, and $d(b,\tilde{z}) \geqslant d(b,a)+d(a,\tilde{z})-t$ (take $x=b$ and $y=a$ in (16.7.5)). Now $d(a,z)$ can take values as close to $\rho(r)$ as we like, say larger than $\rho(r)-t$. We get $\rho(s) \geqslant d(b,\tilde{z}) \geqslant d(b,a)+d(a,\tilde{z})-t \geqslant d(b,a)+d(a,z)-2t \geqslant d(b,a)+\rho(r)-3t$, proving the inequality $\rho(s) \geqslant d(b,a)+\rho(r)$. □

**Theorem 16.7.7.** *Let $G$ be an abelian group and let $d$ be a translation-invariant metric on $G$. Assume that $d$ is almost lower regular for the triangle inequality. Then for all $r,s \in \mathbf{R}$ with $r>0$, $B_<(a,r)$ is contained in $B_<(b,s)$ if and only if $d(a,b)+\sigma(r) \leqslant \sigma(s)$.*

**Proof.** By the definition of $\sigma$, $B_<(a,r)=B_<(a,\sigma(r))$ (cf. (16.7.3)). The inequality $d(a,b)+\sigma(r) \leqslant \sigma(s)$ implies that $B_<(a,\sigma(r)) \subset B_<(b,\sigma(s))$; hence that $B_<(a,r) \subset B_<(b,s)$.

To prove the other implication we assume that $B_<(a,r) \subset B_<(b,s)$. This means that $d(x,b) \geqslant \sigma(s)$ implies $d(x,a) \geqslant \sigma(r)$. For any given $t>0$ we can find $x$ such that $\sigma(s) \leqslant d(x,b) \leqslant \sigma(s)+t$ and a point $\tilde{x}$ such that $d(x,b) \leqslant d(\tilde{x},b) \leqslant d(x,b)+t$ and $d(\tilde{x},b)+t \geqslant d(\tilde{x},a)+d(a,b)$ (take $y=a$ and $z=b$ in formula (16.7.8) in the definition of almost lower regularity).

We obtain $\sigma(s) \geqslant d(x,b)-t \geqslant d(\tilde{x},b)-2t \geqslant d(b,a)+d(a,\tilde{x})-3t \geqslant d(b,a)+\sigma(r)-3t$. Here $d(a,\tilde{x}) \geqslant \sigma(r)$ since, by hypothesis, $\tilde{x} \notin B_<(a,r) = B_<(a,\sigma(r))$.

If we now let $t$ tend to zero we obtain $\sigma(s) \geqslant d(b,a)+\sigma(r)$, which concludes the proof. □

### 16.7.4 *Strict and non-strict balls*

Since the distances we have constructed are in general not lower regular for the triangle inequality it is desirable to impose only upper regularity. Now upper regularity was used for the inclusion of non-strict balls, lower regularity for the inclusion of strict balls. But, luckily, for the finitely generated distances all strict balls are non-strict! We shall now study metrics with this property.

We consider metric spaces $X$ such that the set $\{d(x,c);\ x \in X\}$ of all distances from a given point $c \in X$ is a discrete subset of $\mathbf{R}$. This is equivalent to the statement that the set $\{d(x,c);\ x \in X,\ d(x,c) \leqslant t\}$ is finite for every real number $t$. The main examples are of course the finitely generated distances in any abelian group.

If the set $\{d(x,c);\ x \in X\}$ of all distances from a fixed point $c$ is discrete, the set is denumerable and either finite or equal to a set $\{r_j;\ j \in \mathbf{N}\}$. In the latter case we shall choose the indices so that $r_0 = 0$ and $r_j < r_{j+1}$ for all $j \in \mathbf{N}$. A sphere $S(c,r)$ is nonempty if and only if $r = r_j$ for some $j$. The functions $\rho$, $\sigma$, $\rho_-$ and $\sigma_+$ defined by (16.7.1) and (16.7.2) satisfy

$$\begin{aligned} &\rho(r) = r_j \text{ and } \sigma_+(r) = r_{j+1} \text{ when } r \in [r_j, r_{j+1}[\,, \\ &\sigma(r) = r_{j+1} \text{ and } \rho_-(r) = r_j \text{ when } r \in \,]r_j, r_{j+1}]\,. \end{aligned} \tag{16.7.14}$$

**Proposition 16.7.8.** *Let $X$ be a metric space where the set of all distances from a given point $c$ is discrete. Then all non-strict balls with center at $c$ are strict and all strict balls with center at $c$ are non-strict. More exactly,*

$$B_{\leqslant}(c,r) = B_{<}(c,\sigma_+(r)) \text{ and } B_{<}(c,r) = B_{\leqslant}(c,\rho_-(r)).$$

**Proof.** We consider the case when the set of distances is infinite—easy modifications will take care of the case of finitely many distances. It is convenient in the proof to define $r_{-1} = -\infty$. Given any $r \in \mathbf{R}$ we choose $j \in \mathbf{N}$ so that $r \in [r_{j-1}, r_j[$. Then $B_{\leqslant}(c,r) = B_{<}(c,s)$ for all $s \in \,]r_{j-1}, r_j]$. In particular $B_{\leqslant}(c,r) = B_{<}(c,r_j) = B_{<}(c,\sigma_+(r))$.

Similarly, given any $r \in \mathbf{R}$ we choose $j \in \mathbf{N}$ such that $r \in \,]r_{j-1}, r_j]$; then $B_{<}(c,r) = B_{\leqslant}(c,s)$ for all $s \in [r_{j-1}, r_j[$. In particular, $B_{<}(c,r) = B_{\leqslant}(c,r_{j-1}) = B_{\leqslant}(c,\rho_-(r))$ for all $r \in \mathbf{R}$. □

**Theorem 16.7.9.** *Let $G$ be an abelian group and $d$ a translation-invariant metric in $G$. Assume that $d$ is upper regular for the triangle inequality and that the set of all distances is discrete. Then for all $r, s \in \mathbf{R}$ with $r > 0$, $B_{<}(a,r)$ is contained in $B_{<}(b,s)$ if and only if $d(a,b) + \rho_-(r) \leqslant \rho_-(s)$.*

**Proof.** We know that the strict ball $B_{<}(a,r)$ is equal to the non-strict ball $B_{\leqslant}(a,\rho_-(r))$. Theorem 16.7.6 shows that the inclusion

$$B_{\leqslant}(a,\rho_-(r)) \subset B_{\leqslant}(b,\rho_-(s))$$

is equivalent to the inequality $d(a,b) + \rho(\rho_-(r)) \leqslant \rho(\rho_-(s))$. However, we always have $\rho(\rho_-(r)) = \rho_-(r)$. This completes the proof. □

**Remark 16.7.10.** A final remark on the prime distances. The elements of the symmetric finite set $P$ will be denoted by $p_0 = 0$ and $p_j = -p_{-j}$, for $j = 1, \dots m$, and the values of the prime distance function $F$ at $\pm p_j =$

$p_{\pm j}$, $j = 1, \ldots, m$, by $a_j > 0$. It is of interest to consider the functions $\varphi, \psi \colon \mathbf{R}_! \to \mathbf{R}_!$ defined by

$$\varphi(r) = \sup_{n_j \in \mathbf{N}} \left(\sum n_j a_j;\ \sum n_j a_j < r\right) \text{ and}$$

$$\psi(r) = \sup_{n_j \in \mathbf{N}} \left(\sum n_j a_j;\ \sum n_j a_j \leqslant r\right).$$

The supremum in the definition of $\rho$ is over all distances from the origin that can occur; these are all of the form of a finite sum $\sum_1^m n_j a_j$ with $n_j \in \mathbf{N}$, which shows that $\rho(r) \leqslant \psi(r) \leqslant r$. Also $\rho_-(r) \leqslant \varphi(r) < r$. However, it may happen that $\rho(r) < \psi(r)$ for some values of $r$, because a certain sum $\sum n_j a_j$, although it is allowed in the definition of an infimal convolution $F \sqcap \cdots \sqcap F$, never occurs as an infimum.

We note that we have $\psi(r) = \rho(r)$ for $\mathbf{Z}^n$ with the $l^\infty$- or the $l^1$-metric. Indeed, here all $a_j$ are equal to 1 and all integers can occur as distances $\sum n_j a_j = \sum n_j$. The same is true for the example in $\mathbf{Z}^2$ we considered with $P$ consisting of nine points and $a = F(1,0)$ and $b = F(1,1)$ provided $a \leqslant b \leqslant 2a$; these inequalities guarantee that all numbers of the form $ma + nb$, $m, n \in \mathbf{N}$, do occur as distances. □

## 16.8 Distance transforms in normed vector spaces

In this section we shall calculate the distance transforms of some subsets of normed spaces. For definitions of the terms used here we refer the reader to Section 9.16.

In all what follows we may take $E$ as $\mathbf{R}^n$. However, the statements for more general spaces are really not more difficult to prove.

### 16.8.1 *The distance transform of a convex set*

In particular we shall look at the case of a convex subset $A$ of a normed vector space. The distance transform of a ball $B = B_\leqslant(a, r)$ is $\mathbf{dt}_B(x) = \|x - c\| - r$ (Exercise 16.10). Another simple convex set is a half-space, and we shall now determine its distance transform:

**Proposition 16.8.1.** *Let $Y$ be a closed half-space in a normed space $E$, defined by an inequality $\xi(x - a) \leqslant 0$ for some continuous linear form $\xi \in E'$, $\xi \neq 0$. Then its distance transform is $\mathbf{dt}_Y(x) = \eta(x - a)$, where $\eta = \xi/\|\xi\|'$.*

**Proof.** We may as well define $Y$ by the inequality $\eta(x-a) \leqslant 0$. We note that $\|\eta\|' = 1$.

By definition $\mathbf{dt}_Y(x) = \inf_{\eta(y)\leqslant\eta(a)} \|x-y\|$ when $x \notin Y$. Without loss of generality we may assume that $x = 0$ and $\eta(a) = -1$; this can be achieved by a change of variable. We shall thus have to prove that

$$\inf_{y\in E} \big(\|y\|;\ \eta(y) \leqslant -1\big) = 1. \tag{16.8.1}$$

This is a well-known fact. To prove it, we note that $\sup_{\|z\|\leqslant 1} \eta(z) = 1$ by the definition of the dual norm, and that, given any $t$ with $0 < t < 1$, we can take $z_t$ of norm at most 1 so that $\eta(z_t) \geqslant 1 - t$. Then $y_t = -z_t/\eta(z_t)$ satisfies $\eta(y_t) = -1$ and $\|y_t\| \leqslant 1/(1-t)$. Hence the infimum in (16.8.1) is at most $1/(1-t)$; thus at most 1.

On the other hand, the infimum is at least 1, for $\|y\| \geqslant |\eta(y)| \geqslant 1$. This proves the formula for $\mathbf{dt}_Y$ in the complement of $Y$.

We note that for open half-spaces the formula is the same. For $x \in Y$ we can therefore use the symmetry $\mathbf{dt}_{\complement Y} = -\mathbf{dt}_Y$. □

If $Y$ is a half-space which is not closed (equivalently, defined by a discontinuous linear functional), then $\mathbf{dt}_Y = 0$ identically.

**Theorem 16.8.2.** *Let $A$ be any convex set in a normed vector space $E$. Then its distance transform is*

$$\mathbf{dt}_A = \sup_Y \big(\mathbf{dt}_Y;\ Y \textit{ is a closed half-space containing } A\big).$$

**Proof.** We first note that the theorem is trivially true if $A = \varnothing$ or $A = E$. Assume now that $A \neq \varnothing, E$.

For every set $Y$ containing $A$ we have $\mathbf{dt}_Y \leqslant \mathbf{dt}_A$, so the supremum in the statement of the theorem can never exceed $\mathbf{dt}_A$.

If $x \notin \overline{A}$ we consider the open ball $B_<(x,r)$, where $r = \mathbf{dt}_A(x) > 0$. This ball and $A$ are disjoint. By Theorem 9.15.2 there is a hyperplane separating the two. This means that there is a continuous linear form $\xi$ such that $\xi(y) \leqslant \xi(z)$ for all $y \in A$ and all $z \in B_<(x,r)$. Thus the closed half-space $Y$ defined by $\xi(y) \leqslant \inf_{z\in B_<(x,r)} \xi(z)$ contains $A$, and $\mathbf{dt}_Y(x) \geqslant r = \mathbf{dt}_A(x)$. This shows that the supremum in the statement of the theorem attains $\mathbf{dt}_A$ at the point $x$.

Now take $x \in A$. Then there is an open ball $B_<(x,r)$ with maximal radius $r = -\mathbf{dt}_A(x) \geqslant 0$ contained in $A$. (Note that $r = 0$ is allowed this time.) Take $s = r + t > r$, where $t$ is arbitrarily small. There exists a point $b$ in $B_\leqslant(x,s) \smallsetminus A$. By the Hahn–Banach theorem (Theorem 9.5.2) there is a closed hyperplane passing through $b$ with $A$ on one side. The corresponding half-space $Y$ solves our problem, for $-\mathbf{dt}_Y(x)$ cannot be larger than $d(x,b) \leqslant s$, so that $\mathbf{dt}_Y(x) \geqslant -s = \mathbf{dt}_A(x) - t$. Since $t$ is arbitrary, this proves the equality in $A$, and by continuity also in its closure $\overline{A}$. We already proved the equality in the complement of $\overline{A}$, so we are done. □

**Corollary 16.8.3.** *The distance transform of an arbitrary convex subset of a normed vector space is a convex function.*

**Proof.** The distance transform of a half-space is an affine function as we have seen. Every affine function is convex, and the supremum of any family of convex functions is convex. □

To prove that the distance transform of a convex set $A$ is convex in $A$ and also convex on every segment contained in $\complement A$ is not difficult. It is not so easy to prove convexity on a segment $[a,b]$ with $a \in A$ and $b \notin A$; this is why we have used the half-spaces here.

### 16.8.2 *Distance transforms under duality*

We shall now relate the discussion on convex sets in Subsection 16.8.1 to the Fenchel transformation and the support function.

**Proposition 16.8.4.** *The distance transform of a closed convex subset $A$ of a normed vector space $E$ is equal to*

$$\mathbf{dt}_A(x) = \sup_{\eta \in E'} \big(\eta(x) - H_A(\eta);\ \|\eta\|' = 1\big), \qquad x \in E.$$

**Proof.** In view of the Hahn–Banach theorem (Theorem 9.5.2), the distance transform at a point $x$ is equal to the supremum of all values $\mathbf{dt}_Y(x)$ when $Y$ varies in the family of all closed half-spaces containing $A$ (Theorem 16.8.2). However, every such half-space contains a minimal half-space, and the minimal half-spaces $Y = \{y;\ \eta(y) \leqslant \alpha\}$ are precisely those for which $\alpha = H_A(\eta)$, thus with distance transform $\mathbf{dt}_Y(x) = \eta(x) - \eta(a) = \eta(x) - H_A(\eta)$ for some $a \in \overline{A} \smallsetminus \mathbf{relint}(A)$. □

**Proposition 16.8.5.** *If $A$ is a nonempty closed convex subset of a normed vector space $E$, the Fenchel transform $\widetilde{\varphi}$ of its distance transform $\varphi = \mathbf{dt}_A$*

*is*

$$\widetilde{\varphi}(\xi) = \begin{cases} H_A(\xi), & \|\xi\|' = 1; \\ +\infty, & \|\xi\|' > 1; \\ \textit{the largest convex minorant of } H_A\big|_{S'}, & \|\xi\|' < 1. \end{cases}$$

*Here $S'$ is the unit sphere for the dual norm $\|\cdot\|'$. If $A$ is empty, then $\widetilde{\varphi}$ is $-\infty$ identically.*

**Proof.** Let $g$ be equal to $H_A$ on the unit sphere $S'$ and $+\infty$ elsewhere, and $h$ equal to $H_A$ on the closed unit ball of $E'$ and equal to $+\infty$ elsewhere. It follows from Proposition 16.8.4 that the Fenchel transform of $g$ is equal to $\mathbf{dt}_A$. The transform of $\varphi = \mathbf{dt}_A$ is therefore the second Fenchel transform of $g$: $\widetilde{\varphi} = \widetilde{\widetilde{g}}$. Theorem 9.17.7 tells us that $\widetilde{\widetilde{g}}$ is the largest lower semicontinuous convex minorant of $g$. Since $h \leqslant g$ we have $\widetilde{\widetilde{h}} \leqslant \widetilde{\widetilde{g}}$, and we know that $\widetilde{\widetilde{h}} = h$ since $h$ is convex, lower semicontinuous, and never takes the value $-\infty$; $h = \widetilde{\widetilde{h}} \leqslant \widetilde{\widetilde{g}} \leqslant g$. Therefore, since $h$ and $g$ agree on $S'$, we have $\widetilde{\widetilde{g}} = g = H_A$ on $S'$. Outside the unit ball $B'$ of the dual space $E'$ (in particular in $E^\star \smallsetminus E'$) we have $g = h = +\infty$, so $\widetilde{\varphi} = \widetilde{\widetilde{g}} \geqslant \widetilde{\widetilde{h}} = +\infty$ there. Finally we note that, since $g$ is plus infinity in the open unit ball, $\widetilde{\varphi} = \widetilde{\widetilde{g}}$ is in that ball as described in the statement of the proposition.

All this holds if $A$ is nonempty; otherwise we easily see that $\widetilde{\varphi} = H_A = -\infty$ identically. □

## 16.9 Notes

### 16.9.1 *Sources*

The first use of a weighted distance transform was made by Hilditch & Rutovitz (1969). Barrow et al. (1977) were the first to use the term *chamfer*. (Gunilla Borgefors, personal communication 2019 September 27.)

For hexagonal grids see (Borgefors 1989).

In my lecture notes (2002, 2004) I forgot to make the hypothesis about the Lipschitz constant in Proposition 16.4.4. Su et al. (2012) corrected my omission.

The concepts of upper and lower regularity for the triangle inequality were introduced in my notes (2002; 2004), however only for translation-invariant distances. The generalization presented here comes easily. The concept of inner regularity for the triangle inequality was introduced by Su et al. (2012:865, Definition 2). All three concepts are different (see my notes from 2002 or 2004 and Su et al. 2012). The weakened forms with almost regularity were introduced by Guo et al. (2019).

Su et al. (2012) studied the Lipschitz constant of distance transforms in relation to regularity properties of metrics, and Guo et al. (2019) pursued this study, introducing the three kinds of almost regular distances.

Theorems 16.7.6 and 16.7.7 on the calculus of balls were proved in my notes (2002) under the hypotheses of upper and lower regularity, respectively; their generalizations assuming only almost regularity are due to Guo et al. (2019).

In my paper (1996) I had all the $F_j$ in Theorem 16.5.2 equal to a single function $F$. Nagy et al. (2018) used two different prime distances, and it is easy to let them all be different. See also (Nagy et al. 2019).

In addition to the many articles mentioned in Section 16.5, see works by Thiel (2001), Sintorn & Borgefors (2002), Banon (2005), Schmitt (2005), Fouard & Gedda (2006), Strand et al. (2006), Talbot (2006), Lindblad et al. (2009), Strand et al. (2011), Strand & Normand (2012), Ćurić et al. (2014), Linnér & Strand (2014), Ilić et al. (2015), Lindblad & Sladoje (2015), and Strand (2016).

Often distances must be measured inside certain sets, not in the plane as a whole. This gives rise to the so-called geodesic distances, studied by Lantuejoul & Maisonneuve (1984).

A stochastic model describing how a forest fire as well as other processes can spread, thus a study of the quench function, is presented by Jean Serra in his article (2009).

For distance transforms on the trihexagonal grid, which involves a tessellation consisting of both triangles and hexagons, see (Kovács et al. 2017); for a study of the triangular grid, see (Kovács et al. 2020). For distance transforms in three dimensions, see (Nagy & Strand 2006) and (Strand et al. 2016).

Molchanov & Terán (2003) define and study a distance transform for gray-level images.

### 16.9.2 *Terms used*

The adjective *distant* is of Latin origin. The prefix *dis-* or *di-* means 'apart, separate'. The other part of the word comes from the present participle of the verb *stare* 'to stand', implying that *distant* means 'to be standing apart'.

It seems that the first use of the term *quench function* is in a report from 1965 by William E. Hartnett.

Borgefors (1984:324, 1986:345) calls $F(x-y)$ the *local distances*; Verwer (1991:672) adopted this term. Starovoitov (1995:501) calls the function $F$ used above the *prime distance function*. The resulting metric, $(x, y) \mapsto d_0(x-y)$, was called a *quasi-Euclidean* distance by Montanari (1968) and a *chamfer distance* by Borgefors (1984:326). The latter term has then been used by, e.g., Verwer (1991:672) and Marchand-Maillet & Sharaiha (2000:21). It has also been called a *finitely generated distance*, for instance in my paper (1996).

However, Borgefors herself now prefers the terms *weighted distance* (used for instance by Borgefors 2005 and Abdelkader et al. 2019), and *integer weighted distance*, and refers to them as *constructed by chamfering* (Gunilla Borgefors, personal communication 2019 September 16); the latter term is derived from the method of calculating distance transforms by passing a mask twice over the image. Nevertheless, the term *chamfer distance* as a short term for the metrics constructed in Corollary 16.5.4 is now quite common.

## 16.10 Exercises

16.1. (a) Prove that, with the notation of Theorem 16.3.7, we have $\theta(A) = 1$ and $\Theta(A) = 3$ when $X = \mathbf{Z}^n$ with the $l^\infty$-metric and $A \neq \varnothing, \mathbf{Z}^n$.

(b) Determine the quantities $\theta(A)$ and $\Theta(A)$ when $X = \mathbf{Z}^n$ with the $l^1$-metric.

16.2. Prove Proposition 16.4.2.

16.3. Prove that the distance transform of a ball $B_\leqslant(0, r)$, $r \geqslant 0$, in a normed vector space is given by $\mathbf{dt}_{B_\leqslant(0,r)}(x) = \|x\| - r$.

16.4. Let $A$ be the filled square $\{x \in \mathbf{R}^2;\ \|x\|_\infty \leqslant 1\}$. Calculate the distance transforms $\mathbf{dt}_A$ when the distance is given by $d(x, y) = \|x - y\|_p$ for (a): $p = 1$; (b): $p = 2$. (For $p = \infty$, we get the result by solving Exercise 16.3 above.)

16.5. Let $B$ be the Euclidean closed disk of radius $r$, viz.

$$B = B_\leqslant(0, r) = \{x \in \mathbf{R}^2;\ \|x\|_2 \leqslant r\}.$$

Calculate the distance transforms $\mathbf{dt}_B$ when the distance is given by $d(x, y) = \|x - y\|_p$ for (a): $p = 1$; (b): $p = \infty$. (For $p = 2$, see

Exercise 16.3.)

16.6. (a) Given a subset $A$ of an abelian group with a translation-invariant metric, show that $\mathbf{dt}_{A+B_<(0,r)} = \mathbf{dt}_A - r$ holds in $\complement(A + B_<(0, r))$ (cf. Proposition 16.3.13).

(b) Is the statement in (a) true in $\mathbf{Z}^2$?

(c) Show that, in $\mathbf{R}^n$ with the $l^2$-metric, $\mathbf{dt}_{A+B_<(0,r)} = \mathbf{dt}_A - r$ holds everywhere if $A$ is convex.

16.7. Give an example which shows that the inclusion

$$\varepsilon_{B_\leqslant(0,r)} \supset \{\mathbf{dt}_A < -r\}$$

in (16.4.1) can be strict.

16.8. Let $X$ be a metric space with metric $d$. Define a metric $d_\gamma$ in $X \times \mathbf{R}$ by

$$d_\gamma\big((x, s), (y, t)\big) = \gamma d(x, y) + |s - t|, \qquad (x, s), (y, t) \in X \times \mathbf{R},$$

where $\gamma$ is any positive number. Prove that, for any subset $A$ of $X$,

$$\mathbf{dt}_{\mathbf{epi}(\mathbf{dt}_A)}(x, s) = \mathbf{dt}_A(x) - s, \qquad (x, s) \in X \times \mathbf{R},$$

if $\gamma \geqslant 2$ ($\gamma \geqslant 1$ in a normed vector space with distance $d(x, y) = \|x - y\|$).

Show by examples that this is not necessarily so if $\gamma < 2$ ($\gamma < 1$ in a normed vector space).

16.9. *Square pixels.* Assume that the centers of square pixels are placed at the points $\mathbf{Z}^2$ with integer coordinates in $\mathbf{R}^2$. Since $\mathbf{Z}^2$ is group, we can use infimal convolution. Define a function $F: \mathbf{R}^2 \to [0, +\infty]$ by $F(0) = 0$, $F(\pm 1, 0) = F(0, \pm 1) = a > 0$, $F(\pm 1, \pm 1) = b > 0$ (four points with value $b$), and $F(x) = +\infty$ at all other points. Then let $h = \lim F \sqcap F \sqcap \cdots \sqcap F$ as the number of terms tends to infinity.

For all $a, b > 0$ there is a constant $C$ depending on $a, b$ such that

$$C^{-1} \leqslant \frac{h(x)}{\|x\|_2} \leqslant C, \qquad x \in \mathbf{Z}^2.$$

Show that the smallest possible value of $C$ is

$$c = \lambda(h, \|\cdot\|_2) = \sqrt[4]{4 - 2\sqrt{2}} \approx 1.04038,$$

and that it is attained when $a$ and $b = a\sqrt{2}$ take the values (16.6.1).

16.10. *Hexagonal pixels.* The centers of hexagonal pixels form a triangular pattern: if we identify $\mathbf{R}^2$ with $\mathbf{C}$ for convenience they can be placed at the points $p + q\omega$, $p, q \in \mathbf{Z}$, where $\omega = \frac{1}{2} + \frac{1}{2}\mathrm{i}\sqrt{3}$. (See also Subsection 15.6.5 and Exercise 7.3.)

Define a function $F\colon \mathbf{C} \to [0, +\infty]$ by $F(0) = 0$, $F(\pm 1) = F(\pm\omega) = F(\pm\overline{\omega}) = 1$, and $F(x) = +\infty$ at all other points. Then let $h = \lim F \sqcap F \sqcap \cdots \sqcap F$ as the number of terms tends to infinity. Show that

$$1 \leqslant \frac{h(x)}{\|x\|_2} \leqslant \frac{2}{\sqrt{3}} \approx 1.1547, \qquad x \in \mathbf{Z} + \mathbf{Z}\omega.$$

Thus $\lambda(h, \|\cdot\|_2) \leqslant \sqrt{2}/\sqrt[4]{3} \approx 1.07457$.

16.11. *Triangular pixels.* The barycenters of the triangular pixels form a hexagonal pattern. In Figure 1.6 on page 12 they have the coordinates $(x, \sqrt{3}(y + 1/6))$ for $(x, y) \in \mathbf{Z}^2$, $x$ even, and $(x, \sqrt{3}(y - 1/6))$ for $x$ odd. These hexagonally placed points do not form a subgroup of $\mathbf{R}^2$, since $(0, \sqrt{3}/6) + (1, -\sqrt{3}/6) = (1, 0)$, which is not a barycenter. Therefore infimal convolution cannot be applied directly as in the case of square or hexagonal pixels. But we can equally well use $(x, \sqrt{3})$, $(x, y) \in \mathbf{Z}^2$, as addresses for the pixels—they form an abelian group.[7]

Prove that with the distance from $(0, 0)$ to $(1, 0)$ and the distance form $(0, 0)$ to $(1, \sqrt{3})$ equal to the respective Euclidean distances in the finitely constructed distance $h$, we get $\lambda(h, \|\cdot\|_2) = \sqrt[4]{2} \approx 1.1892$.

16.12. Prove that in an abelian group with a translation-invariant metric which is upper regular for the triangle inequality, $B_{\leqslant}(a, r) = B_{\leqslant}(b, s)$ implies that $a = b$ provided $r, s \geqslant 0$.

16.13. Prove that the finitely generated distance in $\mathbf{Z}^2$ defined by

$$F(\pm 1, 0) = F(0, \pm 1) = 3, \quad F(\pm 1, \pm 1) = 4$$

is not lower regular for the triangle inequality.

16.14. Calculate the values of the functions $\rho$, $\sigma$, $\rho_-$ and $\sigma_+$ for $0 \leqslant r \leqslant 3$ when the space is $\mathbf{Z}^2$ and the distance is given by the $l^2$-norm. (For definitions of the functions, see equations (16.7.1) and (16.7.2).)

[7]Another solution is to use only an even number of steps; see (Abdalla & Nagy 2019). See also (Abuhmaidan & Nagy 2020) for translations in the triangular grid.

# Chapter 17

# Skeletonizing

## 17.1 Introduction

SKELETONS are sets within a given set which are thin in some sense but nevertheless carry meaningful information about its parent set. There are several related concepts, now to be studied.

## 17.2 Defining skeletons, medial axes, and nervures

If $A$ is any subset of a metric space $X$, then its interior $A^\circ$ is the union of all open balls contained in $A$. This is typically the union of a very large family of sets. We would like to describe $A^\circ$ as the union of a smaller family. It is obvious that if we have two balls contained in $A$, $B_<(a,r)$ and $B_<(b,s)$, and one is contained in the other, then we may throw away the smaller ball without changing the union. In fact, for every ball $B_<(a,r)$ in the union, we may throw away all balls strictly contained in that ball without changing the union. This leads to the concept of a maximal ball. A maximal ball must be retained, but all balls strictly contained in a maximal ball may be dispensed with.

The importance of skeletons in applications is due to the fact that they are thin in some sense but nevertheless retain essential information about an object, for instance its general shape, and that, given the skeleton and the distance transform at the points in the skeleton, we can reconstruct the whole object. Typically we save memory when listing only the skeleton and the quench function—the two are combined in the concept of internal medial locus, to be defined in Definition 17.2.5.

If $a$ is the center of a maximal strict ball $B_<(a,r)$ contained in a set $A$, then necessarily $r = -\mathbf{dt}_A(a)$. In fact, when we defined the distance transform $\mathbf{dt}_A(a)$ at a point $a \in A$, we looked at all balls with center $a$ contained in $A$ and we took the largest such ball. Then we kept the center fixed. There is a largest ball with center $a$, which in particular is maximal among these balls. By way of contrast, when we define the skeleton we shall vary both the center and the radius and look at all balls contained in $A$, regardless of their centers. We shall now give a name to the centers of maximal balls.

**Definition 17.2.1.** Let $A$ be a subset of a metric space $X$. We define the ***skeleton*** of $A$, denoted by $\mathbf{skel}(A)$, as the set of all centers of maximal nonempty strict balls contained in $A$. □

To define the skeleton we have to choose a distance. The concept is of interest both for $X = \mathbf{R}^n$ and $X = \mathbf{Z}^n$.

The definition means that $a \in \mathbf{skel}(A)$ if and only if there exists a number $r > 0$ such that $B_<(a,r) \subset A$ and such that if a ball $B_<(b,s)$ is contained in $A$ and contains $B_<(a,r)$, then $B_<(b,s) = B_<(a,r)$. The skeleton may be empty: think of a set with empty interior or of a half-space in $\mathbf{R}^n$. A half-space contains lots of balls, but there are no maximal balls. So obviously we need to investigate whether there exist maximal balls—and whether there are enough of them in the formation of the interior of $A$. For this we shall need the Kuratowski–Zorn lemma.

There is another possibility, viz. to consider maximal closed balls contained in a closed set. See (Matheron 1988c), in particular page 218, for a discussion of the various properties of this somewhat different concept.

To proceed we shall need several definitions related to a given metric:

**Definition 17.2.2.** Given a set $A$ in a metric space $X$ with distance function $d$ and a point $x \in X$, we define the ***proximity set of*** $A$ ***at*** $x$, denoted by $\mathbf{prox}_A(x)$, as the set of all points $y \in \overline{A}$ such that $d(x,y) \leqslant d(x,z)$ for all $z \in \overline{A}$. □

**Definition 17.2.3.** Given a set $A$ in a metric space $X$ with distance function $d$ and a point $x$, we define the ***multiplicity set of*** $A$ ***at*** $x$, denoted by $\mathbf{mult}_A(x)$, as the set of all points $x \in X$ such that $\mathbf{prox}_A(x)$ consists of more than one point. □

Thus $\mathbf{prox}_A$ and $\mathbf{mult}_A$ are functions $X \to \mathscr{P}(X)$; the first has its values in $\overline{A}$. If $a \in \overline{A}$, then $\mathbf{prox}_A(a) = \{a\}$ and $\mathbf{mult}_A(a)$ is empty.

If $A$ is convex in a Euclidean space, then $\mathbf{mult}_A(x)$ is empty for every $x$. However, if we use the $l^1$-norm or the $l^\infty$-norm, then a convex set may well have a nonempty multiplicity set $\mathbf{mult}_A(x)$ for some $x \notin \overline{A}$.

**Definition 17.2.4.** Let $A$ be a subset of $\mathbf{R}^n$ provided with a metric. We define the ***medial axis***, of $A$, denoted by $\mathbf{medial}(A)$, as the set if all points $a \in A$ such that $a$ belongs to $\mathbf{mult}_{\complement(A^\circ)}(a) = \mathbf{mult}_{\partial A}(a)$. □

Thus $\mathbf{medial}(A)$ is the set of all points $a$ in $A$ such that $\mathbf{prox}_{\partial A}(a)$, defined using the chosen distance, has more than one point.

Also this definition depends on the norm chosen. In $\mathbf{R}^2$ with the $l^1$- or $l^\infty$-norm, the medial axis can be all of $A^\circ$.

**Definition 17.2.5.** The ***internal medial locus*** is the set of pairs $(c, r) \in A \times \mathbf{R}$, where $c$ is the center of a maximal inscribed and nonempty strict ball in the set $A$, and $r$ the radius of the ball. □

**Definition 17.2.6** (Rivière 1987:435; 1996:9–10). Let $A$ be a subset of $\mathbf{R}^n$. We define the ***nervure*** of $A$, to be denoted by $\mathbf{nervure}(A)$, as the set of all points $a$ in $A$ such that there are no points $b$ and $p \in \mathbf{prox}_{\partial A}(b)$ such that $a$ belongs to a segment $[b, p]$ with $b \neq a \neq p$. □

Since $\mathbf{prox}_{\partial A}(b)$ depends on the distance chosen, so does the nervure.

The three sets just defined are closely related. We shall now compare them.

## 17.3 Comparing medial axes, nervures, and skeletons

**Theorem 17.3.1.** *In a Euclidean space $E$,*

$$\mathbf{medial}(A) \subset \mathbf{nervure}(A) = \mathbf{skel}(A)$$

*for every subset $A$ of $E$.*

**Example 17.3.2.** The medial axis and the skeleton may be different. Let $A$ be the set in $X = \mathbf{R}^2$ defined by $x_2 \geqslant x_1^2$. If we use the $l^2$-norm, then it is easy to see that $(0, a_2)$ belongs to $\mathbf{skel}(A)$ iff $a_2 \geqslant \frac{1}{2}$ and that $(0, a_2)$ belongs to $\mathbf{medial}(A)$ iff $a_2 > \frac{1}{2}$.

If instead we use the $l^\infty$-norm, then $\mathbf{medial}(A) = \mathbf{skel}(A)$, and $a$ belongs to this set iff $a = (0, a_2)$ with $a_2 > 0$.

For the $l^1$-norm, $\mathbf{medial}(A) = \mathbf{skel}(A)$, and $a$ belongs to this set iff either $a = (0, a_2)$ with $a_2 \geqslant 1$ or $a$ belongs to a piece of an algebraic curve of degree four joining $(0, 1)$ and $(\frac{1}{2}, \frac{1}{4})$ or its mirror image in the $x_2$ axis. □

**Example 17.3.3.** If we go to a discrete version of the example above, things are different. So let now the space be $X = \mathbf{Z}^2$, and take $A$ defined by $x_2 \geqslant x_1^2$. Then the medial axis and the skeleton consist of all points $a = (0, a_2)$ with $a_2 \in \mathbf{N}$, equally for the three metrics $l^2$, $l^\infty$ and $l^1$. □

**Example 17.3.4.** There are easy examples, for instance with a finite space $X$, which show that in general none of the inclusions $\mathbf{medial}(A) \subset \mathbf{skel}(A)$, $\mathbf{medial}(A) \supset \mathbf{skel}(A)$ holds.

Take $X = \{(0,0), (1,0), (2,0), (-1,1), (-1,-1)\}$ with the $l^2$-metric and $A = \{(0,0), (1,0), (2,0)\}$. Then $(0,0)$ belongs to $\mathbf{medial}(A)$ but not to $\mathbf{skel}(A)$.

If we take $X = \mathbf{Z}$ and $A = [0, 2m+1]_\mathbf{Z}$, then

$$\mathbf{skel}(A) = \{m-1, m+1\} \text{ while } \mathbf{medial}(A) \text{ is empty.}$$

But the choice of $X$ is also important: if we take $X = \mathbf{N}$, then the skeleton of the set $A$ as above becomes the set $[0, m+1]_\mathbf{Z}$ and the medial axis is empty. □

**Proof of Theorem 17.3.1.** (1) $\mathbf{medial}(A) \subset \mathbf{nervure}(A)$. Take a point $a \in \mathbf{medial}(A)$ and two points $x$ and $y$ on the boundary of $A$ such that $d(a,x) = d(a,y)$. We have to prove that there is no point $b \neq a$ such that $a$ belongs to the segment $[b,p]$, with $p \in \mathbf{prox}_{\partial A}(b)$.

So assume that there is such a point $b$, and define $r = d(a,x)$ and $q = d(b,a)$. We shall prove that $d(b,x) < d(b,p)$. Since $a \in [b,p]$, it is clear that $d(b,p) = d(b,a) + d(a,p) \geqslant q + r$. It is also clear that $d(b,x) \leqslant d(b,a) + d(a,x) = q + r$. But equality in the triangle inequality can occur only when the three points are on a straight line; in this case we must have $a \in [p,x]$. Then $p$ cannot belong to $\mathbf{prox}_{\partial A}(b)$, which is required in the definition. This impossibility proves that there is no point $b$ with the properties mentioned; in other words, that $a \in \mathbf{nervure}(A)$.

(2) $\mathbf{nervure}(A) \subset \mathbf{skel}(A)$. If $a \notin \mathbf{skel}(A)$, there exist a number $r$ and a point $p \in \partial A$ with $d(a,p) = r$ such that the open ball $B_<(a,r)$ is not maximal. This implies that there is an open ball $B_<(b,s)$ strictly containing the first ball. Then necessarily $a$ must belong to the segment $[b,p]$ so that $a \notin \mathbf{nervure}(A)$.

(3) $\mathbf{nervure}(A) \supset \mathbf{skel}(A)$. Conversely, if $a \notin \mathbf{nervure}(A)$, there are points $b$ and $p \in \mathbf{prox}_{\partial A}(b)$, $b \neq a \neq p$, such that $a \in [b,p]$. Then $B_<(a,r) \subset B_<(b,s)$, where $r = d(a,p)$ and $s = r + d(a,b) > r$, which proves that $B_<(a,r)$ is not maximal: $a \notin \mathbf{skel}(A)$. □

## 17.4 Existence of skeletons

We recall the definition of an inductive order and the Kuratowski–Zorn lemma, both mentioned in Section 9.5. The latter will be used in the proof of Corollary 17.4.4.

**Theorem 17.4.1.** *Let $\mathbf{Z}^n$ be equipped with a metric which is either inherited from a norm on $\mathbf{R}^n$ or a finitely generated distance in the sense of Definition 16.5.5. Let $A$ be a finite subset. Then the set of all nonempty strict balls contained in $A$ is inductively ordered.*

**Proof.** Let us consider the set $A_M = \mathbf{Z}^n \cap \bigcup_{(c,r)\in M} B_<(c,r)$ of a family of strict balls contained in $A$, where $M$ is a subset of $\mathbf{Z}^n \times \mathbf{R}$. Assume that the family is totally ordered, i.e., for any two pairs $(a,r),(b,s) \in M$, either $B_<(a,r)$ is a subset of $B_<(b,s)$ or conversely. Clearly $A_M$, being a subset of $A$, is finite, which implies that it is equal to one of the balls $B_<(c,r)$ with $(c,r) \in M$. We are done. □

In $\mathbf{R}^n$ things are less simple:

**Theorem 17.4.2.** *Let $A$ be a bounded set in a finite-dimensional normed vector space $E$. Then the set of open balls contained in $A$ is inductively ordered.*

**Proof.** Let as before $A_M = \bigcup_{(c,r)\in M} B_<(c,r)$ be a union of open balls contained in $A$, where $M$ is a subset of $E \times \mathbf{R}$. We assume that the family of these balls is totally ordered. Define $r^*$ as the supremum of all numbers $r$ such that there exists a $c \in E$ such that $(c,r) \in M$. Since $A$ is bounded, this supremum is necessarily finite. For every $j = 1,2,3,\dots$ there exists a number $r_j > r^* - 1/j$ and a point $c_j$ in $A$ such that $(c_j, r_j) \in M$. Unless one of the radii is equal to $r^*$ we may also choose the $r_j$ so that $r_{j+1} > r_j$. Since the sequence of centers $(c_j)$ is bounded, it has a converging subsequence; let us change notation so that $(c_j)$ itself is converging. Let its limit be $c^*$. We claim that $A_M$ is equal to the ball $B_<(c^*, r^*)$.

We shall prove first that $A_M$ is contained in $B_<(c^*, r^*)$. Let $x \in A_M$; there exists $(c,r) \in M$ such that $d(x,c) < r$. Define $t = r - d(x,c) > 0$. Then the ball $B_<(x,t)$ is contained in $B_<(c,r)$. If $r = r^*$, then we must have $B_<(c,r) = B_<(c^*, r^*)$, in particular $x$ belongs to $B_<(c^*, r^*)$. If, on the other hand $r < r^*$, then we take $k$ so large that $r_k > r$, and we must have $B_<(c,r) \subset B_<(c_k, r_k)$ since the opposite inclusion is impossible and one of them must hold by hypothesis. Thus $B_<(x,t) \subset B_<(c_j, r_j)$ for all $j \geqslant k$, which implies, on letting $j$ tend to infinity, that $B_<(x,t) \subset B_\leqslant(c^*,r^*)$. But then $x$ belongs also to the open ball $B_<(c^*,r^*)$ and we are done. (Note that this part of the proof is not valid in a general metric space.)

In the other direction, let us prove that $B_<(c^*,r^*)$ is contained in $A_M$. Take any point $x \in B_<(c^*,r^*)$. Then $d(x,c^*) < r^*$, and we may define $t = r^* - d(x,c^*) > 0$. We then have $d(x,c_j) \leqslant d(x,c^*) + d(c^*,c_j) = r^* - t + d(c^*,c_j)$, where the right-hand side is less than $r_j$ for large $j$. Hence $x$ belongs to $B_<(c_j,r_j)$, and therefore to $A_M$. (This part of the proof is valid in any compact metric space.) □

If the norm is Euclidean, it is enough to assume that $A$ does not contain a half-space. Also, for any given norm in $\mathbf{R}^n$, it is enough to assume that $A$ does not contain a cone of a certain aperture.

**Example 17.4.3.** One might think that the result should hold in any compact metric space. However, simple examples show that this is not so. Define a compact metric space $X$ as consisting of the segment $A = [-1,1]\times\{0\}$ in $\mathbf{R}^2$ and adjoin to it the point $(0,1)$. The metric shall be that induced by the usual Euclidean metric in $\mathbf{R}^2$. Consider now the open balls $B_<(c_j,r_j)$, where $c_j = (1/j,0)$ and $r_j = 1$, $j = 1,2,\ldots$. Note that $B_<(c_j,r_j) = ]-1+1/j,1] \subset A$. The sequence of balls is increasing with $j$, and its union is the segment $A_M = ]-1,1] \times \{0\}$. However, this segment is not an open ball in $X$. What is worse: it is not contained in an open ball in $A$. Hence the family of open balls in $A$ is not inductively ordered. □

There is a well-known trick to get rid of the hypothesis that $A$ must not contain any half-space. Indeed, a half-space is a limiting case of a ball, and if we compactify $\mathbf{R}^n$ by adding a point at infinity, then $\mathbf{R}^n \cup \{\infty\}$ can be regarded as the $n$-sphere. The balls in $\mathbf{R}^n$ then become spherical caps on the sphere, and so do the half-spaces. But these spherical caps are actually balls for a suitable metric on the sphere. So $\mathbf{R}^n \cup \{\infty\}$ is a metric space and the open balls contained in any given subset are inductively ordered—the proof is very much like the one we have done in $\mathbf{R}^n$. The compactification allows us to define a generalized skeleton, which may contain $\infty$, but we need of course to modify the quench function.

**Corollary 17.4.4.** *Let $A$ be a bounded subset of a finite-dimensional normed vector space, or a bounded subset of $\mathbf{Z}^n$, where $\mathbf{Z}^n$ is provided with a metric as in Theorem 17.4.1. The union of all open balls with center $c$ belonging to the skeleton and radius equal to $-\mathbf{dt}_A(c)$ is equal to the inte-*

*rior of A. In particular, if A has interior points, then the skeleton of A is nonempty.*

**Proof.** Take any point $x \in A^\circ$. The ball $B_<(x,t)$ is contained in $A^\circ$ for some small positive $t$. By the Kuratowski–Zorn lemma (Theorem 9.5.1) and Theorem 17.4.1 or 17.4.2, respectively, there is a maximal ball $B_<(c,r)$ containing $B_<(x,t)$ and contained in $A$. Thus $c \in \mathbf{skel}(A)$ and $x \in B_<(c,r)$, with $r = -\mathbf{dt}_A(c)$. □

In any metric space where the conclusion of Theorem 17.4.2 holds, we have $A^\circ = \bigcup_{c \in \mathbf{skel}(A)} B_<(c, -\mathbf{dt}_A(c))$. Here $-\mathbf{dt}_A(c) = (\mathbf{dt}_A(c))^-$ is the quench function evaluated at $c$. Knowledge of $\mathbf{skel}(A)$ and the restricion of $\mathbf{dt}_A$ to $\mathbf{skel}(A)$ is equivalent to knowing $A^\circ$. This shows how we can reconstruct $A^\circ$ from $\mathbf{skel}(A)$ and the quench function. However, it is sometimes not necessary to use all the points in the skeleton, e.g., when $A$ is the union of two disks; see Exercise 17.3.

## 17.5 Properties of skeletons

In some sense the skeleton is a thin set. For instance, it is easy to prove that a skeleton in $\mathbf{R}^n$ has no interior points (cf. Exercise 17.1).

The skeleton has, generally speaking, bad continuity properties.

**Example 17.5.1.** Let $D = B_<(0,1)$ be the open unit disk in $\mathbf{R}^2$. Its skeleton is just the origin. Then add a small open disk $D_t = B_<((1,0),t)$ with center at $(1,0)$ and radius $t > 0$. The skeleton of the new set $A = D \cup D_t$ is the entire segment $[0,1] \times \{0\}$ for all small positive $t$. Thus a very small change in the set causes the skeleton to grow. Note that here it is not necessary to use all the points in the skeleton to reconstruct $A$: it suffices to take the disks with centers at 0 and $(1,0)$. Even more dramatic is perhaps the growth in the skeleton when we remove a small closed disk: consider $D \smallsetminus \overline{D_t}$. □

In $\mathbf{Z}^2$ the continuity properties are of course different, but also in this case, a small change can cause points to appear far from the original skeleton; see Exercise 17.6.

**Example 17.5.2.** Let $A = [-m,m]_\mathbf{Z} \times [-m,m]_\mathbf{Z}$ be a large square in $\mathbf{Z}^2$. Its skeleton for the $l^\infty$-metric is just the origin. If we add a single point $(m+1,0)$ to $A$, the skeleton of the new set is $\{0,(m+1,0)\}$. What happens if we remove a corner point? See Exercise 17.5. □

The skeleton of a set $A$ in $\mathbf{R}^2$ need not be a closed set, even if $A$ has a smooth boundary.

**Example 17.5.3.** Let first $C = A + B_<(0,1) \subset \mathbf{R}^n$, the dilation of the segment $A = [(-1,1),(1,1)]$ by the open unit disk for the $l^2$-norm. This is an open set and its skeleton is $\mathbf{skel}(C) = A$. Then modify $C$ as follows. The lower boundary of $C$ is just the segment $[(-1,0),(1,0)]$, so that $C$ is locally defined by the inequality $y > 0$ near this segment. We replace it by an inequality $y > \varphi(x)$, where $\varphi(x) = \sum_1^\infty c_j \psi(2^j x - 1)$, $\psi$ being an even, nonpositive function in $C_0^\infty$ with $\psi(0) = -1$, with support contained in $[-\frac{1}{3}, \frac{1}{3}]$ and such that $\psi'(x) \geqslant 0$ for $x \geqslant 0$. For a suitable choice of positive constants $c_j$, the function $\varphi$ is in $C^\infty$, and the skeleton of the new open set $V$ so defined contains segments $\{2^{-j}\} \times [a,b]$ for $j = 1,2,3,\ldots$ and a suitable choice of $a$ and $b > a$. But no point on the limiting segment $\{0\} \times [a,b]$ belongs to the skeleton. □

These bad continuity properties led Christian Lantuéjoul (1978, 1980) to modify the definition of the skeleton in order to obtain a better stability.

We shall now give a characterization of the points in the skeleton.

**Theorem 17.5.4.** *Let $E$ be a normed space with metric given by the norm: $d(x,y) = \|x - y\|$. Let $A$ be a nonempty proper subset of $E$, fix a point $c$ in the interior of $A$, and define $h(x) = d(x,c) + \mathbf{dt}_A(x)$, $x \in E$. Then $c$ belongs to the skeleton of $A$ if and only if $h$ has a minimum only at $c$.*

**Proof.** If $B_<(c,r) \subset B_<(x,s) \subset A$, where $r = -\mathbf{dt}_A(c) > 0$, then $s \geqslant d(c,x) + r$. This implies that $h(x) = d(x,c) + \mathbf{dt}_A(x) \leqslant d(x,c) - s \leqslant d(c,c) - r = h(c)$. If $c$ is the only point where $h$ attains its minumum, we must have $x = c$ and it follows that $B_<(c,r)$ is maximal, hence that $c \in \mathbf{skel}(A)$.

Conversely, assume that $c$ is in the skeleton and that $x$ is a point where $h(x) \leqslant h(c)$. Then $d(c,x) - s \leqslant -r$, where we define $r = -\mathbf{dt}_A(c)$, $s = -\mathbf{dt}_A(x)$. This implies that $B_<(x,s)$ contains $B_<(c,r)$. Since $c$ is in the skeleton of $A$, the two balls must be equal, which implies that $x = c$. Therefore the infimum of $h$ is attained at $c$ and only there. □

Thanks to the calculus of balls developed in Section 16.7 we can generalize this result to other groups. In a normed space of positive dimension, the open ball of radius $r = -\mathbf{dt}_A(a)$ is the interior of the closed ball of the same radius and the same center. In a group where the set of distances is discrete, the open ball $B_<(a,r)$ can be described as the closed ball of radius

$\rho_-(r)$. Since the conditions for working with closed balls are more easily satisfied than those for open balls, we will get a more applicable result if we replace the function $x \mapsto d(x,c) + \mathbf{dt}_A(x)$ by $x \mapsto d(x,c) - \rho_-(-\mathbf{dt}_A(x))$; to wit:

**Theorem 17.5.5.** *Let $G$ be an abelian group with a translation-invariant metric $d$ which is upper regular for the triangle inequality and such that the set of all distances is discrete. Let $A$ be a nonempty proper subset of $G$, fix a point $c \in A$, and define $h(x) = d(x,c) - \rho_-(-\mathbf{dt}_A(x))$, $x \in G$. Then $c$ belongs to the skeleton of $A$ if and only if $h$ has a minimum only at $c$.*

**Proof.** If $B_<(c,r) \subset B_<(x,s) \subset A$, where $r = -\mathbf{dt}_A(c)$, then by Theorem 16.7.9, $\rho_-(s) \geqslant d(c,x) + \rho_-(r)$. This implies that $h(x) = d(x,c) - \rho_-(-\mathbf{dt}_A(x)) \leqslant d(x,c) - \rho_-(s) \leqslant d(c,c) - \rho_-(r) = h(c)$. If $c$ is the only point where $h$ attains its minumum, we must have $x = c$ and it follows that $B_<(c,r)$ is maximal, hence that $c \in \mathbf{skel}(A)$.

Conversely, assume that $c$ is in the skeleton and that $x$ is a point where $h(x) \leqslant h(c)$. Then $d(c,x) - \rho_-(s) \leqslant -\rho_-(r)$, where we define $r = -\mathbf{dt}_A(c)$, $s = -\mathbf{dt}_A(x)$. This implies that $B_\leqslant(x,\rho_-(s))$ contains $B_\leqslant(c,\rho_-(r))$. But $B_\leqslant(x,\rho_-(s)) = B_<(c,s)$ and $B_\leqslant(c,\rho_-(r)) = B_<(c,r)$ by Proposition 16.7.8. Since $c$ is in the skeleton of $A$, the two balls must be equal, which implies that the difference $x - c$ between their centers belongs to a bounded subgroup, thus that $x = c$. Therefore the infimum of $h$ is attained at $c$ and only there. □

## 17.6 Notes

### 17.6.1 *Sources*

Harry Blum started the study of skeletons with his seminal paper (1967). There we find the expressions *grass fire* (p. 363), *medial axis function* (p. 366), *skeleton* (p. 366, footnote 1). See also his paper (1973). Other early papers are those by Pfaltz & Rosenfeld (1967), Rosenfeld & Pfaltz (1968), Hilditch (1969) and Hilditch & Rutovitz (1969).

Three pioneering texts on skeletons are the doctoral thesis of Christian Lantuéjoul (1978), the reports by Georges Matheron (1978b, 1981), later published as two chapters in the 1988 book edited by Jean Serra: (Matheron 1988c, 1988d).

For later results and surveys of the concepts studied in the present chapter, see, e.g., publications by Borgefors (1984, 1986, 1989, 1994, 1996a,

1996b, 2003, 2005), Borgefors et al. (2008), Hulin & Thiel (2006) and Saha et al. (2017).

The definition of *internal medial locus* appears in (Pizer et al. 2008:8).

Matheron (1978b:33; 1988c:238) proved that the closures of the three sets, the medial axis, the nervure, and the skeleton are equal.

The inclusion $\mathbf{medial}(A) \subset \mathbf{skel}(A)$ in Theorem 17.3.1 was proved by Serra (1982:383).

The inclusion $\mathbf{medial}(A) \subset \mathbf{nervure}(A)$ in Theorem 17.3.1 was proved by Alain Rivière (1987:345, 1996:10).

As mentioned above, a skeleton is of Lebesgue measure zero. On the other hand, its closure need not be of Lebesgue measure zero (cf. Exercise 17.2). These results are mentioned by Serra (1982:378) and Matheron (1988c:218). It is also stated there that it is unknown whether the skeleton has Lebesgue measure zero, and whether the interior of its closure is empty. For the topology of a skeleton, see (Matheron 1978b); for the negligibility of a definition of the skeleton which is different from the usual one, see (Matheron 1988d).

The questions just mentioned were answered by Alain Rivière (1996:12), who proved that the nervure of an open set $\Omega$ in a Euclidean space has Lebesgue measure zero. He also proved that the medial axis of an open set $\Omega$ can be dense; hence also $\mathbf{nervure}(\Omega) = \mathbf{skel}(\Omega)$ can be dense in $\Omega$. In his paper (2001: Théorème 2), Rivière proved that the skeleton can be dense in some convex open set in $\mathbf{R}^n$, $n \geqslant 2$. See also Exercise 17.2.

Example 17.5.3 is essentially taken from (Matheron 1988c:219). Theorem 17.5.4 was proved in $\mathbf{R}^n$ by Matheron (1988c:225).

Eckhardt and Latecki (1996) exhibit a method of thinning a digital image to reach the skeleton. For a method to calculate distance transforms and skeletons in three dimentions, see (van Dortmont et al. 2006). For a variant of the skeleton construction using Voronoi diagrams, see, e.g., (Langer et al. 2019).

Imiya et al. (2003) introduced a transformation called *digital curvature flow*, a motion controlled by the curvature at the boundary of an object and which can be used to obtain the skeleton of a planar digital object.

### 17.6.2 *Terms used*

The terms used in this area vary a lot. Some authors define medial axis to be equal to the skeleton; others define them as two different concepts. Some authors describe thinning as a procedure resulting in a skeleton without formally defining the latter.

It seems that the term *skeleton* was first used in this context by Rosenfeld & Pfaltz (1966). Rivière (1996) used the French term *nervure*$_{(f)}$; we have called it by the English word *nervure* here. Sometimes it is called *rib.*

Matheron (1988c) defines skeletons using maximal open balls. He also discusses the use of maximal closed balls.

We have here followed Jean Serra, who defines the skeleton of an open set with the help of maximal open balls (1982:376), and the medial axis (1982:382) as we have done in Section 17.2.

Najman & Talbot (2008:40) define *axe médian* with the help of maximal closed balls. In Euclidean spaces, *l'axe médian* is called *squelette*$_{(m)}$.

## 17.7 Exercises

17.1. Show that in $\mathbf{R}^n$ with a Euclidean metric every skeleton has an empty interior.

17.2. Find an open set $U$ in $\mathbf{R}^2$ such that the closure of its skeleton with respect to the usual topology of $\mathbf{R}^2$ is of Lebesgue measure equal to that of $U$ minus a given positive number.
*Hint:* Take $U$ as the unit disk, and add to it denumerably many triangles with two vertices on the circumference and the third vertex outside $U$ and with rational argument.

17.3. Determine the skeleton of the union of two intersecting open disks in $\mathbf{R}^2$ for the $l^2$-norm.

17.4. Show that if we use the $l^\infty$-distance, there exists a bounded open connected set in $\mathbf{R}^2$ such that its skeleton is not connected.

17.5. Let the space be $\mathbf{Z}^2$ with the $l^\infty$-metric. What is the skeleton of $([-2,2]_{\mathbf{Z}} \times [-2,2]_{\mathbf{Z}}) \smallsetminus \{(2,2)\}$, a square with a point removed?

17.6. Let the space be $\mathbf{Z}^2$ with the $l^\infty$-metric. What is the skeleton of $([-2,2]_{\mathbf{Z}} \times [-2,2]_{\mathbf{Z}}) \cup \{(3,-1),(3,0),(3,1)\}$, a square with three points added?

17.7. We define, following Serra (1982:376) the ***r*-skeleton**, associated to the radius $r > 0$, as the set $\mathbf{skel}_r(A)$ of points in the skeleton where the quench function $-\mathbf{dt}_A$ takes the value $r$. Since $r$ is positive, it follows that $\mathbf{skel}(A)$ is a subset of $A$ and equal to the (disjoint) union $\bigcup_{r>0} \mathbf{skel}_r(A)$.

Assume now that we are in a normed vector space with distance $d(x, y) = \|x - y\|$. We consider only open sets $A$ that are nonempty and bounded. (Actually we could allow unbounded sets such that the radii of all maximal balls are bounded.)

(a) Show that the reconstruction of the interior of a bounded set $A$ takes the form

$$A^\circ = \bigcup_{r>0} \big(\mathbf{skel}_r(A) + B_<(0, r)\big).$$

(b) Show that we can also reconstruct the dilations and erosions by balls from the $r$-skeletons for bounded open sets $A$:

$$\delta_{B_<(0,r)}(A) = \bigcup_{s>0} \big(\mathbf{skel}_s(A) + B_<(0, r + s)\big), \qquad r > 0;$$

$$\varepsilon_{B_\leqslant(0,r)}(A) = \bigcup_{s>r} \big(\mathbf{skel}_s(A) + B_<(0, s - r)\big), \qquad r > 0.$$

(c) Find some conditions on a set $A$ which guarantee that the skeleton of its dilation by a ball is the same as that of $A$:

$$\mathbf{skel}\big(\delta_{B_<(0,r)}(A)\big) = \mathbf{skel}\big(A + B_<(0, r)\big) = \mathbf{skel}(A), \qquad r > 0.$$

(d) Prove that the skeleton of an erosion is simply obtained by deleting a certain part of the skeleton:

$$\mathbf{skel}\big(\varepsilon_{B_\leqslant(0,r)}(A)\big) = \bigcup_{s>r} \mathbf{skel}_s(A), \qquad r > 0.$$

(e) Show that

$$\alpha_{B_\leqslant(0,r)}(A) = \bigcup_{s>r} \big(\mathbf{skel}_s(A) + B_<(0, s)\big). \qquad r > 0.$$

(f) Show that

$$\mathbf{skel}_r(A) = \bigcap_{s>0} \left(\varepsilon_{B_<(0,r)}(A) \smallsetminus \alpha_{B_\leqslant(0,s)}\big(\varepsilon_{B_<(0,r)}(A)\big)\right), \qquad r > 0.$$

It follows that

$$\mathbf{skel}(A) = \bigcup_{r>0} \bigcap_{s>0} \left(\varepsilon_{B_<(0,r)}(A) \smallsetminus \alpha_{B_\leqslant(0,s)}\big(\varepsilon_{B_<(0,r)}(A)\big)\right).$$

This result is due to Lantuéjoul (1977).

# Chapter 18

# Solutions

THIS APPENDIX contains solutions (sometimes just answers) to some of the exercises mentioned at the end of the chapters. (If an exercise does not get a comment it is either too easy or too difficult.)

## Chapter 2: Sets, mappings, and order relations

2.3. The implication (B) $\Rightarrow$ (A) is easy.

(A) & (C) $\Rightarrow$ (B). If $x \leqslant y$, then we get $f(x) \leqslant f(y)$. If $x \neq y$, then we get $f(x) \neq f(y)$, thus, together with the former conclusion, that $f(x) < f(y)$.

(D) $\Rightarrow$ (C). If $f(x) = f(y)$ we conclude that $x \leqslant y$ and that $x \geqslant y$, thus that $x = y$, since $X$ is now ordered.

(B) $\not\Rightarrow$ (C). Consider a set $X$ with four elements $\mathbf{0}$, $a$, $b$, $\mathbf{1}$ and with the order relations $\mathbf{0} < a < \mathbf{1}$, $\mathbf{0} < b < \mathbf{1}$. Define $f(x) = x$ for $x = \mathbf{0}, a, \mathbf{1}$ and $f(b) = a$. Then $f$ is stricty increasing but not injective.

(B) & (C) $\not\Rightarrow$ (D). Take $X = \mathbf{N}$ with the order $j < j + 1$ for $j \geqslant 1$ and $0 < 2$ while 0 and 1 are not comparable. Define $f\colon X \to X$ by $f(0) = 0$ and $f(j) = j+1$ for $j \geqslant 1$. Then $f$ is strictly increasing, (B), and injective, (D), but not coincreasing, since $0 = f(0) \leqslant f(1) = 2$ while $0 \nleqslant 1$

(D) $\not\Rightarrow$ (A). Take $X = \mathbf{N}$ and introduce the usual order except that $1 < 0$ and 0 has no majorants. Define $f\colon X \to X$ by $f(0) = 0$, $f(x) = x + 1$ for $x \geqslant 1$. Then $X$ has an order, $f$ is coincreasing, (D), but not incresing, since $1 < 0$ but $f(1) = 2 \nleqslant 0 = f(0)$.

2.4. It is easy to see that $f \circ g$ is a cleistomorphism. But $g \circ f$ is not, since the convex hull of the closed set $\{(x, y) \in \mathbf{R}^2;\ y(1 + x^2) \geqslant 1\}$ is the strict upper half plane, thus not closed.

2.5. Let $L$ be the lattice with four elements $\mathbf{0}, a, b, \mathbf{1}$ and with the order relations $\mathbf{0} < a < \mathbf{1}$ and $\mathbf{0} < b < \mathbf{1}$. Define $\alpha(\mathbf{0}) = \alpha(b) = \mathbf{0}$, $\alpha(a) = \alpha(\mathbf{1}) = a$; and $\beta(\mathbf{0}) = \beta(a) = \mathbf{0}$, $\beta(b) = \beta(\mathbf{1}) = b$. Both $\alpha$ and $\beta$ are anoiktomorphisms and $\alpha \vee \beta = \mathbf{id}_L$, so $\mathbf{invar}(\alpha \vee \beta) = L$ while $\mathbf{1}$ does not belong to $\mathbf{invar}(\alpha) \cup \mathbf{invar}(\beta)$.

## Chapter 3: Morphological operations: Set-theoretical duality

3.1. (a) The dilation is the union of two translates of $A$; the erosion the intersection of two translates of $A$.

(b) The dilation is an open disk of radius $R + r$. The erosion is an open disk of radius $R - r$ if $R > r$ and the empty set if $R \leqslant r$.

(c) The dilation is a translate of $A + A$. The erosion is the singleton set $\{-b\}$ if $B = A + b$.

(d) The dilation is equal to $A - A + b$ if $B = \check{A} + b$. The erosion is

$$\varepsilon_B(A) = \{x;\ x - A + b \subset A\} = \complement\,(A + \complement A - b);$$

cf. Theorem 3.2.5. We note that if $A$ is symmetric, i.e., if $\check{A} = A$, then we are as in (c) above. In particular, $\varepsilon_B(A) = \{-b\}$.
If $A$ is not symmetric, then there exists an element $c \in A \smallsetminus \check{A}$. We note that, since $c \in A$,

$$\varepsilon_B(A) = \bigcap_{a \in A} (a - b + A) \subset c - b + A$$

if $B = \check{A} + b = b - A$. If we assume that $-b$ belongs to $\varepsilon_B(A)$, we get $0 \in c + A$, which contradicts the assumption that $c$ does not belong to $\check{A}$. We conclude that $-b \notin \varepsilon_B(A)$. (If $A$ is unbounded, the situation is different.)

3.2. (a) The question is whether we have an equivalence

$$\delta_B(A) = C \Leftrightarrow A = \varepsilon_B(C).$$

We do not have an implication $\Rightarrow$: Take $A = \mathbf{Q}$, $B = [-1, 1]$ and $C = \mathbf{R}$. Then $\delta_B(A) = C$ but $A = \mathbf{Q} \neq \mathbf{R} = \varepsilon_B(C)$. Do we have an implication $\Leftarrow$? No: take $A = \mathbf{R}$, $B = \varnothing$, $C = \mathbf{R}$. Then $\varepsilon_B(C) = A$, while $\delta_B(A) = \varnothing \neq C$.

(b) Is the equivalence $\varepsilon_B(A) \subset C \Leftrightarrow A \subset \delta_B(C)$ true? No. Take $A = C = \mathbf{R}$, $B = \emptyset$. Then $\varepsilon_B(A) = C$, while $\delta_B(C) = \emptyset \neq A$. For the other direction we can take $A = B = C = \emptyset$ in $\mathbf{R}$. Then $\delta_B(C) = A$, while $\varepsilon_B(A) = \mathbf{R} \not\subset C$. So none of the implications holds in general.

If $B$ is a closed ball and $A$ and $C$ are open balls, all nonempty, we do have equivalence.

3.3. If $\varepsilon_B(A) \subset A$ for all $A$, then in particular $0 \in \varepsilon_B(B) \subset B$. In the other direction, if $0 \in B$, then $\varepsilon_B(A) \subset \varepsilon_{\{0\}}(A) = A$.

3.4. The Minkowski sum $[a,b] + [c,d]$ is a quadrilateral with vertices at $a+c$, $a+d$, $b+d$ and $b+c$—and a segment if $a = b$ or $c = d$.

3.5. In general a filled hexagon—sometimes degenerate.

3.6. A rectangular hexahedron, i.e., a block all sides of which are rectangles.

3.7. A prism with a hexagonal base (see 3.5).

3.8. A zonotope must have an even number of vertices. To see this we note that this is true for the sum of one or two segments, and that, when we add a segment to a finite sum of segments, the number of vertices is either unchanged or increased by 2.

3.9. See the explanation in 3.8 above.

3.10. If a tetrahedron in $\mathbf{R}^3$ were a zonohedron, then also each face would be a sum of segments, and they must all be situated in the plane of the face. We can now refer to 3.8 above.

3.11. A zonotope is always convex, so a non-convex quadrilateral cannot be a sum of segments.

3.12. (a) A cube has 8 vertices, 12 edges and 6 faces. And $8 - 12 + 6 = 2$, says Euler. The polyhedron in question is the convex hull of the cube $[0,1]^3$ and its translation by the vector $(1,1,1)$, which is the cube $[1,2]^3$. It has, counting with increasing values of $x_1 + x_2 + x_3$, $1+6+6+1 = 14$ vertices, $3+6+6+6+3 = 24$ edges, and $3+6+3 = 12$ faces. (Euler: $14 - 24 + 12 = 2$.)

(b) The polyhedron in this case is the image under a bijective affine mapping of that in (a). Indeed, we can map the unit normal vectors of the tetrahedron to $(1,0,0)$, $(0,1,0)$, $(0,0,1)$ and $(-1,-1,-1)$. This reduces the problem to that already treated in (a)—the fact that we translate here by the vector $(-1,-1,-1)$ instead of by $(1,1,1)$ makes no difference.

3.13. The four vertices of one square should be added to the four vertices of the other. This gives rise to sixteen points, but if the sides are not parallel, eight of these belong to the convex hull of the eight others. The statement about side lengths follow from the fact that the sides in the octagon are just translations of the sides in one of the original squares.

If the sides are parallel, only four points are necessary to span the sum.

3.14. The sum $r_0T_0 + r_\gamma T_\gamma$ is equal to the triangle $(r_0 + r_\gamma)T_0$ if $\gamma$ belongs to $\frac{2}{3}\pi\mathbf{Z}$, i.e., when the sides are parallel.

If $\gamma$ does not belong to $\frac{2}{3}\pi\mathbf{Z}$, the sum $r_0T_0 + r_\gamma T_\gamma$ is a hexagon with vertices at the point obtained from adding one of the three points $r_0(\cos(2j\pi/3), \sin(2j\pi/3))$, $j = 0, 1, 2$, to one of the three points $r_\gamma(\cos(\gamma + 2k\pi/3), \sin(\gamma + 2k\pi/3))$, $k = 0, 1, 2$. (We get nine sums, but three of the points belong to the convex hull of the six others.)

Define $\gamma' \in [0, \pi/3]$ as the distance from $\gamma \in \mathbf{R}$ to $\frac{2}{3}\pi\mathbf{Z}$. If $\gamma' \neq 0$, the sum is a hexagon with three vertices with the distance to the origin equal to

$$R_1(\gamma) = \sqrt{r_0^2 + r_\gamma^2 + 2r_0r_\gamma \cos\gamma'},$$

and the three others at the distance

$$R_0(\gamma) = \sqrt{r_0^2 + r_\gamma^2 + r_0r_\gamma(\sqrt{3}\sin\gamma' - \cos\gamma')}.$$

The latter are closer to the origin except when $\gamma' = \pi/3$; then we have equality and the hexagon is regular. The largest value of $R_0(\gamma)$ is $r_0 + r_\gamma$ and occurs when $\gamma' = 0$ (when the hexagon is reduced to a triangle); the smallest value is $\sqrt{(r_0 + r_\gamma)^2 - r_0r_\gamma}$ and occurs when $\gamma'$ is maximal, i.e., when $\gamma' = \pi/3$.

3.15. (a) We get $3 + 3 = 6$ vertices, $3 + 6 + 3 = 12$ edges and $1 + 6 + 1 = 8$ faces (and $6 - 12 + 8 = 2$).

(b) We get $3 + 6 = 9$ vertices, $3 + 6 + 6 = 15$ edges and $1 + 6 + 1 = 8$ faces (and $9 - 15 + 8 = 2$).

3.16. The vector sum of $\mathbf{cvxh}(A)$ and $\mathbf{cvxh}(B)$, where $A$ and $B$ are finite sets in the plane, is equal to $\mathbf{cvxh}(A + B)$ with a maximum of

**card**$(A) \cdot$ **card**$(B)$ vertices. When $A$ has three elements which are not collinear, we can always assume that $A = \{(0,0),(1,0),(0,1)\}$. Taking $B = \{(0,0),b,c\}$ (we may assume that $a = (0,0)$), we study the convex set

$$C = \mathbf{cvxh}(A) + \mathbf{cvxh}(B)$$
$$= \mathbf{cvxh}(\{(0,0),(1,0),(0,1)\}) + \mathbf{cvxh}(\{(0,0),b,c\}).$$

We can get three vertices: we always have

$$\mathbf{cvxh}(\{a,b,c\}) + \mathbf{cvxh}(\{a,b,c\}) = \mathbf{cvxh}(\{2a,2b,2c\}).$$

We can get four vertices: take $b = (1,0)$ and $c = (0,2)$. Then $C$ is equal to the convex hull of $\{(0,0),(2,0),(1,2),(0,3)\}$ which are the four vertices.

We can have five vertices: take $b = (1,0)$ and $c = (1,1)$. Then $C$ is equal to the convex hull of the five points $(0,0)$, $(2,0)$, $(2,1)$, $(1,2)$, $(0,1)$, which are the five vertices.

We can get six vertices: take $b = \{(-1,1)\}$ and $c = \{(0,1\}$. Then $C$ is equal to the convex hull of the six points $(0,0)$, $(1,0)$, $(1,1)$, $(0,2)$, $(-1,2)$, $(-1,1)$, which are the six vertices.

3.17. It is enough to take the square as $Q$, the convex hull of the four points $(0,0)$, $(1,0)$, $(1,1)$ and $(0,1)$, and the triangle as $\mathbf{cvxh}(\{(0,0),b,c\})$ for various choices of $b$ and $c$. We can get five, six or seven vertices.

Five if $b = (1,0)$, $c = (0,1)$; the vertices are then $(0,0)$, $(2,0)$, $(2,1)$, $(1,2)$, $(0,2)$.

Six if $b = (1,1)$, $c = (-1,1)$; the vertices are $(0,0)$, $(1,0)$, $(2,1)$, $(2,2)$, $(-1,2)$, $(-1,1)$. +nn Seven if $b = (-1,1)$, $c = (1,2)$; the vertices are in this case $(0,0)$, $(1,0)$, $(2,2)$, $(2,3)$, $(1,3)$, $(-1,2)$, $(-1,1)$.

3.18. For unions we can take $B = \{-1,0,1\}$, $A_1 = 4\mathbf{Z}$ and $A_2 = 4\mathbf{Z}+1$. Then $2 \notin \kappa(A_1) \cup \kappa(A_2) = A_1 \cup A_2$ while $\kappa(A_1 \cup A_2) = \mathbf{Z}$.

For intersections we can take $B = \{-1,0,1\}$, $A_1 = 2\mathbf{Z}$ and $A_2 = 2\mathbf{Z}+1$. Then $\kappa_B(A_1 \cap A_2) = \varnothing$, while $\kappa_B(A_1) \cap \kappa_B(A_2)$ is equal to $\mathbf{Z}$ since $\delta_B(A_1) = \delta_B(A_2) = \mathbf{Z}$.

3.19. (a) If $C \subset B$ we clearly have $\varphi \leqslant \mathbf{id}$. Conversely, $\varphi(B) \subset B$, implies that $C \subset \varphi(B) \subset B$.

We note that $\varepsilon_B(G \smallsetminus \{0\}) = -\complement B$, so that $\varphi(G \smallsetminus \{0\}) = C - \complement B$, a set which is contained in $G \smallsetminus \{0\}$ if and only if $C \cap \complement B$ is empty, equivalently $C \subset B$.

(b) We have $\varphi(B) = C$, so if $C = B$ we get $\varphi \circ \varphi = \varphi$. Conversely, if $\varphi$ is idempotent, then, since $\varphi(B) = C$, in particular we must have $\varphi(C) = \{x;\ x+B \subset C\}+C = C$, which under the assumptions made can only happen if $B \subset C$.

3.20. We can choose $f(x) = \mathrm{e}^x$ for $x \in \mathbf{R}$ and $g$ as the constant identically zero. Then $\mathbf{epi}^{\text{finite}}(f \sqcap g)$ is the closed upper half plane in $\mathbf{R}^2$, while $\mathbf{epi}^{\text{finite}}(f) + \mathbf{epi}^{\text{finite}}(g)$ is the open upper half plane.

3.21. Take $G = \mathbf{N}$ with addition as the operation, $z = 1$, $A = \{0\}$, $B = \{1\}$. Then $S_z(\varepsilon_B(A)) = \varnothing \neq \{0\} = \varepsilon_B(S_z(A))$.

With $g = \mathbf{N}$ and multiplication as the operator we can take $z = 2$, $A = \{1\}$, $B = \{2\}$ and obtain $S_z(\varepsilon_B(A)) = \varnothing \neq \{1\} = \varepsilon_B(S_z(A))$.

## Chapter 4: Complete lattices

4.1. We can simply define $\delta(x) = \mathbf{0}$ for all $x$ in a complete lattice with at least two elements. Then $\delta(\mathbf{1}) = \mathbf{0} \neq \mathbf{1}$.

4.2. The invariance set of $\eta_\diamond$ must be equal to the invariance set of $\eta$, so $\eta_\diamond(x) = x$ for $x = \mathbf{0}, b, \mathbf{1}$. Then $\eta_\diamond(c)$ and $\eta_\diamond(a)$ are easily seen by inspection to be equal to $\mathbf{0}$. We know that $\eta_\diamond \leqslant \eta_\alpha \leqslant \eta_{(1)}$ (Theorem 4.3.10), which fixes the value of $\eta_\alpha(x)$ for $x \neq c$, since for these values of $x$, the values $\eta_\diamond(x)$ and $\eta_{(1)}(x)$ are the same. So only $x = c$ remains to be considered before the whole table is complete. We have $\eta_\diamond(c) = \mathbf{0} < c = \eta_{(1)}(c)$, so a priori the values $\eta_\alpha(c) = \mathbf{0}, a, c$ are possible. Choosing $a$ does not give us an idempotent mapping, but the values $\mathbf{0}$ and $c$ do give idempotent and increasing mappings. Since $\mathbf{0} < c$, the maximal anoiktomorphism minimizing $\eta$ is given by the choice $c$.

## Chapter 5: Inverses and quotients of mappings

5.1. With $g$ as the constant mapping with value equal to $b$ we get $g_{[-1]}(y) = \mathbf{1}$ if $y \geqslant_M b$ and equal to $\mathbf{0}$ otherwise, while $g^{[-1]}(y) = \mathbf{0}$ if $y \leqslant_M b$ and equal to $\mathbf{1}$ otherwise.

5.2. We have $\inf g = (\mathbf{R}^n)^\perp = \{0\}$ and $\sup g = \varnothing^\perp = \mathbf{R}^n$. Example 5.2.2 shows that $g_{[-1]}(B) = \mathbf{R}^n$ if $0 \in B$, otherwise equal to $\varnothing$. The upper inverse $g^{[-1]}$ is equal to the empty set identically.

5.3. Take $L = \{\mathbf{0}, b, c, \mathbf{1}\}$ with the order relations $\mathbf{0} < b < \mathbf{1}$ and $\mathbf{0} < c < \mathbf{1}$, and $M = \{\mathbf{0}, a, b, c, \mathbf{1}\}$ with the order relations $\mathbf{0} < a < b < \mathbf{1}$ and $\mathbf{0} < a < c < \mathbf{1}$. Define $g(x) = x$ for $x = \mathbf{0}, b, \mathbf{1}$ and $g(c) = a$. We get $g_{[-1]}(b) = \mathbf{1} > b = g^{[-1]}(b)$ and $g_{[-1]}(c) = c < \mathbf{1} = g^{[-1]}(c)$, so the two inverses are not comparable. (This mapping is injective and strictly increasing, but if we give up injectivity, we can get a simpler example taking $M = L$.)

5.4. Take $X = L = \{\mathbf{0}, a, \mathbf{1}\}$ with the order relations $\mathbf{0} < a < \mathbf{1}$, and $Y = \{\mathbf{0}\}$, and $\varphi(\mathbf{0}) = \mathbf{0}$, $\varphi(a) = \varphi(\mathbf{1}) = \mathbf{1}$. Then $\big(f^{\rightarrow}(\mathbf{id})\big)(\mathbf{0}) = \mathbf{0}$ and $\big(f_{\rightarrow}(\mathbf{id})\big)(\mathbf{0}) = \mathbf{1}$, so that the interval of the range of $\varphi$ is $[\mathbf{0}, \mathbf{1}]$, and $a$ belongs to this interval but is not in the range of $\varphi$.

## Chapter 6: Structure theorems for mappings

6.1. In the definition of the lower quotient,

$$(\alpha/_{\star}\alpha)(y) = \bigvee_x (\alpha(x);\ \alpha(x) \leqslant y),$$

we can take $x = \alpha(y)$ to show that the quotient is $\geqslant \alpha(y)$. And if $\alpha(x) \leqslant y$, then also $\alpha(x) \leqslant \alpha(y)$, proving that the quotient is $\leqslant \alpha(y)$. Thus we have equality for every $y$.

6.2. The dual to Theorem 6.2.1 is: *For any mapping $f\colon X \to M$ the upper quotient is a cleistomorphism.* In the definition of the upper quotient,

$$(\kappa/^{\star}\kappa)(y)) = \bigwedge_x (\kappa(x);\ \kappa(x) \geqslant y),$$

we can take $x = \kappa(y)$ to prove that the value is $\leqslant \kappa(y)$. If $\kappa(x) \geqslant y$, then also $\kappa(x) \geqslant \kappa(y)$, proving that the quotient is $\geqslant \kappa(y)$.

6.3. If $L$ totally ordered, we have either $\eta(x) \leqslant x$ or $\eta(x) \geqslant x$. In both cases we find that $(\eta \circ \eta_{(1)})(x) = \eta(x)$ and that $(\eta \circ \eta^{(1)})(x) = \eta(x)$.

6.4. Both $\eta^{(1)} \circ \eta_{(1)}$ and $\eta_{(1)} \circ \eta^{(1)}$ are equal to $\eta$ if $\eta$ is a strong ethmomorphism. In fact,

$$\eta_{(1)} \circ \eta^{(1)} = (\eta \circ \eta^{(1)}) \wedge \eta^{(1)} = (\eta^{(1)} \circ \eta) \wedge \eta^{(1)} = \eta,$$

where the first equality comes from (6.5.1). Similarly $\eta^{(1)} \circ \eta_{(1)} = \eta$.

6.5. We can take the lattice as $\{\mathbf{0}, a, b, \mathbf{1}\}$ with the order relations $\mathbf{0} < a < \mathbf{1}$ and $\mathbf{0} < b < \mathbf{1}$ and define $f_1$ as the constant mapping $L \to L$ with value $a$ and $f_2$ as the constant mapping with value $b$. Then, with $g(\mathbf{0}) = \mathbf{0}$ and $g(x) = \mathbf{1}$ for other values of $x$ we get $g(a \wedge b) = g(\mathbf{0}) = \mathbf{0} < \mathbf{1} = g(a) \wedge g(b)$. If we instead take $g(\mathbf{1}) = \mathbf{1}$ and $g(x) = \mathbf{0}$ for the other values of $x$, we get $g(a \vee b) = g(\mathbf{1}) = \mathbf{1} > \mathbf{0} = g(a) \vee g(b)$.

6.6. In Example 6.5.2 we obtain $\eta_{(1)}(x) = \eta(x)$ if $x$ is comparable to $p$ and equal to $\eta(x) \wedge x = p \wedge x < \eta(x)$ otherwise. Similarly $\eta^{(1)}(x) = \eta(x)$ if $x$ is comparable to $p$ and equal to $\eta(x) \vee x = p \vee x > \eta(x)$ otherwise.

We get $(\eta \circ \eta_{(1)})(x) = \eta(x)$ if $x$ is comparable to $p$ and equal to $p \wedge x$ otherwise. By duality, $\eta \circ \eta^{(1)} = \eta(x)$ if $x$ is comparable to $p$ and equal to $p \vee x$ otherwise.

For the compositions in the other order we get

$$(\eta_{(1)} \circ \eta)(x) = \eta(x) \wedge x = x$$

if $x$ is comparable to $p$ and equal to $\eta(p) \wedge p = p$ if not. By duality, $(\eta^{(1)} \circ \eta)(x) = \eta(x) \vee x = x$ if $x$ is comparable to $p$ and equal to $\eta(p) \vee p = p$ otherwise.

6.7. We have $\eta_{(1)}(x) = x$ if $x$ is comparable to $p$, otherwise equal to $p \wedge x$. For the composition $\eta_{(1)} \circ \eta_{(1)}$ we get $(\eta_{(1)} \circ \eta_{(1)})(x) = x \wedge \eta(x) = x$ if $x$ is comparable to $p$ and equal to $\eta_{(1)}(p \wedge x) = (p \wedge x) \wedge \eta(p \wedge x) = p \wedge x$ otherwise. So the two agree for all $x$: $\eta_{(1)}$ is idempotent. By duality, so is $\eta^{(1)}$.

6.8. We can take $L$ as the lattice with five elements $\mathbf{0}$, $p$, $q$, $r$, $\mathbf{1}$ satisfying the order relations $\mathbf{0} < p < r < \mathbf{1}$ and $\mathbf{0} < q < r < \mathbf{1}$, and then define $\eta(x) = x$ for $x \neq p, r$, and $\eta(p) = q$, $\eta(r) = \mathbf{1}$. Then we find that $\eta^{(1)}(p) = r < \mathbf{1} = (\eta^{(1)} \circ \eta^{(1)})(p)$. (We can note that $\eta^{(1)} \circ \eta^{(1)} \circ \eta^{(1)} = \eta^{(1)} \circ \eta^{(1)}$, so $\eta^{(1)} \circ \eta^{(1)}$ is idempotent, thus equal to $\eta^{\kappa}$.)

**Chapter 7: Digitization**

7.1. If there is a point $a \in \dot{X} \smallsetminus P$, we see that $F(\{a\}) = \{a\} \cap P = \emptyset$ while $f_*(\{a\}) = \{f(a)\} \neq \emptyset$ for any mapping $f\colon X \to X$.

7.2. (a) For the $l^2$-norm, the Voronoi cell $V(p)$ is equal to the half plane defined by $x \cdot p \geqslant 0$; the cell $V(-p)$ the half plane defined by $x \cdot p \leqslant 0$.

(b) For the $l^\infty$-norm, let us assume that $0 \leqslant p_2 \leqslant p_1$; this is permissible in view of the symmetry.

(b′) Assume first that $0 = p_2 < p_1$. Then the cells $V(p)$ and $V(-p)$ are

$$V(p) = \{x;\ x_1 \geqslant 0\} \cup \{x;\ |x_2| \geqslant p_1 + |x_1|\} \text{ and}$$
$$V(-p) = \{x;\ x_1 \leqslant 0\} \cup \{x;\ |x_2| \geqslant p_1 + |x_1|\}.$$

In particular, $V(p) \cap V(-p)$ contains the two quadrants $\{x;\ |x_2| \geqslant p_1 + |x_1|\}$.

(b″) Assume next that $0 < p_2 = p_1$. The the Voronoi cells are $V(p) = \{x;\ x_1 + x_2 \geqslant 0\}$ and $V(-p) = \{x;\ x_1 + x_2 \leqslant 0\}$, thus as in 7.2 (a).

(b‴) Assume finally that $0 < p_2 < p_1$. The Voronoi cell $V(p)$ is then the set defined by

$$\begin{aligned} &x_1 \geqslant -p_1 + p_2 - x_2 && \text{when } x_2 \leqslant -p_1 + p_2; \\ &x_1 \geqslant 0 && \text{when } -p_1 + p_2 \leqslant x_2 \leqslant p_1 - p_2; \\ &x_1 \geqslant p_1 - p_2 - x_2 && \text{when } p_1 - p_2 \leqslant x_2. \end{aligned}$$

The Voronoi cell $V(-p)$ is the set defined by replacing $x_1 \geqslant$ by $x_1 \leqslant$ above. The length of the vertical segment in $V(p) \cap V(-p)$ is $2(p_1 - p_2) > 0$.

(c) For the $l^1$-norm, we assume again that $0 \leqslant p_2 \leqslant p_1$.

(c′) Consider first the situation when $0 = p_2 < p_1$. Then the Voronoi cells are $V(p) = \{x;\ x_1 \geqslant 0\}$ and $V(-p) = \{x;\ x_1 \leqslant 0\}$, thus as in 7.2 (a) but smaller than in 7.2 (b′).

(c″) Next assume that $0 < p_2 = p_1$. Then $V(p)$ and $V(-p)$ are like in (b″).

(c‴) Finally, assume that $0 < p_2 < p_1$. Then $V(p)$ is given by the inequalities

$$\begin{aligned} &x_1 \geqslant p_2 && \text{when } x_2 \leqslant -p_2; \\ &x_1 \geqslant -x_2 && \text{when } -p_2 \leqslant x_2 \leqslant p_2; \\ &x_1 \geqslant -p_2, && \text{when } p_2 \leqslant x_2. \end{aligned}$$

The Voronoi cell $V(-p)$ is the set defined by replacing $x_1 \geqslant$ by $x_1 \leqslant$ above. The length of the segment in $V(p) \cap V(-p)$ where $x_1 + x_2 = 0$ is $2p_2\sqrt{2} > 0$.

7.3. (a) For the $l^2$-metric, the Voronoi cell $V(0)$ is the convex hull of the six points $\pm\theta, \pm\overline{\theta}, \pm 1/\sqrt{3}$, where $\theta = \frac{1}{2} + \frac{1}{2}\mathrm{i}/\sqrt{3}$, a smaller version of the hexagon with vertices in $\pm\mathrm{i}, \pm\omega, \pm\overline{\omega}$ turned 30° and scaled down by a factor of $1/\sqrt{3}$. The other Voronoi cells are translations of $V(0)$.

(b) For the $l^\infty$-norm, the Voronoi cell $V(0)$ is the octagon defined by the inequalities $|\operatorname{Re} z| \leqslant \frac{1}{2}$, $|\operatorname{Im} z| \leqslant \frac{1}{4}\sqrt{3}$, $|\operatorname{Re} z \pm \operatorname{Im} z| \leqslant \frac{1}{2}\sqrt{3}$; cf. 7.2 (b) above, where the length of the segment (vertical there but horizontal here) is $2(p_1 - p_2) = \frac{1}{2}(\sqrt{3} - 1)$.

(c) For the $l^1$-norm, the Voronoi cell $V(0)$ is an octagon with sides $\operatorname{Re} z = \pm\frac{1}{2}$ and $\operatorname{Re} z + \operatorname{Im} z = \pm\frac{1}{4}(1 + \sqrt{3})$ and $\operatorname{Re} z - \operatorname{Im} z = \pm\frac{1}{4}(1 + \sqrt{3})$. The length of the sides with slope ±45° is $2\sqrt{2}p_2 = 1/\sqrt{2}$.

## Chapter 8: Digital straightness and digital convexity

8.1. The right-most point $a = (5,2)$ in $A$ satisfies $a_2 \leqslant \frac{1}{2}a_1$, so the point $(3,6/5)$ is on the segment $[(0,0),a]$ and does not belong to $A + B^{\infty}_{<}(0,1)$.
The right-most point $b = (5,3)$ in $B$ satisfies $b_2 \geqslant \frac{1}{2}b_1$, so the point $(2,6/5)$, which is on the segment $[(0,0),b]$, does not belong to $B + B^{\infty}_{<}(0,1)$.

8.2. If we require that $D_1D_1f$ be zero for a function $f\colon \mathbf{Z} \to \mathbf{Z}$, the only straight lines we get are those of equation $y = ax+b$ with $a \in \mathbf{Z}$, and, after switching the coordinates, $y = x/a - b/a$ with $a \in \mathbf{Z}$, $a \neq 0$. So the line of equation $2x = 3y$ with slope $2/3$ is not accepted.

8.3. If we require only that $|D_1D_1f| \leqslant 1$, we find solutions like

$$f(x) = \tfrac{1}{2}x(x-1), \qquad x \in \mathbf{Z},$$

hardly acceptable as a straight line. We can also combine these arcs to get waves as large as we want.

8.4. For every $x \neq 0$ there is a unique integer $m$ such that $q^m \leqslant |x| < q^{m+1}$. If $f(qx) = f(x)$ for every $x$, then $f(x) = f(q^{-m}x)$, where $1 \leqslant q^{-m}x < q$. Writing $X = \left]-q,-1\right] \cup \{0\} \cup \left[1,q\right[$, we get a solution to the problem for any function $F\colon \to \mathbf{R}$ by taking $f(x) = F(q^{-m}x)$, $x \neq 0$, for the integer $m$ such that $q^{-m}x$ belongs to $X$, and $f(0) = F(0)$.
When $f$ is only defined on the integers, the same argument works, but $F$ is now only defined on $X_0 = X \cap \{kq^m;\ k,m \in \mathbf{Z}\}$.

## Chapter 9: Convexity in vector spaces

9.1. If $A$ is convex and $a,b \in A$ then also $\frac{1}{2}(a+b) \in A$, so that $A+A \subset 2A$. The set $A_1 = [0,1] \cap \mathbf{Q}$ of rational points in $[0,1]$ is not convex, but satisfies $2A_1 = A_1 + A_1$. The same is true of the minimal example $A_2 = \{j2^{-k};\ j = 0,1,\dots,2^k,\ k \in \mathbf{N}\}$.
If $A + A \subset 2A$ we see that all points of the form $(1 - j2^{-k})a + j2^{-k}b$, $j = 0,1,\dots,2^k$, $k \in \mathbf{N}$, and with $a,b \in A$ belong to $A$ and they are dense in $\overline{A}$.

9.2. The Minkowski sum of two convex sets is convex—easily proved. So $\mathbf{cvxh}(A) + \mathbf{cvxh}(B)$ is convex and contains $A + B$, hence also its convex hull.
Strict inclusion cannot occur: we do have equality. We see

from Proposition 9.17.3 that the two sets $\mathbf{cvxh}(A) + \mathbf{cvxh}(B)$ and $\mathbf{cvxh}(A + B)$ have the same support function. This implies that their closures are the same. If we apply this to finite subsets $A_j$ and $B_j$, we see that their convex hulls are closed and the set $\mathbf{cvxh}(A_j) + \mathbf{cvxh}(B_j)$ is therefore equal to $\mathbf{cvxh}(A_j + B_j)$. Given two arbitrary sets $A$ and $B$ we can now choose finite subsets $A_j$ and $B_j$ such that $\mathbf{cvxh}(A_j)+\mathbf{cvxh}(B_j)$ converges to $\mathbf{cvxh}(A)+\mathbf{cvxh}(B)$ and $\mathbf{cvxh}(A_j+B_j)$ converges to $\mathbf{cvxh}(A+B)$. This finishes the proof.

9.3. For the function identically equal to zero, the strict finite epigraph and finite epigraph are different.

For the function $f(x) = \mathrm{e}^{-x^2}$ defined for $x \in \mathbf{R}$, the convex hull of the finite epigraph is the open upper half plane, while its convex envelope is zero and its finite epigraph is the closed upper half plane.

9.4. The equality $\kappa_B(A) = A$ holds for all $B$ if $A$ is empty, but not when $B$ is empty. So suppose that both are nonempty and that $n = 1$. We then have $A = [a_0, a_1]$ and $b = [b_0, b_1]$ for some real numbers $a_0 \leqslant a_1$, $b_0 \leqslant b_1$, and it is evident that

$$\kappa_B(A) = \{x;\ x + [b_0, b_1] \subset [a_0 + b_0, a_1 + b_1]\} = [a_0, a_1].$$

In higher dimensions we project the sets on a one-dimensional subspace, which is tantamount to using the support function. We find that

$$H_{A+B} = H_A + H_B \text{ and that } H_{\kappa_B(A)} = H_A + H_B - H_B = H_A,$$

proving that $\kappa_B(A) = A$.

9.5. We see that $\delta_B(A) = \delta_A(B)$ is the convex hull of the union of the four disks $B^2_{\leqslant}(c, r)$, $c = (\pm 1, \pm 1)$.

For the erosions we get that $\varepsilon_B(A)$ is the square $[-1 + r, 1 - r]^2$ if $0 \leqslant r \leqslant 1$ and empty otherwise, while $\varepsilon_A(B)$ is the intersection of the four disks $B^2_{\leqslant}(c, r)$, $c = (\pm 1, \pm 1)$ (which is empty if $r > 1$). Finally, $\alpha_B(A)$ equals the convex hull of the four balls $B^2_{\leqslant}(c, 1 - r)$, $c = (\pm 1, \pm 1)$, if $0 \leqslant r \leqslant 1$ and empty otherwise, and $\alpha_A(B) = \varnothing$ if $r < \sqrt{2}$ and equal to $B \cap \left\{x;\ \|x\|_\infty \leqslant \sqrt{r^2 - 1}\right\} \subset B$ if $r \geqslant \sqrt{2}$.

9.6. We use the partial Fenchel transform in the $y$ variable of any function $\varphi$ of two variables $x \in \mathbf{R}^m$ and $y \in \mathbf{R}^n$:

$$\varphi^\diamond(x, \eta) = \sup_{y \in \mathbf{R}^n} \left(y\eta - \varphi(x, y)\right), \qquad (x, \eta) \in \mathbf{R}^m \times \mathbf{R}^n,$$

and its second transform

$$\varphi^{\circ\circ}(x,y) = \sup_{\eta\in\mathbf{R}^n} \left(y\eta - \varphi^{\circ}(x,\eta)\right), \qquad (x,y)\in\mathbf{R}^m\times\mathbf{R}^n.$$

Take now $\varphi(x,y) = g(x,y) = -y^2/f(x)$. Then the second transform is

$$g^{\circ\circ}(x,y) = \sup_{\eta\in\mathbf{R}} \left(y\eta + \tfrac{1}{4}\eta^2 f(x)\right) = -\frac{y^2}{f(x)} = g(x,y).$$

This is a consequence of the general theory for the Fenchel transformation and can also be verified directly.

9.7. We have

$$\frac{-x}{f(y)}\log\left[\frac{-x}{f(y)}\right] = \sup_{\xi\in\mathbf{R}}\left[\frac{-x}{f(x)}\xi - \mathrm{e}^{\xi-1}\right];$$

$$x\log\left[\frac{-x}{f(y)}\right] = \sup_{\xi\in\mathbf{R}}\left[x\xi + f(y)\mathrm{e}^{\xi-1}\right],$$

where, for every $\xi$, the first term to the right is a linear function of $x$, thus of $(x,y)$, and the second term $f(y)\mathrm{e}^{\xi-1}$ is a convex function of $y$, thus also of $(x,y)$.

9.8. We note that, for all real numbers $a$,

$$a^4 = \sup_{t\in\mathbf{R}}\left(4at^3 - 3t^4\right), \text{ yielding } -\frac{f^4}{g^3} = \sup_{t\in\mathbf{R}}\left(4t^3 f + 3t^4 g\right)$$

and proving all we want.

9.9. We get $f_t(x) = t\|x\|_2^2$ for $\|x\|_2 \leqslant 1/(2t)$ and $f_t(x) = \|x\|_2 - 1/(4t)$, for $\|x\|_2 > 1/(2t)$. And $\mathbf{grad}\, f = 2tx$ when $\|x\|_2 \leqslant 1/(2t)$; $\mathbf{grad}\, f = x/\|x\|_2$ when $\|x\|_2 > 1/(2t)$, thus continuous.

## Chapter 10: Discrete convexity

10.1. The implications (A) $\Leftarrow$ (B) and (B) $\Leftarrow$ (C) follow from (9.13.2).

To see that all three are equvalent, it suffices to prove that (A) implies (C). Assuming that (A) is true, we have for any positive number $c$,

$$\mathbf{epi}^{\text{finite}}\left(\mathbf{cvxe}(f)\right)\cap Y \subset \mathbf{epi}^{\text{finite}}_{\text{strict}}\left(\mathbf{cvxe}(f-c)\right)\cap Y \subset \mathbf{epi}^{\text{finite}}_{\text{strict}}(f-c).$$

Taking the intersection over all $c>0$ we obtain

$$\mathbf{epi}^{\text{finite}}\left(\mathbf{cvxe}(f)\right)\cap Y \subset \bigcap_{c>0}\mathbf{epi}^{\text{finite}}_{\text{strict}}(f-c) = \mathbf{epi}^{\text{finite}}(f).$$

So (C) follows.

10.2. Take $A$ consisting of $(0,0)$ and all points $a = (a_1, 1)$, $a_1 \in \mathbf{Z}$. Then the convex hull of $A$ contains the strip defined by $0 < x_2 \leqslant 1$, so that $(1,0)$ belongs to its closure, but for every finite subset of $A$, $(1,0)$ does not belong the closed convex hull of the subset.

10.3. Take $f(x_1, x_2) = -1/x_1$ when $(x_1, x_2) \in \mathbf{Z}^2$ and $x_1 \leqslant -1$, $x_2 = 0$, equal to $1/x_1$ when $x_1 \geqslant 1$, $x_2 = 1$, and equal to $+\infty$ for all other values of $x \in \mathbf{Z}^2$. Then for every restriction to a finite subset of $\mathbf{Z}^2$, the convex envelope of the restriction is $\geqslant s$ for some positive number $s$ in the strip $0 < x_2 < 1$, but $(\mathbf{cvxe}(f))(x) = 0$ in this strip. In fact, on the segment $[a, b] = [(a_1, 0), (b_1, 1)]$ the convex envelope of $f$ tends to $-1/a_1$ as we approach $a$ and to $1/b_1$ as we approach $b$. Fixing now $x_1$ and $-a_1 < x_1$ and letting $b_1$ tend to $+\infty$, we see that the value of the convex envelope on the segment $[(x_1, 0), (x_1, 1)]$ tends to zero as $x_2 \searrow 0$. Then fixing $x_1$ and $b_1 > x_1$ and letting $a_1$ tend to $-\infty$, we find that the value of the convex envelope on the segment $[(x_1, 0), (x_1, 1)]$ tends to zero also as $x_2 \nearrow 0$. Thus the convex envelope is zero on $[(x_1, 0), (x_1, 1)] \smallsetminus \{(x_1, 0), (x_1, 1)\}$. Since $x_1$ is arbitrary, we are done.

## Chapter 11: Discrete convexity in two dimensions

11.1. The set $A = \{0, a\}$ is $\mathbf{Z} \times \mathbf{Z})$-convex if and only if $a_1$ and $a_2$ are coprime.

11.2. With $\gamma = 2/5$ in Proposition 11.13.2 it is clear that the set $A$ in (11.13.1) must be periodic with period 5. Explicitly, we see that $A = (5\mathbf{Z} + 2) \cup (5\mathbf{Z} + 4)$, so that $A \cap [0, 7] = \{2, 4, 7\}$, implying that both $4 - 2 = 2 = \lfloor 1/\gamma \rfloor$ and $7 - 4 = 3 = \lceil 1/\gamma \rceil$, which are elements of $I_\gamma$, actually do occur as distances.

11.3. (a) We have

$$\frac{f(q) - f(p) - 1}{q - p} < \gamma = \frac{F(q) - F(p)}{q - p} < \frac{f(q) - f(p) + 1}{q - p},$$

implying that, as $q - p$ tends to $+\infty$,

$$\limsup \left[ \frac{f(q) - f(p)}{q - p} - \frac{1}{q - p} \right] \leqslant \gamma \leqslant \liminf \left[ \frac{f(q) - f(p)}{q - p} + \frac{1}{q - p} \right].$$

This proves that the limit of $(f(q) - f(p))/(q - p)$ exists and is equal to $\gamma$.

(b) If the values $f(x)$ are known only for $x$ in a finite interval $[a,b]_{\mathbf{Z}}$, we obtain only

$$\max_{p,q\in[a,b]_{\mathbf{Z}}} \frac{f(q)-f(p)-1}{q-p} < \gamma < \min_{p,q\in[a,b]_{\mathbf{Z}}} \frac{f(q)-f(p)+1}{q-p}.$$

In particular,

$$\left|\gamma - \frac{f(b)-f(a)}{b-a}\right| < \frac{2}{b-a}.$$

## Chapter 12: Three problems in discrete convexity

12.1. The marginal function $h$ of $f(x,y)=\sqrt{x^2+y^2}-(\sqrt{2}-1)\,y$, $(x,y)\in\mathbf{Z}^2$, satisfies $h(0)=f(0,0)=0$, $h(1)=f(1,0)=f(1,1)=1$ and $h(2)=f(2,1)=\sqrt{5}-\sqrt{2}+1<2$, which shows that it is not midpoint convex.

12.2. We have $(\mathbf{cvxe}(g))(x,y)=G(x,y)$ when $x\geqslant 0$, $0\leqslant y\leqslant 1$ and $x=ky$ for some integer $k$, and in the slice $\{(x,y)\in\mathbf{R}^2;\ 0\leqslant y\leqslant 1,\ x\geqslant 0\}$ it is affine between the segments $[(0,0),(k,1)]$. This implies that all points $(k,1)\in\mathbf{Z}^2$, $k\in\mathbf{N}$, are $g$-neighbors of the origin.

## Chapter 13: Duality of convolution operators

13.1. (a) With $L=\mathscr{P}(\mathscr{F}(G,\mathbf{R}))$ we have $\mathbf{0}_L=\varnothing$ and $\mathbf{1}_L=\mathscr{F}(G,\mathbf{R})$. Since $\Gamma$ is decreasing, Example 5.2.2 tells us that

$$\inf\Gamma=\Gamma(\mathbf{1}_L)=\mathscr{F}_{\text{finite}}(G,\mathbf{R}) \text{ and } \sup\Gamma=\Gamma(\mathbf{0}_L)=\mathscr{F}(G,\mathbf{R}).$$

For the lower inverse we get $\Gamma_{[-1]}(\mathscr{H})=\mathbf{1}_L=\mathscr{F}(G,\mathbf{R})$ if $\mathscr{H}\supset\mathscr{F}_{\text{finite}}(G,\mathbf{R})$ and equal to $\mathbf{0}_L=\varnothing$ otherwise.

For the upper inverse we have $\Gamma^{[-1]}(\mathscr{H})=\mathbf{0}_L=\varnothing$ for all sets $\mathscr{H}$.

(b) For $\Gamma^0$ we note that

$$\inf\Gamma^0=\Gamma^0(\mathbf{1}_L)=\{0\} \text{ and } \sup\Gamma^0=\Gamma^0(\mathbf{0})=\mathscr{F}(G,\mathbf{R}).$$

The lower inverse is equal to $(\Gamma^0)_{[-1]}(\mathscr{H})=\mathscr{F}(G,\mathbf{R})$ when $\mathscr{H}\supset\{0\}$ and equal to $\varnothing$ otherwise.

For the upper inverse we see that $(\Gamma^0)^{[-1]}(\mathscr{H})=\varnothing$ for all $\mathscr{H}$.

(c) Here $\inf\Gamma_0^{\gtrless 0}=\{0\}$ and $\sup\Gamma_0^{\gtrless 0}=\mathscr{F}_{\text{finite}}(G,\mathbf{R}^n)$. This gives $(\Gamma_0^{\gtrless 0})_{[-1]}(\mathscr{H})=\mathscr{F}_{\text{finite}}(G,\mathbf{R}^n)$ when $\mathscr{H}\supset\{0\}$ and equal to $\varnothing$ otherwise.

The upper inverse $(\Gamma_0^{\gtrless 0})^{[-1]}(\mathscr{H})$ is always equal to $\varnothing$.

(d) We get $\left(\Gamma_1^{\geqslant 0}\right)_{[-1]}(\mathscr{H}) = \mathscr{F}(G, \mathbf{R}^n)$ when $\mathscr{H} \supset \{0\}$ and equal to $\varnothing$ otherwise.

The upper inverse $\left(\Gamma_1^{\geqslant 0}\right)^{[-1]}(\mathscr{H})$ is always equal to $\varnothing$.

13.2. We find that $\Psi(h) = \Psi(\chi_{2\mathbf{Z}+1}) = \{(a,b) \in \mathbf{Z} \times \mathbf{Z};\ ab \in 2\mathbf{Z}\}$.

13.3. Elementary, Dear Reader: *To be and not to be, that is the answer* (published by Piet Hein the day after a referendum in Denmark).

13.4. Ursula's Duality:

*Light is the left hand of darkness*
*and darkness the right hand of light.* (LeGuin 1977:159)

Note also: *An answer is the left hand of a question.*

## Chapter 14: Topology

14.1. Take $X = \mathbf{Q}$, the set of all rational numbers, with the topology induced by that of $\mathbf{R}$. Its components are all the singletons sets, which are closed but not open. If you want a space with exactly one non-open component, take $X = \{0\} \cup \{1/k;\ k \in \dot{\mathbf{N}}\}$.

14.2. Easy because of the equivalence $f^*(B) \supset A \Leftrightarrow B \supset f_*(A)$ (see Example 5.2.3).

14.3. That $\kappa = f^* \circ \lambda \circ f_*$ is increasing and larger than the identity in $\mathscr{P}(X)$ is easy to see. Idempotency follows from the fact that $f_* \circ f^* = \mathbf{id}_Y$.

14.4. That $\lambda = f_* \circ \kappa \circ f^*$ need not be extensive is clear since $f^*(Y \smallsetminus \mathbf{im}(f)) = \varnothing$, so that also $\lambda(Y \smallsetminus \mathbf{im}(f))$ is empty, a set which is strictly contained in $Y \smallsetminus \mathbf{im}(f)$ as soon as $f$ is not surjective.

Take now $X = Y = [0,2]$ with the topology inherited from $\mathbf{R}$ and define

$$f(x) = \begin{cases} 0, & x \in [0,1] \cap \mathbf{Q}; \\ 1, & x \in ([0,1] \smallsetminus \mathbf{Q}) \cup \,]1,2[\,; \\ 2, & x = 2. \end{cases}$$

Then $\lambda(\{0\}) = \{0,1\}$ while $\lambda(\lambda(\{0\})) = \{0,1,2\}$. So this $\lambda$ is not idempotent.

14.5. Because of the formula $f^*(Y \smallsetminus B) = Y \smallsetminus f^*(B)$, we see that $f\colon X \to Y$ is continuous iff the inverse image of every closed subset of $Y$ is closed in $X$. For the closure operators $\kappa_X$ and $\kappa_Y$, continuity of $f$ is equivalent to the inequality $\kappa_X \leqslant \kappa_Y$.

14.6. With $X = \{a, b, c\}$ we order the possible topologies according to the number $s$ of open singleton sets and the number $t$ of open two-sets; we know that $0 \leqslant s, t \leqslant 3$, so a priori there are 16 possibilities. There are no topologies with the eight combinations

$$(s, t) = (0, 2), (0, 3), (1, 3), (2, 0), (2, 3), (3, 0), (3, 1), (3, 2).$$

For the other eight, we give below the number of different topologies, then the number of topologies that arise after renaming the points.

$(s, t) = (0, 0)$. 1, (1): $\tau = \{\varnothing, X\}$, the chaotic topology, which is connected.

$(s, t) = (0, 1)$. 1, (3): $\{\varnothing, \{a, b\}, X\}$ and two more, connected.

$(s, t) = (1, 0)$. 1, (3): the dual to $(0, 1)$, i.e., $\{X, \{c\}, \varnothing\}$, and two more, connected.

$(s, t) = (1, 1)$. 2, (9): $\{\varnothing, \{a\}, \{a, b\}, X\}$ and five more, all connected; $\{\varnothing, \{a\}, \{b, c\}, X\}$ and two more, not connected ($\{a\}$ is both open and closed).

$(s, t) = (1, 2)$. 1, (3): $\{\varnothing, \{a\}, \{a, b\}, \{c, a\}, X\}$ and two more, connected;

$(s, t) = (2, 1)$. 1, (3): the dual to $(1, 2)$, i.e., $\{X, \{b, c\}, \{c\}, \{b\}, \varnothing\}$, and two more, connected;

$(s, t) = (2, 2)$. 1, (6): $\{\varnothing, \{a\}, \{b\}, \{a, b\}, \{c, a\}, X\}$ and five more, none of them connected ($\{b\}$ is both open and closed);

$(s, t) = (3, 3)$. 1, (1): $\mathscr{P}(X)$ the discrete topology, not connected.

It is clear from this list that the nine topologies mentioned first are all different: none of them can be transformed to another one by permutation. So there are 9 different topologies. If we add all permutations, we get instead $1 + 3 + 3 + (6 + 3) + 3 + 3 + 3 + 1 = 29$ topologies, less than half of the number $2^{2^m - m} = 64$ (cf. the remark on Sierpiński space on page 309).
Of the nine different topologies in the list, $1 + 1 + 1 + (1 + 0) + 1 + 1 + 0 + 0 = 6$ are connected. With all permutations, they are $1 + 3 + 3 + (6 + 0) + 3 + 3 + 0 + 0 = 19$.

14.7. We study $\kappa(A) = \overline{A}$, $\alpha(A) = A^\circ$ and their compositions $\eta = \alpha \circ \kappa$ and $\theta = \kappa \circ \alpha$. The inverses $\kappa_{[-1]}$ and $\alpha^{[-1]}$ are both equal to the identity mapping. The other two are constant mappings given by

$$\kappa^{[-1]}(B) = \varnothing, \quad \alpha_{[-1]}(B) = \mathbf{R}^n \text{ for all } B.$$

Because $\kappa_{[-1]}$ and $\alpha^{[-1]}$ are equal to the identity mapping, the four second inverses $(\kappa_{[-1]})^{[-1]}$, $(\kappa_{[-1]})_{[-1]}$, $(\alpha^{[-1]})^{[-1]}$ and $(\alpha^{[-1]})^{[-1]}$, $(\kappa_{[-1]})_{[-1]}$ are also equal to identity mapping.

Two second inverses are constant:

$$(\kappa^{[-1]})_{[-1]}(A) = \mathbf{R}^n \text{ and } (\alpha_{[-1]})^{[-1]}(A) = \varnothing \text{ for all } A.$$

The two remaining second inverses are

$$(\kappa^{[-1]})^{[-1]}(A) = \mathbf{R}^n \text{ when } A \neq \varnothing \text{ and equal to } \varnothing \text{ when } A = \varnothing;$$

$$(\alpha_{[-1]})_{[-1]}(A) = \varnothing \text{ for } A \neq \mathbf{R}^n \text{ and equal to } \mathbf{R}^n \text{ for } A = \mathbf{R}^n.$$

The inverses of $\eta$ and $\theta$ are all constant: $\eta_{[-1]}(B) = \theta_{[-1]}(B) = \mathbf{R}^n$ and $\eta^{[-1]}(B) = \theta^{[-1]}(B) = \varnothing$ for all $B$.

Their second inverses are easily found since $\eta_{[-1]} = \theta_{[-1]} = \alpha_{[-1]}$ and $\eta^{[-1]} = \theta^{[-1]} = \kappa^{[-1]}$.

## Chapter 15: The Khalimsky topology

15.1. (a) All four mappings $\{0,1\} \to \{0,1\}$ are Lip-1, but $x \mapsto 1-x$ is not continuous. The fraction is 0.75. Similarly for $\{1,2\}$.

(b) Of the $3^3 = 27$ mappings $\{0,1,2\} \to \{0,1,2\}$, 17 are Lip-1 mappings and 11 of these are continuous. These are the three constant mappings and the eight with values at $x$ equal to

$$x, \quad 2-x; \quad x \wedge (2-x), \quad (2-x) \vee x; \quad x \wedge 1, \quad (2-x) \vee 1;$$
$$x \vee 1 \text{ and } (2-x) \wedge 1.$$

There are 5 mappings with $f(0) = 0$ of which 4 are continuous, 7 with $f(0) = 1$ of which 3 are continuous, and 5 with $f(0) = 3$ of which 4 are continuous.

The fraction of the number of continuous mappings over that of Lip-1 mappings is $11/17 \approx 0.647 < 0.75$. The numbers are the same for $\{1,2,3\}$.

(c Of the $4^4 = 256$ selfmappings of $\{0,1,2,3\}$, 68 are Lip-1 mappings and 31 of these are continuous for the Khalimsky topology. The fraction is $31/68 \approx 0.456 < 0.647$. The numbers are the same for $\{1,2,3,4\}$.

For $f(0) = 0,1,2,3$ the numbers are the following.

When $f(0) = 0$, there are 13 Lip-1 mappings, of which 8 are continuous;
when $f(0) = 1$, there are 21 mappings, of which 6 are continuous;
when $f(0) = 2$, also 21, of which 13 are continuous; and, finally,
when $f(0) = 3$, there are 13 Lip-1 mappings, of which 4 are continuous.

(There is symmetry for the Lip-1 mappings but not for the continuous ones here, since the inversion $x \mapsto 3 - x$ is discontinuous.)

15.2. In $\{0,1\}^2$ the open sets are $\varnothing$, $\{(1,1)\}$, $\{(1,1),(1,0)\}$, $\{(1,1),(0,1)\}$, $\{(1,1),(1,0),(0,1)\}$ and $\{0,1\}^2$. A function $\chi_A\colon \{0,1\}^2 \to \{0,1\}$ is continuous if and only if $A$ is open. So there are six continuous mappings $\{0,1\}^2 \to \{0,1\}$, viz. the characteristic functions of the six mentioned open sets.

In $\{0,1,2,3\}$ the open sets are $\varnothing$, $\{1\}$, $\{3\}$, $\{1,3\}$ and $\{0,1,2,3\}$. So there are five continuous mappings $\{0,1,2,3\} \to \{0,1\}$, viz. the two constant functions and the characteristic functions $\chi_{\{1\}}$, $\chi_{\{3\}}$ and $\chi_{\{1,3\}}$.

15.3. (a) We have $\lambda(\bigvee A_j) = \overline{\bigcup A_j} = \bigvee A_j$ since $\bigvee A_j$ is by definition the smallest closed set containing all the $A_j$. For $\kappa$, the question is whether we have $\bigcup A_j$ taken in $\mathscr{C}(X)$ is equal to $\overline{\bigcup A_j}$. It is equivalent to ask whether any union of closed set is closed, and that any intersection of open sets is open. This is the very definition of a smallest-neighborhood space.

(b) For $\kappa\colon \mathscr{P}(\mathbf{R}^n) \to \mathscr{P}(\mathbf{R}^n)$ and $\lambda\colon \mathscr{P}(\mathbf{R}^n) \to \mathscr{C}(\mathbf{R}^n)$, $\kappa_{[-1]}$ is the identity in $\mathscr{P}(\mathbf{R}^n)$ and $\lambda_{[-1]}$ is equal to the inclusion mapping of $\mathscr{C}(\mathbf{R}^n)$ into $\mathscr{P}(\mathbf{R}^n)$.

The upper inverses yield only the empty set. This is clear since $\kappa^{[-1]}(B)$ is a subset of $\mathbf{Q}^n$ as well as of $\mathbf{R}^n \smallsetminus \mathbf{Q}^n$.

(c) We obtain that $\kappa_{[-1]}(B)$ is the interior of $B$ taken in the topology dual to the Khalimsky topology of $\mathbf{Z}^n$. To see this, note that the closure $\overline{\{a\}}$ of a singleton set $\{a\} \subset A$ for the Khalimsky topology shall be contained in $B$ ... but this closure is equal to the smallest neighborhood $N(a)$ for the dual topology, so $\kappa_{[-1]}(B)$ is the union of all open subsets of $B$ for the dual topology.

For $\lambda_{[-1]}(B)$ the same observation holds, but since now $B$ is open for the dual topology, the interior is equal to $B$, so $\lambda_{[-1]}$ is just the inclusion mapping of $\mathscr{C}(\mathbf{R}^n)$ into $\mathscr{P}(\mathbf{R}^n)$.

For the upper inverses, we note that $\kappa^{[-1]}(B) \subset B \cap (2\mathbf{Z}+1)^n$ since $B$ is dense in $B$ and $(2\mathbf{Z}+1)^n$ is dense in $\mathbf{R}^n$. Actually we have equality here. For $\lambda^{[-1]}$ the same thing holds.

15.4. (a) For any subset $A$ of $\mathbf{Z}$, let us write $A_{\text{even}} = A \cap (2\mathbf{Z})$ and $A_{\text{odd}} = A \cap (2\mathbf{Z}+1)$ for the even and the odd numbers in $A$, respectively. Then

$$\kappa(A) = \overline{A} = A_{\text{even}} \cup (A_{\text{odd}} + \{-1,0,1\}) \text{ and}$$
$$\alpha(A) = A^{\circ} = A_{\text{odd}} \cup \{x \in A_{\text{even}};\ x + \{-1,0,1\} \subset A\}.$$

(b) It is clear that $\kappa$ is a dilation and that $\alpha$ is an erosion. Then $\kappa_{[-1]}$ is an erosion. It is however not equal to $\alpha$, but equal to the corresponding anoiktomorphism and erosion for the dual topology of $\mathbf{Z}$, the one for which the subsets of $2\mathbf{Z}$ are open and the subsets of $2\mathbf{Z}+1$ are closed.

15.5. We have to find a mapping $f\colon \mathbf{R}^2 \to \mathbf{Z}^2$ such that the set $f^*(\{p\})$ is closed for every point $p$ such that $\{p\}$ is closed and $f^*\{(p)\}$ is open for $\{p\}$ open.

We can define $f(x) = a$, $f(x) = b$ and $f(x) = c$ according to the following recipe. First we take $f(x) = a$ if $x = a \in (2\mathbf{Z}) \times (2\mathbf{Z})$, so that $\{a\}$ is closed. Second take $f(x) = b$ for $x \in \mathbf{R}^2$ such that $\|x-b\|_\infty < 1$ and $\{b\}$ is open. Thus $f^*(\{p\}) = \{p\}$ if $\{p\}$ is closed and $f^*(\{p\}) = B^\infty_<(p,1)$, an open set, if $\{p\}$ is open. These sets do not cover all of $\mathbf{R}^2$; the complement of their union is neither closed nor open and consists of crosses. So, third, to points $x$ in such a cross we associate its center $c$, which is a mixed point, thus $f(x) = c$.

15.6. For the Wyse–Marcus topology, the singleton sets $\{(0,0)\}$ and $\{(1,1)\}$ are closed; the singleton sets $\{(1,0)\}$ and $\{(0,1)\}$ are open. So, if we have a mapping $f\colon \mathbf{R}^2 \to \mathbf{Z}^2$ defining this topology as a quotient topology, then $f^*(\{(0,0)\})$ and $f^*(\{(1,1)\})$ are closed and $f^*(\{(1,0)\})$ and $f^*(\{(0,1)\})$ are open. The four sets are pairwise disjoint. The set-theoretical difference $U = [0,1]^2 \smallsetminus \big(f^*(\{(0,0)\}) \cup f^*(\{(1,1)\})\big)$ is open and contains the points $(0,1)$ and $(1,0)$. Since the closed sets $f^*(\{(0,0)\})$ and $f^*(\{(1,1)\})$ are disjoint, the infimum of $\|x-y\|_2$ when $f(x) = (0,0)$ and $f(y) = (1,1)$ is positive, the open set $U$ is not too narrow, and it can be seen that $(0,1)$ and $(1,0)$ can be joined by a polygon in $U$ and therefore belong to the same component of $U$, contrary to the assumption that the sets $f^*(\{(0,1)\})$ and

$f^*(\{(1,0)\})$ are disjoint. The absence of mixed points explain this impossibility.

The proof for the hexagonal tessellation is the same. Here the singleton sets $\{0\}$ and $\{2-\omega\}$ are closed in $\mathbf{C}$ and the sets $\{1\}$ and $\{\omega\}$ are open. The four sets $f^*(\{0\})$, $f^*(\{2-\omega\})$, $f^*(\{1\})$, $f^*(\{\omega\})$ are pairwise disjoint and give rise to the same kind of contradiction as in the case of the Wyse–Marcus topology.

## Chapter 16: Distance transforms

16.1. First convince yourself that there are points $p \in A$ and $q \notin A$ such that $p_1 = q_1$ or $|p_1 - q_1| = 1$. In both cases we can find $a \in A$ and $b \notin A$ such that $\|a-b\|_1 = 1$. Then it is easy to see that the values of $\theta$ and $\Theta$ are as indicated for the $l^\infty$-norm, 16.1 (a), and that the values are the same for the $l^1$-norm, 16.1 (b).

16.3. It is easy to see that $\mathbf{dt}_B(0) = -r$, denoting by $B$ the ball $B_\leqslant(0,r)$ for the given norm. For $x \neq 0$ we define $y = rx/\|x\|$, and find that $\mathbf{dt}_B(x) \leqslant \|x-y\|$ for $x \notin B$ as well as $-\mathbf{dt}_B(x) \leqslant \|x-y\|$ for $x \in B$. Equality remains to be proved. For $x \notin B$ we find a hyperplane separating $B_\leqslant(x, \|x-y\|)$ and $B$, which proves that $\mathbf{dt}_B(x) = \|x-y\|$ in the complement of $B$. For $x \in B$, we find that any point $z \notin B$ must have a distance to $x$ at least equal to $\|x-y\|$; hence $-\mathbf{dt}_B(x) = \|x-y\| = r - \|x\|$.

16.4. (a) $\mathbf{dt}_A$, $p=1$. We have $\mathbf{dt}_A(x) = \|x\|_1 - 2$ when

$$|x_1| \wedge |x_2| = \|x\|_1 - \|x\|_\infty \geqslant 1$$

and equal to $\|x\|_\infty - 1$ when $|x_1| \wedge |x_2| \leqslant 1$.

(b) $\mathbf{dt}_A$, $p=2$. When $|x_1| \wedge |x_2| \geqslant 1$, then $\mathbf{dt}_A(x)$ is the minimum of the four numbers $\|x - (\pm 1, \pm 1)\|_2$; when $|x_1| \wedge |x_2| \leqslant 1$, then $\mathbf{dt}_A(x) = \|x\|_\infty - 1$.

16.5. (a) $\mathbf{dt}_B$, $p=1$. We have

$$\mathbf{dt}_B(x) = \left(\|x\|_\infty - \sqrt{r^2 - (|x_1| \wedge |x_2|)^2}\right) \vee \left(\|x\|_1 - r\sqrt{2}\right)$$

when $x \notin B$, and

$$\|x\|_\infty - \sqrt{r^2 - (|x_1| \wedge |x_2|)^2}$$

when $x \in B$.

(b) $\mathbf{dt}_B$, $p = \infty$. Turn everything by 45° in 16.5 (a)! Explicitly, we have $\mathbf{dt}_B(x) = \|x\|_\infty - r$ when $|x_1 - x_2| \geqslant r$ and equal to

$$\tfrac{1}{2}\|x\|_1 - \sqrt{r^2 - |x_1 - x_2|^2/4}$$

when $|x_1 - x_2| \leqslant r$.

16.6. (a) From Proposition 16.3.13 we see that

$$\mathbf{dt}_{A+B}(x) = (\mathbf{dt}_A \sqcap \mathbf{dt}_B)(x) \leqslant \mathbf{dt}_A(x) + \mathbf{dt}_B(0) = \mathbf{dt}_A(x) - r,$$

where we have written $B$ for $B_<(0, r)$.

On the other hand, given any $x \notin A+B$ and any positive number $t$, there exists an element $c \in A+B$ with $d(x, c) \leqslant \mathbf{dt}_{A+B}(x)+t$. There exists a point $a \in A$ such that $d(c, a) \leqslant r + t$. Combining the two, we find that $\mathbf{dt}_A(x) \leqslant d(x, a) \leqslant d(x, c) + d(c, a) \leqslant \mathbf{dt}_{A+B}(x) + t + r + t$. Since $t$ is arbitrarily small, we get $\mathbf{dt}_{A+B}(x) \geqslant \mathbf{dt}_A(x) - r$; hence equality.

(b) No. In $\mathbf{Z}$ we can take $A = [-a, a]$ and $B = B_<(0, r) = [-r + 1, r - 1]$ with $a, r \in \dot{\mathbf{N}}$. Then for $x > a + r$ we have $\mathbf{dt}_{A+B}(x) = x - a - r + 1 = \mathbf{dt}_A(x) - r + 1 > \mathbf{dt}_A(x) - r$. This works also in $\mathbf{Z}^2$.

(c) is easy.

16.7. Take $X = \mathbf{R}^n$ and $d(x, y) = \|x - y\|_2$ as the distance, $A = B_<(0, s)$. Then $x \in \varepsilon_{B_\leqslant(0,r)}(A)$ if and only if $\|x\|_2 \leqslant s - r$, and $\mathbf{dt}_A(x) < -r$ if and only if $\|x\|_2 < s - r$, a strictly smaller set.

16.8. Writing $B$ for $\mathbf{epi}^{\text{finite}}(\mathbf{dt}_A)$, we have, if $x \notin A$ and $(x, s) \notin B$,

$$\mathbf{dt}_B(x, s) = \inf_{y \in X} \left(\gamma d(x, y) + |s - \mathbf{dt}_A(y)|\right) \leqslant \gamma d(x, x) + \mathbf{dt}_A(x) - s.$$

Since $|\mathbf{dt}_A(x) - \mathbf{dt}_A(y)| \leqslant 2d(x, y)$, we see that the infimum is actually attained for $y = x$ when $\gamma \geqslant 2$

In a normed space we have $|\mathbf{dt}_A(x) - \mathbf{dt}_A(y)| \leqslant d(x, y)$, so the same argument works for $\gamma \geqslant 1$.

In the other cases (i.e., when $x \notin A$, $(x, s) \in B$; $x \in A$, $(x, s) \notin B$; $x \in A$, $(x, s) \in B$), similar calculations can be made.

If $\gamma < 2$ and $A = \{0\} \subset \mathbf{Z}$, we see that $dt_A(x) = |x|$ for $x \in \mathbf{Z} \smallsetminus \{0\}$, that $\mathbf{dt}_A(0) = -1$, and that, for $(x, s) = (1, -1)$

$$\mathbf{dt}_B(x, s) = d_\gamma((1, -1), (0, -1)) = \gamma < 1 - (-1) = \mathbf{dt}_A(x) - s.$$

If $\gamma < 1$ and $A = \{0\}$ in the vector space $\mathbf{R}$, we see that $\mathbf{dt}_A(x) = |x|$ for all $x \in \mathbf{R}$ and that $d_B(x, 0) = d(x, 0) = \gamma x < x = \mathbf{dt}_A(x) - s$ when $x > 0$.

16.9. *Square pixels.* By symmetry it is enough to consider points $(x, y)$ with $0 \leqslant y \leqslant x$. We get $\|(x,y)\|_2 = \sqrt{x^2+y^2} = x\sqrt{1+t^2}$, where $t = y/x$ if $x > 0$ and $t = 0$ if $x = 0$. The finitely constructed distance is $f(x, y) = a(x - y) + by = x((1 - t)a + tb)$. The quotient

$$\frac{f(x,y)}{\|(x,y)\|_2} = \frac{(1-t)a+tb}{\sqrt{1+t^2}}, \qquad 0 \leqslant t \leqslant 1,$$

assumes its minimum, equal to $a \wedge (b/\sqrt{2})$, for $t = 0$ or $t = 1$, and its maximum, equal to $\sqrt{2a^2 - 2ab + b^2}$ for $t = b/a - 1$. If $b \geqslant \sqrt{2}a$, the quotient $f(x, y)/\|(x, y)\|_2$ is equal to $\sqrt{b^2/a^2 - b/a + 2}$ (smallest when $b = \sqrt{2}a$). If $b \leqslant \sqrt{2}a$, then the quotient is equal to $\sqrt{4a^2/b^2 - 4a/b + 2}$ (smallest when $b = \sqrt{2}a$). The smallest value is therefore obtained when $b = \sqrt{2}a$ and is $\sqrt{4 - 2\sqrt{2}}$, proving the claim for $\lambda(f, \|\cdot\|_2)$.

We have $a \leqslant \Lambda(f, \|\cdot\|_2) \leqslant a\sqrt{4 - 2\sqrt{2}}$. The best choice of $a$ is when $a^2\sqrt{4 - 2\sqrt{2}} = 1$, yielding the value $a_0$ in (16.6.1) on page 362.

16.10. *Hexagonal pixels.* By symmetry it is enough to consider points in the complex plane of the form $p + q\omega$ with $p, q \geqslant 0$. We define $t = q/p$ when $p > 0$ and $t = 0$ when $p = 0$. The absolute value of the complex number $p + q\omega$ is equal to $p\sqrt{1 + t + t^2}$, and the finitely constructed distance of $p + q\omega$ to the origin is $p(1 + t)$. The quotient $Q(t) = (1 + t)/\sqrt{1 + t + t^2}$ has infimum equal to 1 and supremum equal to $2/\sqrt{3}$. This can be found as in Exercise 16.9 by considering $t$ to be a real variable and finding its derivative $Q'(t) = \frac{1}{2}(1-t)(1+t+t^2)^{-3/2}$, which vanishes for $t = 1$. So $1 \leqslant Q \leqslant 2/\sqrt{3}$ is the result.

16.11. *Triangular pixels.* By symmetry it is enough to consider points $(x, \sqrt{3}y)$ with $x, y \geqslant 0$. We get $\|(x, \sqrt{3}y)\|_2 = \sqrt{x^2 + 3y^2} = x\sqrt{1 + 3t^2}$, where $t = y/x$ if $x > 0$ and $t = 0$ if $x = 0$. The finitely constructed distance is $f(x, y) = x(1 + \sqrt{3}t)$. The quotient

$$\frac{f(x,y)}{\|(x,y)\|_2} = \frac{1+\sqrt{3}t}{\sqrt{1+3t^2}}, \qquad t \geqslant 0,$$

assumes its minimum, equal to 1, for $t = 0$ and tends to 1 as $t \to +\infty$. Again, this can be found as in Exercises 16.9 and 16.10 by considering $t$ to be a real variable and finding its derivative, which vanishes for $t = 1/\sqrt{3}$. So $\lambda(f, \|(x, y)\|_2)$ is equal to $\sqrt[4]{2}$.

16.12. We look at the balls $B_{\leqslant}(a,r)$ and $B_{\leqslant}(b,s)$ and shall prove that equality implies equality of their centers provided the distance function is upper regular and translation invariant. We thus know that to $a$, $b$ and $z$ we can find $\tilde{z}$ such that

$$d(a,z) \leqslant d(a,\tilde{z}) = d(a,b) + d(b,\tilde{z}) = d(a,b) + d(b,z),$$

in particular, if $z \in B_{\leqslant}(b,s) = B_{\leqslant}(a,r)$, we have $\rho(r) \geqslant d(a,\tilde{z}) = d(a,b) + d(b,z)$. We now take the supremum over all $z \in B_{\leqslant}(b,s)$ and obtain $\rho(r) \geqslant d(a,b) + \rho(s)$. But by hypothesis $\rho(r) = \rho(s)$, forcing $d(a,b)$ to be zero.

16.13. We shall see that the distance defined is not lower regular, i.e., that we cannot always find $\tilde{x}$ such that $d(\tilde{x},z) = d(\tilde{x},y) + d(y,z)$. Take now $x = (0,0)$, $y = (1,0)$, $z = (1,1)$. Then the triangle inequality has the form $4 = d(\tilde{x},z) = d(\tilde{x},y) + d(y,z) = d(\tilde{x},y) + 3$, but there is no point $\tilde{x}$ satisfying this equality.

16.14. We see that the possible values of $d(x,0)$ in the interval $[0,3]$ are $r_0 = 0$, $r_1 = 1$, $r_2 = \sqrt{2}$, $r_3 = 2$, $r_4 = \sqrt{5}$, $r_5 = \sqrt{8}$, $r_6 = 3$.

From (16.7.14) we see that the value of $\rho$ when $r_j < r \leqslant r_{j+1}$ is $r_j$ and those of $\sigma_+$ is $r_{j+1}$. When $r_j \leqslant r < r_{j+1}$, the value of $\sigma$ is $r_{j+1}$ while the value of $\rho_-$ is $r_j$.

## Chapter 17: Skeletonizing

17.1. If a point $c$ is the center of a ball $B_<(c,r)$ which is maximal among the nonempty strict balls contained in $A$, then we must have $r > 0$ and there must exist a point $x$ belonging to the boundary of $B_<(c,r)$ and to the boundary of $A$. Then a point $y$ on the segment $[c,x]$ different from $c$ and $x$ cannot belong to the skeleton of $A$; no ball with center $y$ can be maximal.

17.2. We start with the unit disk $U_0 = B_<((0,0),1)$ for the $l^2$-norm in $\mathbf{R}^2$ and add small triangles as follows, essentially following Matheron (1988c:234–235). Let $(\theta_k)_{k=1}^{\infty}$ be a sequence containing all rational numbers in $[0,2\pi]$. For each $k = 1,2,\dots$, add an open triangle with one vertex at $((1+h_k)\cos\theta_k, (1+h_k)\sin\theta_k)$ and the other two vertices at $(\cos(\theta_k - \tau_k), \sin(\theta_k - \tau_k))$ and $(\cos(\theta_k + \tau_k), \sin(\theta_k + \tau_k))$, where the $\tau_k$ are nonnegative numbers with sum at most equal to a given positive number $\tau$. We take $h_k$ positive if $\tau_k$ does not belong to any of the intervals $[\theta_j - \tau_j, \theta_j + \tau_j]$ for $j = 1,\dots,k-1$; otherwise we take $h_k = 0$. (This is to avoid interference between the triangles. This interference happens only in a set of arguments of Lebesgue measure at most $2s$.) The set $U_k$ consisting of $U_0$ with the triangles of index $i = 1, \dots, k$ is contained in $B_<((0,0), 1 + \sum h_k)$. We let $U$ be the union of all the $U_k$. We can take the $h_k$ so small that $U_k$ is a convex polygon, as remarked by Jean Serra (personal message 2020 January 20), implying that the union $U$ is convex. Thus we should take $h_k \leqslant 1/\cos\tau_k - 1$. Then the skeleton of $U$ contains the segment

$$[(0,0), ((1+h_k)\cos\theta_k, (1+h_k)\sin\theta_k)]$$

for all $k$ for which we have taken $h_k$ positive, thus with the set of arguments being dense in a subset of $[0, 2\pi]$ of Lebesgue measure larger than that of $U$ minus twice the sum of the $\tau_k$, which is smaller than $2s$. Then also the two-dimensional Lebesgue measure of the skeleton is close to that av $U$.

17.3. If $B_<(a,r)$ and $B_<(b,s)$ are two open disks for the $l^2$-norm such that the distance between their centers satisfies $\|a-b\|_2 < r+s$, then the skeleton of their union (which is connected) is the segment $[a,b]$.

17.4. Take two disks $B_<(a,2)$ and $B_<(b,1)$ for the $l^\infty$-norm with centers $a = (0,0)$ and $b = (2,0)$. Then the skeleton of their union consists of the point $(0,0)$ and the segment $[(1,0),(2,0)]$. (The maximal balls are $B_<(a,2)$ and $B_<(c,1)$, where $c = (c_1, 0)$, $1 \leqslant c_1 \leqslant 2$.) No point strictly between $(0,0)$ and $(1,0)$ belongs to the skeleton.

17.5. For the $l^\infty$-metric, the skeleton of $[-2,2]^2_{\mathbf{Z}}$ as a subset of $\mathbf{Z}^2$ is just the singleton set $\{(0,0)\}$. (The maximal strict ball is $B_<((0,0),3) = {]-3,3[}^2 \cap \mathbf{Z}^2 = [-2,2]^2_{\mathbf{Z}}$.) When we remove the corner point $(2,2)$, the skeleton becomes larger, equal to the eight-set

$$([-1,1]_{\mathbf{Z}} \times [-1,1]_{\mathbf{Z}}) \smallsetminus \{(1,1)\}.$$

17.6. When we have added the three points $\{(3,-1),(3,0),(3,1)\}$, the skeleton is equal to $\{(0,0),(2,0)\}$.

17.7. The tasks presented in (a) and (b) are easily done.

(c) For the result on dilations it suffices to assume that $A$ is convex; cf. Exercise 16.6 (c) on page 378.

It is easy to find an example of a non-convex set such that the equality does not hold. We can take $A = B_<(-2,1) \cup B_<(2,1) \subset \mathbf{R}$ with $\mathbf{skel}_A = \{-2,2\}$. For $r > 1$ we have

$$\delta_{B_<(0,r)}(A) = A + B_<(0,r) = B_<(0, r+3) = {]-3-r, 3+r[},$$

an open interval with $\mathbf{skel}_{A+B_<(0,r)} = \{0\} \neq \{-2,2\} = \mathbf{skel}_A$.

(d) The result on erosions follows from the equivalence

$$B_<(x,r) \subset B_<(y,r+t) \Leftrightarrow x \in B_<(y,t).$$

(Proposition 16.4.2 on sublevel sets of distance transforms might help.)

The assertions in (e) and (f) are not difficult to prove once we have (c) and (d).

# Bibliography


Abdalla, Mohsen; Nagy, Benedek. 2020. Mathematical morphology on the triangular grid: the strict approach. *SIAM J. Imaging Sci.* **13**, no. 3, 1367–1385. Cited on page 379.

Abdelkader, Ahmed; Arya, Sunil; da Fonseca, Guilherme D.; Mount, David M. 2019. Approximate nearest neighbor searching with non-Euclidean and weighted distances. *Proceedings of the Thirtieth Annual ACM-SIAM Symposium on Discrete Algorithms*, pp. 355–372. Philadelphia, PA: SIAM. Cited on page 377.

Abuhmaidan, Khaled; Nagy, Benedek. 2020. Bijective, non-bijective and semi-bijective translations on the triangular plane. *Mathematics* 2020 *8*(1), 29. Cited on page 379.

Ahlberg, Axel W.; Lundqvist, Nils; Sörbom, Gunnar. 1966. *Latin-svensk ordbok*, 941 pp. Stockholm: Svenska Bokförlaget; Bonniers. (Second edition; several later editions.) Cited on page 19.

Alexandroff, Paul (Aleksandrov, P. S.). 1935. Sur les espaces discrets. *C. R. Acad. Sci. Paris* **200**, 1649–1651. Cited on pages 317, 318.

Alexandroff, Paul (Aleksandrov, P. S.). 1937 Diskrete Räume. *Mat. Sb.* **2** (44), 501–519. Cited on pages 315, 318.

Andres, Eric; Dutt, Mousumi; Biswas, Arindam; Largeteau-Skapin, Gaelle; Zrour, Rita. 2019. Digital two-dimensional bijective reflection and associated rotation. **In:** (Couprie et al., Eds., 2019:3–14). Cited on page 7.

Andres, Eric; Largeteau-Skapin, Gaelle. 2016. Digital surfaces of revolution made simple. **In:** (Normand et al., Eds., 2016:244–255). Cited on page 7.

Angulo, Jesus. 2007. Morphological colour operators in totally ordered lattices based on distances: Application to image filtering, enhancement and analysis. *Computer Vision and Image Understanding* **107**, 56–73. Cited on page 79.

Angulo, Jesus. 2017. Convolution in (max,min)-algebra and its role in mathematical morphology. **In:** *Advances in Imaging and Electron Physics*, **203**, 1–66. Cited on page 51.

Angulo, Jesús. 2019. Hierarchial Laplacian and its spectrum in ultrametric image processing. **In:** (Burgeth et al., Eds., 2019:29–40). Cited on page 52.

Aptoula, E.; Lefèvre, S. 2007. A comparative study on multivariate mathematical morphology. *Pattern Recognition* **40**, 2914–2929. Cited on page 79.

Artstein-Avidan, Shiri; Milman, Vitali. 2009. The concept of duality in convex analysis, and the characterization of the Legendre transform. *Ann. of Math.* **169**, 661–674. Cited on page 193.

Asplund, Edgar. 1960. Comparison between plane symmetric convex bodies and parallelograms. *Math. Scand.* **8**, 171–180. Cited on page 94.

Avkan, Aydın; Nagy, Benedek; Saadetoğlu, Müge. 2020. Digitized rotations of 12 neighbors on the triangular grid. *Ann. Math. Artif. Intell.* **88**, no. 8, 833–857. Cited on page 7.

Baclawski, Kenneth; Björner, Anders. 1979. Fixed points in partially ordered sets. *Advances in Mathematics* **31**, 263–287. Cited on page 326.

Bailey, Donald G. 2004. An efficient Euclidean distance transform. **In:** (Klette & Žunić, Eds., 2004:394–408). Cited on page 355.

Balázs, Péter; Brunetti, Sarah. 2016. A measure of $Q$-convexity. **In:** (Normand et al., Eds., 2016:219–230). Cited on page 256.

Balázs, Péter; Brunetti, Sarah. 2017. A new shape descriptor based on a $Q$-convexity measure. **In:** (Kropatsch et al., Eds., 2017:267–378). Cited on page 256.

Banon, Gerald Jean Francis. 2005. Regular metric: Definition and characterization in the discrete case. **In:** (Ronse et al., 2005:249–258). Cited on page 376.

Banon, Gerald Jean Francis; Barrera, Junior. 1993. Decomposition of mappings between complete lattices by mathematical morphology. *Signal Process.* **30**, 299–327. Cited on page 83.

Banon, Gerald Jean Francis; Barrera, Junior; Braga-Neto, Ulisses de Mendonça, Eds. 2007. *Mathematical Morphology and its Applications to Signal and Image processing. Proceedings of the $8^{th}$ International Symposium on Mathematical Morphology, Rio de Janeiro, RJ, Brasil, October 10–13, 2007*, xii + 475 pp. São José dos Campos, SP: MCT/INPE.

Barneva, Renata P.; Brimkov, Valentin E.; Koroutchev, Kostadin; Koroutcheva, Elka, Eds. 2011. *Advances in Image Analysis and Applications*, xi + 133 pp. Singapore: Research Publishing.

Barbu, Viorel; Precupanu, Teodor. 2012. *Convexity and optimization in Banach spaces*, xii + 368 pp. Springer Monographs in Mathematics. Dortrecht: Springer. Fourth edition. Cited on page 199.

Barrow, H[arry] G.; Tenenbaum, J[ay] M.; Bolles, R[obert] C.; Wolf, H[elen] C. 1977. Parametric correspondence and chamfer matching: Two new techniques for image matching. **In:** *Proceedings of the 5th International Joint Conference on Artificial Intelligence*, pp. 659–663. Cambridge, MA. Cited on page 375.

Bédaride, Niclas; Domenjoud, Eric; Jamet, Damien; Rémy, Jean-Luc. 2010. On the number of balanced words of given length and height over a two-letter alphabet. *Discrete Math. Theor. Comput. Sci.* **12**, no. 3, 41–62. Cited on page 142.

Bell, John L[ane]. 2006. *The Continuous and the Infinitesimal In Mathematics and Philosophy*, 352 pp. Milano: Polimetrica. Cited on pages 7, 17.

Benediktsson, Jón Atli; Channussot, Jocelyn; Najman, Laurent; Talbot, Hugues, Eds. 2015. *Mathematical Morphology and Its Applications to Signal and Image Processing. 12th International Symposium, ISMM 2015, Reykjavik, Iceland, May 27–29, 2015. Proceedings*, XIV + 754 pp. Lecture Notes in Comput. Sci., 9082.

Berthé, Valérie. 2009. Discrete geometry and symbolic dynamics. **In:** (Passare, Ed., 2009:81–110). Cited on page 142.

Bilodeau, Michel; Meyer, Fernand; Schmitt, Michel, Eds. 2005. *Space, structure and randomness. Contributions in honor of Georges Matheron in the field of geostatistics, random sets and mathematical morphology*, xviii + 395 pp. Lecture Notes in Statistics, 183. New York, NY: Springer.

Birkhoff, Garrett. 1933. On the combination of subalgebras. *Proc. Camb. Philos. Soc.* **29**, part IV, 441–464. Cited on page 78.

Birkhoff, Garrett. 1940. *Lattice Theory*, v + 155 pp. New York, NY: American Mathematical Society. Cited on pages 35, 35, 67, 78.

Birkhoff, Garrett. 1948. *Lattice Theory.* New York, NY: American Mathematical Society. Revised Edition. Cited on page 35.

Birkhoff, Garrett. 1995. *Lattice Theory*, xiii + 420 pp. American Mathematical Society Colloquium Publications, vol. 25, Third edition, eighth printing. Cited on pages 67, 112.

Birkhoff, Garret; Mac Lane, Saunders. 1957. *A Survey of Modern Algabra*, xi + 472 pp. New York, NY: The MacMillan Company. Cited on page 35.

Birkhoff, George D[avid]. 1911. General theory of linear difference equations. *Trans. Amer. Math. Soc.* **12**, no. 2, 243–284. Cited on page 163.

Bloch, Isabelle; Maître, Henri. 1995. Fuzzy mathematical morphologies: a comparative study. *Pattern Recognition* **28**, no. 9, 1341–1387. Cited on page 102.

Blum, Harry. 1967. A transformation for extracting new descriptors of shape. **In:** Wathen-Dunn, W., Ed., *Models for the Perception of Speech and Visual Form*, pp. 362–380. Cambridge, MA: M.I.T. Press. Cited on page 389.

Blum, Harry. 1973. Biological shape and visual science (part I). *Journal of Theoretical Biology* **18**, issue 2, 205–287. Cited on page 389.

Blyth, T. S. 2005. *Lattices and Ordered Algebraic Structures*, X + 303 pp. Springer. Cited on pages 35, 67.

Blyth, T. S.; Janowitz, M. F. 1972. *Residuation Theory*, ix + 379 pp. Oxford et al.: Pergamon Press. Cited on pages 35, 36, 67.

Boman, Jan. 1967. Differentiability of a function and of its compositions with functions of one variable. *Math. Scand.* **20**, 249–268. Cited on page 95.

Bombelli, Luca; Lee, Joohan; Meyer, David; Sorkin, Rafael D. 1987. Space-time as a causal set. *Physical Review Letters* **59**, 521–524. Cited on page 8.

Bonnesen, T.; Fenchel, W. 1934. *Theorie der konvexen Körper.* Berlin: Verlag von Julius Springer. Cited on page 199.

Borgefors, Gunilla. 1984. Distance transformations in arbitrary dimensions. *Comput. Vision Graphics Image Process.* **27**, 321–345. Cited on pages 363, 390.

Borgefors, Gunilla. 1986. Distance transformations in digital images. *Comput. Vision Graphics Image Process.* **34**, 344–371. Cited on pages 363, 390.

Borgefors, Gunilla. 1989. Distance transformations on hexagonal grids. *Pattern Recognition Letters* **9**, 97–105. Cited on pages 375, 390.

Borgefors, Gunilla. 1991. Another comment on "A note on 'Distance transformations in digital images'." *Comput. Vision Graphics Image Process.* **54**, 301–306. Cited on page 363.

Borgefors, Gunilla. 1994. Applications using distance transforms. **In:** Arcelli, Carlo; Cordella, Luigi P.; Sanniti di Baja, Gabriella, Eds., *Aspects of Visual Form Processing*, pp. 83–108. Singapore: World Scientific. Cited on pages 355, 390.

Borgefors, Gunilla. 1996a. On digital distance transforms in three dimensions. *Computer Vision and Image Understanding* **64**, 368–376. Cited on page 390.

Borgefors, Gunilla. 1996b. Discretization – problems and solutions. **In:** Maître, H[enri].; Zinn-Justin, J[ean]. Eds., *Les Progrès du Traitement des Images. Progress in Picture Processing*, pp. 1–80. (Les Houches, Session LVIII, 1992.) Amsterdam: Elsevier Science B.V. Cited on page 390.

Borgefors, Gunilla. 2003. Weighted digital distance transforms in four dimensions. *Discrete Appl. Math.* **125**, no. 1, 161–176. Cited on page 390.

Borgefors, Gunilla. 2005. Digital distance transforms in 2D, 3D, and 4D. **In:** Chen, C. H.; Wang, P. S. P., Eds., *Handbook of Pattern Recognition and Computer Vision*, pp. 157–176. 3rd Edition. Singapore: World Scientific. Cited on pages 377, 390.

Borgefors, Gunilla; Nyström, Ingela; Sanniti di Baja, Gabriella. 2008. Discrete skeletons from distance transforms in 2D and 3D. **In:** (Siddiqi & Pizer 2008:155–190). Cited on page 390.

Borwein, Jonathan M.; Giladi, Ohad. 2016. Some remarks on convex analysis in topological groups. *J. Convex Anal.* **23**, no. 2, 313–332. Cited on page 199.

Bourbaki, N[icolas]. 1954. *Éléments de mathématique. Théorie des ensembles.* Chapitres I et II. Paris: Hermann & C[ie]. Cited on page 24.

Bourbaki, N[icolas]. 1961. *Éléments de mathématique. Topologie générale.* Chapitres 1 & 2. Paris: Hermann. Third edition. Cited on pages 131, 311, 312, 313.

Bourbaki, N[icolas]. 1963. *Éléments de mathématique. Théorie des ensembles.* Chapitre 3. Ensembles ordonnés, Cardinaux, Nombres Entiers. Paris: Hermann. Second edition. Cited on pages 22, 27, 35, 170.

Boyd, Stephen; Vandenberghe, Lieven. 2009. *Convex Optimization.* Cambridge: Cambridge University Press. Cited on page 199.

Brimkov, Valentin; Coeurjolly, David; Klette, Reinhard. 2007. Digital planarity—A review. *Discrete Appl. Math.* **155**, 468–495. Cited on page 142.

Brimkov, Valentin E.; Maimone, Angelo; Nordo, Giorgio. 2006. On the notion of dimension in digital spaces. **In:** (Reulke et al., Eds., 2006:241–252). Cited on page 343.

Blondin-Massé, Alexandre; Brlek, Srečko; Garon, Ariane; Labé, Sébastien. 2009. Christoffel and Fibonacci tiles. **In:** *Discrete Geometry for Computer Imagery*, pp. 67–78. Lecture Notes in Comput. Sci., 5810. Berlin: Springer. Cited on page 9.

Bruckstein, Alfred M. 1991. Self-similarity properties of digitized straight lines. **In:** *Vision geometry* (Hoboken, NJ, 1989), pp. 1–20. Contemporary Mathematics, 119. Providence, RI.: Amer. Math. Soc. Cited on pages 142, 144.

Brunetti, Sara; Balázs, Péter; Bodnár, Péter; Szűcs, Judit. 2019. A spatial convexity descriptor for object enlacement. **In:** (Couprie et al., Eds., 2019:330–342). Cited on page 256.

Brüning, E[rwin A. K.]; Nagamachi, S[higeaki]. 2004. Relativistic quantum field theory with a fundamental length. *J. Math. Physics* **45**, no. 6, 2199–2231. Cited on page 8.

Burgeth, Bernhard; Kleefeld, Andreas; Naegel, Benoît; Passat, Nicolas; Perret, Benjamin, Eds. 2019. *Mathematical Morphology and Its Applications to Signal and Image Processing. 14th International Symposium, ISMM 2019; Saarbrücken, Germany, July 8–10, 2019; Proceedings.* Lecture Notes in Comput. Sci., 11564.

Buzer, Lilian. 2006. Digital line recognition, convex hulls, thickness, a unified and logarithmic technique. **In:** (Reulke et al., Eds., 2006:189–198). Cited on page 256.

*The Cabinet Dictionary of the English Language, Etymological, Explanatory, and Pronouncing.* 1871. viii + 968 pp. London; Glasgow: William Collins, Sons, and Company. Cited on page 19.

Callender, Clifton. 2013. Sturmian canons. **In:** Yust, J.; Wild, J.; Burgoyne, A., Eds., *Mathematics and Computation in Music: Fourth International Conference, MCM 2013*, pp. 64–75. Lecture Notes in Comput. Sci., 7937. Lecture Notes in Artificial Intelligence. Heidelberg: Springer. Cited on page 256.

Callender, Clifton. Web site: `www.cliftoncallender.com`; accessed 2019-12-05. Cited on page 256.

Campbell, Paul J. 1978. The origin of "Zorn's lemma". *Historia Math.* **5**, no. 1, 77–89. Cited on page 200.

Capraro, Valerio. 2013. A notion of continuity in discrete spaces and applications. *Appl. Gen. Topol.* **14**, no. 1, 61–72. Cited on page 332.

Cavallin, Chr[istian]. 1885. *Latinskt skol-lexikon*, 786 + 712 columns. Stockholm: F. & G. Beijers Förlag. Cited on page 19.

Cegrell U[rban]; Kiselman, C[hrister] O[scar]. 1970. Några anmärkningar om summabiliteter [Some remarks on summabiblity methods]. *Nordisk matematisk tidskrift* **18**, 127–136. Cited on page 288.

Chan, Tony F.; Shen, Jianhong (Jackie). 2005. *Image Processing and Analysis. Variational, PDE, Wavelet, and Stochastic Methods.* Philadelphia, PA: Society of Industrial and Applied Mathematics. Cited on page 61.

Chassery, Jean-Marc; Coeurjolly, David; Sivignon, Isabelle. 2006. Duality and geometry straightness, characterization and envelope. **In:** (Kuba et al., Eds., 2006:1–16). Cited on page 303.

Chassery, Jean-Marc; Sivignon, Isabelle. 2013. Optimal covering of a straight line applied to discrete convexity. **In:** Gonzalez-Diaz, Rocio; Jimenez, Maria-Jose; Medrano, Belen [González-Díaz, Rocío; Jiménez, María-José; Medrano, Belén], Eds. *Discrete Geometry for Computer Imagery. 17th IAPR International Conference, DGCI 2013, Seville, Spain, March 20–22. Proceedings*, pp. 1–10. Lecture Notes in Comput. Sci., 7749. Cited on page 155.

Chattopadhyay, Sudipta. 2019. An analysis of Zeno's view regarding motion. *International Journal of Trend in Scientific Research and Development* **3**(2), 17–21. Cited on page 5.

Ciesielski, Krzysztof. 1997. *Set theory for the working mathematician*, xii + 236 pp. London Mathematical Society Student Texts, 39. Cambridge: Cambridge University Press. Cited on pages 170, 136.

Choquet, Gustave. 1955. Theory of capacities. *Ann. Inst. Fourier,* Grenoble 5 (1953/1954), 131–295. Cited on page 231.

Church, Alonzo. 1936. An unsolvable problem of elementary number theory. *Amer. J. Math.* **58**, 345–363. Cited on page 136.

Coeurjolly, David; Brimkov, Valentin. 2006. Computational aspects of digital plane and hyperplane recognition. **In:** (Reulke et al., Eds., 2006:291–306). Cited on page 256.

Cohn, P. M. 1981. *Universal algebra,* xv + 412 pp. Second edition. Dortrecht; Boston, MA: D. Reidel Publishing Co. Cited on page 112.

Colbert, Edwin H[arris]. 1966. *Evolution of the Vertebrates: A History of the Backboned Animals Through Time.* New York, NY: John Wiley & Sons, Inc. Cited on page 1.

Čomić, Lidija. 2019. Convex and concave vertices on a simple closed curve in the triangular grid. **In:** (Couprie et al., Eds., 2019:397–408). Cited on page 9.

Coppel, W[illiam] A[ndrew]. 1998. *Foundations of Convex Geometry,* XIV + 222 pp. Cambridge: Cambridge University Press. Cited on pages 196, 199.

Couprie, Michel; Cousty, Jean; Kenmochi, Yukiko; Mustafa, Nabil, Eds. 2019. *Discrete Geometry for Computer Imagery. 21st IAPR International Conference, DGCI 2019, Marne-la-Vallée, France, March 26–28, 2019. Proceedings,* XIV + 496 pp. Springer.

Crombez, Loïc; de Fonseca, Guilherme D.; Gérard, Yan. 2019. Efficient algorithms to test digital convexity. **In:** (Couprie et al., Eds., 2019:409–419). Cited on page 229.

Ćurić, Vladimir; Lindblad, Joakim; Sladoje, Nataša; Sarve, Hamid; Borgefors, Gunilla. 2014. A new set distance and its application to shape registration. *Pattern Anal. Appl.* **17**, no. 1, 141–152. Cited on page 376.

Das, P. P.; Chatterji, B. M. 1988. Knight's distance in digital geometry. *Pattern Recognition Letters* **7**, 215–226. Cited on page 359.

Davey, B. A.; Priestely, H. A. 2003. *Introduction to Lattices and Order*, xii + 298 pp. Cambridge: Cambridge University Press. Second Edition. Cited on pages 78, 94, 112.

Debled, Isabelle. 1995. *Étude et reconnaissance des droites et plans discrets.* Doctoral Thesis, 209 pp. Strasbourg: Université Louis Pasteur. Cited on page 144.

Deng, T.-Q. 2000. *Fuzzy logic and Mathematical Morphology.* Report PNA-R0014. Amsterdam: CWI. Cited on page 104.

Deng, T.-Q.; Heijmans, H. J. A. M. 2000. *Grey-scale morphology based on fuzzy logic.* Report PNA-R0012. Amsterdam: CWI. Cited on pages 103, 104.

Domenjoud, Eric; Jamet, Damien; Toutant, Jean-Luc. 2009. On the connecting thickness of arithmetical discrete planes. **In:** *Discrete Geometry for Computer Imagery*, pp. 362–372. Lecture Notes in Comput. Sci., 5810. Berlin: Springer. Cited on page 222.

Domenjoud, Eric; Laboureix, Bastien; Vuillon, Laurent. 2019. Facet connectedness of arithmetic discrete hyperplanes with non-zero shifts. **In:** (Couprie et al., Eds., 2019:38–50). Cited on page 222.

Dorst, L.; Duin, R. P. W. 1984. Spirograph theory: A framework for calculations on digitized straight lines. *IEEE Transactions on Pattern Analysis and Machine Intelligence* **6**, no. 5, 632–638. Cited on page 164.

van Dortmont, M. A. M. M.; van de Wetering, H. M. M.; Telea, A. C. 2006. Skeletonization and distance transforms of 3D volumes using graphics hardware. **In:** (Kuba et al., Eds., 2006:617–629). Cited on page 390.

Dubreil, P[aul]; Dubreil-Jacotin, M. L. [Maire-Louise]. 1964. *Leçons d'algèbre moderne.* Second edition. Paris: Dunod. Cited on pages 36, 72, 78.

Eckhardt, Ulrich. 2001. Digital lines and digital convexity. *Digital and Image Geometry*, pp. 209–228. Lecture Notes in Comput. Sci., 2243. Springer. Cited on page 144.

Eckhardt, U.; Latecki, L. 1996. Invariant thinning and distance transform. **In:** *Theoretical foundations of computer vision (Dagstuhl, 1994)*, pp. 21–36, Comput. Suppl., 11, Vienna: Springer. Cited on page 390.

Eckhoff, Jürgen. 2016. Problems in discrete geometry. **In:** *Convexity and discrete geometry including graph theory*, pp. 269–273. Springer Proc. Math. Stat., 148, [Cham]: Springer. Cited on page 199.

Einstein, A[lbert]. 1905. Über einen die Erzeugung und Verwandlung des Lichtes betreffenden heuristischen Gesichtspunkt. *Annalen der Physik*, **322**(6), 132–148. Cited on page 7.

Emerson, W[illiam] R.; Greenleaf, F[rederick] P. 1969. Asymptotic behavior of products $C^p = C + \cdots + C$ in locally compact abelian groups. *Trans. Amer. Math. Soc.* **145**, 171–204. Cited on page 200.

Ernst, Bruno. 2018. *The Magic Mirror of M. C. Escher*, 116 pp. Cologne: Taschen. Cited on page 10.

Esbelin, Henri-Alex; Malgouyres, Rémy. 2016. Sparse convolution-based digital derivatives, fast estimation for noisy signals and approximation results. *Theoret. Comput. Sci.* **624**, 2–24. Cited on page 164.

Escher, M[aurits] C[ornelis]. 1967. *The Graphic Work of M. C. Escher.* New York, NY: Ballantine Books. (Tenth printing 1976.) Cited on page 10.

Euclid. 1956. *The Thirteen Books of Euclid's Elements.* Translated from the text of Heiberg, with introduction and commentary by Sir Thomas L. Heath. Volume I: Introduction and Books I, II. New York, NY: Dover Publications, Inc. Cited on pages v, 168.

Everett, C. J. 1944. Closure operators and Galois theory in lattices. *Trans. Amer. Math. Soc.* **55**, 514–525. Cited on page 36.

Favati, Paola; Tardella, Fabio. 1990. Convexity in nonlinear programming. *Ricerca operativa*, **53**, 3–44. Cited on pages 209, 222, 236, 268.

Federspiel, Michel. 1992. Sur l'origine du mot σημεῖον en géométrie. *Revue des Études grecques.* Publication de l'Association pour l'Enseignement des Études grecques, Tome **105**, 385–405. Cited on page 5.

Federspiel, Michel. 1998. Sur un emploi de *sèmeion* dans les mathématiques grecques. **In:** *Sciences exactes et sciences appliquées à Alexandrie.* Actes du Colloque International de Saint-Étienne (6–8 juin 1996), pp. 55–78. Saint Étienne: Université de Saint-Étienne. Cited on page 5.

Fenchel, Werner. 1949. On conjugate convex functions. *Canad. J. Math.* **1**, 73–77. Cited on page 199.

Fenchel, W. 1952. A remark on convex sets and polarity. **In:** *Communications du Séminaire de l'Université de Lund. Tome supplémentaire dédié à Marcel Riesz*, pp. 82–89. Lund: C. W. K. Gleerup. Cited on page 171.

Fenchel, W. 1953. *Convex cones, sets, and functions.* Notes of lectures at Princeton University, Spring Term 1951. 155 pp. Cited on page 199.

Fenchel, W. 1983. Convexity through the ages. **In:** Gruber, Peter M.; Wills, Jörg M., Eds., 1983. *Convexity and Its Applications*, pp. 120–130. Basel et al.: Birkhäuser Verlag. Cited on page 199.

Fouard, Céline; Gedda, Magnus. 2006. An objective comparison between gray weighted distance transforms and weighted distance transforms on curved spaces. **In:** (Kuba et al., Eds., 2006:259–270). Cited on page 376.

Fradelizi, Matthieu; Madiman, Mokshay; Marsiglietti, Arnaud; Zvavitch, Artem. 2018. The convexification effect of Minkowski summation. *EMS Surv. Math. Sci.* **5**, 1–64. Cited on page 200.

Franzén, Torkel. 2004. *Inexhaustibility. A Non-Exhaustive Treatment*, xi + 251 pp. Wellesley, MA: A K Peters, Ltd.; Association for Symbolic Logic. Cited on page 136.

Franzén, Torkel. 2005. *Gödel's Theorem. An Incomplete Guide to Its Use and Abuse*, x + 172 pp. Wellesley, MA: A K Peters, Ltd. Cited on page 136.

Franzén, Torkel. 2006. The popular impact of Gödel's incompleteness theorem. *Notices Amer. Math. Soc.* **53**, no. 4, 440–443. Cited on page 136.

Fujishige, Satoru; Murota, Kazuo. 2000. Notes on L-/M-convex functions and the separation theorems. *Math. Program.* **88**, no. 1, Ser. A, 129–146. Cited on pages 221, 303.

Ganter, Bernhard; Wille, Rudolf. 1999. *Formal Concept Analysis: Mathematical Foundations*, x + 284 pp. Berlin et al.: Springer-Verlag. Cited on page 67.

Ghosh, Pijush K.; Kumar, K. Vinod. 1998. Support function representation of convex bodies, its application in geometric computing, and some related representations. *Computer Vision and Image Understanding* **72**, 379–403. Cited on page 200.

Gierz, G.; Hofmann, K. H.; Keimel, K.; Lawson, J. D.; Mislove, M.; Scott, D. S. 1980. *A Compendium of Continuous Lattices*, XX + 371 pp. Berlin et al.: Springer-Verlag. Cited on pages 92, 117.

Gierz, G.; Hofmann, K. H.; Keimel, K.; Lawson, J. D.; Mislove, M.; Scott, D. S. 2003. *Continuous Lattices and Domains*, xxxvi + 591 pp. Cambridge: Cambridge University Press. Cited on pages 35, 67, 94, 112.

Goldstein, Rebecca. 2005. *Incompleteness. The Proof and Paradox of Kurt Gödel*, 296 pp. New York, NY; London: W. W. Norton & Company. Cited on page 136.

Gonzalez, Rafael C.; Woods, Richard E. 1993. *Digital Image Processing*, xvi + 716 pp. Addison-Wesley Publishing Company. Cited on pages 62, 63.

Gradšteĭn, Izrail′ Solomonovič; Ryžik, Iosif Moiseevič. 1962. *Tablicy integralov, summ, ryadov i proizvedeniĭ.* Moscow: Gosudarstvennoe izdatel′stvo fizikomatematičeskoĭ literatury. Cited on page 138.

Grätzer, G. 1998. *General Lattice Theory*, xx + 663 pp. Basel: Birkhäuser Verlag. (Second printing 2003.) Cited on pages 35, 67, 78.

Gruber, Peter M. 1993a. History of convexity. **In:** (Gruber & Wills, Eds., 1993:1–15). Cited on page 199.

Gruber, Peter M. 1993b. The space of convex bodies. **In:** (Gruber & Wills, Eds., 1993:301–318). Cited on page 199.

Gruber, P. M.; Wills, J. M., Eds. 1993. *Handbook of Convex Geometry. Volume A, B.* Volume A: lxvi + 735 pp.; volume B: pp. i–lxvi and 737–1438. Amsterdam: North-Holland Publishing Co. Cited on page 199.

Guo, Qi; Su, XunLi; Guo, JinFeng. 2019. Almost regular metrics on groups and Lipschitz-continuity of distance transforms. *J. Math. Imaging Vision.* Cited on page 376.

Haas, A.; Matheron, G.; Serra, J. 1967. Morphologie mathématique et granulométrie en place. *Annales des Mines* **11**, 736–753; **12**, 767–782. Cited on page 61.

Hadwiger, H[ugo]. 1950. Minkowskische Addition und Subtraktion beliebiger Punktmengen und die Theoreme von Erhard Schmidt. *Math. Z.* **53** (3), 210–218. Cited on page 61.

Hadwiger, H[ugo]. 1957. *Vorlesungen über Inhalt, Oberfläche und Isoperimetrie*, xiii + 213 pp. Berlin et al.: Springer-Verlag. Cited on page 61.

Halimskiĭ, E. D. [Efim Khalimsky]. 1970. Applications of connected ordered topological spaces in topology. Conference of Math. Departments of Povolsia. Cited on pages 332, 342.

Halimskiĭ, E. D. [Efim Khalimsky]. 1977. *Uporyadočennye topologičeskie prostranstva*, 92 pp. Kiev: Naukova Dumka. Cited on pages 332, 342.

Hanner, Olof. 1949–1950. Connectedness and convex hull. Seminar on convex sets, pp. 35–40. Princeton, NJ: The Institute for Advanced Study.

Hanner, Olof; Rådström, Hans. 1951. A generalization of a theorem of Fenchel. *Proc. Amer. Math. Soc.* **2**, 589–593.

Hardy, G. H. 1949. *Divergent Series.* Oxford: At the Clarendon Press. Cited on page 288.

Hartnett, William E. 1965. *Plane figures: Their skeletons and quench functions.* Tech. Rep. No. 60429 SR-3. Carlisle, MA: Parke Math. Laboratories. Cited on page 377.

Hausdorff, Felix. 1914. *Grundzüge der Mengenlehre*, 476 pp. Leipzig: Verlag Veit & Co. Cited on page 131.

Heijmans, Henk J. A. M. 1994. *Morphological Image Operators*, xv + 509 pp. Boston, MA: Academic Press. Cited on page 67.

Heijmans, Henk J. A. M. 1995. Mathematical morphology: A modern approach in image processing based on algebra and geometry. *SIAM Review* **37**, no. 1, 1–36. Cited on page 13.

Heijmans, Henk J. A. M.; Keshet, Renato. 2002. Inf-semilattice approach to self-dual morphology. *J. Math. Imaging Vision* **17**, 55–80. Cited on page 79.

Heijmans, Henk J. A. M.; Maragos, Petros. 1997. Lattice calculus of the morphological slope transform. *Signal Process.* **59**, 17–42. Cited on page 112.

Heijmans, H. J. A. M.; Molchanov, Ilya S. 1998. Morphology on convolution lattices with applications to the slope transform and random set theory. *J. Math. Imaging Vision* **8**, no. 3, 199–214. Cited on page 112.

Heijmans, Henk J. A. M.; Roerdink, Jos B. T. M., Eds. 1998. *Mathematical Morphology and its Applications to Image and Signal Processing*, ix + 442 pp. Dortrecht et al.: Kluwer Academic Publishers. Cited on page 79.

Heijmans, H. J. A. M.; Ronse, C. 1990. The algebraic basis of mathematical morphology. I. Dilations and erosions. *Comput. Vision Graphics Image Process.* **50**, 245–295. Cited on pages 67, 94, 112.

Herman, Gabor T. 1998. *Geometry of Digital Spaces*, x + 216 pp. Boston, MA, et al.: Birkhäuser. Cited on pages 15, 342.

Hilbert, David; Ackermann, Wilhelm. 1928. *Grundzüge der theoretischen Logik.* Springer-Verlag. Cited on page 135.

Hilditch, C. Judith. 1969. Linear skeletons from square cupboards. **In:** Meltzer, Bernard; Michie, Donald, Eds., *Machine Intelligence. 4*, pp. 403–420. New York, NY: American Elsevier Publishing Co., Inc. Cited on page 389.

Hilditch, C. Judith; Rutovitz, Denis. 1969. Chromosome recognition. *Annals of the New York Academy of Sciences* **157**(1), 339–364. Cited on pages 359, 375, 389.

Hiriart-Urruty, Jean-Baptiste; Lemaréchal, Claude. 1993. *Convex Analysis and Minimization Algorithms I. Fundamentals*, VXII + 417 pp. Berlin et al.: Springer-Verlag. Cited on page 199.

Hiriart-Urruty, Jean-Baptiste; Lemaréchal, Claude. 2001. *Fundamentals of Convex Analysis*, X + 259 pp. Berlin et al.: Springer. Cited on page 199.

Hofstadter, Douglas R. 1979; 1999. *Gödel, Escher, Bach: an Eternal Golden Braid*, 23 + xxi + 777 pp. New York, NY: Basic Books. Cited on page 10.

Hofstadter, Douglas R. 1985. *Metamagical Themas: Questing for the Essence of Mind and Pattern*. New York, NY: Basic Books. Cited on page vii.

Hörmander, Lars. 1983. *The Analysis of Linear Partial Differential Operators. I. Distribution Theory and Fourier Analysis*, ix + 391 pp. Grundlehren der Mathematischen Wissenschaften, 256. Berlin et al.: Springer-Verlag. Cited on page 201.

Hörmander, Lars. 1994. *Notions of Convexity*. Boston, MA, et al.: Birkhäuser. Cited on page 199.

Hossenfelder, Sabine. 2013. Minimal length scale scenarios for quantum gravity. *Living Reviews in Relativity* **16**:2. Cited on page 8.

Hübler, Albrecht; Klette, Reinhard; Voss, Klaus. 1981. Determination of the convex hull of a finite set of planar points within linear time. *Elektron. Informationsverarb. Kybernet.* **17**, no. 2–3, 121–139. Cited on pages 144, 256.

Hulin, Jérôme; Thiel, Edouard. 2006. Chordal axis on weighted distance transforms. **In:** (Kuba et al., 2006:271–282). Cited on page 390.

Hung, S. H. Y.; Kasvand, T. 1984. On the chord property and its equivalences. **In:** *Proceedings of the 7th International Conference on Pattern Recognition*, pp. 116–119. Cited on page 142.

Ilić, Vladimir; Lindblad, Joakim; Sladoje, Nataša. 2015. Precise Euclidean distance transforms in 3D from voxel coverage representation. *Pattern Recognition Letters* **65**, 184–191. Cited on page 376.

Imiya, Atsushi. 2011. Adjacent topology on nonsquare and noncubic grids. **In:** (Barneva et al., 2011:1–12). Cited on page 343.

Imiya, Atsushi; Saito, Masahiko; Tatara, Ken; Nakamura, Kiwamu. 2003. Digital curvature flow and its application for distances. *J. Math. Imaging Vision* **18**, 55–68. Cited on page 390.

Ioffe, A. D.; Tikhomirov, V. M. 1968. Duality of convex functions and extremum problems (in Russian). *Uspekhi Mat. Nauk* **23**, no. 6, 51–116. Cited on page 199.

Iwata, Satoru. 2010. Submodular functions: Optimization and approximation. **In:** Bathia, Rajendra, Ed., 2010. *Proceedings of the International Congress of Mathematicians. Hyderabad, August 10–27, 2010*, pp. 2943–2963. New Delhi: Hindustan Book Agency. Cited on page 260.

Jamet, Damien; Toutant, Jean-Luc. 2006. On the connectedness of rational arithmetic discrete hyperplanes. **In:** (Kuba et al., Eds., 2006:223–234.) Cited on page 222.

Jamet, Damien; Toutant, Jean-Luc. 2009. Minimal arithmetic thickness connecting discrete planes. *Discrete Appl. Math.* **157**, no. 3, 500–509. Cited on page 222.

Jansirani, N.; Raykumar Dare, V.; Subramanian, K. G. 2011. Sturmian arrays. **In:** (Barneva et al., 2011:93–98). Cited on page 142.

Jourlin, M.; Laget, B. 1988. Convexity and symmetry; Part 1. **In:** (Serra, Ed., 1988:343–357). Cited on page 222.

Kandel, Eric R[ichard]. 2012. *The Age of Insight: The Quest to Understand the Unconscious in Art, Mind, and Brain, from Vienna 1900 to the Present.* New York, NY: Random House, Inc. Cited on page 2.

Kaufmann, A.; Gupta, M. M. 1988. *Fuzzy Mathematical Models in Engineering and Managements Science.* Amsterdam: North-Holland. Cited on page 113.

Khalimsky, Efim; Kopperman, Ralph; Meyer, Paul R. 1990. Computer graphics and connected topologies on finite ordered sets. *Topol. Appl.* **36**, 1–17. Cited on pages 332, 333, 341, 342.

Kim, Chul E.; Rosenfeld, Azriel. 1982. Digital straight lines and convexity of digital regions. *IEEE Trans. Pattern Anal. Mech. Intell.* **PAMI-4**, no. 2, 149–153. Cited on page 228.

Kiran, Bangalore Ravi; Serra, Jean. 2013. Optima on Hierarchies of partitions. **In:** *Mathematical Morphology and Its Applications to Signal and Image Processing, 11th International Symposium, ISMM 2013, Uppsala, Sweden, May 27–29, 2013.* Lecture Notes in Comput. Sci., 7883, pp. 147–158. Cited on page 79.

Kiselman, Christer O. 1969. Prolongement des solutions d'une équation aux dérivées partielles à coefficients constants. *Bull. Soc. Math. France* **97**, 329–356. Cited on pages 31, 36.

Kiselman, Christer O. 1996. Regularity properties of distance transformations in image analysis. *Computer Vision and Image Understanding* **64**, no. 3, 390–398. Cited on pages 355, 376, 377.

Kiselman, Christer O. 2000. Digital Jordan curve theorems. **In:** Borgefors, Gunilla; Nyström, Ingela; Sanniti di Baja, Gabriella. *Discrete Geometry for Computer Imagery, 9th International Conference, DGCI 2000, Uppsala, Sweden, December 13–15, 2000*, pp. 46–56. Lecture Notes in Comput. Sci., 1953. Springer. Cited on page 342.

Kiselman, Christer O. 2002a. A semigroup of operators in convexity theory. *Trans. Amer. Math. Soc.* **354**, no. 5, 2035–2053. Cited on page 200.

Kiselman, Christer O. 2002b. *Digital Geometry and Mathematical Morphology*, 78 pp. Lecture Notes, Uppsala University. Available at `www.cb.uu/~kiselman`; accessed 2021-03-26. Cited on page xii.

Kiselman, Christer O. 2003. La geometrio de la komputila ekrano. **In:** Lipari, Michela, Ed., *Internacia Kongresa Universitato*, pp. 1–12. Rotterdam: Universala Esperanto Asocio. 83 pp. Cited on pages 9, 343.

Kiselman, Christer O. 2004a. *Digital Geometry and Mathematical Morphology*, 95 pp. Lecture Notes, Uppsala University. Available at `www.cb.uu/~kiselman`; accessed 2021-03-26. Cited on page xii.

Kiselman, Christer O. 2004b. Convex functions on discrete sets. **In:** (Klette & Žunić, Eds., 2004:443–457). Cited on pages 245, 256.

Kiselman, Christer O. 2005. Subharmonic functions on discrete structures. **In:** *Harmonic Analysis, Signal Processing, and Complexity. Festschrift in Honor of the 60th birthday of Carlos A. Berenstein*, pp. 67–80. Progress in Mathematics, Vol. 238. Sabadini, Irene; Struppa, Daniele C.; Walnut, David F., Eds., xi + 162 pp. Boston, MA, et al.: Birkhäuser. Cited on pages 302, 303.

Kiselman, Christer O. 2007a. Division of mappings between complete lattices. **In:** (Banon et al., Eds., 2007:27–38). Cited on pages 35, 112.

Kiselman, Christer O. 2007b. *Trois problèmes en convexité digitale : minima locaux, fonctions marginales et hyperplans séparants.* Lecture at GeoNet VI, June 12, 2007. Available at `www.cb.uu.se/~kiselman`; accessed 2021-03-26. Cited on page 286.

Kiselman, Christer O. 2008a. Minima locaux, fonctions marginales et hyperplans séparants dans l'optimisation discrète. *C. R. Acad. Sci. Paris Sér. I* **346**, 49–52. Cited on pages 222, 222, 276, 280, 286, 303.

Kiselman, Christer O. 2008b. Datorskärmens geometri. [The geometry of the computer screen. In Swedish.] **In:** Helenius, Ola; Wallby, Karin, Eds., *Människor och matematik – läsebok för nyfikna* [*People and Mathematics—a reading book for the curious*], pp. 211–229. Göteborg: Nationellt centrum för matematikutbildning, NCM, 390 pp. Cited on page 343.

Kiselman, Christer O. 2010a. Characterizing digital straightness by means of difference operators. **In:** Cris Luengo; Milan Gavrilovic, Eds., *Proceedings SSBA 2010. Symposium on Image Analysis, Uppsala, March 11–12, 2010*, pp. 15–18. Cited on page 163.

Kiselman, Christer O. 2010b. Inverses and quotients of mappings between ordered sets. *Image and Vision Computing* **28**, 1429–1442. Cited on pages 35, 112.

Kiselman, Christer O. 2010c. Local minima, marginal functions, and separating hyperplanes in discrete optimization. **In:** Bhatia, Rajendra, Ed., *Abstracts: Short communications; Posters.* International Congress of Mathematicians, Hyderabad, August 19–27, 2010, pp. 572–573. Cited on page 286.

Kiselman, Christer O. 2011. Characterizing digital straightness and digital convexity by means of difference operators. *Mathematika* **57**, 355–380. Cited on pages 163, 222.

Kiselman, Christer O. 2014a. Questions inspired by Mikael Passare's mathematics. *Afrika Matematika* **25**, issue 2, 271–288. Cited on page xi.

Kiselman, Christer O. 2014b. Pierre Lelong 1912–2011. *Nordisk Matematisk Tidskrift* (*Normat*) **60**, no. 2, 70–81. Cited on page x.

Kiselman, Christer O. 2015a. Euclid's straight lines. *Nordisk Matematisk Tidskrift. Normat* **60**, no. 4, 145–169. Cited on pages 5, 168.

Kiselman, Christer O. 2015b. Estimates for solutions to discrete convolution equations. *Mathematika* **61**, issue 02, 295–308. Cited on page 163.

Kiselman, Christer O. 2015c. Lars Hörmander—some early memories. **In:** Boman, Jan; Sigurdsson, Ragnar; Lerner, Nicolas; Demailly, Jean-Pierre; Atiyah, Michael; Treves, François; Helgason, Sigurdur; Bony, Jean-Michel; Kiselman, Christer O.; Broström, Sofia, pp. 904–905. *Notices of the American Mathematical Society* **62**, no. 8, 890–907. (Translation into Chinese published in *Mathematical Advances in Translation*, 2017, no. 2, 120–121.) Cited on page x.

Kiselman, Christer Oscar. 2019a. Werner Fenchel, a pioneer in convexity theory and a migrant scientist. *Normat. Nordisk matematisk tidskrift* **61**, no. 2–4, 133–152. Cited on pages x, 199, 201.

Kiselman, Christer Oscar. 2019b. Duality of convolution operators: A tool for shape analysis. Accepted on 2021 July 14 for publication in the Springer Series *Mathematics and Visualization*. Cited on pages 222, 303.

Kiselman, Christer Oscar. 2020. André Martineau: Some memories. *Complex Analysis and Operator Theory* **14**, issue 7. Published online on 2020-09-14. Cited on page x.

Kiselman, Christer Oscar. 2021. Which convexity properties are preserved under linear mappings? Manuscript, 13 pp.

Kiselman, Christer; Mouwitz, Lars. 2008. *Matematiktermer för skolan* [*Mathematical terms for school use*], 312 pp. Göteborg: Göteborgs universitet, National Center for Mathematics Education. Cited on page 19.

Kiselman, Christer O.; Samieinia, Shiva. 2010. Convexity of marginal functions in the discrete case. **In:** (Samieinia 2010d, Paper VI). Cited on pages 62, 302.

Kiselman, Christer O.; Samieinia, Shiva. 2017. Convexity of marginal functions in the discrete case. **In:** Andersson, Mats; Boman, Jan; Kiselman, Christer; Sigurdsson, Ragnar, Eds., *Analysis Meets Geometry. A Tribute to Mikael Passare*, pp. 287–309. Basel: Birkhäuser. Cited on pages 62, 302, 304.

Kjeldsen, Tinne Hoff. 2010. History of convexity and mathematical programming: Connections and relationships in two episodes of research in pure and applied mathematics of the 20th century. **In:** *Proceedings of the International Congress of Mathematicians, Hyderabad, August 19–27, 2010*, volume 4, pp. 3233–3257. Hindustan Book Agency.

Klein, Felix. 1895. Ueber eine geometrische Auffassung der gewöhnliche Kettenbruchentwicklung. *Nachrichten von der Königl. Gesellschaft der Wissenschaften zu Göttingen, Mathematisch-Physikalische Klasse*, pp. 357–359. [Also in (Klein 1922:209–211).] Cited on page 143.

Klein, Felix. 1922. *Gesammelte mathematische Abhandlingen*. Volume 2. Berlin: Verlag von Julius Springer. Available at `https://gdz.sub.uni-goettingen.de`; accessed 2020-01-01.

Klette, Reinhard; Rosenfeld, Azriel. 2004. *Digital Geometry. Geometric Methods for Digital Picture Analysis*, xviii + 656 pp. Amsterdam et al.: Elsevier. Cited on pages 17, 63.

Klette, Reinhard; Žunić, Jovisa, Eds. 2004. *Combinatorial Image Analysis. 10th International Workshop, IWCIA 2004. Auckland, New Zealand, December 1–3, 2004. Proceedings*, XII + 760 pp. Lecture Notes in Comput. Sci., 3322. Berlin; Heidelberg: Springer-Verlag.

Koné, Adama Arouna. 2016. *Géométrie digitale utilisée pour la discrétisation et le recouvrement optimal des objects euclidiens*, vi + 114 pp. Doctoral Thesis presented and approved on 2016 January 14 at l'Université des Sciences, des Techniques et des Technologies de Bamako (USTTB). Cited on pages xii, 304.

Koné, Adama Arouna. 2017. Covering a Euclidean line or hyperplane by dilations of its discretization. *Vietnam J. Math.* **45**, no. 3, 351–368. Cited on page 155.

Kong, T. Yung. 2003. The Khalimsky topologies are precisely those simply connected topologies on $\mathbf{Z}^n$ whose connected sets include all $2n$-connected sets but no $(3^n - 1)$-disconnected sets. *Theoret. Comput. Sci.* **305**, 221–235. Cited on page 342.

Kong, T. Y.; Rosenfeld, A. 1989. Digital topology: Introduction and Survey. *Comput. Vision Graphics Image Process.* **48**, 357–393. Cited on page 342.

Kong, Yung; Kopperman, Ralph; Meyer, Paul R. 1991. A topological approach to digital topology. *Amer. Math. Monthly* **98**, 901–917. Cited on pages 315, 340, 342.

Kopperman, R. D.; Kronheimer, E. H.; Wilson, R. G. 1998. Topologies on totally ordered sets. *Topol. Appl.* **90**, 165–185. Cited on page 342.

Kovács, Gergely; Nagy, Benedek; Vizvári, Béla. 2017. Weighted distances on the trihexagonal grid. **In:** (Kropatsch et al., Eds., 2017:82–93). Cited on page 376.

Kovács, Gergely; Nagy, Benedek; Vizvári, Béla. 2020. On disks of the triangular grid: An application of optimization theory in discrete geometry. *Discrete Appl. Math.* **282**, 136–151. Cited on page 376.

Kovalevsky, Vladimir. 2006. Axiomatic digital topology. *J. Math. Imaging Vision* **26**, no. 1–2, 41–58. Cited on page 342.

Kovalevsky, Vladimir A. 2008. *Geometry of Locally Finite Spaces. Computer Agreeable Topology and Many Modern Algorithms for Computer Imagery.* Berlin: Editing House Dr. Baerbel Kovalevski. Cited on page 342.

Kovalevsky, Vladimir; Schulz, Henrik. 2004. Convex hulls in a 3-dimensional space. **In:** (Klette & Žunić, Eds., 2004:176–196).

Kriegl, Andreas; Michor, Peter W. 1997. *The convenient setting of global analysis.* Mathematical Surveys and Monographs, 53. Providence, RI: American Mathematical Society. Cited on page 96.

Kriegl, A.; Nel, L. D. 1990. Convenient vector spaces of smooth functions. *Math. Nachr.* **147**, 39–45. Cited on page 96.

Kronheimer, E. H. 1992. The topology of digital images. *Topol. Appl.* **46**, 279–303. Cited on page 342.

Kropatsch, Walter G.; Artner, Nicole M.; Janusch, Ines, Eds. 2017. *Discrete Geometry for Computer Imagery. 20th IAPR International Conference, DGCI 2017. Vienna, Austria, September 19–21, 2017. Proceedings*, XI + 400 pp. Springer.

Kuba, Attila; Nyúl, László; Palágyi, Kálmán, Eds. 2006. *Discrete Geometry for Computer Imagery. 13th International Conference, DGCI 2006. Szeged, Hungary, October 25–27, 2006. Proceedings*, XIII + 688 pp. Lecture Notes in Comput. Sci., 4245.

Kudryavtseva, Ganna; Mazorchuk, Volodymyr. 2009. On Kiselman's semigroup. *Yokohama Math. J.* **55**, no. 1, 21–46. Cited on page 200.

Kuroš, Aleksandr Gennadievič. 1962. *Lekcii po obščeĭ algebre*, 396 pp. Moscow: Gosudarstvennoe izdatel′stvo fiziko-matematičeskoĭ literatury. Cited on page 112.

Lachaud, Jacques-Olivier. 2021. An alternative definition for digital convexity. **In:** Lindblad, Joakim; Malmberg, Filip; Sladoje, Nataša, Eds. *Discrete Geometry and Mathematical Morphology. First International Joint Conference, DGMM 2021. Uppsala, Sweden, May 24–27, 2021. Proceedings*, pp. 269–282. Lecture Notes in Comput. Sci., 12708. Cited on page 205.

Langer, Maximilian; Gabdulkhakova, Aysylu; Kropatsch, Walter G. 2019. Non-centered Voronoi skeletons. **In:** (Couprie et al., Eds., 2019:355–366). Cited on page 390.

Lantuéjoul, Christian. 1977. *Sur le modèle de Johnson–Mehl généralisé.* Internal report. Fontainebleau: Centre de Morphologie mathématique. Cited on page 392.

Lantuéjoul, Christian. 1978. *La squelettisation et son application aux mesure topologiques des mosaïques cristallines.* Doctoral Dissertation. Paris: École des Mines. Cited on pages 388, 389.

Lantuéjoul, C. 1980. Skeletonization in quantitative metallography. **In:** Haralick, R. M.; Simon, J. C., Eds., *Issues in Digital Image Processing*, pp. 107–135. NATO Advanced Study Institutes Series E, Applied Science, No. 34. Alphen aan den Rijn: Sijthoff & Noordhoff. Cited on page 388.

Lantuejoul, C[hristian]; Maisonneuve, F[rancis]. 1984. Geodesic methods in quantitative image analysis. *Pattern Recognition* **17**, no. 2, 177–187. Cited on page 376.

LeGuin, Ursula K[roeber]. 1977. *The Left Hand of Darkness.* London: Cox & Wyman Ltd. (First published in 1969.) Cited on pages 304, 407.

Lévi, Robert. 1927. Théorie de l'action universelle et discontinue. *J. Phys. Radium* **8**, 182–198. Cited on page 6.

Lin, Huibao; Si, Jennie; Abousleman, Glen P. 2008. Orthogonal rotation-invariant moments for digital image processing. *IEEE Trans. Image Process.* **17**, no. 3, 272–282. Cited on page 7.

Lindblad, Joakim; Ćurić, Vladimir; Sladoje, Nataša. 2009. On set distances and their application to image registration. **In:** *Proceedings of the 6th International Symposium on Image and Signal Processing and Analysis (ISPA), IEEE, Salzburg, Austria, September 2009*, pp. 449–454. Cited on page 376.

Lindblad, Joakim; Sladoje, Nataša. 2015. Exact linear time Euclidean distance transforms of grid line sampled shapes. **In:** (Benediktsson et al., 2015:645–656). Cited on page 376.

Linder, C. W.; Walberg, C. A. 1862. *Svenskt-Grekiskt Lexikon.* Upsala: Lundequistska Bokhandeln. Cited on page 19.

Linnér, Elisabeth; Strand, Robin. 2014. Anti-aliased Euclidean distance transform on 3D sampling lattices. **In:** Barcucci, Elena; Frosini, Andrea; Rinaldi, Simone, Eds., *Discrete Geometry for Computer Imagery. 18th IAPR International Conference, DGCI 2014, Siena, Italy, September 10–12, 2014. Proceedings*, pp. 88–98. Lecture Notes in Comput. Sci., 8668, XI + 423 pp. Cham: Springer. Cited on page 376.

Lo Bello, Anthony. 2013. *Origins of Mathematical Words. A Comprehensive Dictionary of Latin, Greek, and Arabic Roots*, xv + 350 pp. Baltimore, MD: The Johns Hopkins University Press. Cited on page 19.

Locher, J. L., Ed. 1991. *The World of M. C. Escher.* New York, NY: Abradale Press; Harry N. Abrams. Cited on page 10.

Lothaire, M. 2002. *Algebraic Combinatorics on Words.* Cambridge: Cambridge University Press. Cited on pages 142, 256.

Lovász, L. 1983. Submodular functions and convexity. **In:** Bachem, A.; Grötschel, M.; Korte, B., Eds., *Mathematical Programming — The State of the Art*, pp. 235–257. Berlin: Springer-Verlag. Cited on page 231.

Maclagan, Diane; Sturmfels, Bernd. 2015. *Introduction to Tropical Geometry*, xii + 363 pp. Providence, RI: American Mathematical Society. Cited on page 13.

Maimonides, Moses. 1904. *A Guide to the Perplexed.* Translated from the original Arabic text by Michael Friedlaender. New York, NY: E. P. Dutton. Fourth revised edition. Available at `www.sacred-texts.com/jud/gfp/gfp083.htm`; accessed 2018-11-25. Cited on pages 6, 17.

Maimonides, Moses. 1963. *The Guide of the Perplexed*, Volume 1. Translated and with an Introduction and Notes by Shlomo Pines. Introductory Essay by Leo Strauss, 384 pp. Chicago: University of Chicago Press. Cited on page 6.

Malgouyres, Rémy. 1997. A definition of surfaces of $\mathbf{Z}^3$. A new 3D discrete Jordan theorem. *Theoret. Comput. Sci.* **186**, no. 1–2, 1–41. Cited on page 343.

Mandelbrojt, Szolem. 1939. Sur les fonctions convexes. *Comptes Rendus hebdomadaires des Séances de l'Académie des Sciences* **209**, 977–978. Séance du 27 décembre 1939. Cited on page 199.

Maragos, Petros. 2017. Dynamical systems on weighted lattices: general theory. *Math. Control Signals Syst.* **29**:21. Cited on page 78.

Maragos, Petros. 2019. Tropical geometry, mathematical morphology and weighted lattices. **In:** (Burgeth et al., Eds., 2019:3–15). Cited on page 13.

Maragos, Petros; Charisopoulos, Vasileios; Theodosis, Emmanouil. 2021. Tropical geometry and machine learning. *Proceedings of the IEEE* **109**, no. 5, 728–755. Cited on page 13.

Maragos, Petros; Theodosis, Emmanouil. 2019. Tropical geometry and piecewise-linear approximation of curves and surfaces on weighted lattices. Accepted on 2021 July 14 to be published in the Springer Series *Mathematics and Visualization.* `arXiv:1912.03891`, 2019 December 09. Cited on page 13.

Marchand-Maillet, Stéphane; Sharaiha, Yazid M. 2000. *Binary Digital Image Processing. A Discrete Approach*, xxv + 251 pp. San Diego et al.: Academic Press. Cited on pages 18, 377.

Matheron, Georges. 1965. *Étude théorique des granulométriques.* Report N-54 published by École des Mines. Cited on pages 17, 36.

Matheron, G[eorges]. 1967. *Éléments pour une théorie des milieux poreux.* Paris: Masson et C[ie]. Cited on pages 36, 61.

Matheron, G[eorges]. 1975. *Random Sets and Integral Geometry*, xxiii + 261 pp. Cited on pages 29, 31, 34, 36, 56, 57, 58, 60, 61, 62, 63, 67, 315.

Matheron, G[eorges]. 1978a. Les érosions infinitésimales, et les mesures de surface. **In:** (Miles & Serra, Eds., 1978:251–269). Cited on page 17.

Matheron, G[eorges]. 1978b. *Quelques propriétés topologiques du squelette.* Internal report N-560, 43 pp. Fontainebleau: Centre de Morphologie mathématique. Cited on pages 389, 390.

Matheron, G[eorges]. 1981. *Sur la négligeabilité du squelette, et la continuité absolue des érosions*, 26 pp. Internal report N-724. Fontainebleau: Centre de Morphologie mathématique. Cited on page 389.

Matheron, G[eorges]. 1988a. Dilations on topological spaces. **In:** (Serra, Ed., 1988:59–70). Cited on page 61.

Matheron, G[eorges]. 1988b. Filters and lattices. **In:** (Serra, Ed., 1988:115–140). Cited on pages 36, 121, 123.

Matheron, G[eorges]. 1988c. Examples of topological properties of skeletons. **In:** (Serra, Ed., 1988:217–238). Cited on pages 382, 389, 390, 391, 415.

Matheron, G[eorges]. 1988d. On the negligibility of the skeleton and the absolute continuity of erosions. **In:** (Serra, Ed., 1988:239–256). Cited on pages 389, 390.

Matheron, G[eorges]. 1996. *Treillis compacts et treillis coprimaires*, 183 pp. Fontainebleau: École Nationale Supérieure des Mines de Paris. Cited on page 78.

Matheron, G[eorges]; Serra, J[ean]. 1988a. Strong filters and connectivity. **In:** (Serra, Ed., 1988:141–157). Cited on page 123.

Matheron, G[eorges]; Serra, J[ean]. 1988b. Convexity and symmetry: Part 2. **In:** (Serra, Ed., 1988:359–375). Cited on page 222.

Matheron, G[eorges]; Serra, J[ean]. 2002. The birth of mathematical morphology. **In:** (Talbot & Beare, Eds., 2002:1–16). Cited on page 11.

Mejlbo, Lars C. 1968. Om konvergente, uendlige række og deres sum [On convergent, infinite sequences and their sum]. *Nordisk matematisk tidskrift* **16**, 92–103. Cited on page 288.

Melin, Erik. 2004. How to find a Khalimsky-continuous approximation of a real-valued function. **In:** (Klette & Žunić, Eds., 2004:351–365). Cited on page 343.

Melin, Erik. 2005a. Digital straight lines in the Khalimsky plane. *Math. Scand.* **96**, no. 1, 49–62. Cited on page 343.

Melin, Erik. 2005b. Extension of continuous functions in digital spaces with the Khalimsky topology. *Topology Appl.* **153**, no. 1, 52–65. Cited on page 343.

Melin, Erik. 2007. Digital surfaces and boundaries in Khalimsky spaces. *J. Math. Imaging Vision* **28**, no. 2, 169–177. Cited on page 343.

Melin, Erik. 2008a. Continuous digitization in Khalimsky spaces. *J. Approx. Theory* **150**, no. 1, 96–116. Cited on page 343.

Melin, Erik. 2008b. *Digital Geometry and Khalimsky spaces.* Doctoral Thesis defended on 2008 February 29. Uppsala: Uppsala University, Department of Mathematics. Cited on page xi.

Melin, Erik. 2008c. Continuous extension in topological digital spaces. *Appl. Gen. Topol.* **9**, no. 1, 51–66. Cited on page 343.

Melin, Erik. 2009. Digital Khalimsky manifolds. *J. Math. Imaging Vision* **33**, no. 3, 267–280. Cited on page 343.

Michor, Peter. 1984. A convenient setting for differential geometry and global analysis. *Cahiers Topologique Géom. Différentielle* **25**, no. 1, 63–109; no. 2, 113–178. Cited on page 96.

Miles, R[oger] E.; Serra, J[ean], Eds. 1978. *Geometrical Probability and Biological Structures: Buffon's 200th Anniversary. Proceedings of the Buffon Biocentenary Symposium on Geometrical Probability, Image Analysis, Mathematical Stereology, and Their Relevance to the Determination of Biological Structures, Held in Paris, June 1977*, XII + 338 pp. Lecture Notes in Biomathematics 23. Berlin et al.: Springer-Verlag.

Millén, J. A. 1853. *Grekiskt och svenskt hand-lexikon öfver Nya Testamentets skrifter.* Örebro: N. M. Lindhs Bokhandel. Cited on page 19.

Miller, Bruce L. 1971. On minimizing nonseparable functions defined on the integers with an inventory application. *SIAM J. Appl. Math.* **21**, no. 1, 166–185. Cited on pages 220, 268.

Minkowski, Hermann. 1903. Volumen und Oberfläche. *Math. Ann.* **57**, 447–495. [Also in (Minkowski 1911a:230–276).] Cited on page 199.

Minkowski, Hermann. 1910. *Geometrie der Zahlen.* Leipzig; Berlin: B. G. Teubner. [The first 240 pages were published in 1896.] Cited on page 199.

Minkowski, Hermann. 1911a. *Gesammelte Abhandlungen.* Volume 2. Leipzig; Berlin: B. G. Teubner.

Minkowski, Hermann. 1911b. *Theorie der konvexen Körper, insbesondere Begründung ihres Oberflächenbegriffs.* **In:** (Minkowski 1911a:131–229). Cited on page 199.

Molchanov, Ilya S.; Terán, Pedro. 2003. Distance transforms for real-valued functions. *J. Math. Anal. Appl.* **278**, no. 2, 472–484. Cited on page 376.

Montanari, U[go]. 1968. A method for obtaining skeletons using a quasi-Euclidean distance. *J. Assoc. Comput. Mach.* **15**, 600–624. Cited on pages 359, 377.

Moore, Eliakim Hastings. 1910. *Introduction to a Form of General Analysis.* New Haven, CT: Yale University Press. Cited on page 79.

Moreau, Jean-Jacques. 1963. Remarques sur les fonctions à valeurs dans $[-\infty, +\infty]$ définies sur un demi-groupe. *C. R. Acad. Sci. Paris* **257**, 3107–3109. Cited on page 112.

Moreau, J. J. Convexity and duality. 1966. **In:** *Functional Analysis and Optimization*, pp. 145–169. New York, NY: Academic Press. Cited on page 199.

Moreau, J. J. 1966–1967. Fonctionelles convexes. *Séminaire Jean Leray. Sur les équations aux dérivées partielles*, n° 2, pp. 1–108. Cited on page 199.

Moreau, J. J. Sous-différentiabilité. 1967. **In:** *Proc. Colloquium on Convexity (Copenhagen, 1965)*, pp. 185–201. Copenhagen: Københavns Univ. Mat. Inst. Cited on page 199.

Moreau, Jean-Jacques. 1970. Inf-convolution, sous-additivité, convexité des fonctions numériques. *J. Math. Pures et Appl.* **49**, 109–154. Cited on pages 34, 112.

Morse, Marston; Hedlund, Gustav A. 1940. Symbolic dynamics II. Sturmian trajectories. *Amer. J. Math.* **62**, no. 1, 1–42. Cited on pages 142, 250, 255.

Murota, Kazuo. 1996. Convexity and Steinitz's exchange property. *Advances in Mathematics* **124**, 272–311. Cited on page 221.

Murota, Kazuo. 1998. Discrete convex analysis. *Math. Programming* **83**, 313–371. Cited on page 221.

Murota, Kazuo. 2003. *Discrete Convex Analysis*, xxii + 389 pp. Philadelphia, PA: Society for Industrial and Applied Mathematics (SIAM). Cited on pages 144, 152, 214, 221, 222, 240, 262.

Murota, Kazuo. 2009. Recent developments in discrete convex analysis. *Research trends in combinatorial optimization*, pp. 219–260. Berlin: Springer. Cited on page 221.

Murota, Kazuo; Shioura, Akiyoshi. 1999. M-convex function on generalized polymatroid. *Math. Oper. Res.* **24**, 95–105. Cited on pages 221, 303.

Murota, Kazuo; Shioura, Akiyoshi. 2001. Relationship of M/L-convex functions with discrete convex functions by Miller and Favati–Tardella. *Discrete Appl. Math.* **115**, 151–176. Cited on pages 221, 221, 303.

Murota, Kazuo; Shioura, Akiyoshi. 2004. Conjugacy relationship between M-convex and L-convex functions in continuous variables. *Math. Program.* **101**, no. 3, Ser. A, 415–413. Cited on page 303.

Nagy, Benedek; Strand, Robin. 2006. Approximating Euclidean distance using distance based on neighbourhood sequences in non-standard three-dimensional grids. **In:** (Reulke et al., Eds., 2006:89–100). Cited on page 376.

Nagy, Benedek; Strand, Robin; Normand, Nicolas. 2018. Distance functions based on multiple types of weighted steps combined with neighborhood sequences. *J. Math. Imaging Vision* **60**(8), 1209–1219. Cited on page 376.

Nagy, Benedek; Strand, Robin; Normand, Nicolas. 2019. Distance transform based on weight sequences. **In:** *Discrete Geometry for Computer Imagery. 21st IAPR International Conference, DGCI 2019. Marne-la-Vallée, France, March 26–28, 2019. Proceedings*, pp. 62–74. Cited on page 376.

Najman, Laurent; Pesquet, Jean-Christophe; Talbot, Hugues. 2015. When convex analysis meets mathematical morphology on graphs. **In:** (Benediktsson et al., 2015:473–484). Cited on page 79.

Najman, Laurent; Talbot, Hugues. 2008. Introduction à la morphologie mathématique. **In:** (Najman & Talbot, Eds., 2008:19–45). Cited on pages 63, 391.

Najman, Laurent; Talbot, Hugues, Eds. 2008. *Morphologie mathématique 1. Approches déterministes*, 260 pp. Paris: Lavoisier. Cited on page 18.

Najman, Laurent; Talbot, Hugues, Eds. 2010. *Mathematical Morphology. From Theory to Applications*, 507 pp. London; Hoboken, NJ: ISTE Ltd; John Wiley & Sons. Cited on page 18.

Ngo, Phuc; Kenmochi, Yukiko; Debled-Rennesson, Isabelle; Passat, Nicolas. 2017. Convexity-preserving rigid motions of 2D digital objects. **In:** (Kropatsch et al., Eds., 2017:69–81). Cited on page 7.

Ngo, Phuc; Passat, Nicolas; Kenmochi, Yukiko; Debled-Rennesson, Isabelle. 2019. Geometric preservation of 2D digital objects under rigid motions. *J. Math. Imaging Vision* **61**, no. 2, 204–223. Cited on page 7.

Normand, Nicolas; Guédon, Jeanpierre; Autrusseau, Florent, Eds. 2016. *Discrete Geometry for Computer Imagery. 19th IAPR International Conference, DGCI 2016. Nantes, France, April 18–20, 2016. Proceedings*, IX + 451 pp. Lecture Notes in Comput. Sci., 9647. Springer.

Noyel, Guillaume. 2019. A link between the multiplicative and additive functional Asplund's metrics. **In:** (Burgeth et al., Eds., 2019:41–53). Cited on page 94.

Noyel, Guillaume; Jourlin, Michel. 2017. Double-sided probing by map of Asplund's distances using logarithmic image processing in the framework of mathematical morphology. **In:** Angulo, Jesús; Velasco-Forero, Santiago; Meyer, Fernand, Eds., *Mathematical Morphology and Its Applications to Signal and Image Processing; 13th International Symposium, ISMM 2017, Fontainebleau, France, May 15–17, 2017; Proceedings*, pp. 408–420. Lecture Notes in Comput. Sci., 10225. Cited on page 94.

Noyel, Guillaume; Jourlin, Michel. 2020. Functional Asplund metrics for pattern matching robust to variable lighting conditions. *Image Anal. Stereol.* **39**, no. 2, 53–71.

Nörlund, Niels Erik. 1924. *Vorlesungen über Differenzenrechnung*, IX + 551 pp. Berlin: Verlag von Julius Springer. Cited on page 163.

Ore, Oystein. 1944. Galois connexions. *Trans. Amer. Math. Soc.* **55**, 493–513. Cited on pages 35, 92, 112.

*Oxford English Dictionary. The definitive record of the English language.* Available at `www.oed.com`; accessed 2019-12-03. Cited on page 19.

Passare, Mikael. 2009. Christer Kiselman's mathematics. **In:** (Passare, Ed., 2009:9–26). Cited on page x.

Passare, Mikael, Ed. 2009. *Complex Analysis and Digital Geometry. Proceedings from the Kiselmanfest, 2006*, pp. 81–110. Acta Universitatis Upsaliensis, vol. 86, 364 pp. Uppsala: Uppsala University. ISSN 0502-7454, ISBN 978-91-554-7672-4.

Passat, Nicolas; Kenmochi, Yukiko; Ngo, Phuc; Pluta, Kacper. 2019. Rigid motions in the cubic grid: a discussion on topological issues. **In:** (Couprie et al., Eds., 2019:127–140). Cited on page 7.

Petermann, Eike. 1979. On a method of constructing categories. *J. Pure Appl. Algebra* **15**, 271–281. Cited on page 96.

Pfaltz, J. L.; Rosenfeld, A. 1967. Computer representation of planar regions by their skeletons. *Comm. ACM* **10**, 119–125. Cited on page 389.

Pfaltz, J. L.; Šlapal, Josef. 2013. Transformations of discrete closure systems. *Acta Math. Hungar.* **138**, no. 4, 386–405. Cited on page 35.

Pizer, Stephen; Siddiqi, Kaleem; Yushkevich, Paul. 2008. Introduction. **In:** (Siddiqi & Pizer 2008:1–34). Cited on page 390.

Pluta, Kacper; Romon, Pascal; Kenmochi, Yukiko; Passat, Nicolas. 2017. Honeycomb geometry: Rigid motions on the hexagonal grid. **In:** (Kropatsch et al., Eds., 2017:33–45). Cited on page 7.

Pytheas Fogg, N. 2002. *Substitutions in Dynamics, Arithmetics and Combinatorics.* Lecture Notes in Math., 1794. Springer-Verlag. Cited on page 142.

Rådström, Hans. 1952. Convexity and norm in topological groups. *Ark. mat.* **2**, no. 7, 99–137. Cited on pages 173, 199.

Ragnemalm, I. 1993. The Euclidean distance transform in arbitrary dimensions. *Pattern Recognition Letters* **14**, 883–888. Cited on page 355.

Renaudineau, Arthur. 2021. Raconte moi ... la géométrie tropicale. *la Gazette des Mathématiciens*, no. 167, 51–59. Cited on page 13.

Reulke, Ralf; Eckardt, Ulrich; Flach, Boris; Knauer, Uwe; Polthier, Konrad, Eds. 2006. *Combinatorial Image Analysis. 11th International Workshop, IWCIA 2006. Berlin, Germany, June 19–21, 2006. Proceedings*, XII + 482 pp. Springer.

Reveilles, J. P. [Reveillès, Jean-Pierre]. 1991. *Géométrie discrète, calcul en nombres entiers et algorithmique*, 251 pp. Thèse d'État, soutenue le 20 décembre 1991. Strasbourg: Université Louis-Pasteur. Cited on page 256.

Rivière, Alain. 1987. Nervure d'un ouvert d'un espace euclidien. *C. R. Acad. Sci. Paris Sér. I.* **304**, no. 14, 435–438. Cited on pages 383, 390.

Rivière, Alain. 1996. Nervure d'un ouvert d'un espace euclidien. *J. Sci. Univ. Tehran Int. Ed.* **1**, no. 1, 1–24. Cited on pages 383, 390.

Rivière, Alain. 2001. Dimension de Hausdorff de la nervure. *Geom. Dedicata* **85**, no. 1–3, 217–235. Cited on page 390.

Roerdink, J. B. T. M. 1996. Computer vision and mathematical morphology. *Theoretical foundations of computer vision (Dagstuhl, 1994)*, 131–148. Comput. Suppl., 11. Vienna: Springer. Cited on page 79.

Rockafellar, R. Tyrrell. 1970, 1997. *Convex Analysis.* Princeton, NJ: Princeton University Press. Cited on pages 112, 199.

Ronse, Christian. 1990. Why mathematical morphology needs complete lattices. *Signal Processing* **21**, 129–154. Cited on pages 67, 68.

Ronse, C.; Heijmans, H. J. A. M. 1991. The algebraic basis of mathematical morphology. II. Openings and closings. *Comput. Vision Graphics Image Process.* **54**, no. 1, 74–97. Cited on page 67.

Ronse, C.; Heijmans, H. J. A. M. 1998. A lattice-theoretical framework for annular filters in morphological image processing. *Applicable Algebra in Engineering, Communications and Computing* **9**, 45–89. Cited on page 67.

Ronse, Christian; Najman, Laurent; Decencière, Etienne, Eds. 2005. *Mathematical Morphology: 40 Years On*, xvii + 490 pp. Dortrecht: Springer.

Ronse, Christian; Serra, Jean. 2008. Fondements algébriques de la morphologie. **In:** (Najman & Talbot, Eds., 2008:49–96). Cited on pages 11, 67, 79.

Ronse, Christian; Serra, Jean. 2010. Algebraic foundations of morphology. **In:** (Najman & Talbot, Eds., 2010:35–80). Cited on pages 11, 67, 79.

Rosenfeld, Azriel. 1974. Digital straight line segments. *IEEE Transactions on Computers*, **c-23**, no. 12, 1264–1269. Cited on pages 128, 134, 142, 145, 154, 154, 164, 228.

Rosenfeld, Azriel. 1979. Digital topology. *Amer. Math. Monthly* **86**, 621–630. Cited on page 340.

Rosenfeld, Azriel; Kim, Chul E. 1982. How a digital computer can tell whether a line is straight. *Amer. Math. Monthly* **89**, no. 4, 230–235. Cited on page 164.

Rosenfeld, Azriel; Klette, Reinhard. 2001. Digital straightness. *Electronic Notes in Theoretical Computer Science* **46**, 32 pp. Cited on page 142.

Rosenfeld, A.; Pfaltz, J. L. 1966. Sequential operations in digital picture processing. *Journal of the ACM* **13**, 471–494. Cited on page 391.

Rosenfeld, A.; Pfaltz, J. L. 1968. Distance functions in digital pictures. *Pattern Recognit.* **1**, 33–61. Cited on page 389.

Rovelli, Carlo. 2016. *Reality Is Not What It Seems*, xv + 255 pp. Translated by Simon Carnell and Erica Segre. London: Allen Lane; Penguin Books. Cited on page 8.

Rovelli, Carlo. 2019. *The Order of Time*, 214 pp. Translated by Erica Segre and Simon Carnell. UK: Penguin Books. Cited on page 17.

Rovelli, Carlo; Vidotto, Francesca. 2015. *Covariant Loop Quantum Gravity: An Elementary Introduction to Quantum Gravity and Spinfoam Theory*, xii +254 pp. Cambridge: Cambridge University Press. Cited on page 8.

Roy, Anthony J.; Stell, John G. 2003. Convexity in discrete space. **In:** Kuhn, W.; Worboys, M. F.; Timpf, S., Eds., *COSIT 2003*, pp. 253–269. Lecture Notes in Comput. Sci., 2825. Berlin; Heidelberg: Springer. Cited on page 196.

Saha, P. K.; Borgefors, Gunilla; Sanniti di Baja, Gabriella, Eds. 2017. *Skeletonization: Theory, Methods, and Applications*, 392 pp. London et al.: Academic Press (Elsevier). Cited on page 390.

Samieinia, Shiva. 2007. *Digital straight line segments and curves.* Licentiate Thesis presented at Stockholm University, 2007 September 25. Available at `www2.math.su.se/reports/2007/6/`; accessed 2018-07-12. Cited on page 142.

Samieinia, Shiva. 2010a. Chord properties of digital straight line segments. *Math. Scand.* **106** no. 2, 169–195. Cited on page 142.

Samieinia, Shiva. 2010b. The number of continuous curves in digital geometry. *Port. Math.* **67**, no. 1, 75–89. Cited on page 142.

Samieinia, Shiva. 2010c. The number of Khalimsky-continuous functions on intervals. *Rocky Mountain J. Math.* **40**, no. 5, 1667–1687. Cited on page 321.

Samieinia, Shiva. 2010d. *Digital Geometry, Combinatorics, and Discrete Optimization.* Doctoral Thesis published on 2010 December 15 and defended on

2011 January 21. Stockholm: Stockholm University, Department of Mathematics. Cited on pages xi, 221.

Samieinia, Shiva. 2011. The number of Khalimsky-continuous functions between two points. **In:** Aggarwal, Jake K.; Barneva, Reneta P.; Brimkov, Valentin E.; Koroutchev, Kostadin N.; Korutcheva, Elka R., Eds., *Combinatorial Image Analysis: 14th International Workshop, IWCIA 2011. Madrid, Spain, May 23–25, 2011. Proceedings*, pp. 96–106. Lecture Notes in Comput. Sci., 6636. Heidelberg et al.: Springer. Cited on page 321.

Samieinia, Shiva. 2014. Discrete convexity built on differences. *Port. Math.* **71**, no. 1, 25–37. Cited on page 221.

Sapiro, Guillermo. 2001. *Geometric Partial Differential Equations and Image Analysis*, xxvi + 385 pp. (Reprinted 2006.) Cambridge: Cambridge University Press. Cited on page 139.

Schaefer, Helmut H[einrich]. 1971. *Topological Vector Spaces*, XI + 294 pp. New York, NY et al.: Springer-Verlag. Cited on page 93.

Schaeffer, Luke; Shallit, Jeffrey. 2016. Closed, palindromic, rich, privileged, trapezoidal, and balanced words in automatic sequences. *Electron. J. Combin.* **23**, no. 1. Paper 1.25, 19 pp. Cited on page 142.

Schmitt, Michel. 2005. Ubiquity of the distance function in mathematical morphology. **In:** (Bilodeau et al., Eds., 2005:349–368). Cited on page 376.

Schwartz, Laurent. 1954. Sur l'impossibilité de la multiplication des distributions. *C. R. Acad. Sci. Paris* **239**, 847–848. Cited on page 288.

Schwartzman, Steven. 1994. *The Words of Mathematics. An Etymological Dictionary of Mathematical Terms Used in English.* Washington, DC: The Mathematical Association of America. Cited on page 19.

Serra, Jean. 1966. Remarques sur une lame mince de minerai lorrain. *Bulletin of the BRGM* [Bureau de Recherches géologiques et minières], nr. 6, 1–36, December 1966. Cited on page 61.

Serra, J[ean]. 1978. "One, two, three, ... infinity" **In:** (Miles & Serra, Eds., 1978:137–153). Cited on page 12.

Serra, J[ean]. 1982. *Image Analysis and Mathematical Morphology*, xviii + 610 pp. London et al.: Academic Press. Cited on pages 17, 36, 62, 63, 67, 128, 390, 391.

Serra, Jean. 1986a. *Éléments de théorie pour l'optique morphologique.* Thèse de Doctorat d'État. Paris: Université Pierre et Marie Curie. Cited on page 79.

Serra, J. 1986b. Introduction to mathematical morphology. *Comput. Vision Graphics Image Process.* **35**(3), 283–305. Cited on page 79.

Serra, Jean. 1987. *Thickenings, thinnings.* Technical Report 39. Fontainebleau: École Nationale Supérieure des Mines de Paris, Centre de Morphologie mathématique. Cited on pages 79, 79.

Serra, J[ean]. 1988a. Mathematical morphology for complete lattices. **In:** (Serra, Ed., 1988:13–35). Cited on pages 79, 118.

Serra, J[ean]. 1988b. Mathematical morphology for Boolean lattices. **In:** (Serra, Ed., 1988:37–58). Cited on page 72.

Serra, J[ean]. 1988c. Examples of structuring functions and their use. **In:** (Serra, Ed., 1988:71–99). Cited on page 63.

Serra, J[ean]. 1988d. Introduction to morphological filters. **In:** (Serra, Ed., 1988:101–114). Cited on pages 35, 79.

Serra, J[ean]. 1988e. Alternating sequential filters. **In:** (Serra, Ed., 1988:203–216). Cited on page 61.

Serra, Jean. 1988f. Principles of mathematical morphology. **In:** Chung, C. F.; Fabbri, A.G.; Sinding-Larsen, R., Eds., *Quantitative Analysis of Mineral and Energy Resources*, pp. 237–254. NATO Advances Science Institutes Series C: Mathematical and Physical Sciences, Volume 223. Dortrecht et al.: D. Reidel Publishing Company. Cited on page 12.

Serra, Jean. 1989. Boolean random functions. *Journal of Microscopy.* **156**, 41–63. Cited on page 72.

Serra, Jean. 1998. Connectivity on complete lattices. *J. Math. Imaging Vision.* **9** (3), 231–251. Cited on page 79.

Serra, Jean. 2001. *Lecture Notes on Morphological Operators.* Institut Mittag-Leffler, Lecture Notes no. 1, 2000/2001. Cited on page 119.

Serra, Jean. 2002. Viscous lattices. **In:** (Talbot & Beare 2002:79–89). Cited on page 79.

Serra, Jean. 2005a. Viscous lattices. *J. Math. Imaging Vision* **22**, no. 2–3, 269–282. Cited on page 79.

Serra, Jean. 2005b. A few words about Georges Matheron. **In:** (Bilodeau et al., Eds., 2005:ix–xii). Cited on page 11.

Serra, Jean. 2005c. Morphological operators for the segmentation of colour images. **In:** (Bilodeau et al., Eds., 2005:223–255). Cited on page 79.

Serra, Jean. 2006. A lattice approach to image segmentation. *J. Math. Imaging Vision*, **24**, 1, 83–130. Cited on page 67.

Serra, Jean. 2007. Digital Steiner sets and Matheron semi-groups. **In:** (Banon et al., Eds., 2007:75–86). Cited on page 34.

Serra, Jean. 2009. The random spread model. **In:** (Passare, Ed., 2009:283–310). Cited on page 376.

Serra, Jean. 2010. Digital Steiner sets and Matheron semi-groups. *Image and Vision Computing* **28**, 1452–1459. Cited on page 34.

Serra, Jean, Ed. 1988. *Image Analysis and Mathematical Morphology. Volume 2: Theoretical Advances*, xvi + 411 pp. London et al.: Academic Press. Cited on pages 17, 67.

Shakespeare, William. 1868. *The Works of William Shakspere*, iv + 764 pp. Edited by Charles Knight. London: George Routledge and Sons, The Broadway, Ludgate; New York, NY: 416 Broome Street. Cited on pages 17, 304, 407.

Siddiqi, Kaleem; Pizer, Stephen M., Eds. 2008. *Medial Representations: Mathematics, Algorithms and Applications.* Springer Science + Business Media B.V.

Simon, Barry. 2014. *Convexity. An Analytic Viewpoint.* Cambridge: Cambridge University Press. Cited on page 199.

Singer, Ivan. 1984/85. Extensions of functions of 0-1 variables and applications to combinatorial problems. *Numer. Funct. Anal. Optim.* **7**, no. 1, 23–62. Cited on page 221.

Singer, Ivan. 1997. *Abstract Convex Analysis*, xix + 491 pp. New York, NY et al.: John Wiley & Sons, Inc. Cited on pages 36, 67, 79, 87, 91, 107, 194.

Sintorn, Ida-Maria; Borgefors, Gunilla. 2002. Weighted distance transforms for images using elongated voxel grids. *Discrete Geometry for Computer Imagery (Bordeaux, 2002)*, pp. 244–254. Lecture Notes in Comput. Sci., 2301. Berlin: Springer. Cited on page 376.

Sladoje, Nataša. 2005. *On Analysis of Discrete Spatial Fuzzy Sets in 2 and 3 Dimensions.* Doctoral Thesis, defended 2005 November 25. Acta Universitatis Agriculturae Sueciae no. 2005:112. Uppsala: Swedish University of Agricultural Sciences, Faculty of Forest Science. Cited on pages 103, 113.

Šlapal, Josef. 2009. Convenient closure operators on $\mathbb{Z}^2$. **In:** Wiederhold, Petra; Barneva, Renata P., Eds. *Combinatorial Image Analysis*, pp. 425–436. Lecture Notes in Comput. Sci., 5852. Cited on page 343.

Šlapal, Josef. 2013. Topological structuring of the digital plane. *Discrete Mathematics and Teoretical Computer Science* **15**, no. 2, 165–176. Cited on page 343.

Šlapal, Josef. 2014. Jordan curves in the digital plane. *Bull. Malays. Math. Soc.* (2), **37**, 295–310. Cited on page 343.

Šlapal, Josef. 2018. Galois connections between sets of paths and closure operators in simple graphs. *Open Mathematics* **16**, no. 1. 1573–1581. Cited on page 343.

Šlapal, Josef. 2020a. Digital Jordan curves with respect to a closure operator. *AIP Conference Proceedings 2293, 420090.* Cited on page 343.

Šlapal, Josef. 2020b. A 3D digital Jordan-Brouwer separation theorem. *Comput. Appl. Math.* **39**, Paper No. 211. Cited on page 343.

Šlapal, Josef. 2021a. A convenient graph connectedness for digital imagery. **In:** *High Performance Computing in Science and Engineering.* Cited on page 343.

Šlapal, Josef. 2021b. Realization of categorical idempotent closure operators by preorders. *Quaestiones Mathematicae* 2021: 1–10. Cited on page 343.

Smith, Peter. 2013. *An introduction to Gödel's theorems*, xvi +388 pp. Cambridge: Cambridge University Press. Second edition. Cited on page 136.

Soille, Pierre. 2004. *Morphological Image Analysis. Principles and Applications*, xvi + 391 pp. Berlin et al.: Springer. Second edition. Cited on page 18.

Solvang, Ragnar. 2002. *Matematisk etymologi med historiske kommentarer* [*Mathematical etymology with comments on history*]. Oslo: N. W. Damm & Søn. Cited on page 19.

Starovoitov, Valery. 1995. Toward a distance transform generalization. **In:** Borgefors, Gunilla, Ed., *Proceedings of the 9th Scandinavian Conference on Image Analysis*, pp. 499–506. Uppsala: Uppsala University. Cited on page 355.

Starr, Ross M. 1969. Quasi-equilibria in markets with non-convex preferences (Appendix 2: The Shapley–Folkman theorem, pp. 35–37). *Econometrica*, **37**(1), 25–38. Cited on page 200.

Stoltenberg-Hansen, Viggo; Lindström, Ingrid; Griffor, Edward R. 1994. *Mathematical Theory of Domains*, xii + 349 pp. Cambridge: Cambridge University Press. Cited on page 67.

Strand, Robin. 2016. Minimal paths by sum of distance transforms. **In:** (Normand et al., Eds., 2016:349–358). Cited on page 376.

Strand, Robin; Nagy, Benedek; Borgefors, Gunilla. 2011. Digital distance functions on three-dimensional grids. *Theoret. Comput. Sci.* **412**, no. 15, 1350–1363. Cited on page 376.

Strand, Robin; Nagy, Benedek; Fouard, Céline; Borgefors, Gunilla. 2006. Generating distance maps with neighborhood sequences. **In:** (Kuba et al., Eds., 2006:295–307). Cited on page 376.

Strand, Robin; Normand, Nicolas. 2012. Distance transform computation for digital distance functions. *Theoret. Comput. Sci.* **448**, 80–93. Cited on page 376.

Strömberg, Thomas. 1996. The operation of infimal convolution. *Dissertationes Math. (Rozprawy Mat.)* **352**, 58 pp. Cited on page 112.

Su, XunLi; Guo, JinFeng; Guo, Qi. 2012. The discrete degree of metric spaces and the Lipschitz-continuity of distance transforms. *J. Math. Anal. Appl.* **389**, 863–870. Cited on page 376.

Taisbak, Christian Marinus. 2003. ΔΕΔΟΜΕΝΑ: *Euclid's Data or The Importance of Being Given.* The Greek Text translated and explained by Christian Marinus Taisbak. [Δεδομένα, Dedoména.] Copenhagen: University of Copenhagen, Museum Tusculanum Press. Cited on page 5.

Talbot, Hugues. 2006. Elliptical distance transforms and applications. **In:** (Kuba et al., Eds., 2006:320–330). Cited on page 376.

Talbot, Hugues. 2016. Discrete calculus, optimisation and inverse problems in imaging. **In:** (Normand et al., Eds., 2016:18–27). Cited on page 112.

Talbot, Hugues; Beare, Richard. 2002. *Mathematical Morphology. Proceedings of the VI$^{th}$ International Symposium—ISMM 2002. Sydney, 3–5 April 2002*, xii + 441 pp. Collingwood VIC: CSIRO Publishing.

Tarski, Alfred. 1931. Sur les ensembles définissables de nombres réels I. *Fundamenta Mathematica* **17**, 210–239. [Also published in translation in (Tarski 1956:110–142) with the title On definable sets of real numbers.] Cited on page 135.

Tarski, Alfred. 1948. *A Decision Method for Elementary Algebra and Geometry*, iii + 60 pp. RAND Corporation, Santa Monica, CA. [Mentioned in Tarski 1931.] Cited on page 135.

Tarski, Alfred. 1951. *A decision method for elementary algebra and geometry*, iii + 63 pp. Berkeley, CA; Los Angeles, CA: University of California Press. Second edition. Cited on page 135.

Tarski, Alfred. 1955. A lattice-theoretical fixpoint theorem and its applications. *Pacific J. Math.* **5**, 285–309. Cited on page 326.

Tarski, Alfred. 1956. *Logic, Semantics, Metamathematics. Papers from 1923 to 1938*, xiv + 471 pp. Translated by J. H. Wodger. Oxford: At the Clarendon Press. Cited on pages 22, 135.

Thelin, Nils B[ertil]. 2014. *On the Nature of Time. A Biopragmatic Perspective on Language, Thought and Reality*, 477 pp. Acta Universitatis Upsaliensis. Studia Slavica Upsaliensia 48. Uppsala: Uppsala University. Cited on page 5.

Thibault, Lionel. 2016. Jean Jacques Moreau: A selected review of his mathematical works. *J. Convex Anal.* **23**, no. 1, 5–22. Cited on page 199.

Thibault, Yohan; Kenmochi, Yukiko; Sugimoto, Akihiro. 2008. Computing admissible rotation angles from rotated digital images. *Combinatorial image analysis*, pp. 99–111. Lecture Notes in Comput. Sci., 4958, Berlin: Springer. Cited on page 7.

Thiel, Edouard. 2001. *Géométrie des distances de chanfrein*, 100 pp. Mémoire scientifique; Habilitation à diriger des recherches. Marseille: Université de la Méditerranée (Aix-Marseille II). Cited on page 376.

Tucker, A. W. 1936. Cell spaces. *Ann. of Math.* **37**, 92–100. Cited on pages 36, 317.

Turing, A. M. 1937. On Computable Numbers, with an Application to the Entscheidungsproblem. *Proc. London Math. Soc.* **s2-42**, Issue 1, 230–265. Cited on page 136.

Uscka-Wehlou, Hanna. 2007. Digital lines with irrational slopes. *Theoret. Comput. Sci.* **377**, no. 1–3, 157–169. Cited on pages 142, 256.

Uscka-Wehlou, Hanna. 2008a. Continued fractions and digital lines with irrational slopes. **In:** *Discrete Geometry for Computer Imagery*, pp. 93–104. Lecture Notes in Comput. Sci., 4992. Berlin: Springer. Cited on pages 142, 256.

Uscka-Wehlou, Hanna. 2008b. A Run-hierarchical description of upper mechanical words with irrational slopes using continued fractions. **In:** *Proceedings of 12th Mons Theoretical Computer Science Days (Mons, Belgium), 27–30 August 2008.* Cited on page 256.

Uscka-Wehlou, Hanna. 2009a. Two equivalence relations on digital lines with irrational slopes. A continued fraction approach to upper mechanical words. *Theoret. Comput. Sci.* **410**, no. 38–40, 3655–3669. Cited on pages 142, 256.

Uscka-Wehlou, Hanna. 2009b. Run-hierarchical structure of digital lines with irrational slopes in terms of continued fractions and the Gauss map. *Pattern Recognition* **42**, 2247–2254. Cited on pages 142, 144, 256.

Uscka-Wehlou, Hanna. 2009c. Sturmian words with balanced construction. **In:** *Proceedings of Words 2009, the 7th International Conference on Words (Salerno, Italy), 14–18 September 2009.* Cited on page 256.

Uscka-Wehlou, Hanna. 2009d. *Digital Lines, Sturmian Words, and Continued Fractions.* Doctoral Thesis defended on 2009 September 25. Uppsala: Uppsala University, Department of Mathematics. Cited on pages xi, 142, 144, 154, 256.

Uscka-Wehlou, Hanna. 2010. Continued fractions, Fibonacci numbers, and some classes of irrational numbers. *Acta Math. Acad. Paedagog. Nyházi.* (N.S.) **26**, no. 1, 17–30. Cited on page 144.

Verwer, Ben J. H. 1991. Local distances for distance transformations in two and three dimensions. *Pattern Recognition Letters* **12**, 671–682. Cited on pages 361, 362, 377.

Vonnegut, Kurt, Jr. 1974. *The Sirens of Titan: An Original Novel*, 319 pp. New York, NY: Dell. [First published in 1959.] Cited on page 20.

Voss, Klaus. 1991. Coding of digital straight lines by continued fractions. *Comput. Artificial Intelligence* **10**, no. 1, 75–80. Cited on pages 144, 256.

Vuillon, Laurent. 2003. Balanced words. *Bull. Belg. Math. Soc. Simon Stevin* **10**, suppl., 787–805. Cited on page 142.

Ward, Morgan; Dilworth, R. P. 1939. Residuated lattices. *Trans. Amer. Math. Soc.* **45**, 3, 335–354. Cited on page 93.

*Webster's Ninth New Collegiate Dictionary.* 1983. Springfield, MA: Merriam-Webster, Inc. Cited on pages 15, 19.

*Webster's New World Dictionary of American English.* 1994. Neufeldt, Victoria, Editor in Chief. New York, NY et al.: Prentice Hall. Cited on page 19.

Welk, Martin; Breuß, Michael; Sridhar, Vivek. 2019. Matrix morphology with extremal principle. **In:** (Burgeth et al., Eds., 2019:177–188). Cited on page 61.

White, Michael J. 1992. *The Continuous and the Discrete. Ancient Physical Theories from a Contemporary Perspective*, 345 pp. Oxford: Clarendon Press. Cited on page 7.

Whitrow, G[erald] J[ames]. 1990. *Time in History*, xi + 217 pp. Oxford; New York, NY: Oxford University Press. Cited on pages 5, 17.

Wyse, Frank [Oliver]; Marcus, Dan et al. 1970. A special topology for the integers. *Amer. Math. Monthly* **77**, 1119. (Problem proposed by Marcus and solved by him, Wyse and ten other persons.) Cited on pages 319, 340.

Yanchilin, Vasily. 2003. *The Quantum Theory of Gravitation.* Moscow: Editorial USSR.

Yanchilin, Vasily. 2013. Discrete movement as a key to quantum measurements understanding. *Measurements World* **12**, 30–35.

Yüceer, Ümit. 2002. Discrete convexity: convexity for functions defined on discrete spaces. *Discrete Appl. Math.* **119**, no. 3, 297–304. Cited on page 222.

Yüceer, Ümit. 2006. The equivalence of discrete convexity and the classical definition of convexity. *Int. Math. Forum* **1**, no. 5–8, 299–308. Cited on page 222.

Zhang, Yunxiang; Blusseau, Samy; Velasco-Forero, Santiago; Bloch, Isabelle; Angulo, Jesús. 2019. Max-plus operators applied to filter selection and model pruning in neural networks. **In:** *Mathematical morphology and its applications to signal and image processing*, pp. 310–322, Lecture Notes in Comput. Sci., 11564, Cham: Springer. Cited on page 52.

Zong, Chuanming. 1996. *Strange Phenomena in Convex and Discrete Geometry*, x + 158 pp. New York, NY et al.: Springer. Cited on page 199.

# Author Index

# Subject Index

Lightning Source UK Ltd.
Milton Keynes UK
UKHW020931190122
397385UK00001B/43